Hossein Oloomi

LINEAR
SYSTEMS
CONTROL

Dynamics of Physical Circuits and Systems
Lindsay and Katz

Filter Theory and Design: Active and Passive
Sedra and Brackett

Introduction to Linear Systems Analysis
Swisher

Linear Dynamic Systems
Lewis

Linear Systems Control
Sage

**MATRIX SERIES
IN CIRCUITS AND SYSTEMS**

Andrew P. Sage, *Editor*

LINEAR SYSTEMS CONTROL

Andrew P. Sage

School of Engineering and Applied Science
University of Virginia

MATRIX PUBLISHERS, INC., ● Beaverton, Oregon

To LaVerne, Theresa, Karen, Philip

Library of Congress catalog card number: 78-53839

Matrix Publishers, Inc.
Beaverton, Oregon

ISBN: 0-916460-19-3

Illustrations by Scientific Illustrators.
In-house Editor was Merl K. Miller.

CONTENTS

SYSTEM RESPONSE STABILITY AND PERFORMANCE SPECIFICATIONS   4

LINEAR SYSTEMS CONTROL DESIGN BY BODE DIAGRAMS   5

THE ROOT LOCUS METHOD   6

DIGITAL OPTIMAL AND ADAPTIVE LINEAR SYSTEMS CONTROL   7

PREFACE

The purpose of this text is to present a treatment of linear automatic control theory and practice that will serve the two important goals of preparing students for professional practice in automatic control or for further graduate study of advanced topics in automatic control.

This is intended to be a basic and introductory textbook suitable for a first course in automatic control. Although no attempt has been made to avoid mathematics, we have attempted to utilize the minimum amount of mathematics appropriate for a sound understanding and appreciation of the important field of automatic control.

Prerequisites for the course on which this text is based are an introductory course in differential equations and an introductory course in the dynamics of electrical or mechanical circuits. There is perhaps slightly more material in the text than can be thoroughly covered in a one-semester course unless students taking the course have a background such that it is possible to treat Chaps. 2 and 3 as review and refresher material and cover them very quickly. For students without this background the amount of time needed to discuss Chaps. 2 and 3 may be such that some of the more specialized topics in Chap. 7 may have to be eliminated or not covered completely.

Comments from users of this text are most welcome. I will be most pleased to respond to specific questions concerning use of the material and will be very pleased also to receive suggestions and critiques concerning this text.

Andrew P. Sage

Charlottesville, Virginia
April, 1978

1

INTRODUCTION TO LINEAR SYSTEMS CONTROL

This introductory chapter to our textbook in linear systems control has two goals. We will present a very brief development of the field of automatic control. Finally, we will outline and highlight some of the material to be covered in this introductory text in linear systems control.

1.1 HISTORICAL DEVELOPMENT OF SYSTEMS CONTROL

The subject of systems control has been under evolutionary development for more than four thousand years. Although much of the early development of systems control or automatic control was very pragmatic, analytical development has taken place since the industrial revolution and major scientific and mathematical tools have been used to great advantage since World War II.

Every living thing, including humankind, possesses many feedback control systems and most of our daily activities intimately involve the use of feedback principles. Essentially every biological and physiological activity in the human involves the use of feedback. A very simple example of a feedback control system is that of a person driving a car into a garage. The object of the driver is to steer the automobile with a pair of hands in accordance with commands issued by the brain. The brain obtains an input signal from the eye. The eye visually measures the difference between the actual position of the car somewhere outside of the garage and the desired position of the car inside the garage. The mind, somehow, computes an error signal which is the difference between the input signal or desired output which is the car's location inside the garage and the actual output which is the present position of the car. A very small low power signal from the eye causes hand movement on a steering wheel and foot movement on an accelerator pedal and brake. This causes a rather massive powerful device, the automobile, to move in the general direction of the garage. Certain constraints are avoided in the automobile's trajectory into the garage such as rose bushes and the garage door. It is highly desirable to avoid overshooting the desired target and ending up with the automobile in the kitchen. It would be possible for the eye to sense a possible trajectory which needs to be followed in order to put the car from a present position into the garage. The mind could mentally compute an "open-loop" trajectory and steer the car with appropriate feet movement on the accelerator pedal and brake and with the eyes closed. If the human is sufficiently precise in estimation of the trajectory and if the model of the human machine system is sufficiently accurate such that precalculation of desired foot pressure on the accelerator pedal and pressure on the brake pedal with an appropriate time and steering wheel sequence could be mentally calculated with sufficient accuracy, it would certainly be possible to have the final position of the car precisely where one wishes it in the garage. On the other hand if there are any errors in modeling the human machine system or errors in trajectory calculations, then it is quite likely that constraints involving the garage doors, or the rose bush, or the kitchen will be violated, to the ultimate embarrassment of the automobile driver.

Rather than using this "open-loop" philosophy, the driver typically adopts a "closed-loop" philosophy. The eyes are kept open during the movement of the automobile into the garage and the eyes continually sense the difference between the input or desired final state location in the garage and the actual state location or position of the vehicle with respect to the garage. The error signals which ultimately cause foot movement on the accelerator and brake and steering movements by the hand are proportional to the instantaneous error signal detected by the eyes. And so typically, the automobile winds up under closed-loop control in a good safe position within the garage.

However, there can be disturbances acting upon the signal being transmitted to the hand steering the car and the feet controlling the acceleration and braking of the car. These disturbances could be such as a very strong wind, or four extra martinis at a cocktail party, or a recent heated argument with one's consort or business partner. As we know, these disturbances can, upon occasions, lead to errors and occasionally catastrophic errors. It turns out, however, that the errors committed because of disturbances will generally be much less under closed-loop control than under open-loop control.

On the basis of this simple description of the feedback control system, we might state a definition of a "feedback" control system as any system which maintains a nominally prescribed relationship of one system variable to another by the process of comparing functions of these variables and using the relationship to determine feedback control signals. Alternately, we might define a feedback control system as a system in which measurement of the value of the control variable and comparing the value of the control variable with the value of the command result in generation of an established relationship between the control variable and the command such as to actuate the system in accordance with this relationship. When we compare the two alternate methods of controlling the automobile mentioned in our foregoing discussion, an open-loop control with the eyes closed and a closed-loop control in which the eye measures the error between the command input, or desired final state of the car in the garage, and the actual output or present state of the car, it is apparent that there are great intuitive advantages to the use of a feedback control system.

One advantage which we might be tempted to cite is that the energy necessary on the part of the driver is very small compared with the actual energy being expended. But this is an advantage of a control system be it open-loop or closed-loop. The energy or power from the automobile will be the same regardless of whether our eyes are open or closed. Nevertheless, this energy advantage is an inherent one with any control system be it open-loop or closed-loop. There are at least three major advantages for closed-loop control as opposed to open-loop control:

1. The preciseness of the trajectory of the final state depends primarily upon the accuracy of the error detecting equipment and not upon the accuracy of the high power final drive elements in the control system. In our automobile example for instance, one experienced driver can, with relative ease, put a large number of cars with vastly different handling characteristics into the garage when closed-loop control is used. However, if the eyes are closed such that open-loop control results, then sufficient training may well enable the driver to put a Cadillac in a garage from a given initial position, but the same steering and acceleration and braking commands may be totally inadequate to put a Honda in the garage even when starting from the same

position. This is an extremely significant advantage to a closed-loop feedback control system in that it becomes possible to obtain accurate control using very inaccurate high power components.

2. The effects of noises, disturbances, and modelling errors may be ameliorated by use of a closed-loop control as contrasted with an open-loop control.

3. Some output high power elements may have characteristics of information transmission or information structures that make open-loop control infeasible.

It is of interest to present a very brief and incomplete history of feedback control. The Babylonians, circa 2000 B.C., utilized waters from the Euphrates and Tigres Rivers to form a feedback control system in which water flow from the rivers was regulated by opening and closing very crude valves in accordance with the moisture content of the soil. Other early uses of the feedback control also involved water where water was used to indicate the passage of time. Ketesibios of Alexandria discovered the water clock approximately 300 B.C. The water clock of Ketesibios was actually a regulator in which a float-operated valve was regulated by the difference between a desired and actual water level and this valve adjusted the flow of water entering the regulating vessel. A float in the vessel rose and this indicated the passage of time. The primary thing which made this a feedback control system was the regulation of the water flow since, if this was not done, the water flow rate would be quite erratic and the time indication worthless. These water clocks were refined by a number of people and remained the preferred method of accurate time determination until improvements in the mechanical clock, at approximately the midst of the seventeenth century, replaced the water clock. Many of the early principles of the water clock are in use today in other fluid regulation systems.

The first temperature regulator appears to date from an invention of Cornelius Drebbel of Holland approximately 1610 A.D. This is often accepted as the first feedback invention of western civilization. This invention appears based on the fact that the heat quantity ignited in a fire is a function of the quantity of air available. A temperature sensor was used in this device in the form of a thermometer consisting of an alcohol-filled tube inserted in a vessel of mercury. The level of mercury in the tube changes due to thermal expansion and contraction of the alcohol which is heated by the furnace. The tube moves with the expansion and contraction of the alcohol and this regulates a damper which opens and closes a flue regulating air available for the fire. Thus provision for automatic control via feedback of the desired temperature is made. A number of modifications to the basic thermal regulator have, of course, been made but many present systems bear strong resemblance to this original system.

Mechanical regulators came into preeminence during the eighteenth century. These early mechanical feedback control systems included the automatic turning gear for windmills, and the steam engine fly-ball governor of the Englishman James Watt which was a modification of the early windmill regulators of the Dutch and Germans.

Almost simultaneously with invention of mechanisms during the industrial revolution, many of which incorporated feedback control, developments in mathematics were occurring which would have a dramatic influence in improving not only the practice of feedback control but of initiating a theory of feedback control. Mathematicians such as Laplace, Fourier, and Cauchy researched transformation theory, the representation of periodic functions as sums of sinusoidal functions, and the theory of functions of a complex variable. Many of these ideas built upon the work of Newton in the late 1600's. The pioneering work of Laplace, Fourier and Cauchy was apparently not used in feedback control until approximately one century after these mathematicians accomplished their seminal efforts. J. C. Maxwell published, in 1868, his famous work "On Governors" in the Proceedings of the Royal Society of London. This perhaps marked the first formal written work dealing with analysis of control systems involving feedback. It was followed soon thereafter in 1877 with the discovery by Routh of his famous stability criterion. This appeared in "A Treatis on the Stability of a Given State of Motion" which was published in London and followed soon thereafter by the famous paper by A. Hurwitz published in German in 1895, which also concerned linear system stability.

After approximately thirty years these works concerning feedback systems analysis and stability were followed by the classic work of N. Minorsky "Directional Stability of Automatically Steered Bodies" published in the *Journal of the American Society of Naval Engineers* in 1922. H. Nyquist published his famous work concerning "Regeneration Theory" in the *Bell System Technical Journal* in 1932 and another extra-ordinarily useful analytical technique for the analysis of control systems was born. A fundamental paper concerning the "Theory of Servomechanisms" was published in the *Journal of the Franklin Institute* by H. L. Hazen in 1934 at approximately the same time that the paper by H. S. Black "Stabilized Feedback Amplifiers" appeared in the *Bell System Technical Journal*.

Approximately ten years later a number of important texts presenting the theory and practice of servomechanisms or feedback control systems began to appear. The text by A. C. Hall "Analysis and Synthesis of Linear Servomechanisms" published by the Technology Press at M.I.T. in 1943 represents perhaps the first serious effort which concerns feedback control systems theory in servomechanisms. This pioneering text deals with system compensation including the use of impedance matching and lead network equalizers. The exemplary work by LeRoy MacColl "Fundamental Theory of Servomechanisms" published by VanNostrand in 1945 presents a rather comprehensive discussion of linear servomechanisms including the rudiments of sample data or discrete time systems. A fundamental work by R. C. Oldenbourg and H. Sartorius entitled "The Dynamics of Automatic Controls" was published in German in 1944 and translated into English in 1948 by the American Society of Mechanical Engineers. The work presents a reasonably general treatment of linear feedback control systems theory and design and a study of selected nonlinearities. The work "Servomechanism Fundamentals" by H. Lauer, R. Lesnick, and L. E. Matson was published by McGraw-Hill in 1947 and was almost immediately followed by the very excellent classic text concerning linear servomechanisms and feedback control systems "Principles of Servomechanisms" by G. Brown and D. Campbell which was published by John Wiley & Sons in 1948. All of these works utilized the pioneering efforts of the mathematicians Laplace, Fourier, and Cauchy in that the analysis of feedback control systems and servomechanisms was heavily based upon the Laplace and Fourier transforms and upon the complex variable theory of Cauchy.

Feedback control systems engineering made dramatic progress during World War II both in the United States and in other countries. After the war when the veil of classified secrecy was removed the Radiation Laboratory Series of M.I.T. appeared. Several of these works discussed feedback control systems and there rapidly appeared a significant theory of automatic feedback control systems which had demonstrated its value in coping with problems of guided missiles, inertial navigation, auto-pilots, ballistic missile controls, and fire control systems and radar. Also, there had been trained during the period of World War II a significant cadre of engineers and scientists fully capable of further developments in the theory and practice of feedback control systems. The most stimulating of the M.I.T. radiation laboratories series, from the point of view of feedback control systems, was without question the effort of H. M. James, M. B. Nichols, and R. S. Phillips

"Theory of Servomechanisms" published by McGraw-Hill in 1947. This text discusses reasonably detailed design techniques for linear servomechanisms and presents the Nichols transformation of the Nyquist diagram into a form which was used for and is currently used for many linear servomechanism design problems. This text also contained a thorough study of steady state errors but, more importantly, presented three superb chapters devoted to what is probably the first effort to apply stochastic process phenomena for the analysis of random disturbances in feedback control systems. Phillips developed the root mean square criterion for the design of optimum linear systems with fixed structure and adjustable parameters. These works appearing in the 1940's together with the pioneering effort in network analysis of H. W. Bode "Network Analysis and Feedback Amplifier Design" published by Van Nostrand in 1945, the work by Norbert Wiener "The Interpolation, Extrapolation and Smoothing of Stationary Time Series" published by John Wiley & Sons in 1950 (although it had been developed a number of years before during World War II and classified) which presented an advanced mathematical treatment of filtering, and the discovery of the work of A. M. Liapunov "General Problems of the Stability of Movement" published in French translation in 1949 by Princeton University Press from the original version in Russian published in 1892, provide much of the background for the even more rapid developments in control theory and practice which have occurred in the 1950's, 60's, and 70's.

Most of the works in the 1950's contain extensions of the developments in automatic feedback control which occurred during World War II. Although a number of excellent textbooks appeared during the 1950's particularly significant with respect to these refinements is the work of J. G. Truxal "Automatic Feedback Control Systems Synthesis." This work, published by McGraw-Hill in 1955, discusses some of the original s-plane compensation approaches of Gillemin and Truxal and also discusses refined versions of sampled data systems, Wiener filtering, nonlinear systems, and other approaches which had much stimulus during World War II.

Much effort during the early 1960's was devoted to the establishment of a theoretical basis for extensions to the integral square optimization and root mean square error optimization approaches of Phillips and the optimum frequency domain filtering of Wiener. Control systems engineers and applied mathematicians interested in control became much interested in the calculus of variations and extensions to the basic techniques in the calculus of variations by a number of workers including L. S. Pontryagin. These efforts in optimization theory were applied to systems with both deterministic and random disturbances and were applied to both linear and nonlinear systems. A considerable number of journal papers and advanced textbooks describe these contributions, many of which have been motivated by the availability of the ubiquitous large scale digital computer and the more recent microcomputer and microprocessor, in much detail.

1.2   OUTLINE OF THIS TEXTBOOK IN LINEAR SYSTEMS CONTROL

This text in linear systems control, intended as an introductory text, attempts, in six chapters to follow, to discuss those techniques in control systems design and synthesis which appear most useful for practicing professional engineers engaged in the design of control systems. Although our discussions are restricted to linear systems with deterministic inputs, we do emphasize the limitations of a strictly linear theory and the need for both nonlinear and stochastic approaches at appropriate points in the text. We attempt to present both the "classical" and "modern" approach to linear systems control design. Since this is an introductory text and since the "classical" methods have proven themselves over and over again whereas many of the "modern" methods are still evolving from research to practice and since the "modern" theory is considerably more advanced, we unashamedly emphasize the "classical" approach. We do strongly stress the point, however, that there is no conflict whatever between these two approaches and insist that they mutually support and reinforce one another. Thus a considerable portion of our later efforts are devoted to obtaining solutions to the same, or similar, problems using more than one approach.

Chapter 2.   Mathematics for Linear Systems Control

In our next chapter we will examine three very important topics necessary for linear systems control. We present a discussion of linear time invariant differential equations from both the single input single output point of view as well as the linear state space viewpoint. A very convenient tool for the analysis of linear constant coefficient systems with deterministic inputs is that of the Laplace transform. We present a study of that portion of Laplace transform theory particularly focused on those topics which will be needed in our later discussion of design procedures for linear systems control. Finally, we present a detailed discussion of linear systems block diagrams and two approaches for the determination of transfer functions for linear systems from block diagrams.

Chapter 3.   Linear Systems Control Components

Physical control systems typically consist of electrical networks and systems, mechanical networks and systems, thermal systems, fluid systems, and various combinations of these. Many control system engineers on a first design assignment will find, sometimes to their amazement, that it is not the lack of understanding of control systems theory which presents a major impediment to problem solving but uneasiness and lack of confidence with respect to one's ability to model a physical system. Because space limitations and intent considerations do not allow an extensive discussion of physical control system modeling here, we will, in this chapter, consider only a few of the many components used in physical linear control systems. We will obtain models for these physical components and the associated differential equations and transfer functions.

We will examine electrical networks and components, operational amplifiers and analog computers both for use in control systems and for simulation of control systems, mechanical network and components, and thermal and fluid networks and components. This brief introduction to this very important and needed area will be adequate for our primary purposes in this text. With the background in transform mathematics and system control components obtained in this chapter and the previous chapter, we will then be in a position to undertake a study of the analysis design and synthesis of linear feedback control systems.

Chapter 4.   System Response Stability and Performance Specifications

As a preliminary to a study of design methods for linear systems control, we must introduce appropriate analysis techniques. In this chapter we shall consider a number of systems concepts which yield information of considerable value concerning the typical behavior of simple first and second order systems, requirements which must be imposed to insure system stability, and performance criteria and specifications for linear systems. In order to achieve these goals, we will discuss the time and frequency response of simple first and second order systems, some simple system stability criteria for linear systems and will then conclude our presentation by presenting a variety of performance criteria and rule of thumb approximations for linear system design that are based primarily upon the response of typical first and second order systems.

Chapter 5.   Linear System Control Design by Bode Diagrams

In this chapter our efforts and emphasis turns from the analysis of linear control systems to system design. The classic approach to system design involves trial and error repetition or iteration of analysis until convergence upon a set of design specifications is achieved. Thus we will find our study of analysis methods in earlier chapters to be extremely useful in the design process. In this chapter we will present the frequency domain approach to the design of linear control systems. We will determine appropriate design criteria for frequency domain design and relate this to the Nyquist stability and performance characterizations that we have obtained in our previous chapter. This approach to design suggests that we first obtain from a client a set of system specifications which may include such items as speed of response, steady state error, overshoot, and cost, size, and design time specifications. Often many of these will not be in a form directly suitable for an analysis procedure and specifications must then be converted, often by approximations, into a form suitable for design. Often, the specifications will include a fixed plant or set of unalterable components which may well already be in existence but which are not subject to change by the control system designer. A mathematical model, generally in the form of a transfer function, is obtained for the fixed plant. On the basis of experience, a trial form of compensating network is selected with a number of fixed but unspecified parameters. The object of Bode diagram design is to allow specification of these parameters such that a set of performance specifications is achieved. We shall develop several frequency domain approaches to accomplishing this end in this chapter. We will consider series equalization or compensation methods, as well as compensation methods based on minor loop design. We will consider sensitivity effects and the effects of output disturbances upon systems in this chapter. We do not attempt encyclopedic or survey coverage of all known frequency domain design techniques but rather present one approach, in the opinion of the author the simplest and most useful one, in some depth.

Chapter 6. Linear System Control Design by Root Locus

In this chapter we shall present a method of design in the s plane which compliments the Bode diagram frequency domain design approach of the previous chapter. The root locus method was first developed by W. R. Evans and is a method much in use in control systems practice. In this approach to design, an initial form of system compensation is selected generally on the basis of past experience. All parameters but one are typically frozen and a set of rules, known as root locus rules, is used to determine how the closed-loop system behavior changes as the unspecified parameter is changed. The most acceptable value of this parameter is generally found directly on the root locus from the engineer's intuitive knowledge of which of the many closed-loop transfer functions obtainable from several discrete values of the unspecified parameter will result in "best" performance. Analysis is again used to compare the performance of the closed-loop system with the system specifications. If these specifications have been met, the design is satisfactory. If the specifications have not been met, the specified parameters within the compensation network are changed and root locus diagram constructed again and the procedure iterated. Athough this chapter concerns itself primarily with development and use of the root locus approach for the design of linear feedback control systems, the approach to design actually advocated involves a combination of root locus and Bode diagram approaches and so we continually relate locus methods and results to Bode diagram methods and results in this chapter. As in all of our chapters, a rather large number of examples are solved and many exercises and problems are suggested for solution by the interested reader.

Chapter 7. Digital, Optimum, and Adaptive Control

In this chapter we present several extensions to the basic linear control analysis and design procedures that have been found very useful. Widespread use of the digital computer, both as a component in control systems as well as a tool for the simulation of control systems, has made it highly desirable to develop a theory of sampled data or discrete time or digital control. Our first part of this chapter is devoted to extending the continuous time methods of our earlier efforts such that we can apply the Bode diagram and root locus design techniques to discrete time as well as continuous time systems.

Since digital computers are so often used to simulate continuous time control systems, we devote one section of this chapter to the development of digital simulation methods which allow the simulation and modeling of linear and nonlinear continuous time systems on a general or a special purpose digital computer.

The next section in this chapter presents an introductory discussion of optimum linear systems control. The primary difference between the approaches of optimum systems control or "modern" control and the "classical" approaches of our previous two chapters is that the optimum systems control approach involves a cost function or objective function which is typically a time integral of some function of the system error. This cost function or objective function attempts to express, in a single scalar number, a measurement of the quality of system performance. The end result of an optimum systems control procedure is the determination of a control input or a set of compensation network parameters which extremizes, typically minimizes, the performance index or objective function.

We present a discussion both of linear optimum systems control for the case where all system parameters are constant and where the time interval of interest is very large or infinite. A frequency domain approach to optimization turns out to be possible in this case and we develop this infinite time integral square error approach in the frequency domain here. Often, systems are not fundamentally linear and the time interval of operation is often short compared with dominant system time constants. A nonlinear system can be linearized about an operating trajectory which may well be an optimum operating trajectory. This linearization will typically produce a time varying linear system for which the time invariant frequency domain optimization techniques are not applicable. Many controls for nonlinear systems are based upon a "fly by wire" approach in which the object of the linear controller is to cause the system to "fly" a predetermined trajectory. In our concluding portions of the section devoted to optimum linear systems control, we develop the Pontryagin maximum principle and show how one specific solution to this maximum principle can be obtained for a general class of problems. This particular solution involves a linear time invariant system and a performance index or cost function which is the integral of a quadratic function of the input control and output state variables. It turns out that a feedback control solution is possible for this optimization problem and the solution is developed here. It turns out that the solution is one which is readily implemented on a digital computer to determine optimum feedback time varying gain coefficients. We illustrate this linear regulator problem, and an extension of it to the linear servomechanism, with some numerical examples of interest.

The concluding section of this, our last chapter in linear systems control, deals with adaptive control. In many ways, the adaptive control problem is an outgrowth of the optimum control problem where it is desired that a system optimize or tune itself for best performance as the system operates with real time data. We define a performance adaptive system as one which tunes itself to minimize a function of the system error and a parameter adaptive system as one in which system parameters are identified and compensating network parameters adjusted in accordance with a predetermined law as a function of the identified fixed system characteristics. A few simple examples of adaptive control systems are given.

Appendix.   The DELTA Chart

In our analysis and design efforts in linear systems control we are especially concerned with the development of systematic procedures with which to analyze or design linear control systems. The DELTA Chart is a very appropriate graphical method which allows us to expose Decision, Events, Logic, Time, and Activities in a way that can be very readily comprehended. We present DELTA Charts for most of the important analysis and design methods in this text and devote a very brief appendix to an explanation of the DELTA Chart method.

1.3 SUMMARY

This chapter has presented a very brief overview of the historical development of automatic control and a brief discussion of efforts to follow.

1.4 REFERENCES

Those interested in readings in the fascinating history of automatic control should find the following two texts of considerable interest:

Usher, A. P., *A History of Mechanical Inventions*, Harvard University Press, Cambridge, Mass., 1954.

Mayr, O., *The Origins of Feedback Control*, The M.I.T. Press, Cambridge, Mass., 1970.

There are numerous discussions in encyclopedias concerning specific historical developments in automatic control. The recent special issue:

Proceedings of the IEEE, "Special Issue on Two Centuries in Retrospect," Vol. 64, No. 9, September 1976.

surveys historical developments in electrical engineering including automatic control and presents a number of viable references to the history of technology including electrical engineering and automatic control technology.

A vast number of early references to automatic control is contained in the AIEE Report:

"Bibliography on Feedback Control," *Applications in Industry*, October 1954, pp. 430–462.

References to more recent developments in automatic control can be found in our chapters to follow as well as in the references cited in these chapters.

1.5 PROBLEMS

1. Figure P.1 represents an adaptation of the basic AIEE feedback control systems committee standard block diagram for automatic control.

    Discuss the automobile control example which we have discussed in Sec. 1.1 in terms of this figure and show that it has a model which fits this basic block diagram.

2. Think of a simple feedback control system with which you are personally familiar and discuss how your example can be modeled such as to fit the basic block diagram of Fig. P.1.1.

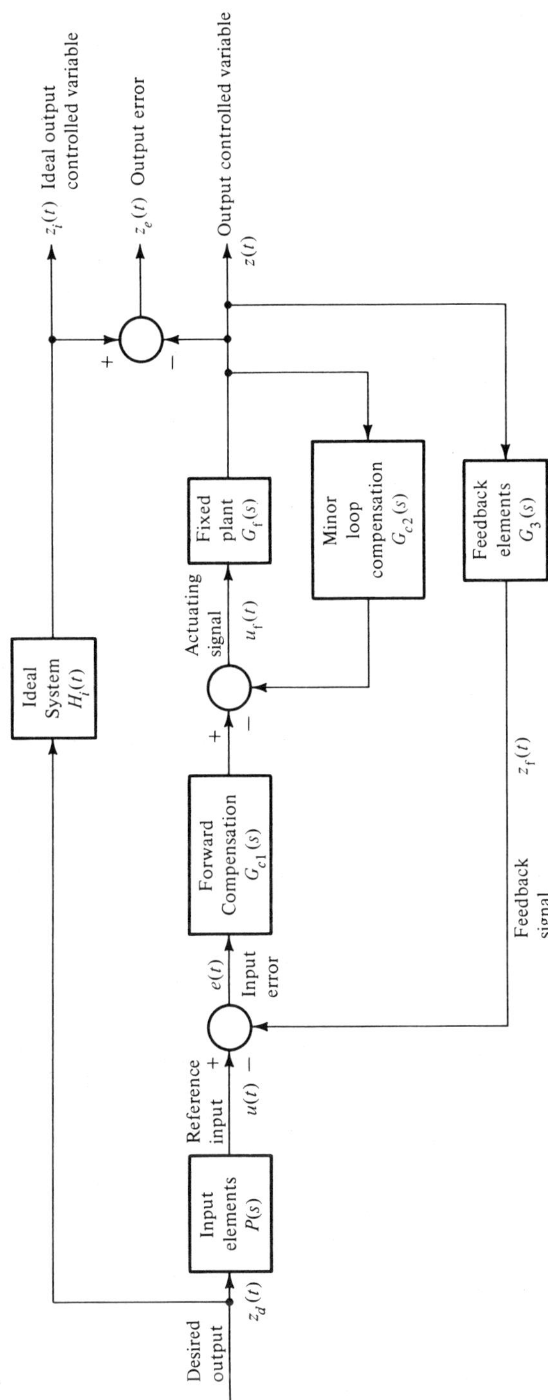

Figure P1.1 General block diagram of linear feedback control system

2
MATHEMATICS FOR LINEAR SYSTEMS CONTROL

In this chapter we will examine three very important topics necessary for linear systems control. We present a discussion of linear time invariant differential equations from both the single input single output point of view as well as the linear state space point of view. A very convenient tool for the analysis of linear constant coefficient systems with deterministic inputs is the Laplace transform. We present a study of Laplace transforms particularly focused on those topics that will be needed in our later discussions of design procedures for linear systems control. The chapter concludes with a presentation of block diagrams and their reduction. This is a particularly important topic since systems are often constructed by interconnecting subsystems and we need to have a convenient representation of information* flow in systems in order to develop useful approaches to the regulation and control of systems.

2.1 LINEAR DIFFERENTIAL EQUATIONS

Most of the systems we study in this text can be modeled mathematically by linear differential equations. The assumption of linearity is a crucial and critical one in many systems control problems and is fundamental to obtaining reasonably simple mathematical results. It is fortunate that, as we have previously indicated, most nonlinear systems can be approximated by linear systems. There are many definitions of

*Information is data of value in decision making.

linearity that can be postulated. Two of these will be sufficient for our purposes here.

Suppose that an initially unexcited system is subjected to an input $u_1(t)$ and an output $z_1(t)$ results. The same initially unexcited system when subjected to an input $u_2(t)$ produces an output $z_2(t)$. The system is linear if and only if the system's response to the input $\alpha_1 u_1(t) + \alpha_2 u_2(t)$ is the output $\alpha_1 z_1(t) + \alpha_2 z_2(t)$ for any constant values α_1 and α_2. This is the *superposition property* of a linear system which asserts that the response to an input, which can be disaggregated into several input components, can be found by finding the response to each of the several inputs and adding these responses.

An alternate definition of linearity is that the system input and output are related by a linear transformation. Although many systems can be adequately described by linear algebraic equations we will be much more concerned here with systems that can be described by *linear differential equations* of the general form

$$a_n \frac{d^n z(t)}{dt^n} + a_{n-1} \frac{d^{n-1} z(t)}{dt^{n-1}} + \ldots + a_1 \frac{dz(t)}{dt} + a_0 z(t)$$

$$= b_m \frac{d^m u(t)}{dt^m} + \ldots + b_1 \frac{du(t)}{dt} + b_0 u(t) \tag{2.1-1}$$

with initial conditions specified for $z(t_0)$, $dz(t_0)/dt_0$, $\ldots$, $d^{n-1} z(t_0)/dt_0^{n-1}$. We may rewrite Eq. (2.1-1) in the somewhat more compact but equivalent form

$$\sum_{i=0}^{n} a_i \frac{d^i z(t)}{dt^i} = \sum_{i=0}^{m} b_i \frac{d^i u(t)}{dt^i} \tag{2.1-2}$$

In either of these representation of a *differential equation of order* n, the a_i and b_i may be time functions. In many of the linear systems control design methods that we shall develop, the coefficients a_i and b_i must be constant, however. A notable exception to this will occur in Chap. 7 in which general time varying linear systems, in which the a_i and b_i coefficients may be time varying, will be considered.

In many cases the *input-output* differential equation of Eqs. (2.1-1) or (2.1-2) will not be the most convenient method to portray the mathematical model of a physical system. A frequency domain representation of these equations will often be more useful for many applications. For others, a *state variable* representation in the time domain will often be more useful. It is possible to show that a differential equation of the form of Eq. (2.1-1) can always be written in the form

$$\frac{d\mathbf{x}(t)}{dt} = \mathbf{A}\,\mathbf{x}(t) + \mathbf{b}u(t)$$

$$z(t) = \mathbf{c}\mathbf{x}(t) + du(t)$$

(2.1-3)

where $\mathbf{A}$ is an n by n matrix, $\mathbf{b}$ and $\mathbf{c}$ are n vectors, and d is a scalar. Each is independent of $\mathbf{x}$, u, and z. An initial condition vector $\mathbf{x}(t_0)$ must be specified in order to solve Eq. (2.1-3). One of the principal advantages to the representation of Eq. (2.1-3) contrasted to that of Eq. (2.1-1) is that the solution to Eq. (2.1-3) may be written, for the case where $\mathbf{A}$, $\mathbf{b}$, $\mathbf{c}$, and d are constant, and $\mathbf{I}$ is the identity matrix, as

$$\mathbf{x}(t) = \phi(t)\mathbf{x}(0) + \int_0^t \phi(t-\tau)\mathbf{b}u(\tau)\,d\tau$$

(2.1-4)

$$z(t) = \mathbf{c}\mathbf{x}(t) + du(t)$$

where

$$\frac{d\phi(t)}{dt} = \mathbf{A}\phi(t), \quad \phi(0) = \mathbf{I}$$

(2.1-5)

or

$$\phi(t) = e^{\mathbf{A}t} = \mathbf{I} + \mathbf{A}t + \frac{\mathbf{A}^2}{2!}t^2 + \dots$$

(2.1-6)

is called the *fundamental matrix* or *transition matrix*. Solutions from Eq. (2.1-4) may generally be determined more easily (on a digital computer) than solutions to Eq. (2.1-1). In addition Eq. (2.1-3) has a number of theoretical uses which are difficult to realize from use of Eq. (2.1-1). For example, the system response to an input $u(t) = \alpha_1 u_1(t) + \alpha_2 u_2(t)$, with $\mathbf{x}(0) = 0$ is easily determined from Eq. (2.1-4) as

$$\mathbf{x}(t) = \int_0^t \phi(t - \tau)\mathbf{b}[\alpha_1 u_1(\tau) + \alpha_2 u_2(\tau)]\, d\tau$$

$$= \alpha_1 \mathbf{x}_1(t) + \alpha_2 \mathbf{x}_2(t)$$

where $\mathbf{x}_1(t)$ and $\mathbf{x}_2(t)$ are the responses to the inputs $u_1(t)$ and $u_2(t)$. Thus we easily see, by use of the state variable representation, that superposition is a property of a linear system.

EXAMPLE 2.1-1. Let us consider the system described by Eq. (2.1-3) where

$$A = \begin{bmatrix} -3 & 0 \\ 1 & -1 \end{bmatrix}, \qquad \mathbf{b} = \begin{bmatrix} 3/2 \\ 0 \end{bmatrix}$$

$$\mathbf{c} = \begin{bmatrix} 1 & 1 \end{bmatrix}, \qquad \mathbf{d} = \begin{bmatrix} 0 \end{bmatrix}$$

Insertion of the various terms in Eq. (2.1-3) leads immediately to

$$\begin{bmatrix} \dot{x}_1 \\ \dot{x}_2 \end{bmatrix} = \begin{bmatrix} -3 & 0 \\ 1 & -1 \end{bmatrix}\begin{bmatrix} x_1 \\ x_2 \end{bmatrix} + \begin{bmatrix} 3/2 \\ 0 \end{bmatrix}u(t)$$

$$z = \begin{bmatrix} 1 & 1 \end{bmatrix}\begin{bmatrix} x_1 \\ x_2 \end{bmatrix}$$

which becomes

$$\dot{x}_1 = -3x_1 + 3/2u \tag{1}$$

$$\dot{x}_2 = x_1 - x_2 \tag{2}$$

$$z = x_1 + x_2 \tag{3}$$

We obtain dz/dt and d^2z/dt^2 from (3) and use (1) and (2) to obtain

$$\dot{z} = \dot{x}_1 + \dot{x}_2 = -2x_1 - x_2 + 3/2u$$

$$\ddot{z} = -2\dot{x}_1 - \dot{x}_2 + 3/2\dot{u} = 5x_1 + x_2 - 3u + 3/2\dot{u}$$

By any of several approaches we find that z, $\dot{z}$, and $\ddot{z}$ may be combined to eliminate x_1 and x_2 such that

$$2\ddot{z} + 8\dot{z} + 6z = 3\dot{u} + 6u \tag{4}$$

which is the input-output differential equation for this problem.

It turns out that it is possible to obtain (1) through (3) from (4) by suitable linear transforms. Equation (4) is the unique input-output representation of Eqs. (1) through (3). However, there are several state variable representations of Eq. (4), (see EXERCISE 2.1-1).

EXERCISE 2.1-1. Show that the matrices

$$\text{(a)} \quad \mathbf{A} = \begin{bmatrix} -1 & 0 \\ 0 & -3 \end{bmatrix}, \quad \mathbf{b} = \begin{bmatrix} 1 \\ 1 \end{bmatrix}, \quad \mathbf{c} = \begin{bmatrix} 3/4 & 3/4 \end{bmatrix}, \quad \mathbf{d} = 0$$

$$\text{(b)} \quad \mathbf{A} = \begin{bmatrix} 0 & 1 \\ -3 & 4 \end{bmatrix}, \quad \mathbf{b} = \begin{bmatrix} 0 \\ 1 \end{bmatrix}, \quad \mathbf{c} = \begin{bmatrix} 3 & 3/2 \end{bmatrix}, \quad \mathbf{d} = 0$$

all lead to the input output differential equation of EXAMPLE (2.1-1).

EXERCISE 2.1-2. Show that

$$\phi(t) = e^{\mathbf{A}t} = \begin{bmatrix} e^{-3t} & 0 \\ 1/2(e^{-t} - e^{-3t}) & e^{-t} \end{bmatrix}$$

is the transition matrix for EXAMPLE (2.1-1) in that this matrix satisfies Eq. (2.1-5) for the parameters of EXAMPLE (2.1-1).

2.2 THE LAPLACE TRANSFORMATION

A convenient way to solve constant coefficient differential equations is to use the Laplace transformation. In addition there is considerable physical insight into the behavior of constant coefficient linear systems that may be gained through use of this very valuable tool. In this section, we will develop some fundamental theorems concerning the Laplace transform and the inverse transform. In addition, we will discuss convolution properties of linear systems and illustrate how a few simple problems in input-output format as well as in state variable format may be solved using the Laplace transformation. We assume that the reader has some previous exposure to the Laplace transformation and operational mathematics or else is interested primarily in learning the use of this important technique for linear systems control problems since there are numerous other works which provide a much more detailed treatment of this important topic.

We begin our presentation with an attempt to develop a relationship between the input and output of a simple linear system such as shown in Fig. (2.2-1). We could write a differential equation in input output form, as given by Eq. (2.1-1), or in state variable form, as given by Eq. (2.1-3), to describe this system. Here we consider a somewhat different approach by breaking the system input $u(t)$ into simpler inputs consisting of sums of impulses and determine the system response to each of these inputs and then use the superposition principle to obtain the system output response due to the actual input.

First we consider a typical system input $u(t)$ as shown in Fig. (2.2-2). We approximate this input by the square pulse train as shown in the figure. Thus we are using the approximation

$$u(t) \approx \sum_{k=-\infty}^{\infty} u(t, t_k, \Delta t) = \sum_{k=-\infty}^{\infty} u(t_k)P(t - t_k, \Delta t)\,\Delta t \qquad (2.2\text{-}1)$$

where $P(t - t_k, \Delta t)$ is a pulse of height $1/\Delta t$, width Δt, and centered at $t = t_k$. This is called a *unit pulse* since its area is unity and is illustrated in Fig. (2.2-3). We note that there is a Δt term multiplying $P(t - t_k, \Delta t)$ in Eq. (2.2-1) such that the actual height of the pulse multiplying $u(t_k)$ in Eq. (2.2-1) is unity. We are particularly interested in representing $P(t - t_k, \Delta t)$ as a unit area pulse since we wish it to become a valid impulse function in the limit as we let Δt approach zero.

We will now define the response of the initially unexcited linear system of Fig. (2.2-1) to a unit pulse $P(t - \tau, \Delta t)$ as $h(t, \tau, \Delta t)$. Here τ is the time the center of the unit pulse occurs and Δt is its width. We do not need to be at all concerned now with the mechanics of how $h(t, \tau, \Delta t)$ is obtained. It is sufficient here to say that it could be obtained by solving Eq. (2.1-1) with the given unit pulse input. Now we let the width of the unit pulse, Δt, approach zero such that we obtain a unit impulse occurring at time τ

$$\delta(t - \tau) \overset{\Delta}{=} \lim_{\Delta t \to 0} P(t - \tau, \Delta t) \qquad (2.2\text{-}2)$$

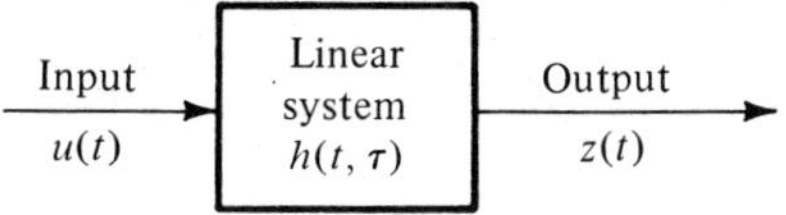

Figure 2.2-1   Linear system block diagram

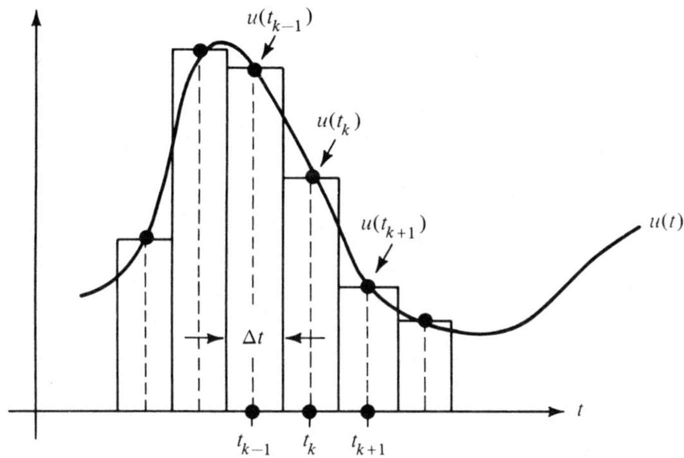

Figure 2.2-2   A typical input $u(t)$ and decomposition of the continuous input into a pulsed input

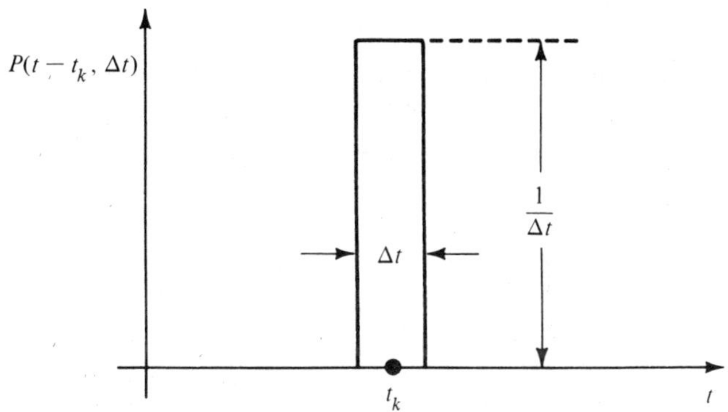

Figure 2.2-3   Definition of a unit pulse

The corresponding system output for this unit impulse input is defined as the system impulse response and is formally given by

$$h(t, \tau) = \lim_{\Delta t \to 0} h(t, \tau, \Delta t) \qquad (2.2\text{-}3)$$

The actual system input time t_k is not just the unit pulse but the unit pulse multiplied by $u(t_k)\Delta t$. Thus the system response to the input portion occurring at time t_k, $u(t, t_k, \Delta t)$ is

$$z(t, t_k, \Delta t) = h(t, t_k, \Delta t)u(t_k)\Delta t$$

The response due the complete pulse train is, by the superposition principle,

$$z(t, \Delta t) = \sum_{k=-\infty}^{\infty} z(t, t_k, \Delta t) = \sum_{k=-\infty}^{\infty} h(t, t_k, \Delta t)u(t_k)\Delta t$$

When Δt becomes zero we obtain the exact system response. From the foregoing equation, this is

$$z(t) = \lim_{\Delta t \to 0} z(t, \Delta t) = \lim_{\Delta t \to 0} \sum_{k=-\infty}^{\infty} h(t, t_k, \Delta t)u(t_k)\Delta t$$

which is just the fundamental definition of the integral such that we have, using Eq. (2.2-3),

$$z(t) = \int_{-\infty}^{\infty} h(t, \tau)u(\tau)\, d\tau \qquad (2.2\text{-}4)$$

This very important integral is called the convolution integral. Although it may not appear to be the case, Eqs. (2.1-1), (2.1-3), and (2.2.4) are equivalent in the sense that knowledge of one is essentially equivalent to knowledge of the other in that we can determine the response of a system to an arbitrary input from either of these relations.

EXAMPLE 2.2-1. We consider a simple differential equation which has as its impulse response to an impulse applied at time zero

$$z(t) = h(t, 0) = te^{-t} \qquad (1)$$

The first two derivatives of the output (or impulse response $h(t, o)$) are easily obtained as

$$\dot{z} = e^{-t} - te^{-t} \tag{2}$$

$$\ddot{z} = \delta(t) - 2e^{-t} + te^{-t} \tag{3}$$

The impulse occurring in $\ddot{z}$ is just the impulse input $u(t) = \delta(t)$ and we easily obtain by eliminating the time expressions in (1), (2), and (3)

$$\ddot{z} + 2\dot{z} + z = u(t) \tag{4}$$

as the system differential equation. This is the unique input output differential equation which corresponds to the impulse response of (1). There are many state variable representations, one of which is obtained from the definitions

$$z = x_1$$

$$\dot{x}_1 = x_2 = \dot{z}$$

and the clearing equation to satisfy (4)

$$\dot{x}_2 = \ddot{z} = u(t) - x_1 - 2x_2$$

as

$$\dot{x} = Ax + bu, \quad z = cx$$

where

$$A = \begin{bmatrix} 0 & 1 \\ -1 & -2 \end{bmatrix}, \quad b = \begin{bmatrix} 0 \\ 1 \end{bmatrix}, \quad c = \begin{bmatrix} 1 & 0 \end{bmatrix}$$

EXERCISE 2.2-1. Determine the differential equation whose unit impulse response, $u(t) = \delta(t)$, is $h(t, 0) = e^{-\alpha t} \sin \omega t$.

$$[Answer: \ddot{z} + 2\alpha\dot{z} + (\alpha^2 + \omega^2)z = \omega u(t)]$$

EXERCISE 2.2-2. Determine the differential equation whose unit impulse response is $h(t, 0) = e^{-t} - te^{-t}$. Find a state variable representation for this system.

One of the uses of the convolution integral of Eq. (2.2-4) is to find the response of a system to a known input. There are some restrictions on any physical system impulse response. We will, in Chap. 4, discuss stability requirements for a linear system which will impose bounds on the impulse response function. Since we are dealing with a physical system, the requirement of causality must be imposed. This requires that the system "cannot laugh before it is tickled" or respond before it is excited. This causality or non-anticipatory requirement insures that the impulse response is zero before the impulse is applied or

$$h(t, \tau) = 0 \qquad t < \tau$$

We can use this information to replace the upper limit in Eq. (2.2-4) by t such that we obtain for the convolution integral of a causal linear system

$$z(t) = \int_{-\infty}^{t} h(t, \tau)u(\tau)\, d\tau \tag{2.2-5}$$

It is often difficult to evaluate this integral even though it is very useful for theoretical purposes. If we are dealing with a linear time invariant system it turns out that the impulse response is a function of the age variable $t - \tau$, or the time after the impulse is applied, only and not of any other t, τ combination. Using

$$h(t, \tau) = h(t - \tau)$$

we obtain for Eq. (2.2-5)

$$z(t) = \int_{-\infty}^{t} h(t - \tau)u(\tau)\, d\tau \tag{2.2-6}$$

Now we let $t - \tau = \eta$ and change the variable of integration in the foregoing to obtain

$$z(t) = \int_{0}^{\infty} h(\eta)u(t - \eta)\, d\eta \tag{2.2-7}$$

which is generally easier to use for actual calculations than Eq. (2.2-6).

EXAMPLE 2.2-2. Let us consider the use of the convolution integral of Eq. (2.2-7) to obtain the response of a simple system to a unit step input. We consider the impulse response

$$h(t, \tau) = \begin{cases} (t - \tau)e^{-(t - \tau)} & t > \tau \\ 0 & \text{otherwise} \end{cases}$$

such that Eq. (2.2-7) becomes

$$z(t) = \int_0^\infty \eta e^{-\eta} u(t - \eta)\, d\eta \tag{1}$$

where $u(t)$ is one for $t > 0$ and 0 for $t < 0$. Thus $u(t - \eta)$ is one for $t - \eta > 0$ and zero for $t - \eta < 0$. The upper limit on Eq. (2.2-7) and (1) may be replaced by t whenever we have an input that is always zero for $t < 0$. We may obtain the system response by evaluating

$$z(t) = \int_0^t \eta e^{-\eta}\, d\eta \qquad t > 0$$
$$= 0 \qquad\qquad t < 0$$

to obtain

$$z(t) = 1 - e^{-t} - te^{-t} \qquad t \geq 0$$
$$= 0 \qquad\qquad t < 0$$

We were able to evaluate the convolution integral in a reasonably simple fashion for this example in order to obtain the system time response. In more difficult situations this determination could become quite complex. Also we are much more interested in the design of linear control systems than their analysis. We now describe a very potent tool, the Laplace transformation, which is generally much easier to use, and much more applicable to the system design problem than is the convolution integral approach.

EXERCISE 2.2-3. What is the time response of the system with impulse response $h(t) = (1 - t)e^{-t}$ to a unit step function occurring at $t = 0$?

$$[\text{Answer: } z(t) = te^{-t}]$$

EXERCISE 2.2-4. What is the time response of the system with impulse response $h(t) = e^{-t} \sin t$ to a unit step input that starts at $t = 0$ and stops at $t = 4$?

We wish to inquire concerning use of some sort of transformation which will make the operations required by the convolution integral of Eq. (2.2-7) easier to perform. It turns out that an integral transformation known as the Laplace transformation is exceptionally useful for this purpose. The Laplace transform pair

$$F(s) = \int_0^\infty f(t)e^{-st}\, dt = \mathcal{L}f(t) \tag{2.2-8}$$

$$f(t) = \frac{1}{2\pi j} \int_{\sigma-j\infty}^{\sigma+j\infty} F(s)e^{st}\, ds = \mathcal{L}^{-1} F(s) \tag{2.2-9}$$

forms the basis for our development. The validity of these relations can be established in a number of ways.*

In the definition of the Laplace transform of Eq. (2.2-8) and the inverse transform of Eq. (2.2-9), the variable s must be considered as a complex variable having a real component σ and an imaginary component ω such that $s = \sigma + j\omega$ where $j = \sqrt{-1}$. Further it is assumed that $f(t) = 0$ for $t < 0$ and that σ is selected such that $\lim_{t\to\infty} f(t)e^{-\sigma t} = 0$ and $\int_0^\infty |f(t)e^{-\sigma t}|\, dt < \infty$. The smallest possible value of σ for which this equality is satisfied is called the abscissa of absolute convergence. It is generally necessary to know complex variable theory in order to evaluate the inverse transform integral of Eq. (2.2-9). Although this is a very valuable subject its development would take us too far afield here and so we shall have little formal use for the inversion integral of Eq. (2.2-9). We shall generally use tables of Laplace transforms in order to obtain the inverse transform. Table (2.2-1) on page 32 presents Laplace transforms of some elementary time functions.

We have demonstrated that the response of a constant coefficient linear system, $z(t)$, to an arbitrary input $u(t)$ is given by the convolution integral

---

*One simple way to do this is to insert Eq. (2.2-8) into Eq. (2.2-9) and note that

$$\frac{1}{2\pi j} \int_{\sigma-j\infty}^{\sigma+j\infty} e^{-s(\eta-t)}\, ds = \delta(\eta - t) \qquad \sigma > 0$$

such that an identity results in that we obtain $f(t) = f(t)$.

$$z(t) = \int_0^\infty h(\tau)u(t-\tau)\,d\tau \qquad (2.2\text{-}10)$$

where $h(\tau)$ is the impulse response of the system. We now seek to determine the Laplace transformation of the system output $z(t)$. Using the definition of the Laplace transform, Eq. (2.2-8), we have

$$Z(s) = \int_0^\infty \left[\int_0^\infty h(\tau)u(t-\tau)\,d\tau \right] e^{-st}\,dt$$

We interchange the order of integration and have

$$Z(s) = \int_0^\infty \left[\int_0^\infty u(t-\tau)e^{-st}\,dt \right] h(\tau)\,d\tau$$

Since $e^{st}e^{-st} = 1$, we can certainly introduce this term into the foregoing which gives us, after a slight rearrangement,

$$Z(s) = \int_0^\infty \left[\int_0^\infty u(t-\tau)e^{-s(t-\tau)}\,dt \right] h(\tau)e^{-s\tau}\,d\tau$$

Now we make the change of variable $t - \tau = \eta$ and note that $u(\eta) = 0$ for $\eta < 0$, as required by a Laplace transformed time function, such that we have

$$Z(s) = \int_0^\infty \left[\int_0^\infty u(\eta)e^{-s\eta}\,d\eta \right] h(\tau)e^{-s\tau}\,d\tau$$

We immediately recognize the first integral as the Laplace transform definition of $U(s)$ and can then easily see that the remaining integral is just the definition of the transfer function $H(s)$, the Laplace transform of $h(t)$. We have just demonstrated that the Laplace transform of the output signal is just the product of the system transfer function and the Laplace transform of the system input. In symbols we have

$$Z(s) = H(s)U(s) \qquad (2.2\text{-}11)$$

This is a very important result as it turns out to be far simpler to take transforms of time function, multiply them and then look up the inverse transform than it is to evaluate convolution integrals. In addition, much physical insight can be obtained by careful inspection of system transfer functions as we shall see many times in our later chapters in linear systems control.

Before we can make extensive use of Eq. (2.2-11) we need to establish a table of Laplace transforms for commonly encountered time functions. The exponential and the singularity functions turn out to have particularly simple Laplace transforms. It is of value to see how a few of these are obtained.

EXAMPLE 2.2-3. The Laplace transform of the impulse $\delta(t - t_1)$, where $0 < t_1$, is easily obtained. From Eq. (2.2-8) we have

$$f(t) = \delta(t - t_1)$$

$$F(s) = \int_0^\infty \delta(t - t_1)e^{-st}\,dt = e^{-st_1}$$

EXAMPLE 2.2-4. The Laplace transform of the time function te^{-at} is also easily obtained. We have

$$f(t) = te^{-at} \qquad t > 0$$

$$F(s) = \int_0^\infty te^{-at}e^{-st}\,dt = \int_0^\infty te^{-(a+s)t}\,dt$$

We use the relation for integration by parts

$$\int u\,dv = uv - \int v\,du$$

and let $u = t$ to obtain $du = dt$. From $dv = e^{-(a+s)t}\,dt$ we obtain $v = -e^{-(a+s)t}/(s + a)$ such that

$$F(s) = te^{-(a+s)t}\Big|_0^\infty - \frac{1}{s + a}\int_0^\infty e^{-(a+s)t}\,dt$$

$$= \frac{1}{(a + s)^2}$$

In this example as well as in evaluating transforms in general we use $\lim_{t\to\infty} e^{-st} = \lim_{t\to\infty} e^{-(\sigma + j\omega)t} = 0$ for any $\sigma > 0$.

EXERCISE 2.2-5. Find the Laplace transform of a) $\sin \omega t,\ t > 0$; b) $e^{-\alpha t}\cos \omega t,\ t > 0$.

It will not be often that the impulse response of a system will be specified initially. It will occur far more often that the differential equation of a system will be given or obtained from physical considerations. Thus we need to be able to take the Laplace function of derivatives and other functions if we are to obtain maximum use of the Laplace transform method. Table (2.2-2) presents the Laplace transform of a number of useful general operational functions.

EXAMPLE 2.2-5.　One of the most useful of these is the Laplace transform of the derivative of a time function. This example will concern obtaining the Laplace transform of time derivatives of functions. Integration by parts easily yields the desired transforms.

$$\text{a) } f(t) = \frac{dg(t)}{dt}, \qquad F(s) = \int_0^\infty \frac{dg(t)}{dt} e^{-st}\, dt$$

We let $du = dg(t)$, $v = e^{-st}$ so that $u = g(t)$, $dv = -se^{-st}\, dt$ and obtain

$$F(s) = g(t)e^{-st}\Big|_0^\infty + s\int_0^\infty g(t)e^{-st}\, dt$$

so that

$$F(s) = -g(0) + sG(s)$$

b) $f(t) = \dfrac{d^2 e(t)}{dt^2}$. We just let $g(t) = de(t)/dt$ in part a) and utilize the results of part a) to yield

$$F(s) = -\frac{de(t)}{dt}\Big|_{t=0} + s\int_0^\infty \frac{de(t)}{dt} e^{-st}\, dt$$

We again integrate by parts and get

$$F(s) = s^2 E(s) - se(0) - \frac{de(t)}{dt}\Big|_{t=0}$$

c) We can easily extend these results to the general case and have

$$f(t) = \frac{d^n g(t)}{dt^n}, \quad F(s) = s^n G(s) - \sum_{i=1}^{n} s^{i-1}\frac{d^{n-i}g(t)}{dt^{n-i}}\Big|_{t=0}$$

TABLE 2.2-1 LAPLACE TRANSFORMS OF TYPICAL TIME FUNCTIONS

$f(t)$	$F(s)$
1	$\dfrac{1}{s}$
$t^n \ (n \geq 0)$	$\dfrac{n!}{s^{n+1}} \left(\dfrac{\Gamma(n+1)}{s^{n+1}} \text{ if } n \text{ not an integer} \right)$
e^{-at}	$\dfrac{1}{s+a}$
$t^n e^{-at}$	$\dfrac{n!}{(s+a)^{n+1}}$
$\cos \omega t$	$\dfrac{s}{s^2 + \omega^2}$
$\sin \omega t$	$\dfrac{\omega}{s^2 + \omega^2}$
$\sin (\omega t + \phi)$	$\dfrac{s \sin \phi + \omega \cos \phi}{s^2 + \omega^2}$
$t^n \cos \omega t$	$\dfrac{n!}{2} \dfrac{(s+j\omega)^{n+1} + (s-j\omega)^{n+1}}{(s^2 + \omega^2)^{n+1}}$
$t^n \sin \omega t$	$\dfrac{n!}{2j} \dfrac{(s+j\omega)^{n+1} - (s-j\omega)^{n+1}}{(s^2 + \omega^2)^{n+1}}$
$\sin \omega_1 t \sin \omega_2 t$	$\dfrac{2\omega_1 \omega_2 s}{[s^2 + (\omega_1 + \omega_2)^2][s^2 + (\omega_1 - \omega_2)^2]}$
$\cos \omega_1 t \cos \omega_2 t$	$\dfrac{s(s^2 + \omega_1^2 + \omega_2^2)}{[s^2 + (\omega_1 + \omega_2)^2][s^2 + (\omega_1 - \omega_2)^2]}$

TABLE 2.2-1 *(cont'd)*

$f(t)$	$F(s)$
$\sin \omega_1 t \cos \omega_2 t$	$\dfrac{\omega_1 (s^2 + \omega_1^2 - \omega_2^2)}{[s^2 + (\omega_1 + \omega_2)^2][s^2 + (\omega_1 - \omega_2)^2]}$
$e^{-at} \sin (\omega t + \phi)$	$\dfrac{(s + a) \sin \phi + \omega \cos \phi}{(s + a)^2 + \omega^2}$

TABLE 2.2-2 LAPLACE TRANSFORMS OF USEFUL GENERAL OPERATIONAL FUNCTIONS

$f(t)$	$F(s)$
$f_1(t) + f_2(t)$	$F_1(s) + F_2(s)$
$\dfrac{df(t)}{dt}$	$sF(s) - f(0)$
$\dfrac{d^2 f(t)}{dt}$	$s^2 F(s) - sf(0) - \dfrac{df(t)}{dt}\bigg|_{t=0}$
$\dfrac{d^n f(t)}{dt^n}$	$s^n F(s) - \displaystyle\sum_{i=1}^{n} s^{i-1} \dfrac{d^{n-1} f(t)}{dt^{n-1}}\bigg|_{t=0}$
$\displaystyle\int_0^t f(\lambda)\, d\lambda = g(t)$	$\dfrac{F(s)}{s} + \dfrac{g(0)}{s}$
$\displaystyle\int_0^t \int_0^t f(\lambda_1)\, d\lambda_1\, d\lambda_2 = \int_0^t g(\lambda_2)\, d\lambda_2 = h(t)$	$\dfrac{F(s)}{s^2} + \dfrac{g(0)}{s^2} + \dfrac{h(0)}{s}$
$af(t)$	$aF(s)$
$f(t/a)$	$aF(as)$

TABLE 2.2-2 *(cont'd)*

$f(t)$	$F(s)$
$f(t-a)$	$e^{-as}F(s)$
$e^{-at}f(t)$	$F(s+a)$
$tf(t)$	$-\dfrac{dF(s)}{ds}$
$\displaystyle\int_0^t f(\lambda)g(t-\lambda)\,d\lambda$	$F(s)G(s)$
$\displaystyle\lim_{t\to\infty} f(t)$	$\displaystyle\lim_{s\to 0} sF(s)$
$\displaystyle\lim_{t\to 0} f(t)$	$\displaystyle\lim_{s\to\infty} sF(s)$

Now we are in a position to take the Laplace transform of the linear differential equation (2.1-1) or (2.1-2) for a general linear constant coefficient system. We assume that the system is initially unexcited such that all initial conditions are zero. We use the results of EXAMPLE 2.2-5 and have, from Eq. (2.1-2),

$$\sum_{i=0}^{n} a_i s^i Z(s) = \sum_{i=0}^{m} b_i s^i U(s)$$

and thus have

$$Z(s) = H(s)U(s) \tag{2.2-11}$$

where

$$H(s) = \frac{\displaystyle\sum_{i=0}^{m} b_i s^i}{\displaystyle\sum_{i=0}^{n} a_i s^i} \tag{2.2-12}$$

is called the system transfer function. We note that this relation is just what we obtained by Laplace transforming the convolution integral.

Now we see a generally easy way to find the transfer function of a system. It will normally be quite a difficult task to find the impulse response of a system directly. A much simpler approach is to take the Laplace transform of the system differential equations to obtain the system transfer function. We can take the inverse of this transfer function to obtain the system impulse response. This will generally require factoring of polynomials in the Laplace variable s and, for other than simple systems, numerical methods will often be needed to accomplish this.

EXAMPLE 2.2-6. We consider the linear system described by

$$2\ddot{z} + 8\dot{z} + 6z = 3\dot{u} + 6u \tag{1}$$

It is desired to obtain the system transfer function and the system impulse response. All initial conditions are zero. We take the Laplace transform of the system input output equation and obtain

$$(2s^2 + 8s + 6)Z(s) = (3s + 6)U(s) \tag{2}$$

such that we have for the system transfer function

$$H(s) = \frac{3(s + 2)}{2(s + 3)(s + 1)} \tag{3}$$

This transfer function is said to have a zero at $s = -2$ and poles at $s = -1$ and $s = -3$. The Laplace transform of a unit impulse $u(t) = \delta(t)$ is $U(s) = 1$ and thus we have to take the inverse transform of (3). This is not in a form such that the more elementary Laplace transform tables will contain an entry so that the time function can be directly determined. We may use the method of partial fractions, which we discuss next, to obtain

$$H(s) = \frac{3/4}{s + 3} + \frac{3/4}{s + 1}$$

and then immediately find the time function corresponding to the impulse response

$$h(t) = 3/4\,(e^{-3t} + e^{-t}) \qquad t > 0$$

EXAMPLE 2.2-7. Laplace transform manipulations are as equally applicable to systems described in state variable form as they are to systems described in input output form. As an example, let us consider the standard state variable representation

$$\dot{\mathbf{x}} = \mathbf{A}\mathbf{x} + \mathbf{b}u, \; z = \mathbf{c}\mathbf{x} + du \tag{1}$$

with

$$\mathbf{A} = \begin{bmatrix} -3 & 0 \\ 1 & -1 \end{bmatrix} \quad \mathbf{b} = \begin{bmatrix} 3/2 \\ 0 \end{bmatrix}$$

$$\mathbf{c} = \begin{bmatrix} 1 & 1 \end{bmatrix} \quad d = 0$$

We take the Laplace transform of (1) to obtain

$$s\mathbf{X}(s) - \mathbf{X}(0) = \mathbf{A}\mathbf{X}(s) + \mathbf{b}U(s) \tag{2}$$

$$Z(s) = \mathbf{c}X(s) + dU(s) \tag{3}$$

We may formally solve (2) for $\mathbf{X}(s)$ as

$$\mathbf{X}(s) = (s\mathbf{I} - \mathbf{A})^{-1}[\mathbf{x}(0) + \mathbf{b}U(s)]$$

We have for the output

$$Z(s) = [\mathbf{c}(s\mathbf{I} - \mathbf{A})^{-1}\mathbf{b} + d]\,U(s) + \mathbf{c}(s\mathbf{I} - \mathbf{A})^{-1}\mathbf{x}(0)$$

The system transfer function for this single input single output system is

$$H(s) = \mathbf{c}(s\mathbf{I} - \mathbf{A})^{-1}\mathbf{b} + d \tag{4}$$

and this is a general result. The expression for the Laplace transformation of the fundamental or transition matrix is

$$\Phi(s) = (s\mathbf{I} - \mathbf{A})^{-1} \tag{5}$$

and this is also a general result.

For the specific parameters assumed for this example we have

$$s\mathbf{I} - \mathbf{A} = \begin{bmatrix} s & 0 \\ 0 & s \end{bmatrix} - \begin{bmatrix} -3 & 0 \\ 1 & -1 \end{bmatrix} = \begin{bmatrix} s+3 & 0 \\ -1 & s+1 \end{bmatrix}$$

The inverse matrix is easily obtained as

$$\Phi(s) = (s\mathbf{I} - \mathbf{A})^{-1} = \frac{1}{(s+1)(s+3)} \begin{bmatrix} s+1 & 0 \\ 1 & s+3 \end{bmatrix}$$

We can take the inverse transform of this to yield

$$\phi(t) = \mathcal{L}^{-1}\Phi(s) = \begin{bmatrix} e^{-3t} & 0 \\ 1/2(e^{-t} - e^{-3t}) & e^{-t} \end{bmatrix}$$

and this is the fundamental or transition matrix for this example. The system transfer function of (4) becomes

$$H(s) = \begin{bmatrix} 1 & 1 \end{bmatrix} \begin{bmatrix} \dfrac{1}{s+3} & 0 \\ \dfrac{1}{(s+1)(s+3)} & \dfrac{1}{s+1} \end{bmatrix} \begin{bmatrix} 3/2 \\ 0 \end{bmatrix}$$

$$= \begin{bmatrix} 1 & 1 \end{bmatrix} \begin{bmatrix} \dfrac{3/2}{s+3} \\ \dfrac{3/2}{(s+1)(s+3)} \end{bmatrix} = \frac{3(s+2)}{2(s+3)(s+1)}$$

and this corresponds to the impulse response

$$h(t) = 3/4(e^{-3t} + e^{-t}) \qquad t > 0$$

EXERCISE 2.2-6. What are the fundamental or transition matrix, input output system transfer function, and impulse response of the state variable systems described in EXERCISE 2.2-1?

For a complex system, it will often occur that the Laplace transform of a system output for a given input will be too complex to be found in a simple table of inverse Laplace transforms. It is a simple matter to show that the transfer function of a linear constant coefficient system with no pure time delays in the system will always be a ratio of rational polynomials in s. In a similar way the Laplace transform of any exponential time function, no matter how complex, will always turn out to be a ratio of polynomials in s. Thus the Laplace transform for essentially all systems of interest* will turn out to be a ratio of rational polynomials in s. Let us look at some illustrative examples to see how we might simplify ratios of polynomials in s and then state some general rules.

EXAMPLE 2.2-8. It is well known in elementary algebra that it is possible to expand ratios of rational polynomials in s into simpler functions. For example, we may write for $\alpha_1 \neq \alpha_2$,

$$F(s) = \frac{\beta_1 s + \beta_0}{(s + \alpha_1)(s + \alpha_2)} = \frac{K_1}{s + \alpha_1} + \frac{K_2}{s + \alpha_2}$$

and multiply both sides of this equation by $(s + \alpha_1)(s + \alpha_2)$ and obtain

$$K_1(s + \alpha_2) + K_2(s + \alpha_1) = \beta_1 s + \beta_0$$

If we let $s = -\alpha_2$ we obtain

$$K_2 = \left. \frac{\beta_1 s + \beta_0}{s + \alpha_1} \right|_{s=-\alpha_2} = \frac{-\beta_1 \alpha_2 + \beta_0}{-\alpha_2 + \alpha_1}$$

If we let $s = -\alpha_1$, we obtain

$$K_1 = \left. \frac{\beta_1 s + \beta_0}{s + \alpha_2} \right|_{s=-\alpha_1} = \frac{-\beta_1 \alpha_1 + \beta_0}{-\alpha_1 + \alpha_2}$$

and we have obtained what is commonly called a partial fraction of the ratio of polynomials. We can easily take the inverse Laplace transform of the first order polynomials in s. Thus we see that the inverse transform of the function $F(s)$ expressed in terms of its partial fraction expansion

---

*An exception to this occurs when there are time delays incorporated within the feedback loop of a system. It is quite difficult to determine the time response of such systems.

$$F(s) = \frac{\beta_1 s + \beta_0}{(s + \alpha_1)(s + \alpha_2)}$$

$$= \frac{1}{\alpha_1 - \alpha_2} \left[\frac{-\beta_0 + \beta_1 \alpha_1}{s + \alpha_1} + \frac{\beta_0 - \beta_1 \alpha_2}{s + \alpha_2} \right]$$

is

$$f(t) = \mathcal{L}^{-1} \frac{\beta_1 s + \beta_0}{(s + \alpha_1)(s + \alpha_2)}$$

$$= \frac{1}{\alpha_1 - \alpha_2} \left[(-\beta_0 + \beta_1 \alpha_1) e^{-\alpha_1 t} + (\beta_0 - \beta_1 \alpha_2) e^{-\alpha_2 t} \right]$$

for $t \geq 0$. The time function is, of course, zero for $t < 0$.

EXAMPLE 2.2-9. When we have repeated roots the procedure to obtain the partial fraction is more complex. For example, we must write for the case of a double root

$$F(s) = \frac{\beta_0}{(s + \alpha_1)^2 (s + \alpha_2)} = \frac{K_1}{(s + \alpha_1)^2} + \frac{K_2}{(s + \alpha_1)} + \frac{K_3}{(s + \alpha_2)}$$

We multiply both sides of this relation by $(s + \alpha_1)^2 (s + \alpha_2)$ and obtain

$$K_1 (s + \alpha_2) + K_2 (s + \alpha_1)(s + \alpha_2) + K_3 (s + \alpha_1)^2 = \beta_0 \tag{1}$$

We can use the same rule we used in the last example to evaluate K_1 and K_3. We obtain

$$K_1 = \frac{\beta_0}{s + \alpha_2} \bigg|_{s=-\alpha_1} = \frac{\beta_0}{\alpha_2 - \alpha_1}$$

$$K_3 = \frac{\beta_0}{(s + \alpha_1)^2} \bigg|_{s=-\alpha_2} = \frac{\beta_0}{(\alpha_1 - \alpha_2)^2}$$

However, this procedure cannot be applied to determine K_2. If we divide both sides of (1) by $(s + \alpha_2)$ and take the derivative of the result with respect to s and evaluate this at $s = -\alpha_1$, we obtain

$$K_1 + K_2(s + \alpha_1) + \frac{K_3(s + \alpha_1)^2}{(s + \alpha_2)} = \frac{\beta_0}{s + \alpha_2}$$

and

$$0 + K_2 + K_3 \left. \frac{2(s + \alpha_2)(s + \alpha_1) - (s + \alpha_1)^2}{(s + \alpha_2)^2} \right|_{s=-\alpha_1} = \left. \frac{-\beta_0}{(s + \alpha_2)^2} \right|_{s=-\alpha_1}$$

such that

$$K_2 = \frac{-\beta_0}{(\alpha_2 - \alpha_1)^2}$$

and thus have determined the partial fraction expansion for this example. The time function has been determined to be

$$f(t) = \frac{\beta_0}{\alpha_2 - \alpha_1} \left[te^{-\alpha_1 t} - \frac{1}{\alpha_2 - \alpha_1}(e^{-\alpha_2 t} - e^{-\alpha_1 t}) \right]$$

EXAMPLE 2.2-10. Complex conjugate poles do not present any particular difficulty in terms of partial fraction expansions. Since complex conjugate poles occur so frequently it is desirable to present an example which treats this important class of problems. We consider

$$F(s) = \frac{\omega_0^2}{s(s^2 + 2\zeta\omega_0 s + \omega_0^2)} = \frac{K_1}{s} + \frac{K_2}{(s + \alpha + j\beta)} + \frac{K_3}{(s + \alpha - j\beta)}$$

where $\alpha = \zeta\omega_0$ and $\beta = \omega_0\sqrt{1 - \zeta^2}$. By repeating the procedure of EX-AMPLE 2.2-8 we obtain

$$K_1 = sF(s)\Big|_{s=0} = 1$$

$$K_2 = (s + \alpha + j\beta)F(s)\Big|_{s=-\alpha-j\beta} = \frac{\omega_0^2}{2j\beta(\alpha + j\beta)}$$

$$K_3 = (s + \alpha - j\beta)F(s)\Big|_{s=-\alpha+j\beta} = \frac{\omega_0^2}{2j\beta(-\alpha + j\beta)}$$

thus

$$F(s) = \frac{1}{s} + \frac{\omega_0^2}{2\beta}\left[\frac{1}{j}\frac{1}{\alpha + j\beta}\frac{1}{s + \alpha + j\beta} + \frac{1}{j}\frac{1}{-\alpha + j\beta}\frac{1}{s + \alpha - j\beta}\right]$$

$$= \frac{1}{s} + \frac{\omega_0}{2\beta}\left[\frac{\exp\{-j[(\pi/2) + \theta]\}}{s + \alpha + j\beta} + \frac{\exp\{j[(\pi/2) + \theta]\}}{s + \alpha - j\beta}\right]$$

where we use the Euler identity

$$e^{jx} = \cos x + j \sin x$$

to obtain

$$\theta = \tan^{-1}\left[\frac{-\beta}{\alpha}\right]$$

We take the inverse Laplace transformation and obtain

$$f(t) = 1 + \frac{1}{\sqrt{1 - \zeta^2}}\, e^{-\zeta\omega_0 t} \sin\left(\sqrt{1 - \zeta^2}\,\omega_0 t - \theta\right)$$

EXAMPLE 2.2-11. As our last example of partial fraction expansions we consider the case where the numerator polynomial is of greater order than the demonimator polynomial. We consider the specific example

$$F(s) = \frac{s^3 + 1}{(s + 1)(s + 2)} \tag{1}$$

As s approaches infinity, $F(s)$ approaches infinity. The numerator is one order greater than the denominator. Thus $F(s)$ must contain a $K_1 s$ term and may also contain a constant K_2 term. So it is reasonable that $F(s)$ is of the form

$$F(s) = K_1 s + K_2 + \frac{K_3}{s+1} + \frac{K_4}{s+2}$$

A long division procedure may be employed to extract the K_1 and K_2 terms from $F(s)$. From (1) we have

$$
\begin{array}{r}
s - 3 \\
s^2 + 3s + 2 \overline{)s^3 + 0s^2 + 0s + 1} \\
\underline{s^3 + 3s^2 + 2s} \\
-3s^2 - 2s + 1 \\
\underline{-3s^2 - 9s - 6} \\
7s + 7
\end{array}
$$

such that we have

$$F(s) = s - 3 + \frac{7s + 7}{s^2 + 3s + 2}$$

We expand the ratio of polynomials in the usual fashion to obtain

$$\frac{7s + 7}{(s + 2)(s + 1)} = \frac{7}{s + 2} + \frac{0}{s + 1}$$

where we note the curious occurrence of a zero coefficient in the partial fraction expansion. The reason for this is that a root of $(s^3 + 1)$ is $s = -1$, and thus there is partial cancellation of numerator and denominator terms in $F(s)$. We should have determined this prior to starting the partial fraction expansion although we see that there is no error committed in this case if we do not do this. In other words we should have rewritten the Laplace transform $F(s)$ as

$$F(s) = \frac{s^2 - s + 1}{s + 2} = s - 3 + \frac{7}{s + 2}$$

The time function for this example consists of a doublet (the time derivative of an impulse), an impulse and a single exponential. The time function is

$$f(t) = \delta'(t) - 3\delta(t) + 7e^{-2t}$$

Now that we have solved some particular examples involving partial fractions of rational polynomials, let us state the general rules for determining partial fraction expansions.

1. Determine the simplest form of the ratio of polynomials. Cancel out any factors common to the numerator and demoninator.

2. If the numerator of the polynomial is greater than or of equal degree to the denominator, use long division to obtain a remainder ratio of polynomials with denominator polynomial of greater degree than numerator polynomial. Thus we expand such that, where the order of $Q(s)$ is m greater than $P(s)$,

$$F(s) = \frac{Q(s)}{P(s)} = \sum_{i=1}^{m} M_i s^i + \frac{Q'(s)}{P(s)} \tag{2.2-13}$$

and further expand $Q'(s)$ by the partial fraction rules to follow.

3. If all roots of $P(s)$ or poles are real and simple then $F(s)$ can be written as

$$F(s) = \frac{Q(s)}{P(s)} = \frac{Q(s)}{\displaystyle\prod_{i=1}^{n}(s + p_i)} \tag{2.2-14}$$

We can expand $F(s)$ in partial fractions as

$$F(s) = \sum_{i=1}^{n} \frac{K_i}{s + p_i} \tag{2.2-15}$$

where K_j is determined from

$$K_j = (s + p_j)F(s)\Big|_{s=-p_j} = \frac{Q(-p_j)}{\displaystyle\prod_{\substack{i=1 \\ i \neq j}}^{n}(s + p_i)} \tag{2.2-16}$$

4. If m of the roots or poles in $F(s)$ are identical, or in equivalent language the pole at p_j is of multiplicity m, then the Laplace transform $F(s)$ is of the form

$$F(s) = \frac{Q(s)}{P(s)} = \frac{Q(s)}{(s + p_k)^m P'(s)} = \frac{Q(s)}{(s + p_k)^m \prod_{i=1}^{n-m}(s + p_i)} \qquad (2.2\text{-}17)$$

The residues of the poles in $P'(s)$, as the partial fraction coefficients of the poles of $P'(s)$ are called, may be determined by applying Rule 2 to the ratio of polynomials $P(s)/Q(s)$. Thus for the non-repeated roots we simply use Eq. (2.2-16) to determine the residues or partial fraction coefficients K_j.

For the repeated root we use the relationships

$$L_m = (s + p_k)^m F(s)\Big|_{s=-p_k}$$

$$L_{m-1} = \frac{d}{ds}\left[(s + p_k)^m F(s)\right]\Big|_{s=-p_k}$$

$$L_{m-2} = \frac{1}{2!}\frac{d^2}{ds^2}\left[(s + p_k)^m F(s)\right]\Big|_{s=-p_k} \qquad (2.2\text{-}18)$$

$$\vdots$$

$$L_{m-\ell} = \frac{1}{\ell!}\frac{d^\ell}{ds^\ell}\left[(s + p_k)^m F(s)\right]\Big|_{s=-p_k}$$

$$\vdots$$

$$L_1 = \frac{1}{(m-1)!}\frac{d^{m-1}}{ds^{m-1}}\left[(s + p_k)^m F(s)\right]\Big|_{s=-p_k}$$

The complete partial fraction expansion is then

$$F(s) = \sum_{i=1}^{n-m}\frac{K_i}{s + p_i} + \sum_{i=1}^{m}\frac{L_i}{(s + p_k)^i} \qquad (2.2\text{-}19)$$

5. Multiple repeated roots are treated one repeated root at a time utilizing the procedure outlined in step four for a single repeated root considering all other roots as belonging in $P'(s)$.

6. Complex conjugate poles pose no particular problem although they should be considered together as we have seen in EXAMPLE 2.2-10.

It is convenient to summarize our procedure for obtaining the inverse Laplace transform by using the partial fraction expansion method. The DELTA chart, described in Appendix A, is a very convenient graphical procedure to accomplish this. Figure (2.2-4) represents a DELTA chart outlining the steps in processing $F(s)$ to obtain $f(t)$ using the partial fraction expansion method. In conjunction with the brief table of inverse Laplace transforms we present as Table 2.2-3, use of this DELTA chart will allow us to find the time function corresponding to essentially any Laplace transform that could represent the output time function of a linear system.

EXERCISE 2.2-7. Find the inverse Laplace transforms of the $F(s)$ functions given.

$$\text{a) } \frac{s^2 + 2}{s + 1} \qquad \text{b) } \frac{s^2 + 2s + 2}{s(s + 1)(s + 2)} \qquad \text{c) } \frac{3s - 4}{s(s + 1)(s + 2)}$$

$$\text{d) } \frac{1}{s^5(s + 1)}$$

[Answers: a) $\delta'(t) - \delta(t) + 3e^{-t}$; c) $7e^{-t} - 2 - 5e^{-2t}$]

EXERCISE 2.2-8. Find the inverse Laplace transforms of the $F(s)$ functions given.

$$\text{a) } \frac{1}{s^2(s + 1)^2(s + 2)^2} \qquad \text{b) } \frac{1}{s(s^2 + 2s + 10)}$$

$$\text{c) } \frac{s}{(s^2 + 4)(s^2 + 2s + 2)^2} \qquad \text{d) } \frac{s + 1}{s(s + 2)^3}$$

$$\left[\begin{array}{l} \text{Answers: a) } -\frac{3}{4} + \frac{t}{4} + te^{-t} + \frac{3}{4}e^{-2t} + \frac{1}{4}te^{-2t}; \\[2mm] \qquad\qquad \text{b) } 0.1 - 0.1055e^{-t}\sin(3t + 71.56°) \\[2mm] \qquad\qquad \text{d) } \frac{1}{8}[1 - e^{-2t} - 2te^{-2t} + 2t^2e^{-2t}] \end{array}\right]$$

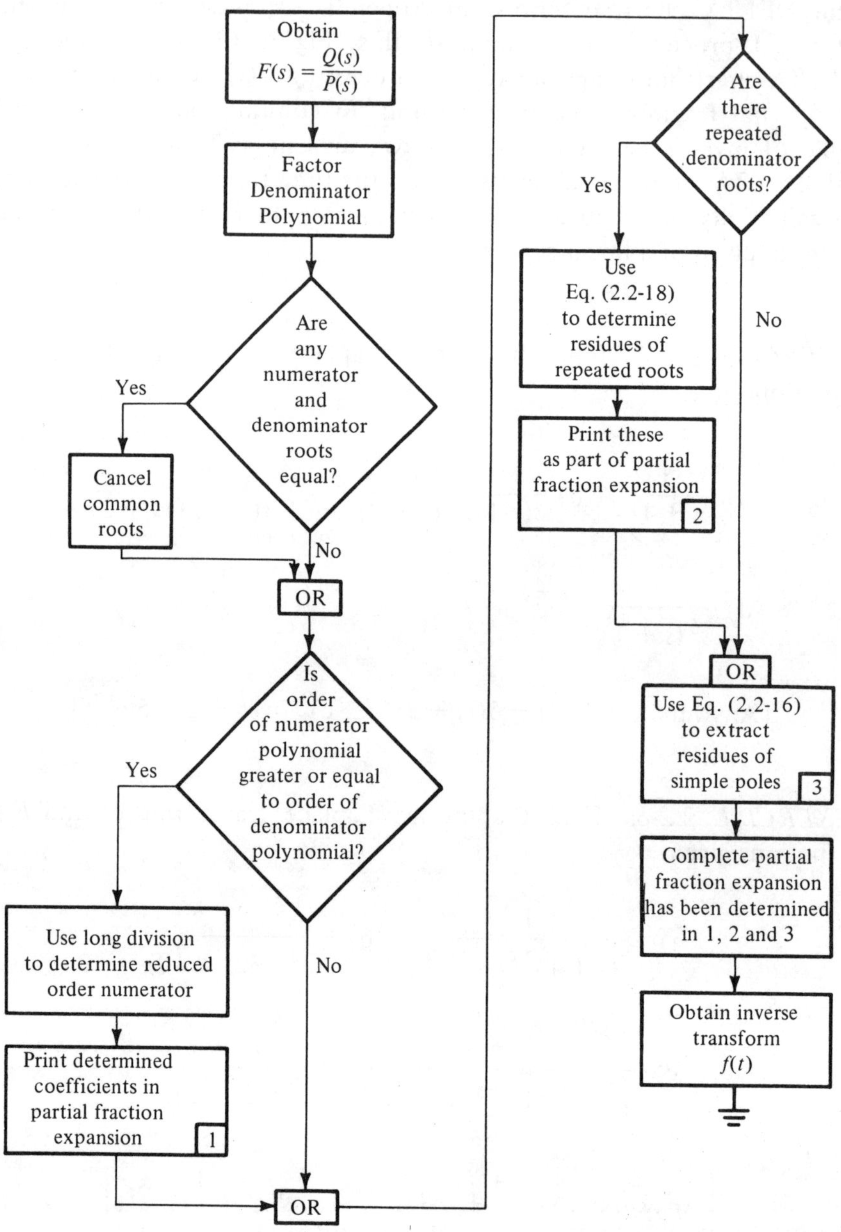

Figure 2.2-4   DELTA Chart of procedure to obtain inverse transform of $F(s)$

TABLE 2.2-3   BRIEF TABLE OF INVERSE LAPLACE TRANSFORMS

$F(s)$	$f(t)$
1	$\delta(t)$
$\dfrac{1}{s}$	1
$\dfrac{1}{s^2}$	t
$\dfrac{1}{s^n},\ n>0$	$\dfrac{1}{(n-1)!}t^{n-1}\left[\dfrac{1}{\Gamma(n)}t^{n-1}\ \begin{array}{l}\text{if } n \text{ not}\\ \text{an integer}\end{array}\right]$
s	$\dfrac{d\delta(t)}{dt}=\delta'(t)$
s^n	$\dfrac{d^n\,\delta(t)}{dt^n}=\delta^n(t)$
$\dfrac{1}{s+a}$	e^{-at}
$\dfrac{1}{(s+a)^n}$	$\dfrac{1}{(n-1)!}t^{n-1}e^{-at}$
$\dfrac{1}{(a+s)s}$	$\dfrac{1-e^{-at}}{a}$
$\dfrac{1}{(a+s)(b+s)}$	$\dfrac{e^{-at}-e^{-bt}}{b-a}$
$\dfrac{c+s}{(a+s)(b+s)}$	$\dfrac{(c-a)e^{-at}-(c-b)e^{-bt}}{b-a}$
$\dfrac{c+s}{(a+s)(b+s)(d+s)}$	$\dfrac{(c-a)e^{-at}}{(b-a)(d-a)}+\dfrac{(c-b)e^{-bt}}{(d-b)(a-b)}+\dfrac{(c-d)e^{-dt}}{(a-d)(b-d)}$

TABLE 2.2-3 *(cont'd)*

$F(s)$	$f(t)$
$\dfrac{1}{s^2(a+s)}$	$\dfrac{at+e^{-at}-1}{a^2}$
$\dfrac{b+s}{s(a+s)^2}$	$\dfrac{bat+(b-a)e^{-at}+(a-b)}{a^2}$
$\dfrac{s}{s^2+\omega^2}$	$\cos\omega t$
$\dfrac{\omega}{s^2+\omega^2}$	$\sin\omega t$
$\dfrac{s+b}{s^2+\omega^2}$	$\dfrac{(b^2+\omega^2)^{\frac{1}{2}}\sin(\omega t+\phi)}{\omega}$ $\phi=\tan^{-1}\dfrac{\omega}{b}$
$\dfrac{s+b}{s(s^2+\omega^2)}$	$\dfrac{b-(b^2+\omega^2)^{\frac{1}{2}}\cos(\omega t+\phi)}{\omega^2}$ $\phi=\tan^{-1}\dfrac{\omega}{b}$
$\dfrac{1}{(s+a)^2+b^2}$	$\dfrac{1}{b}e^{-at}\sin bt$
$\dfrac{s+c}{(s+a)^2+b^2}$	$\dfrac{(a^2+b^2+c^2-2ac)^{\frac{1}{2}}e^{-at}\sin(bt+\phi)}{b}$ $\phi=\tan^{-1}\dfrac{b}{c-a}$

TABLE 2.2-3 *(cont'd)*

$F(s)$	$f(t)$
$\dfrac{1}{s[(s+a)^2 + b^2]}$	$\dfrac{b + (a^2 + b^2)^{\frac{1}{2}} e^{-at} \sin(bt - \phi)}{b(a^2 + b^2)}$ $\phi = -\tan^{-1} \dfrac{b}{a}$

EXERCISE 2.2-9. Find the transfer function for the system described by

$$\frac{d^3 z}{dt^3} + 5\frac{d^2 z}{dt^2} + 8\frac{dz}{dt} + 4z = \frac{du}{dt} + 4u$$

What is the linear system response if the input is a unit step $u(t) = 1$ for $t \geqslant 0$?

2.3 LINEAR SYSTEM BLOCK DIAGRAMS

It is very helpful to have a pictorial representation to help describe the flows of information and physical quantities in a system. We will examine two essentially equivalent ways of accomplishing this in this text. We will examine the use of linear system block diagrams in this section. This is the traditional conventional approach of the control systems engineer. We shall show that this method leads to the same transfer function determination formulae as the flow graph method.

In our previous discussions we have defined the Laplace transform of a system impulse response $h(t)$ as the system transfer function $H(s)$. It is often convenient to display this as a simple block diagram as in Fig. (2.3-1). Although it is proper to call either of these representations a *block diagram* we will usually reserve the use of this term for the frequency domain representation.

Systems are often constructed by interconnecting subsystems. It will be of considerable value therefore to be able to have a convenient

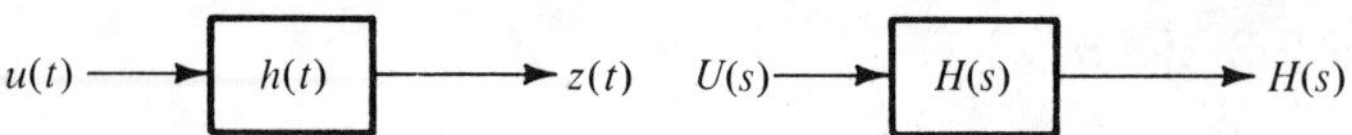

Figure 2.3-1   Simple block diagram representation

method to cope with two (or more) coupled subsystems as indicated in Fig. (2.3-2). We must be very careful to describe precisely what we mean by the interconnection shown in this figure. The easiest way to accomplish this is to use a differential equation representation of the two subsystems. We assume that

$$\sum_{i=0}^{n_1} a_{1i} \frac{d^i z_1}{dt^i} = \sum_{i=0}^{m_1} b_{1i} \frac{d^i u_1}{dt^i} \tag{2.3-1}$$

describes the first system. The second system is assumed to be described by

$$\sum_{i=0}^{n_2} a_{2i} \frac{d^i z_2}{dt^i} = \sum_{i=0}^{m_2} b_{2i} \frac{d^i u_2}{dt^i} \tag{2.3-2}$$

The systems are assumed to be interconnected such that the output from the first system serves as the input to the second system. Thus we assume

$$u_2 = z_1 \tag{2.2-3}$$

There are some physical requirements, involving such things as non-loading considerations, which must be imposed to insure that the output z_1 of the first differential equation can be input to the second differential equation without altering the physical characteristics and hence the differential equation of the first system. Our next chapter will examine some of these modeling considerations.

There are several ways in which we could describe this coupled system. We could use the impulse response characterization to obtain

$$z_1(t) = \int_0^\infty h_1(\tau_1) u_1(t - \tau_1) \, d\tau_1$$

$$z_2(t) = \int_0^\infty h_2(\tau_2) u_2(t - \tau_2) \, d\tau_2$$

$$u_2(t) = z_1(t)$$

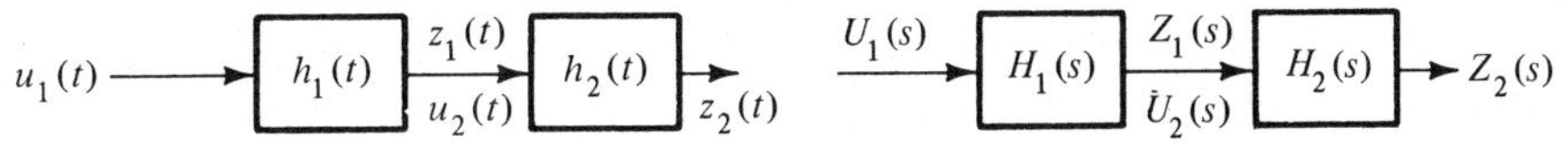

Figure 2.3-2 Interconnection of two subsystems

We can combine these three relations into a single relation which expresses the input-output dependency of the coupled system. We obtain

$$z_2(t) = \int_0^\infty \int_0^\infty h_2(\tau_2) h_1(\tau_1) u_1(t - \tau_1 - \tau_2)\, d\tau_1\, d\tau_2 \qquad (2.3\text{-}4)$$

but this is a difficult integral to evaluate and shows very complex time domain coupling between subsystems 1 and 2. Fortunately, the Laplace transform of this equation yields a very simple result as we will now demonstrate. We use the fundamental definition of the Laplace transform and obtain for Eq. (2.3-4)

$$Z_2(s) = \int_0^\infty \int_0^\infty \int_0^\infty h_2(\tau_2) h_1(\tau_1) u_1(t - \tau_1 - \tau_2) e^{-st}\, d\tau_1\, d\tau_2\, dt$$

We now insert exp $[-s(\tau_1 + \tau_2)]$ exp$[s(\tau_1 + \tau_2)] = 1$ in the inner integral of this expression, interchange the order of integration such that we integrate first over t, then over τ_1 and finally over τ_2. We carefully preserve the exp $[-s(t - \tau_1 - \tau_2)]$ term and obtain

$$Z_2(s) = \int_0^\infty h_2(\tau_2) \left\{ h_1(\tau_1) \left[\int_0^\infty u_1(t - \tau_1 - \tau_2) \exp[-s(t - \tau_1 - \tau_2)]dt \right. \right.$$

$$\left. \left. e^{-s\tau_1}\, d\tau_1 \right\} e^{-s\tau_2}\, d\tau_2 \right.$$

We change variables in the first integral to be evaluated by letting $\eta = t - \tau_1 - \tau_2$ and noting that $u_1(\eta) = 0$ for $\eta < 0$. Thus we recognize the inner integral as just the Laplace transform of $u_1(t)$ which we denote by $U_1(s)$. The other integrals are just the definitions of $H_1(s)$ and $H_2(s)$ such that we have finally and simply

$$Z_2(s) = H_1(s) H_2(s) U_1(s) \qquad (2.3\text{-}5)$$

The transfer function of the interconnected system is just the ratio $Z_2(s)/U_1(s)$ and we see that this is simply the product of the two transfer functions, a very simple and desirable relationship.

A somewhat simpler proof of this product relationship can be obtained by determining the transfer function of the individual subsystems. We have from Eqs. (2.3-1) through (2.3-3)

$$\frac{Z_1(s)}{U_1(s)} = \frac{\displaystyle\sum_{i=0}^{m_1} b_{1i}s^i}{\displaystyle\sum_{i=0}^{n_1} a_{1i}s^i} = \frac{Q_1(s)}{P_1(s)} = H_1(s)$$

$$\frac{Z_2(s)}{U_2(s)} = \frac{\displaystyle\sum_{i=0}^{m_2} b_{2i}s^i}{\displaystyle\sum_{i=0}^{n_2} a_{2i}s^i} = \frac{Q_2(s)}{P_2(s)} = H_2(s)$$

$$U_2(s) = Z_1(s)$$

$$\frac{Z_2(s)}{U_1(s)} = \frac{\dfrac{Z_2(s)}{U_2(s)}}{\dfrac{U_1(s)}{Z_1(s)}} = \frac{Z_2(s)}{U_2(s)}\frac{Z_1(s)}{U_1(s)} = H_2(s)H_1(s)$$

as we have obtained before.

EXAMPLE 2.3-1. We have just developed a very useful method to determine the transfer function of two interconnected systems. It is also of interest to obtain the impulse response of an interconnected system. For the system considered here we can obtain this from Eq. (2.3-4) by letting $u_1(t)$ be an impulse occurring at $t = 0$. For $u_1(t) = \delta(t)$, we obtain from Eq. (2.3-4)

$$h(t) = \int_0^\infty h_2(\tau_2)h_1(t - \tau_2)\,d\tau_2 \tag{1}$$

and we see that the impulse response of the interconnected system is just the convolution of the impulse response of the individual systems.

It will generally be simpler to obtain the transfer function of systems such as these and manipulate these rather than convolve impulse responses.

EXERCISE 2.3-1. Obtain the Laplace transform $H(s) = \mathcal{L}h(t)$ of the impulse response (1) of EXAMPLE 2.3-1. Use the fundamental definition of the Laplace transformation.

Two other elements are needed in order to complete our description of the fundamental linear block diagram elements. These are the summer and pickoff point and are depicted in Fig. (2.3-3). We usually place signs alongside a summer to denote whether inputs to the summer are added or subtracted. If no signs are shown all inputs are generally summed. As was the case with interconnection of subsystems it is assumed that the physical model is such that no loading occurs.

We could devise elements for multiplication and other nonlinear elements such as shown in Fig. (2.3-4). Since we are primarily concerned here with linear systems we shall not explore nonlinear elements in any detail here.

A linear system block diagram is a representation of oriented lines and transfer function boxes, each of which we label with a variable and interconnect using summers and pickoff points. More often than not there will be one or more feedback loops in a typical linear system block diagram. One very simple such diagram, representative of a linear positioning servomechanism, is shown in Fig. (2.3-5a). The presence of a closed loop whereby we can start at point $A(s)$ and move through the system in the feedforward position and return to point $A(s)$ is noted. The name *closed loop system* is used to denote systems of this type in which there are feedback elements such that we may go through a complete path in the forward direction and return to the starting point. We may combine the forward transfer functions G_3 and G_4 and redraw the block diagram as in Fig. (2.3-5b). We desire to obtain a single block, block diagram which represents the transfer function of this system. Such a single block is shown in Fig. (2.3-5c). Using the elements in Fig. (2.3-5b) we have the transfer function relations

$$Z(s) = G_2(s)A(s)$$

$$A(s) = U(s) - G_1(s)Z(s)$$

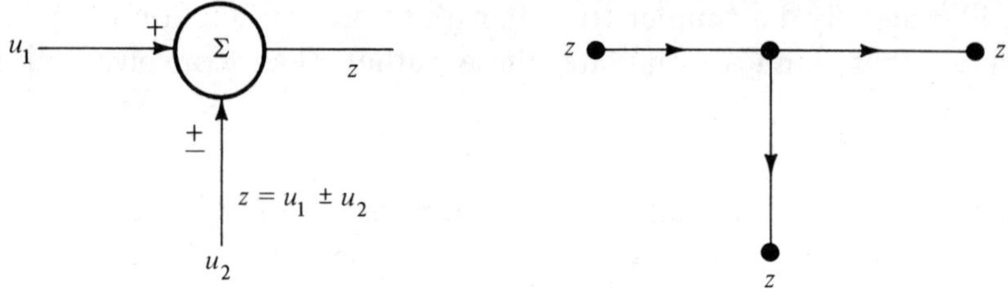

Figure 2.3-3 Summation and pickoff elements

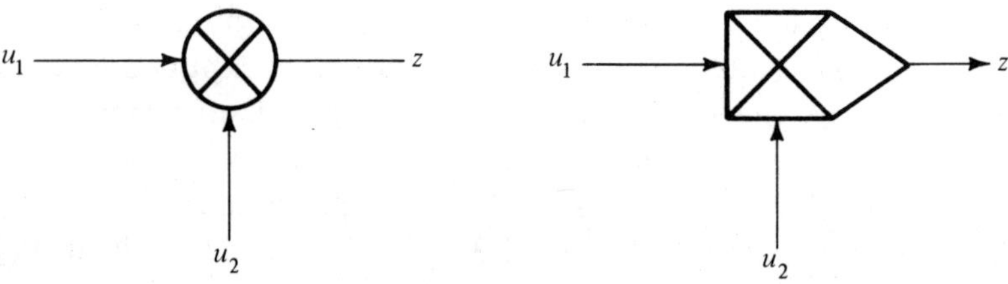

Figure 2.3-4 Alternate multiplier element $(z = u_1 u_2)$ symbols

which we may solve for the desired input output transfer relation as

$$\frac{Z(s)}{U(s)} = H(s) = \frac{G_2(s)}{1 + G_1(s)G_2(s)} \tag{2.3-6}$$

This relationship is very important. We may restate it as follows. *The input output transfer function for a single closed loop system is the forward transfer function divided by one plus the closed loop transfer function.*

Three other rules for block diagram manipulation are helpful. These concern a rule for combining transfer functions in parallel, and two rules for moving summing junctions and take off points around transfer functions. Figure (2.3-6) illustrates these three rules (a, b, and c) as well as the rules for cascading transfer functions and feedback loop reduction that we have already obtained.

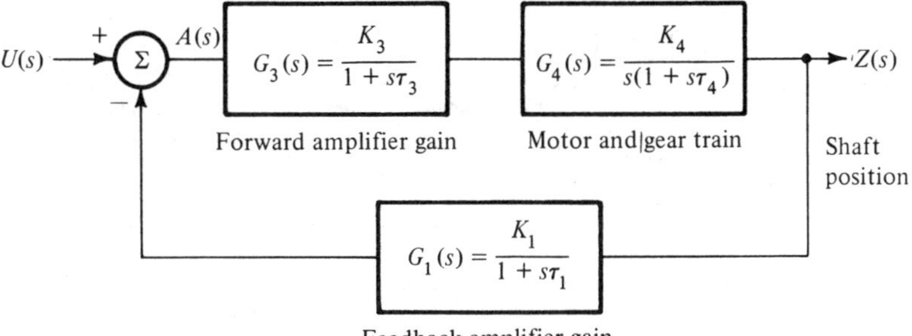

(a) Original block diagram of feedback system

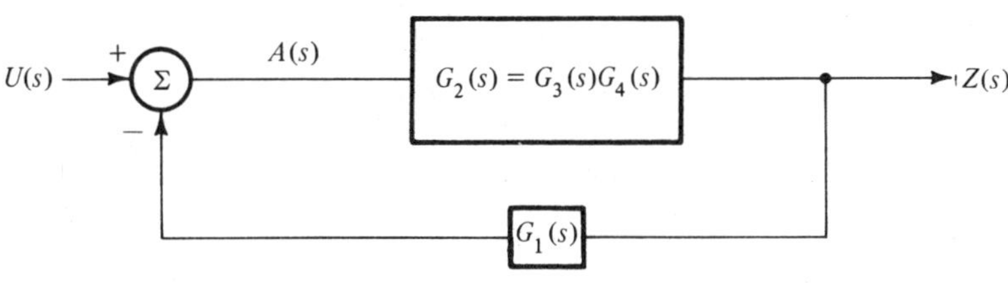

(b) Reduced block diagram

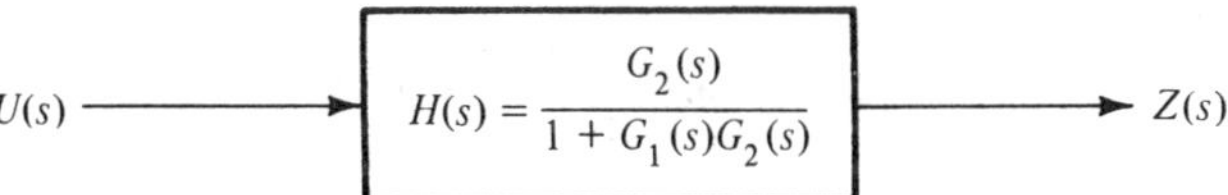

(c) Single block equivalent system

Figure 2.3-5 Simple feedback system and equivalents

EXERCISE 2.3-2. Demonstrate the validity of rules a), b), and c) in Fig. (2.3-6) for block diagram reduction.

We could now proceed to establish general formulae for transfer functions of complex multiple loop feedback systems. To motivate this

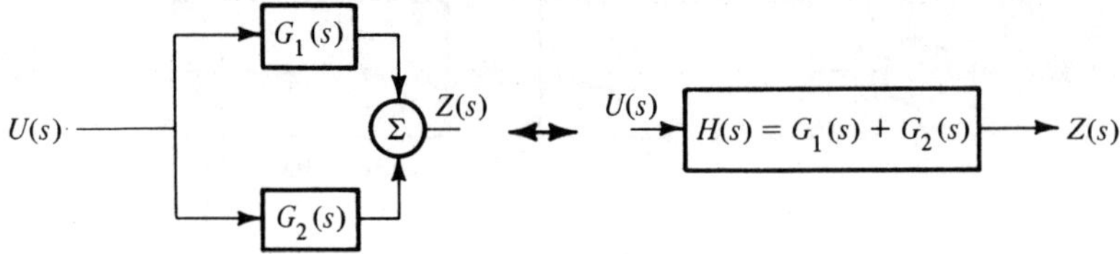

(a) Parallel transfer function

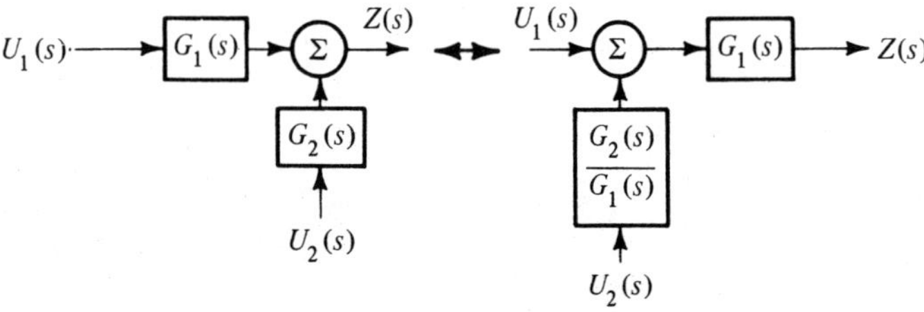

(b) Moving a summing junction

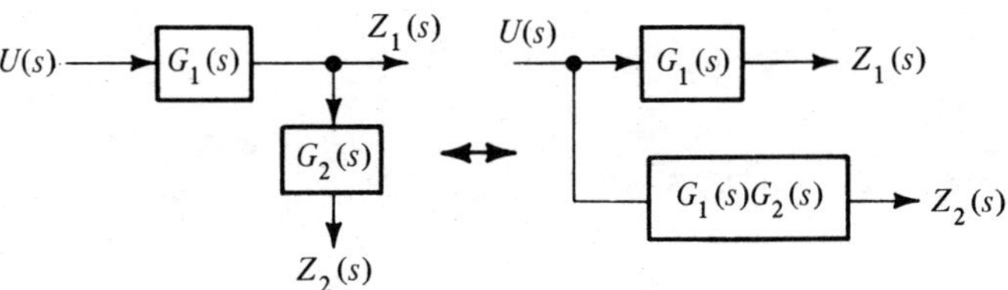

(c) Moving a takeoff point

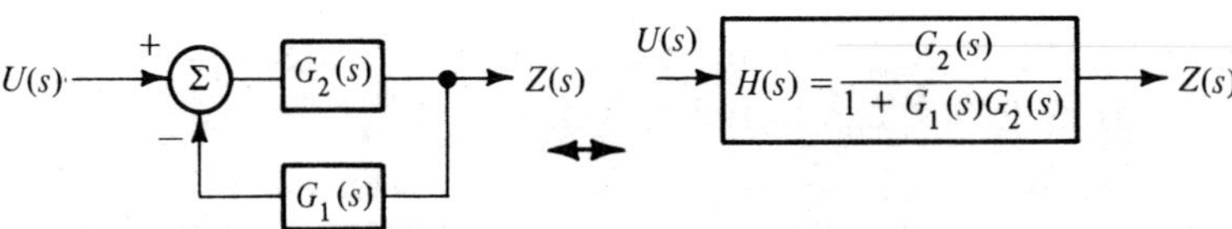

(d) Combining cascade transfer functions

(e) Single loop feedback system

Figure 2.3-6 Basic rules for block diagram reduction

let us consider several examples which establish some preliminary ground rules and then present the more complex general rule.

EXAMPLE 2.3-2. Let us consider the three loop feedback system shown in Fig. (2.3-7a). We desire to obtain the input output transfer function. Each of the single loop feedback systems may be reduced to yield Fig. (2.3-7b) by use of rule e) from Fig. (2.3-6e). The cascade rule (Fig. 2.3-6d) then allows us to represent the system as in Fig. (2.3-7c). Use of the simple feedback rule (Fig. 2.3-6e) then allows us to obtain the final single block transfer function of Fig. (2.3-7d).

We have reduced the complex three loop system to an equivalent transfer function. This is the end of the example as far as block diagram reduction is concerned. With respect to ultimately stating a general formula for multiple loop feedback systems it is desirable to give some further interpretation of the overall system transfer function which is

$$H(s) = \frac{G_1 G_3 G_4}{1 + G_1 G_2 + G_4 G_5 + G_1 G_3 G_4 G_6 + G_1 G_2 G_4 G_5}$$

In this expression we have deleted the (s) arguments for convenience. We notice that the numerator of this expression is just the forward transfer function from $U(s)$ to $Z(s)$. The denominator contains one minus the sum of the closed loop transfer functions (which sum is $-G_1 G_2 - G_4 G_5 - G_1 G_3 G_4 G_6$). Also there is the term $G_1 G_2 G_4 G_5$. This is the product of two of the closed loop transfer functions. We naturally ask why there are no products with the third loop transfer function $(G_1 G_3 G_4 G_6)$. A little introspection leads us to suspect that it is because this loop contains elements in common with other closed loops. Thus it seems that a plausible rule which certainly works for this example is that the overall input output transfer function is the forward transfer function divided by one plus the sum of the loop gains plus the product of loop gains taken two at a time with non-touching nodes or common paths or elements.

This turns out to be a correct rule for this system but not to be a complete rule as we see in our next example in which there is a minor loop feedback loop.

EXAMPLE 2.3-3. We now consider block diagram reduction for the system shown in Fig. (2.3-8) in which there is a minor loop feedback loop. The system shown is reduced by applying our block diagram

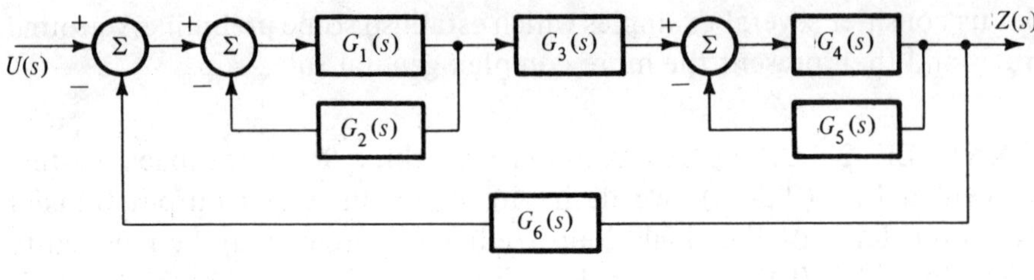

(a) Original system

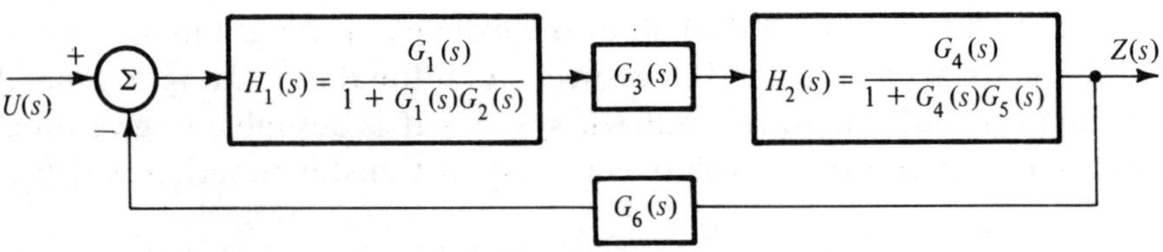

(b) System after reduction of two feedback loops

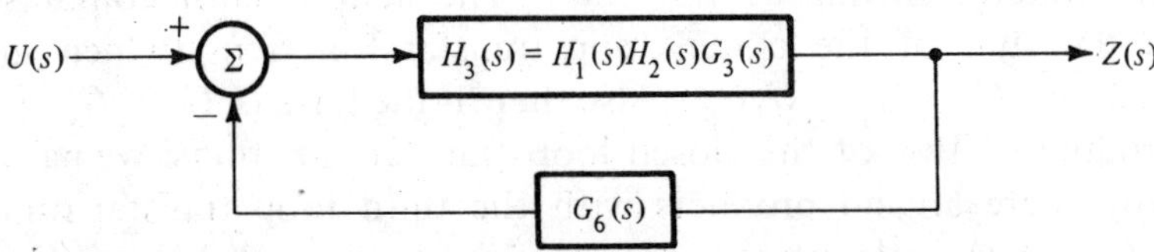

(c) System after application of cascade transfer function rule

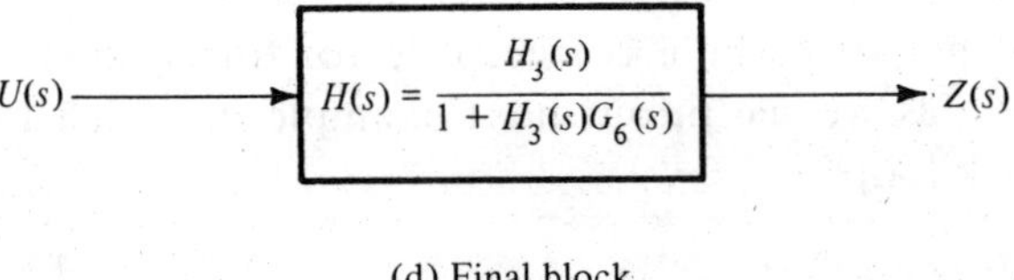

(d) Final block

Figure 2.3-7 Block diagram reduction of a three-loop system

reduction rules. The final transfer function is determined, in terms of the basic transfer functions of Fig. (2.3-8a), as

$$H(s) = \frac{Z(s)}{U(s)} = \frac{G_1 G_2 (1 + G_6 G_7)}{1 + G_2 G_3 G_4 + G_1 G_2 G_3 G_5 + G_1 G_2 G_6 + G_6 G_7 + G_2 G_3 G_4 G_6 G_7 + G_1 G_2 G_3 G_5 G_6 G_7}$$

Inspection of this result indicates that the numerator of this expression is different from that which would have been computed using the rule established in the previous example. Introspection indicates that the numerator here is multiplied by all terms in the denominator that do not have a path element in common with the forward transfer function contained in the numerator.

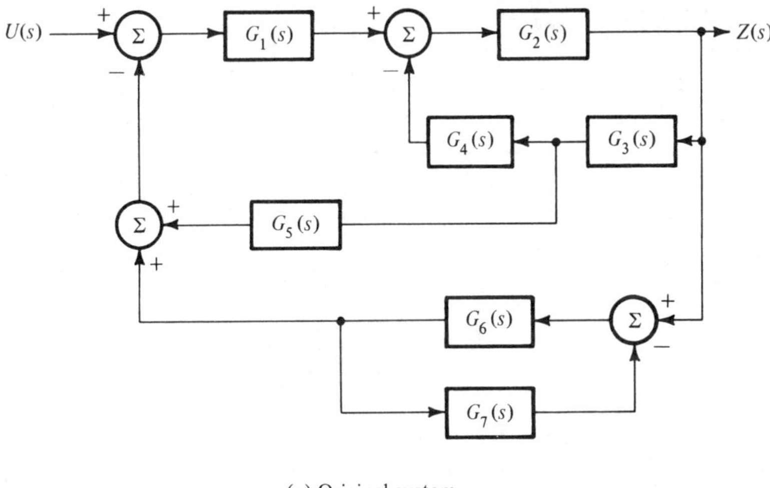

(a) Original system

Figure 2.3-8 System reduced by block diagram reduction

This is still not quite the most general statement of the transfer function formula. Extension of these examples to cases where there are three or more closed loops that have no paths in common would show that the general input output transfer function formula is

$$H(s) = \frac{Z(s)}{U(s)} = \sum_k \frac{G_k(s) \Delta_k(s)}{\Delta(s)} \tag{2.3-7}$$

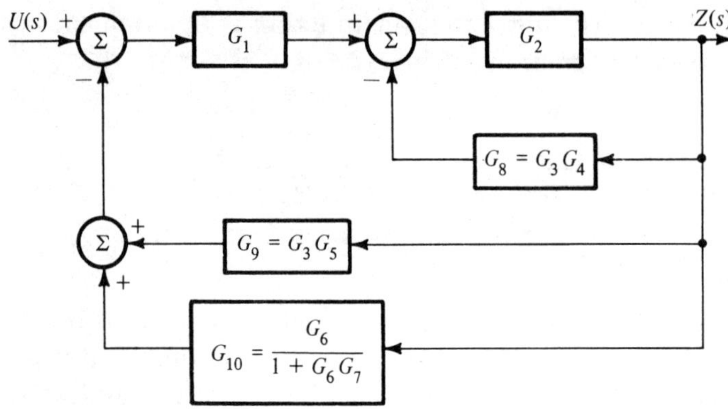

(b) System after moving takeoff point and reducing one feedback loop

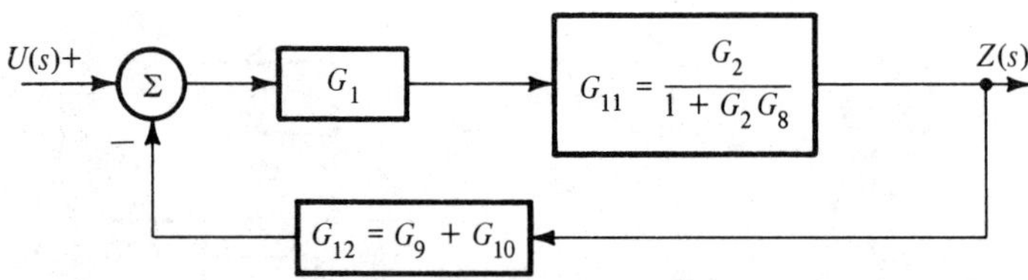

(c) System after reducing another feedback loop and using parallel element rule

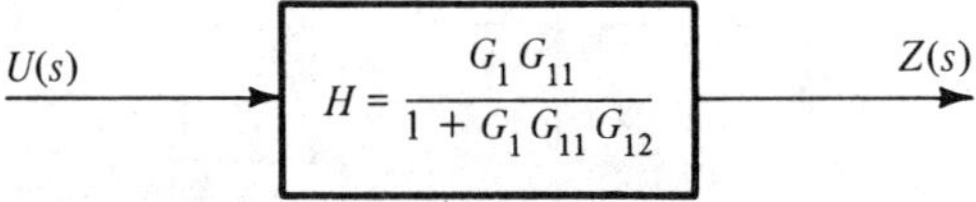

(d) Final reduced block diagram

Figure 2.3-8 System reduced by block diagram reduction (cont'd)

where

1. $g_k(s)$ is the kth forward transfer function through the system from $u(t)$ to $z(t)$. NO feedback loops can be contained in any $g_k(s)$.

2. $\Delta(s) = 1 -$ (sum of all individual closed loop transfer functions) + (sum of products of all closed loop transfer functions

taken two at a time with no common paths or elements) — (sum of products of all closed loop transfer functions taken three at a time with no common paths or elements) + . . .

3. $\Delta_k(s)$ = all terms in Δ that do not have elements or paths in common with an element or path in $g_k(s)$.

4. The summation is taken over all forward transfer function paths in the system.

5. The input output ($u(t)$ and $z(t)$) must be true inputs and outputs and not just nodes within a more complex feedback system.

EXAMPLE 2.3-4. In this example we will use the general transfer function formula of Eq. (2.3-7) to evaluate the input output transfer function of the system shown in Fig. (2.3-9). Direct application of Eq. (2.3-7) yields

$$\Delta = 1 + G_1 + G_2 + G_1 G_2 G_3 + G_4 + G_1 G_2 +$$
$$G_1 G_4 + G_2 G_4 + G_1 G_2 G_3 G_4 + 2G_1 G_2 G_4$$

$$g_k = G_1 G_2$$

$$\Delta_k = 1 + G_4$$

and so we have

$$\frac{Z(s)}{U(s)} = H(s) = \frac{g_k \Delta_k}{\Delta}$$

EXERCISE 2.3-2. Determine the input output transfer function of EXAMPLE 2.3-4 using the block diagram reduction technique.

EXERCISE 2.3-3. Determine the input output transfer function corresponding to the block diagram of Fig. (2.3-10). Use both the block diagram reduction technique as well as the general feedback system transfer function formula. Contrast and compare the advantages of each approach.

It is of interest to determine a block diagram to represent the general input output differential equation (2.1-2). We will accomplish

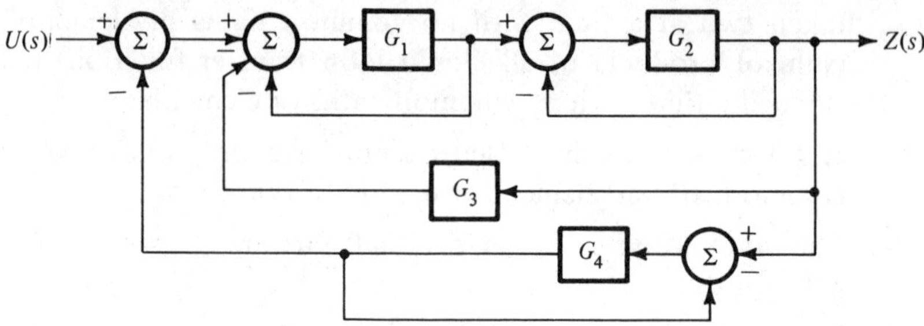

Figure 2.3-9 Block diagram for EXAMPLE 2.3-4

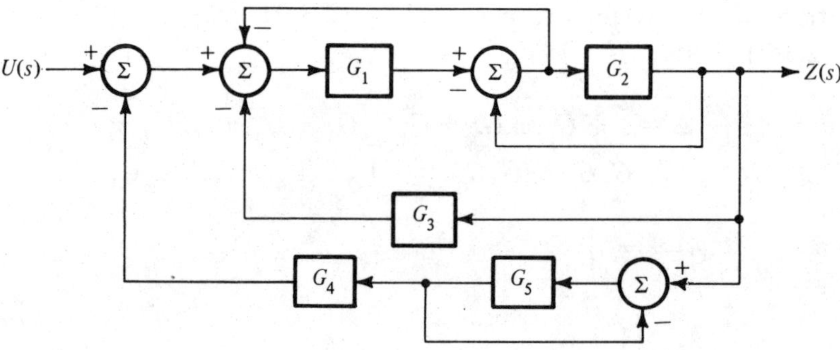

Figure 2.3-10 Block diagram for EXERCISE 2.3-3

this using only integrators for subsystem transfer functions. Thus the resulting block diagram will have an interpretation directly applicable to a state variable representation or to an analog computer simulation. We thus consider the differential equation

$$\sum_{i=0}^{n} a_i \frac{d^i z(t)}{dt} = \sum_{i=0}^{m} b_i \frac{d^i u(t)}{dt^i} \qquad (2.3\text{-}8)$$

for the case where $n \geqslant m$. We define state variables as follows:

$$x_1 = a_n z - b_n u \tag{2.3-9}$$

$$x_2 = a_{n-1} z - b_{n-1} u + \dot{x}_1 \tag{2.3-10}$$

$$x_3 = a_{n-2} z - b_{n-2} u + \dot{x}_2 \tag{2.3-11}$$

$$\vdots$$

$$x_n = a_1 z - b_1 u + \dot{x}_{n-1} \tag{2.3-12}$$

We solve Eq. (2.3-9) for z and obtain

$$z = \frac{x_1 + b_n u}{a_n} \tag{2.3-13}$$

We solve Eq. (2.3-10) for $\dot{x}_1$, Eq. (2.3-11) for $\dot{x}_2, \ldots,$ and Eq. (2.3-12) for $\dot{x}_{n-1}$ to yield

$$\dot{x}_1 = x_2 - a_{n-1} z + b_{n-1} u \tag{2.3-14}$$

$$\dot{x}_2 = x_3 - a_{n-2} z + b_{n-2} u \tag{2.3-15}$$

$$\vdots$$

$$\dot{x}_{n-1} = x_n - a_1 z + b_1 u \tag{2.3-16}$$

Now we differentiate Eq. (2.3-9) and substitute the resulting value of $\dot{x}_1$ into Eq. (2.3-10) to obtain

$$x_2 = a_{n-1} z - b_{n-1} u + a_n \dot{z} - b_n \dot{u}$$

We differentiate this expression and substitute the resulting value of x_2 into Eq. (2.3-11) and obtain

$$x_3 = a_{n-2} z - b_{n-2} u + a_{n-1} \dot{z} - b_{n-1} \dot{u} + a_n \ddot{z} - b_n \ddot{u}$$

We continue this procedure until finally we obtain from Eq. (2.3-12)

$$x_n = \sum_{i=1}^{n} a_i \frac{d^{i-1}z}{dt^{i-1}} - \sum_{i=1}^{m} b_i \frac{d^{i-1}u}{dt^{i-1}}$$

We differentiate this expression again and use Eq. (2.3-8) to obtain

$$\dot{x}_n = b_0 u - a_0 z \tag{2.3-17}$$

Eqs. (2.3-14) through (2.3-17) are the desired state variable equations. However, they are not in the standard form. If we substitute Eq. (2.3-13) into Eqs. (2.3-14) through (2.3-17) we obtain the normal state variable representation

$$\dot{x} = \mathbf{A}x + \mathbf{b}u \tag{2.3-18}$$

$$z = \mathbf{c}x + du \tag{2.3-19}$$

where

$$\mathbf{A} = \begin{bmatrix} -\dfrac{a_{n-1}}{a_n} & 1 & 0 & 0 & \ldots & 0 \\[2ex] -\dfrac{a_{n-2}}{a_n} & 0 & 1 & 0 & \ldots & 0 \\[2ex] -\dfrac{a_{n-3}}{a_n} & 0 & 0 & 1 & \ldots & 0 \\[2ex] \cdot & \cdot & \cdot & \cdot & & \cdot \\ \cdot & \cdot & \cdot & \cdot & & \cdot \\ \cdot & \cdot & \cdot & \cdot & & \cdot \\[1ex] -\dfrac{a_1}{a_n} & 0 & 0 & 0 & \ldots & 1 \\[2ex] -\dfrac{a_0}{a_n} & 0 & 0 & 0 & \ldots & 0 \end{bmatrix} = \left[\begin{array}{c|c} \begin{matrix} -\dfrac{a_{n-1}}{a_n} \\[2ex] -\dfrac{a_{n-2}}{a_n} \\[2ex] -\dfrac{a_{n-3}}{a_n} \\[2ex] \vdots \\[1ex] -\dfrac{a_1}{a_n} \\ \hline \end{matrix} & \begin{matrix} \mathbf{I} \\[6ex] \hline \end{matrix} \\ -\dfrac{a_0}{a_n} & 0 \;\ldots\; 0 \end{array} \right]$$

$$\mathbf{b} = \begin{bmatrix} b_{n-1} - \dfrac{a_{n-1}b_n}{a_n} \\[2ex] b_{n-2} - \dfrac{a_{n-2}b_n}{a_n} \\[2ex] \cdot \\ \cdot \\ \cdot \\[1ex] b_1 - \dfrac{a_1 b_n}{a_n} \\[2ex] b_0 - \dfrac{a_0 b_n}{a_n} \end{bmatrix}, \quad \mathbf{c}^T = \begin{bmatrix} \dfrac{1}{a_n} \\[2ex] 0 \\[1ex] 0 \\[1ex] 0 \\[1ex] \cdot \\ \cdot \\ \cdot \\[1ex] 0 \end{bmatrix}, \quad d = \dfrac{b_n}{a_n} \qquad (2.3\text{-}20)$$

Figure (2.3-11) represents a block diagram of this general differential equation (2.3-8) or a state variable equivalent, Eqs. (2.3-18) through (2.3-20). In obtaining these results we assumed that $m \leqslant n$ since physical systems cannot have dynamics such that there are more finite zeros than finite poles.* For convenience in our discussion we assume $m = n$ but it is a simple matter to set $b_i = 0$, $i = m, m + 1, \ldots, n$ if in fact $m < n$.

EXERCISE 2.3-4.   Use the general input output feedback system transfer function formula to obtain the differential equation corresponding to Fig. (2.3-11).

EXERCISE 2.3-5.   Use block diagram reduction to obtain the transfer function corresponding to Fig. 2.3-11.

EXERCISE 2.3-6.   Recast the input output transfer function

---

*We discuss this point again in Chaps. 5 and 6, specifically Sec. 6.2.

$$\frac{Z(s)}{U(s)} = \frac{1+s}{1+s+s^2}$$

in the form of the block diagram of Fig. (2.3-11).

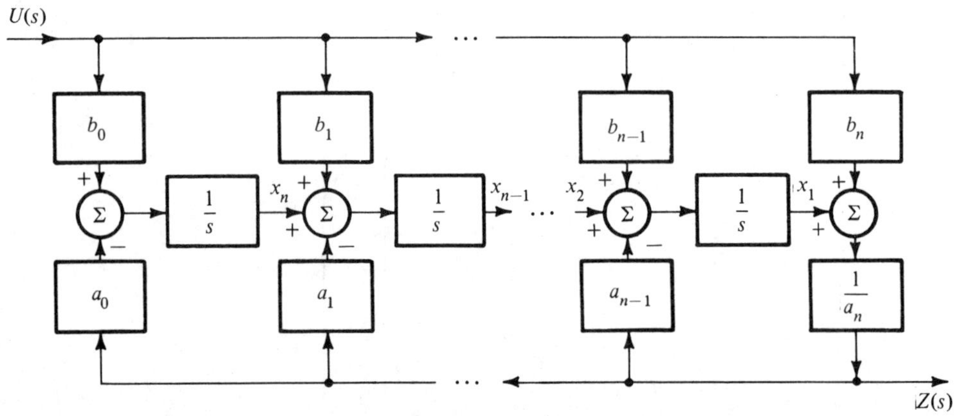

Figure 2.3-11   Block diagram representation of general single input single output linear system

2.4 SUMMARY

In this chapter we have begun an intensive study of linear systems control by considering three closely interrelated topics: linear differential equations, Laplace transforms, and block diagram reduction. These are preliminary tools for the analysis of linear systems. In our next chapter we will examine several important simple systems. Our goal will be to obtain transfer functions describing these systems. We will make much use of the three topics of this chapter in the next chapter and throughout this text and in real world control system investigations.

2.5 REFERENCES

The material in this chapter is very basic and very fundamental to a study of linear systems control. It is discussed in many references in applied mathematics as well as in most introductory control systems texts. As has been suggested earlier, alternate points of view are always helpful in gaining a fuller understanding of the material presented here.

More complete coverage of the differential equation and Laplace transform material is contained in several applied mathematics and linear systems texts. The student should find the coverage in:

Wylie, C. R., Jr. *Advanced Engineering Mathematics*, 2nd Edition, McGraw-Hill Book Co., New York, 1960;

Hildebrand, F. B. *Methods of Applied Mathematics*, Prentice Hall Inc., Englewood Cliffs, New Jersey, 1952;

Hildebrand, F. B. *Advanced Calculus for Engineers*, Prentice Hall Inc., Englewood Cliffs, New Jersey, 1972;

Schwarz, R. J., and Friedland, B. *Linear Systems*, McGraw-Hill Book Co., Inc., New York, 1965;

Gardner, M. F., and Barnes, J. L. *Transients in Linear Systems*, John Wiley and Sons, Inc., New York, 1942;

very useful. The last reference is an extraordinarily well-done classic.

Any beginning control systems text discusses, to some extent, block diagram reduction. The material concerning the general formula for the transfer function of a linear system is adapted from the general gain formula of a linear flow graph as originally presented in:

Mason, S. J. "Feedback Theory - Some Properties of Signal Flow Graphs," *Proc. IRE*, Vol. 41, No. 9, pp. 1144-1156, September 1953;

Mason, S. J. "Feedback Theory - Further Properties of Signal Flow Graphs," *Proc. IRE*, Vol. 44, No. 7, pp. 920-926, July 1956;

and the student may find it interesting to consult these sources for a quite different presentation of this material in a signal flow graph format.

2.6 PROBLEMS

Unless otherwise specified, assume that all systems are initially at rest such that all initial conditions are zero.

1. The differential equations describing a system are

$$\ddot{x}_1 + \alpha_1 \dot{x}_1 + \alpha_2 x_1 + \alpha_3 \dot{x}_2 + \alpha_4 x_2 = u(t)$$

$$\ddot{x}_2 + \beta_1 \dot{x}_1 + \beta_2 x_1 + \beta_3 \dot{x}_2 + \beta_4 x_2 = 0$$

$$z = x_1$$

Find the system transfer function $Z(s)/U(s)$.

2. The differential equations describing a system are

$$\dot{x}_1 = -x_1 + 2x_2$$

$$\dot{x}_2 = -2x_2 - 1.5x_3$$

$$\dot{x}_3 = -x_1 - 10x_2 - 3x_3 + u$$

$$z = x_1$$

(a) Find the state variable matrix coefficients such that the system is described by $\dot{x} = Ax + bu$, $z = cx$ and (b) Find the system transfer function $Z(s)/U(s)$.

3. Find the Laplace transforms of

(a) $f(t) = t^2 e^{-at}$ (b) $f(t) = te^{-t} \cos t$

(c)
$$f(t) = \begin{cases} 0 \text{ for } t < 2 \\ 1 \text{ for } 2 < t < 3 \\ 0 \text{ for } 3 < t \end{cases}$$
     (d) $f(t) = e^t + e^{-t}$

4. What is the impulse response of a system which is described by the differential equation

$$\frac{d^3 z}{dt^3} + 3\frac{d^2 z}{dt^2} + 2\frac{dz}{dt} = u$$

5. Determine three different state variable descriptions of the system in Problem 4. What is the fundamental matrix or state transition matrix for each of these systems?

6. A system has the impulse response

$$h(t) = \frac{5}{2} e^{-t} - 4e^{-2t} + \frac{3}{2} e^{-3t}$$

What is the step response of the system? What is the system transfer function? What is the response if $z(0) = \alpha$, $\dot{z}(0) = \beta$, $\ddot{z}(0) = \gamma$ and $u(t) = 0$?

7. A system transfer function is

$$H(s) = \frac{8(s+1)}{(s+2)^3}$$

What is the response of the system to a unit impulse input? to a unit step input?

8. The function

$$H(s) = s + \cfrac{1}{s + \cfrac{1}{4s + 1}}$$

is expressed as a continued fraction expansion. What is the system transfer function and how is the continued fraction expression obtained from the transfer function? What is the partial fraction expansion for this transfer function and how is it related to the continued fraction expansion?

9. A system, initially at rest, is described by the differential equation

$$\frac{dz}{dt} + az = u(t) - u(t-1), \quad a \geqslant 0$$

(a) What is the impulse response of the system? (b) What is the system transfer function? (c) What is the unit step response of the system?

10. A system, containing a time delay, is described by the differential equation

$$\frac{dz}{dt} + az(t-1) = u(t)$$

(a) What is the impulse response of the system? (b) What is the system transfer function? (c) What is the unit step response of the system?

11. Prepare simple sketches of block diagrams and reduced versions of these block diagrams to illustrate the rules for the following:

(a) interchange of elements, (b) interchange of summation points, (c) moving a summing point ahead of a transfer function element, (d) moving a summing point behind a transfer function element, (e) moving a takeoff point ahead of an element, and (f) inserting a transfer function element in the forward path of a feedback loop.

12. Show that another state variable representation of Eq. (2.3-8) is

$$\dot{x}_1 = x_2 + \alpha_1 u$$

$$\dot{x}_2 = x_3 + \alpha_2 u$$

$$\dot{x}_{n-1} = x_n + \alpha_{n-1} u$$

$$\dot{x}_n = -\frac{a_0}{a_n} x_1 - \frac{a_1}{a_n} x_2 - \ldots - \frac{a_{n-1}}{a_n} x_n + \alpha_n u$$

where the output is

$$z = x_1 + \frac{b_n}{a_n} u$$

Unfortunately there is no simple relationship between the α coefficients and the differential equation coefficients. It turns out that

$$\alpha_1 = \frac{b_{n-1} a_n - a_{n-1} b_n}{a_n^2}$$

$$\alpha_2 = \frac{b_{n-2} a_n - a_{n-2} b_n}{a_n^2} - \frac{a_{n-1}}{a_n} \alpha_1$$

$$\alpha_3 = \frac{b_{n-3}a_n - a_{n-2}b_n}{a_n^2} - \frac{a_{n-1}}{a_n}\alpha_1 - \frac{a_{n-1}}{a_n}\alpha_2$$

$$\vdots$$

$$\alpha_n = \frac{b_0 a_n - a_0 b_n}{a_n^2} - \frac{a_1}{a_n}\alpha_1 - \frac{a_2}{a_n}\alpha_2 - \cdots - \frac{a_{n-2}}{a_n}\alpha_{n-1} - \frac{a_{n-1}}{a_n}\alpha_n$$

(a) Draw a block diagram representing this system. (b) Show that this is the state variable representation corresponding to

$$x_1 = z - \frac{b_n}{a_n}u$$

$$x_2 = \dot{x}_1 - \alpha_1 u$$

$$x_3 = \dot{x}_2 - \alpha_2 u$$

$$\vdots$$

$$x_n = \dot{x}_{n-1} - \alpha_{n-1}u$$

13. For the transfer function

$$\frac{Z(s)}{U(s)} = \frac{1}{(s+1)^4}$$

find a block diagram representation in the form of Fig. (2.3-11). Contrast the associated state variable representation with the phase variable representation which is obtained by lettering

$$z = x_1$$

$$\dot{x}_1 = x_2$$

$$\dot{x}_2 = x_3$$

$$\dot{x}_3 = x_4$$

$$\dot{x}_4 = \gamma_1 x_1 + \gamma_2 x_2 + \gamma_3 x_3 + \gamma_4 x_4 + u$$

14. Repeat P2.6-13 utilizing the state variable representation suggested in P2.12.

15. Repeat EXERCISE 2.3-6 using the state variable representation of P2.6-12.

16. Find the transfer functions

$$\left.\frac{Z_1(s)}{U_1(s)}\right|_{U_2=0} \qquad \left.\frac{Z_1(s)}{U_2(s)}\right|_{U_1=0} \qquad \left.\frac{Z_2(s)}{U_1(s)}\right|_{U_2=0} \qquad \left.\frac{Z_2(s)}{U_2(s)}\right|_{U_1=0}$$

for the system shown in Fig. P2.6-16.

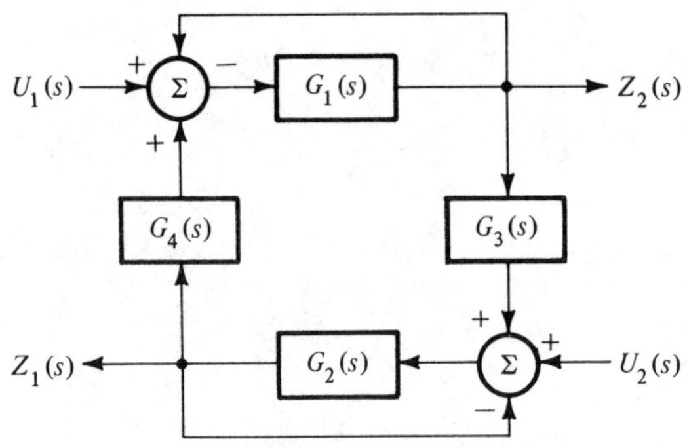

Figure P2.6-16   Block diagram for Problem 16

17. Suppose that, in Fig. P2.6-16

$$G_1(s) = \frac{K}{s} \qquad G_2(s) = \frac{K}{s} \qquad G_3(s) = 1 + T_3 s \qquad G_4(s) = \frac{1 + T_4 s}{1 + T_5 s}$$

Determine a system state variable representation.

3
LINEAR SYSTEMS CONTROL COMPONENTS

We have just completed a study of Laplace transforms and linear differential equations for the analysis of linear control systems. In this chapter we will consider some of the many components used in linear systems. We will obtain models of these physical components and the associated differential equations and transfer functions. With the background in transform mathematics and systems control components obtained in this chapter and the previous chapter we will be in a position to obtain general response of simple first and second order systems, as well as higher order systems, and thus determine desirable performance specifications for linear systems.

In this chapter we will examine several items of interest, namely

electrical networks

operational amplifiers and analog computers

mechanical networks

thermal and fluid networks

and some representative physical examples. This can by no means be considered other than a very brief introduction but it will be adequate for our purposes in this text.

3.1 ELECTRICAL NETWORKS

In this section we shall consider some electrical networks typically used in linear systems control. These networks are typically used as compensation networks to filter and shape the flow of information in control systems such as to enhance their performance in accordance with stated design criteria.

The three basic passive electrical network elements are resistors, capacitors, and inductors. The two basic active electric network elements are the current source and the voltage source. Table 3.1-1 presents these basic elements and the linear relationship between current and voltage in the passive elements.

The passive linear elements may be connected together to form information shaping networks or filters. In later chapters we shall illustrate some of the many very useful properties of these networks in linear systems control. Three methods are often used to analyze linear electrical networks. These are branch current analysis, loop current analysis, and node voltage analysis.* Each method of analysis is developed such as to satisfy Kirchoffs laws which state that:

1. the algebraic sum of the instantaneous voltages around a complete closed path must be zero.

2. the algebraic sum of the instantaneous branch currents into the node of a network must be zero.

EXAMPLE 3.1-1. As an illustration of the three methods of electric circuit analysis let us consider the simple electrical network shown in Fig. (3.1-1). Shown on this fugure are three branch currents ($\mathcal{I}_1$, $\mathcal{I}_2$, and $\mathcal{I}_3$), the two loop currents (i_1 and i_2), and the three node voltages (v_1, v_2, and v_3) needed as dependent variables for our analysis. All voltages are taken with respect to the ground terminal shown.

We use Kirchoffs second or current law in writing, for the branch currents of Fig. (3.1-1)

$$\mathcal{I}_1 + \mathcal{I}_2 + \mathcal{I}_3 = 0 \tag{1}$$

Then we use the first or voltage law of Kirchoffs to obtain, using the current-voltage relationships of Table (3.1-1),

---

*The two terminals of any of the basic electrical elements are called *nodes*. Connection of two or more elements will result in the coincidence of two or more nodes into a single node. Either a single two terminal element or a series connection of such elements such that the same current flows through them is called a *branch*. A number of branches connected together forms a *network*. A *loop* is a closed contour formed by one or more branches in series in some portion of a network.

TABLE 3.1-1 ELEMENTARY ELECTRICAL NETWORK ELEMENTS

Element Name	*Element Symbol*	*Current Voltage Relationships*
Voltage source—v (volts)	$v(t)$ $V(s)$	Independent source
Current source—i (amperes)	$i(t)$ $I(s)$	Independent source
Resistance—R (ohms)	$v(t)$ R $i(t)$	$v(t) = Ri(t)$ $V(s) = RI(s)$
Capacitance—C (farads)	$v(t)$ C $i(t)$	$i = C\dfrac{dv}{dt}$ $v = \dfrac{1}{c}\displaystyle\int_0^t i\,dt + v(0)$ $I(s) = sCV(s) - Cv(0)$
Inductance—L (henrys)	$v(t)$ L $i(t)$	$v = L\dfrac{di}{dt}$ $i = \dfrac{1}{L}\displaystyle\int_0^t v\,dt + i(0)$ $V(s) = sLI(s) - Li(0)$

$$v(t) - \mathcal{I}_1 R_1 + \mathcal{I}_2 R_2 + \frac{1}{C} \int_0^t \mathcal{I}_2 \, dt - v_c(0) = 0 \tag{2}$$

$$v(t) - \mathcal{I}_1 R_1 + \mathcal{I}_3 R_3 + L \frac{d\mathcal{I}_3}{dt} = 0 \tag{3}$$

Equations (1)–(3) constitute a complete set of equations for this simple electrical network obtained by the branch current method of analysis. In order to solve this set of equations for the branch currents it is convenient to use the Laplace transform methods of our previous chapter. We could, for example, obtain the input impedance, $Z_{in}(s) = \dfrac{V(s)}{\mathcal{I}_1(s)}$, of the network.

To obtain a node voltage analysis of this simple electrical circuit we make use of the second or branch current law of Kirchoff to obtain, for the node voltages at nodes 1, 2, and 3,

$$\frac{v_1 - v}{R_1} + \frac{v_1 - v_2}{R_2} + \frac{v_1 - v_3}{R_3} = 0 \tag{4}$$

$$\frac{v_2 - v_1}{R_2} + C \frac{dv_2}{dt} = 0 \tag{5}$$

$$\frac{v_3 - v_1}{R_3} + \frac{1}{L} \int_0^t v_3 \, dt + i_L(0) = 0 \tag{6}$$

We may now take the Laplace transform of this complete set of Eqs. (4)–(6) for node voltage analysis of this electrical network.

We now use Kirchoffs voltage law to obtain, using the loop currents of Fig. (3.1-1), the following set of equations for the sum of the voltages in the two loops of this network:

$$v - R_1 i_1 - R_2(i_1 - i_2) - \frac{1}{C} \int_0^t (i_1 - i_2) \, dt - v_c(0) = 0 \tag{7}$$

$$v_c(0) + \frac{1}{C} \int_0^t (i_1 - i_2) \, dt + R_2(i_1 - i_2) - R_3 i_2 - L \frac{di_2}{dt} = 0 \tag{8}$$

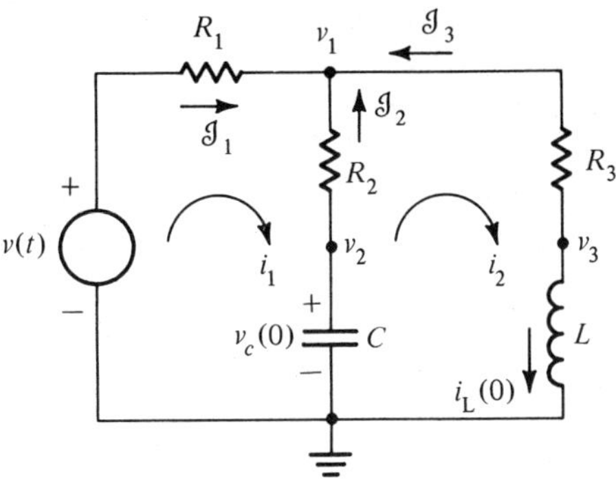

Figure 3.1-1 Electrical network for EXAMPLE 3.1-1

We may now take Laplace transforms of the loop current equations (5) and (6) to obtain

$$V(s) - \left[R_1 + R_2 + \frac{1}{Cs}\right]I_1(s) + \left[R_2 + \frac{1}{Cs}\right]I_2(s) - \frac{v_c(0)}{s} = 0$$

for loop 1 as represented by Eq. (7) and

$$\frac{v_c(0)}{s} + \left[\frac{1}{Cs} + R_2\right]I_1(s) - \left[\frac{1}{Cs} + R_2 + R_3 + Ls\right]I_2(s) + Li_L(0) = 0$$

for Eq. (8). We arrange these in standard matrix form

$$\begin{bmatrix} a_{11}(s) & a_{12}(s) \\ a_{21}(s) & a_{22}(s) \end{bmatrix} \begin{bmatrix} I_1(s) \\ I_2(s) \end{bmatrix} = \begin{bmatrix} U_1(s) \\ U_2(s) \end{bmatrix} \tag{9}$$

or

$$\mathbf{A}(s)\mathbf{I}(s) = \mathbf{U}(s) \tag{10}$$

where

$$a_{11}(s) = (R_1 + R_2)Cs + 1$$

$$a_{12}(s) = -(R_2 Cs + 1)$$

$$a_{21}(s) = -(R_2 Cs + 1)$$

$$a_{22}(s) = LCs^2 + (R_2 + R_3)sC + 1$$

$$U_1(s) = sCV(s) - Cv_c(0)$$

$$U_2(s) = Cv_c(0) - LCsi_L(0)$$

We can now calculate the inverse of the $\mathbf{A}(s)$ matrix such that we obtain

$$\mathbf{I}(s) = \mathbf{A}^{-1}(s)U(s) \tag{11}$$

where

$$\mathbf{A}^{-1}(s) = \frac{1}{\Delta(s)} \begin{bmatrix} a_{22}(s) & -a_{12}(s) \\ -a_{21}(s) & a_{11}(s) \end{bmatrix}$$

with

$$\Delta(s) = Cs\left\{ LC(R_1 + R_2)s^2 + [(R_1 R_2 + R_1 R_3 + R_2 R_3)C + L]s + R_1 + R_3 \right\}$$

In order to carry the analysis much further it is convenient to assume numerical values for the linear electrical network elements of Fig. (3.1-1). We let $C = 0.788675$, $L = 0.633975$, $R_1 = R_2 = R_3 = 1$, to obtain

$$\Delta(s) = 0.788675s(s + 2)(s + 1)$$

and

$$\mathbf{A}^{-1}(s) = \frac{1.267949}{s(s + 1)(s + 2)}$$

$$\times \begin{bmatrix} 0.5(s + 2.275944)(s + 0.878756) & 0.788675(s + 1.267949) \\ 0.788675(s + 1.267949) & 1.57735(s + 0.633945) \end{bmatrix}$$

We could now calculate the response due to any particular input $v(t)$ and any given initial conditions $v_c(0)$ and $i_L(0)$.

Often we will desire to use electrical networks as information shaping networks in control systems such as to produce a desired voltage transfer function $V_o(s)/V_i(s)$ which would have characteristics such as to enhance system response. We will conclude this example by obtaining the transfer function $V_1(s)/V(s)$ for the electrical network of Fig. (3.1-1). We have

$$v_3 = L\frac{di_2}{dt}, \qquad v_1 - v_3 = i_2 R_3$$

such that we obtain

$$v_1 = L\frac{di_2}{dt} + R_3 i_2$$

which has the Laplace transform

$$v_1(s) = (Ls + R_3)I_2(s) - Li_L(0) \tag{12}$$

We now assume that all initial conditions are zero and obtain from Eq. (11)

$$I_2(s) = \frac{-a_{21}(s)sCV(s)}{\Delta(s)}$$

which becomes when inserting Eq. (12) and using the given parameter values

$$\frac{V_1(s)}{V(s)} = \frac{0.5\,(s + 1.267949)\,(s + 1.577349)}{(s + 1)(s + 2)}$$

In later chapters we will be interested in examining transfer functions of this sort with respect to determination of frequency and phase response characteristics.

EXERCISE 3.1-1. Obtain the transfer function $V_3(s)/V(s)$ using the loop current method.

EXERCISE 3.1-2. Obtain the transfer function $V_1(s)/V(s)$ using the node voltage method.

EXERCISE 3.1-3. Obtain the transfer function $V_1(s)/V(s)$ using the branch current method.

EXERCISE 3.1-4. What is the input impedance $Z_{in}(s) = V(s)/I_1(s)$ of this electrical network?

In most linear system control design applications, it is rare to use inductors as part of compensation networks. Cost and the limited linear range of inductors make networks with resistors and capacitors only much more attractive than those with inductors. Table 3.1-2 illustrates some elementary electrical networks commonly used in control system design and their transfer functions. We will have extensive use for these networks and their transfer functions in our later efforts where the asymptotic gain and phase characteristics associated with them will be of much value. The gain of each transfer function is just the magnitude of the transfer function and the phase shift is the angle for a frequency ω

$$\alpha(\omega) = \text{Gain} = |H(j\omega)| = \left| \frac{V_o(s)}{V_i(s)} \right|_{s=j\omega} = \sqrt{H(j\omega)H(-j\omega)} \qquad (3.1\text{-}1)$$

$$\beta(\omega) = \text{Phase Angle} = \underline{/H(j\omega)} \qquad (3.1\text{-}2)$$

where we can always write

$$H(j\omega) = \alpha(\omega)e^{j\beta(\omega)} \qquad (3.1\text{-}3)$$

EXERCISE 3.1-5. Obtain the transfer functions depicted in Table 3.1-2.

In many linear system control applications operational amplifiers are used to obtain desired compensation transfer functions. Also analog computer simulation of systems is often used and the basic element of an analog computer simulation is an operational amplifier, which is a very high gain directly coupled electronic amplifier. It is true that essentially any transfer function for a linear system may be realized using only passive circuit elements (resistors, inductors and capacitors). However, loading problems lead to very difficult network synthesis problems for other than simple transfer functions. Often the purpose of a system simulation is to determine the system responses for a variety of possible

TABLE 3.1-2 COMMONLY USED COMPENSATING NETWORKS

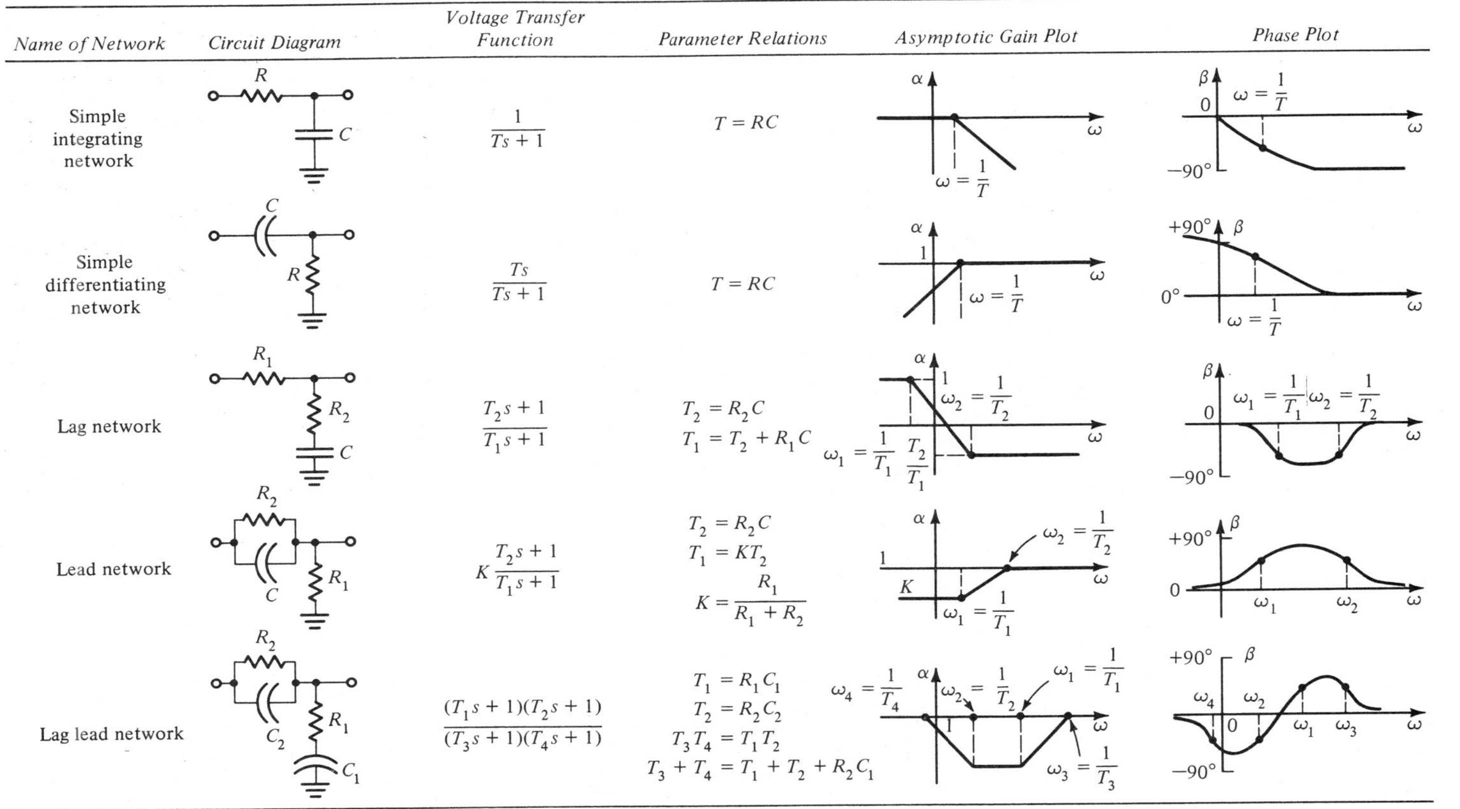

Name of Network	Circuit Diagram	Voltage Transfer Function	Parameter Relations	Asymptotic Gain Plot	Phase Plot
Simple integrating network		$\dfrac{1}{Ts+1}$	$T=RC$		
Simple differentiating network		$\dfrac{Ts}{Ts+1}$	$T=RC$		
Lag network		$\dfrac{T_2 s+1}{T_1 s+1}$	$T_2=R_2 C$ $T_1=T_2+R_1 C$		
Lead network		$K\,\dfrac{T_2 s+1}{T_1 s+1}$	$T_2=R_2 C$ $T_1=KT_2$ $K=\dfrac{R_1}{R_1+R_2}$		
Lag lead network		$\dfrac{(T_1 s+1)(T_2 s+1)}{(T_3 s+1)(T_4 s+1)}$	$T_1=R_1 C_1$ $T_2=R_2 C_2$ $T_3 T_4=T_1 T_2$ $T_3+T_4=T_1+T_2+R_2 C_1$		

variable parameters. The high degree of coupling which exists in passive linear systems would generally lead to very complicated relations between a variable parameter in a physical linear control system and many variable parameters in a linear electrical network model of the system.

The basic diagram for an operational amplifier is shown in Fig. (3.1-2). The various impedances are selected to yield specified transfer functions as we will now indicate. First we will attempt an approximate solution for the transfer function of the operational amplifier system that will depend heavily upon the operational amplifier being a perfect device. We assume that the input impedance of the amplifier is infinite or at least very large compared with the other impedances shown in the figure. Also we assume that the output impedance is zero or at least very small compared with the other impedances shown in the figure. Finally we assume that the gain of the operational amplifier is negative infinity* or at least a very large negative number. When we actually assume that the gain is negative infinity then the node labeled v_g must be a virtual earth or virtual ground since any non-zero voltage would amplify through the infinite gain amplifier to produce an infinite output voltage v_0. With v_g set equal to zero we must have, for no initial conditions associated with the impedances,

$$I_i(s) = \frac{V_i(s)}{Z_i(s)} \qquad i = 1, 2, \ldots, n \tag{3.1-4}$$

and

$$I_0(s) = \frac{V_0(s)}{Z_0(s)} \tag{3.1-5}$$

Since the amplifier is assumed to have infinite input impedance no current can flow into it and Kirchoffs current law yields

$$I_0(s) + \sum_{i=0}^{n} I_i(s) = 0 \tag{3.1-6}$$

We may easily combine Eqs. (3.1-4)–(3.1-6) to obtain

$$V_0(s) = -Z_0(s) \sum_{i=0}^{n} \frac{V_i(s)}{Z_i(s)} \tag{3.1-7}$$

---

*The gain must be negative in order that the overall system may be stable.

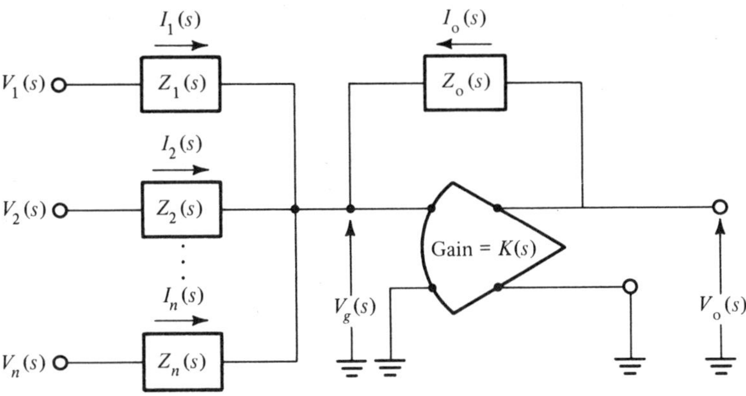

Figure 3.1-2 Operational amplifier system

which is the desired voltage transfer relationship for the operational amplifier.

As we noted previously, a virtual earth assumption is made in our derivation. This assumption is easily removed. Using Kirchoff's laws at nodes $V_1, V_2, \ldots, V_n, V_g$ and V_o, we obtain from the current relationship for infinite amplifier input impedance

$$I_o(s) = -\sum_{i=1}^{n} I_i(s)$$

the corresponding voltage relationships

$$\frac{V_o(s) - V_g(s)}{Z_o(s)} = -\sum_{i=1}^{n} \frac{V_i(s) - V_g(s)}{Z_i(s)} \qquad (3.1\text{-}8)$$

Now we use the gain relationship

$$V_o(s) = K(s)V_g(s)$$

to obtain from Eq. (3.1-8)

$$V_o(s)\left[\frac{1}{Z_o(s)} - \frac{1}{K(s)Z_o(s)} - \frac{1}{K(s)}\sum_{i=1}^{n}\frac{1}{Z_i(s)}\right] = -\sum_{i=1}^{n}\frac{V_i(s)}{Z_i(s)}$$

From this expression the output transfer relationship for the finite gain amplifier

$$V_o(s) = -\frac{Z_o(s)}{1 + \epsilon(s)} \sum_{i=1}^{n} \frac{V_i(s)}{Z_i(s)} \qquad (3.1\text{-}9)$$

where

$$\epsilon(s) = \frac{-1}{K(s)} \left[1 + Z_o(s) \sum_{i=1}^{n} \frac{1}{Z_i(s)} \right] \qquad (3.1\text{-}10)$$

is obtained. We easily see that $\epsilon(s) = 0$ for $K(s) = -\infty$ as we have suggested. Then we see that Eq. (3.1-10) is identical to Eq. (3.1-7).

EXAMPLE 3.1-2. One of the most used operational amplifier systems is the integrator. We use a capacitor for the impedance $Z_o(s)$ and a resistor for the impedance $Z_1(s)$ as indicated in Fig. (3.1-3). We assume that the gain is independent of frequency and obtain from Eqs. (3.1-9) and (3.1-10)

$$\frac{V_o(s)}{V_1(s)} = \frac{-\dfrac{1}{RCs}}{1 - \dfrac{1}{K}\left(1 + \dfrac{1}{RCs}\right)} \qquad (1)$$

For the typical unit gain integrator the RC product is set equal to unity. Suppose that the gain K is not infinitely large but finite, say $K = -1,000$. Equation (1) becomes

$$\frac{V_o(s)}{V_1(s)} = \frac{-10^3}{1 + s(1 + 10^3)} \qquad (2)$$

If this were a perfect integrator the response to a unit impulse input (which we simulate not by attempting to input a unit impulse but by putting an initial condition of one on the capacitor) would be a unit step. For $V_1(s) = 1$, we obtain upon taking the inverse Laplace transform of Eq. (2)

$$V_o(t) = -\frac{10^3}{1 + 10^3} \exp\left[-t/(1 + 10^3)\right]$$

and we see that the response is very nearly a constant. At $t = 50$ seconds we obtain $v_o(50) = -0.9503$ and so we see that there is less than 5 percent error at $t = 50$ seconds. Actually a gain of -10^3 is really quite low. A gain of -10^6 is not at all unreasonable and this would give an output voltage at $t = 50$, for a unit impulse input, of $v_o(50) = -0.9999949$. Even after one hour the output voltage is $v_o(1 \text{ hr}) = -0.99964$ which represents an error of only 0.036 percent. Thus we see that we may obtain an exceptionally good integrator using operational amplifiers, one far better than can be obtained using the simple integrating network of Table 3.1-2.

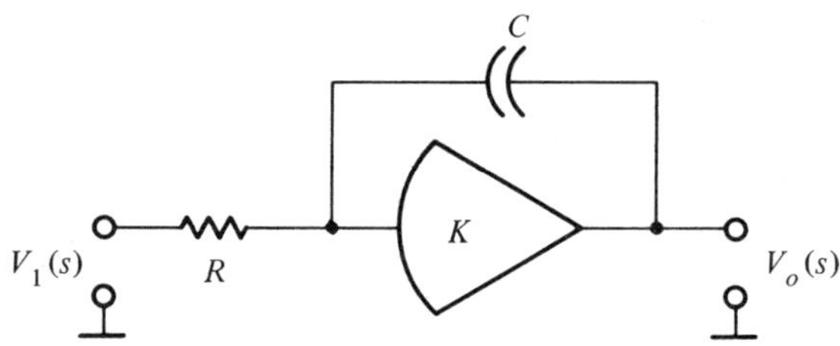

Figure 3.1-3 Operational amplifier integrator

EXERCISE 3.1-6. Contrast and compare results from using the simple integrating network of Table 3.1-2 with those obtained from using the operational amplifier integrator of EXAMPLE 3.1-2. Accomplish this for a unit impulse impulse and a unit step input. What is a reasonable value for T for the simple integrating network?

Table 3.1-3 illustrates the two fundamental operations obtainable by use of operational amplifier circuits. The summer and the summing integrator are the most important for general purpose analog simulation of control systems. Others are especially useful when we desire to simulate compensating network transfer functions. We can obtain desired transfer relations by use of Eq. (3.1-7) and the appropriate Z_o and Z_i.

TABLE 3.1-3 SUMMER AND SUMMING INTEGRATOR REALIZATION USING OPERATIONAL AMPLIFIERS

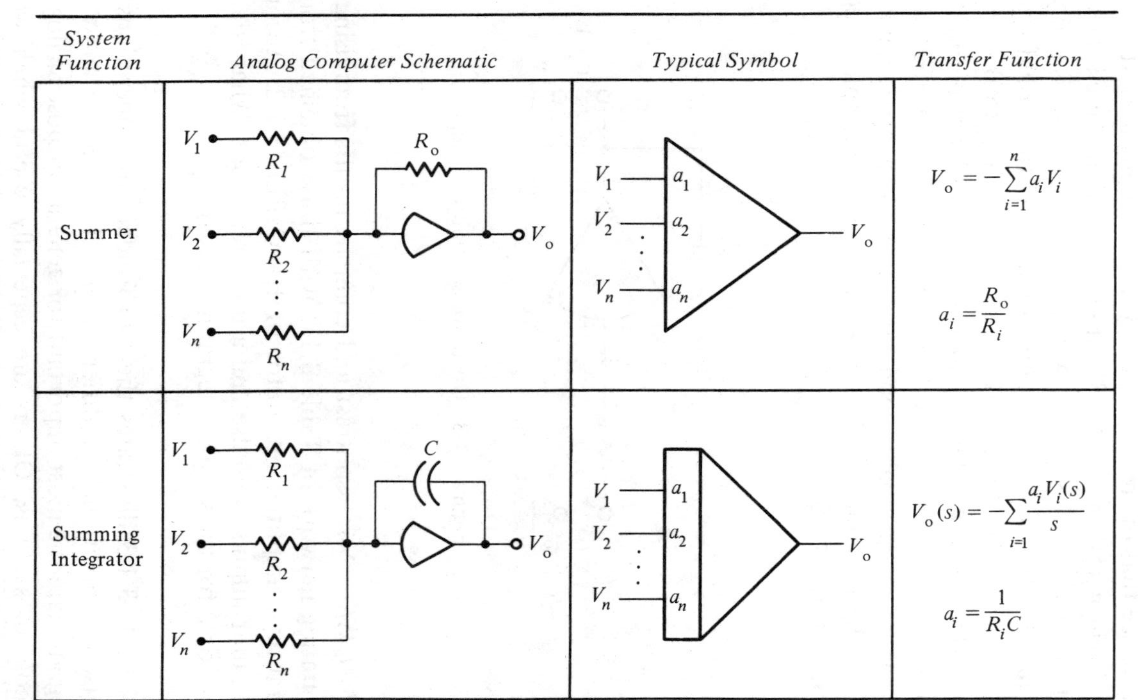

System Function	Analog Computer Schematic	Typical Symbol	Transfer Function
Summer			$V_o = -\sum_{i=1}^{n} a_i V_i$ $a_i = \dfrac{R_o}{R_i}$
Summing Integrator			$V_o(s) = -\sum_{i=1} \dfrac{a_i V_i(s)}{s}$ $a_i = \dfrac{1}{R_i C}$

The summer and summing integrator, in association with a resistance divider for multiplication by a constant less than one, can be used to simulate essentially any linear constant coefficient system. There are other analog computer elements such as multipliers and function generators that are used to simulate nonlinear systems but we will not consider them here as our concern is primarily with linear elements. Simulation of linear systems is relatively straightforward. There are many procedures and we illustrate one of these by means of an example.

EXAMPLE 3.1-3. We desire to determine the analog computer simulation to represent the system described by

$$\frac{d^2 z}{dt^2} + 2\zeta\omega_n \frac{dz}{dt} + \omega_n^2 z(t) = \omega_n^2 u(t)$$

By solving for the highest derivative we obtain

$$\frac{d^2 z}{dt^2} = -\omega_n^2 [z(t) - u(t)] - 2\zeta\omega_n \frac{dz}{dt}$$

Thus we can integrate an assumed $d^2 z/dt^2$ twice to get z and then formulate $d^2 z/dt^2$ by summing up appropriate inputs as indicated in Fig. (3.1-4). This is a complete analog computer simulation diagram. We can insert initial conditions on dz/dt and $z(t)$ by putting appropriate initial voltages on the integrating capacitors. To complete and actually implement the analog computer simulation we must be concerned with such things as time and amplitude scaling. Also it is possible to obtain a simulation with fewer amplifiers. To explore these topics would take us too far afield here.

EXAMPLE 3.1-4. We will now obtain an analog computer representation of the general second order system

$$T_1^2 \frac{d^2 z}{dt^2} + 2\zeta_1 T_1 \frac{dz}{dt} + z = T_2^2 \frac{d^2 u}{dt^2} + 2\zeta_2 T_2 \frac{du}{dt} + u$$

which can be used for several useful purposes including simulation of lag lead networks in which we may more easily change parameters than

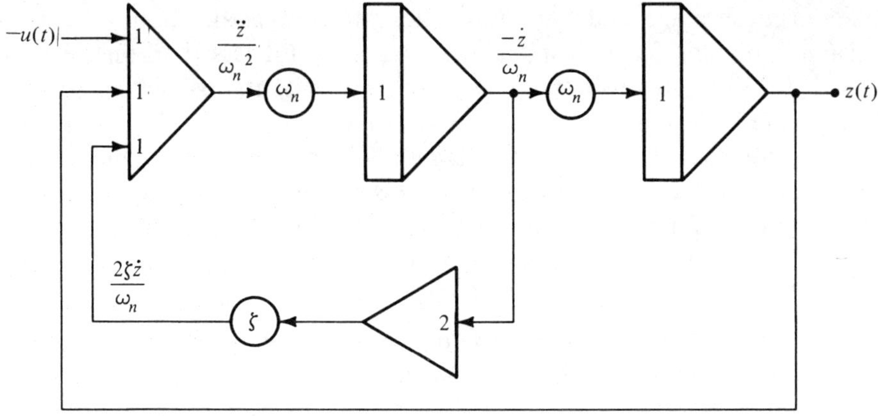

Figure 3.1-4   Analog computer simulation of differential equation

in the passive network we have previously discussed. It is convenient to write this differential equation in transfer function form as

$$\frac{Z(s)}{U(s)} = \frac{T_2^2 s^2 + 2\zeta_2 T_2 s + 1}{T_1^2 s^2 + 2\zeta_1 T_1 s + 1}$$

and divide by s^2 such that there are no "differentiators" involved and we obtain

$$\frac{Z(s)}{U(s)} = \frac{T_2^2\left[1 + \dfrac{2\zeta_2}{sT_2} + \dfrac{1}{s^2 T_2^2}\right]}{T_1^2\left[1 + \dfrac{2\zeta_1}{sT_1} + \dfrac{1}{s^2 T_1^2}\right]} \tag{2}$$

The most convenient way to implement the analog simulation of this is to visualize the general gain formula of Chap. 2 and implement a block diagram using integrators and summers such that application of the general gain formula yields Eq. (2). Slight rearrangement of this block diagram to accommodate the analog computer symbols leads immediately to Fig. (3.1-5) which represents the analog computer simulation for this example.

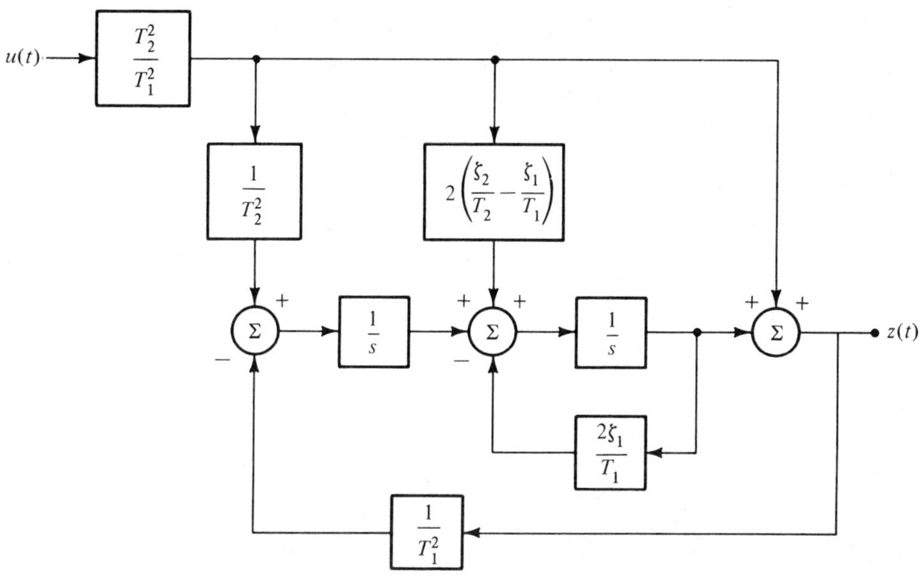

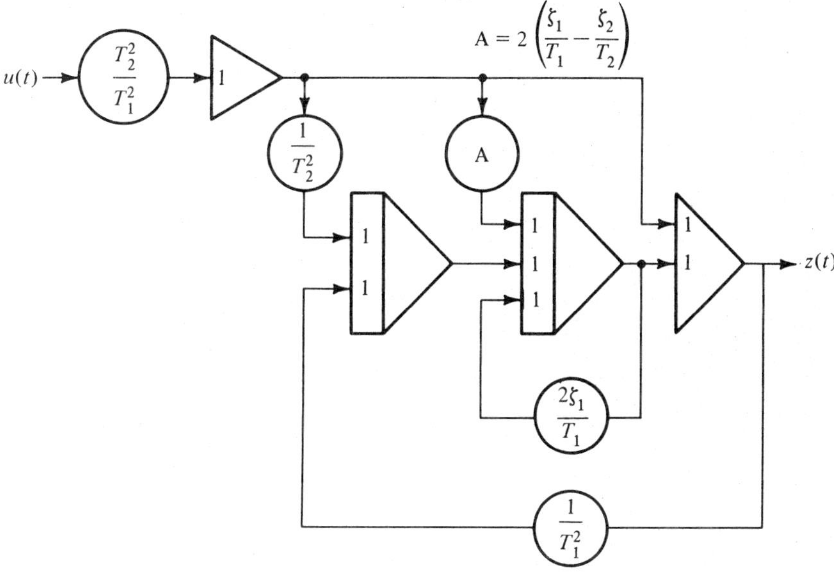

Figure 3.1-5 Block diagram and associated analog computer simulation of general second order system

EXERCISE 3.1-7. Show that the analog computer schematic of Fig. (3.1-6) solves the differential equation of EXAMPLE 3.1-3. What are appropriate voltage divider settings? What is the state variable differential equation solved by this schematic?

EXERCISE 3.1-8. Show that the analog computer schematic of Fig. (3.1-7) can represent a lag lead network. Will this be an easier network from which to determine parameter values than the corresponding network of Table 3.1-2?

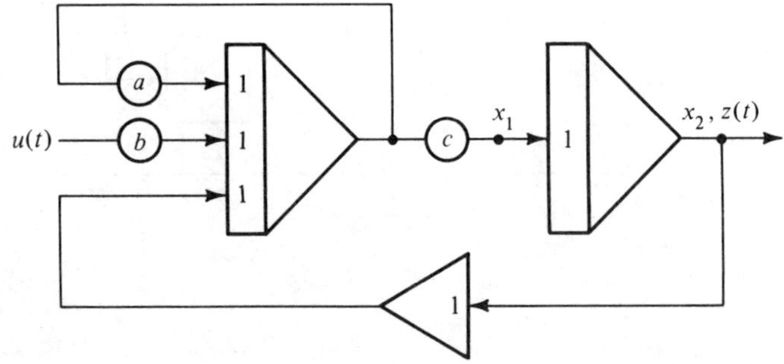

Figure 3.1-6   Analog computer schematic for EXERCISE 3.1-7

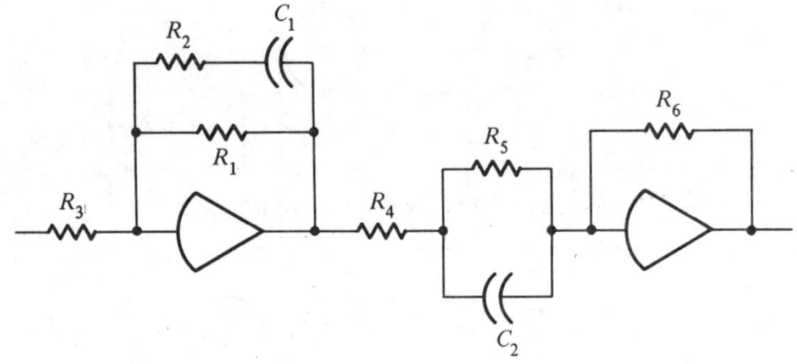

Figure 3.1-7   Lag lead network for EXERCISE 3.1-8

In this section we have examined the most common electrical networks and electronic analog computer networks that are used in control systems design. We now turn our attention to a similar topic— mechanical networks.

3.2 MECHANICAL NETWORKS

We will concern ourselves here with motion of rigid bodies such that the only independent variable we need deal with is time. Lumped constant coefficient mechanical systems or mechanical networks are entirely analogous to electrical networks except that they can be described as translational, rotational or a combination of these. Equations governing the motion of mechanical systems, which may contain active and passive elements, are determined from Newton's law of motion or D'Alembert's principle. Passive mechanical elements are springs, masses, and friction (resistance). The active elements are the various energy sources for mechanical systems.

Table 3.2-1 presents the basic translational and rotational elements and the linear relationship between force and motion in these elements. The passive linear mechanical elements may be connected together to form information shaping networks or filters although it is much more common for the mechanical elements to be part of a plant that is to be controlled. Differential equations may be written using Newton's law which states that when a body is acted upon by forces, it is accelerated in the direction of these forces with a magnitude proportional to the forces and inversely proportional to the mass of the body. It is more convenient for us to use the modification of Newton's law known as D'Alembert's principle which states that the instantaneous sum of the external forces acting on a body and the body's reaction force due to inertia is zero.*

In order to determine the differential equations representing a mechanical system we connect together at a common node all mechanical element terminals, active and passive, that move together and connect to a ground terminal or node all elements that remain stationary with respect to the particular reference frame selected. Then we assign a coordinate to every movable node. Arrows are used to indicate direction for positive values of movement of each coordinate as well as the positive direction of each independent force or velocity. Examples are best used to illustrate the procedure.

EXAMPLE 3.2-1. As a simple example of a mechanical system we consider the one coordinate translational system shown in Fig. (3.2-1a). This figure shows a mass M, constrained by fixed guides with friction

---

*Alternately Lagrange's equations based on energy considerations could be used to obtain equations of motion but we shall not use this method here.

TABLE 3.2-1 ELEMENTARY MECHANICAL NETWORK ELEMENTS

Element Name	Element Symbol	Input-Output Relationship
Translational force source—f (newtons)	$f(t)$ $F(s)$	Independent translational source
Translational velocity source—v (metres per second)	$v(t)$ $V(s)$	Independent translational source
Translational mass—M (kilograms)	M	$f(t) = M\dfrac{dv}{dt}$ $v(t) = \dfrac{1}{M}\displaystyle\int_0^t f(t)\,dt + v(0)$ $F(s) = MsV(s) - Mv(0)$
Translational stiffness—K (newton per metre)	K	$f(t) = K\displaystyle\int_0^t v(t)\,dt + f(0)$ $v(t) = \dfrac{1}{K}\dfrac{df(t)}{dt}$ $sF(s) = KV(s) + f(0)$
Translational friction (newtons per metre per second)	B	$f(t) = Bv(t)$ $F(s) = BV(s)$
Rotational torque source—τ (newton metres)	$\tau(t)$ $\tau(s)$	Independent rotational source

TABLE 3.2-1 *(cont'd)*

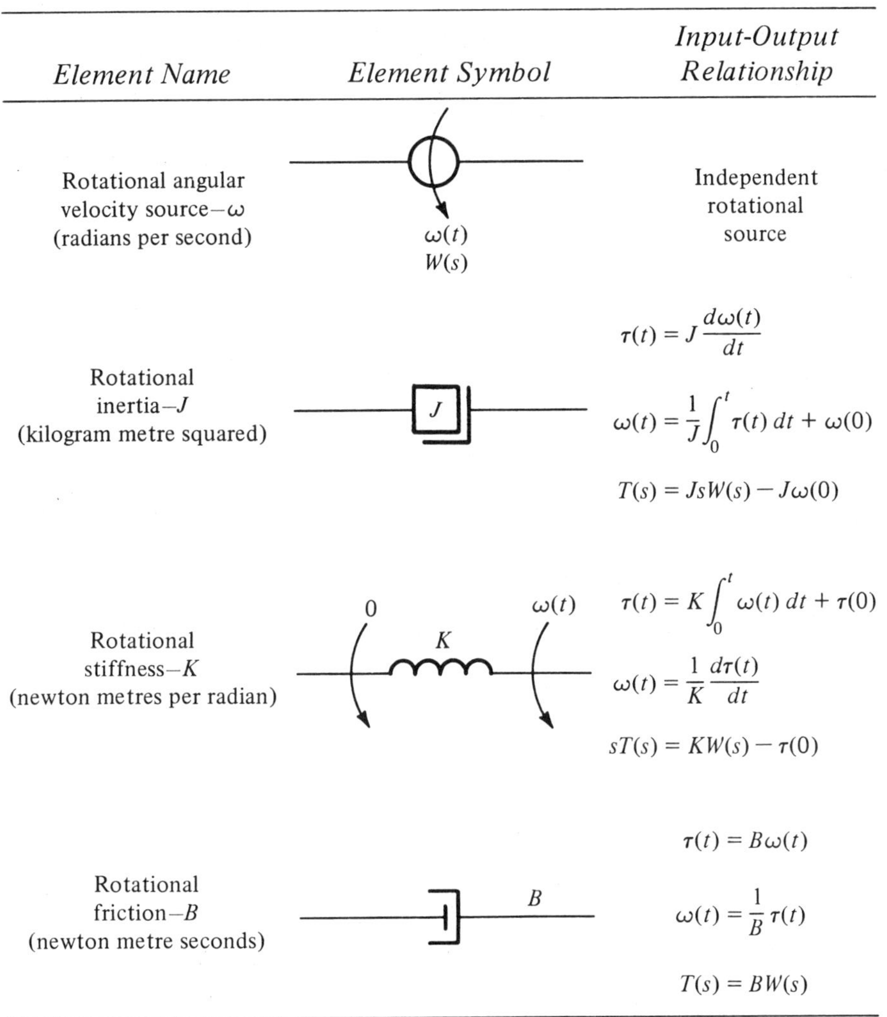

Element Name	Element Symbol	Input-Output Relationship
Rotational angular velocity source—ω (radians per second)	$\omega(t)$ $W(s)$	Independent rotational source
Rotational inertia—J (kilogram metre squared)	J	$\tau(t) = J\dfrac{d\omega(t)}{dt}$ $\omega(t) = \dfrac{1}{J}\displaystyle\int_0^t \tau(t)\,dt + \omega(0)$ $T(s) = JsW(s) - J\omega(0)$
Rotational stiffness—K (newton metres per radian)	$0 \qquad \omega(t)$ K	$\tau(t) = K\displaystyle\int_0^t \omega(t)\,dt + \tau(0)$ $\omega(t) = \dfrac{1}{K}\dfrac{d\tau(t)}{dt}$ $sT(s) = KW(s) - \tau(0)$
Rotational friction—B (newton metre seconds)	B	$\tau(t) = B\omega(t)$ $\omega(t) = \dfrac{1}{B}\tau(t)$ $T(s) = BW(s)$

coefficient B, which is supported by a spring with spring constant K. A driving force $f(t)$ acts in a vertical direction. The mass is at rest at $t = 0$ at a position where the spring supports the force due to gravity which is Mg. The mechanical network corresponding to this single coordinate system is shown in Fig. (3.2-1b). We use the node method of analysis and have, using D'Alembert's principle,

$$f_K(t) + f_B(t) + f_M(t) - f(t) = 0 \tag{1}$$

where the subscript is used to indicate the component upon which the force acts. Using appropriate input-output relations from Table 3.2-1 we have for Eq. (1)

$$M\frac{dv(t)}{dt} + Bv(t) + K\int_0^t v(t)\,dt = f(t) \tag{2}$$

where we note that $Mg = f_K(0)$ since the system is initially in equilibrium. If we desire we can insert the relation $v(t) = dx(t)/dt$ into Eq. (1) to obtain

$$M\frac{d^2x(t)}{dt^2} + B\frac{dx(t)}{dt} + Kx(t) = f(t) \tag{3}$$

where $x(t)$ is the position displacement from the equilibrium $x(0) = 0$. Appropriate initial conditions for this problem are $x(0) = v(0) = 0$ because of our equilibrium assumption.

We may take the Laplace transform of Eq. (3) to obtain the mechanical system transfer function

$$\frac{X(s)}{F(s)} = \frac{1}{Ms^2 + Bs + K} \tag{4}$$

For a given force function, $f(t)$, we may find $F(s)$ and then employ the procedure used in Chap. 2 to find the time response of motion $x(t)$.

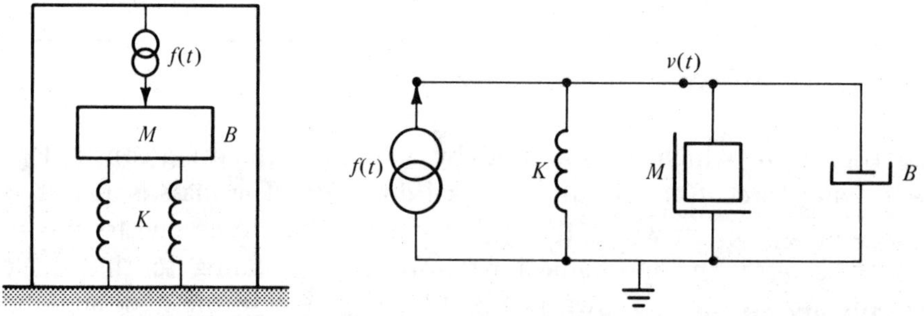

Figure 3.2-1   Single coordinate translational system

EXAMPLE 3.2-2. In this example we consider the two coordinate mechanical system shown in Fig. (3.2-2a) which can be modeled by the mechanical network shown in Fig. (3.2-2b). Each mass is supported on a guidewire and there is friction between each mass and the ground-reference as well as friction between the two mass displacements. We assume that initial conditions are such that the system is initially at rest with no force $f(t)$ applied. This requires $M_1 g = K_1 [x_1(0) - x_2(0)]$ and $M_2 g = K_2 x_2(0) + K_1 [x_2(0) - x_1(0)]$ or $(M_1 + M_2)g = K_2 x_2(0)$. From Fig. (3.2-2b) we obtain, using nodal analysis,

$$f(t) = M_1 \frac{dv_1(t)}{dt} + B_3 v_1(t) + B_1 [v_1(t) - v_2(t)] + K_1 \int_0^t [v_1(t) - v_2(t)]\, dt$$

for the sum of forces at node 1 and

$$0 = B_1 [v_2(t) - v_1(t)] + K_1 \int_0^t [v_2(t) - v_1(t)]\, dt +$$
$$B_2 v_2(t) + K_2 \int_0^t v_2(t)\, dt + M_2 \frac{dv_2(t)}{dt}$$

as the force equation at node 2. We may now take Laplace transforms of these equations and obtain, after multiplication by s,

$$sF(s) = [M_1 s^2 + (B_1 + B_3)s + K_1] V_1(s) - (B_1 s + K_1)V_2(s)$$

$$0 = -(B_1 s + K_1)V_1(s) + [M_2 s^2 + (B_1 + B_2)s + K_1 + K_2] V_2(s)$$

which we may also solve for the velocities $V_1(s)$ and $V_2(s)$. Also we can use the relation $V_1(s) = sX_1(s) - x_1(0)$ and $V_2(s) = sX_2(0) - x_2(0)$ and solve for the distance displacement $x_1(t)$ and $x_2(t)$. The way in which equilibrium is defined in this example $x_1(0)$ and $x_2(0)$ are the spring compressions needed (with zero compression as a reference) to support the masses M_1 and M_2.

EXAMPLE 3.2-3. As an example of a two coordinate rotational mechanical system we consider the system shown in Fig. (3.2-3a) which represents two discs mounted on a flexible shaft with torsional spring constant K. There is torsional friction B_1 and B_2 between each disc and some fixed reference point. The moments of inertia of the two discs are J_1 and J_2. Torque $\tau_1(t)$ is applied to one end of the shaft and torque $\tau_2(t)$ to the other end.

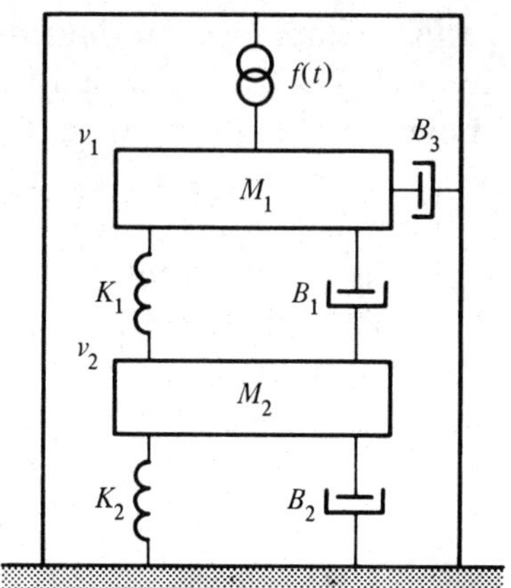

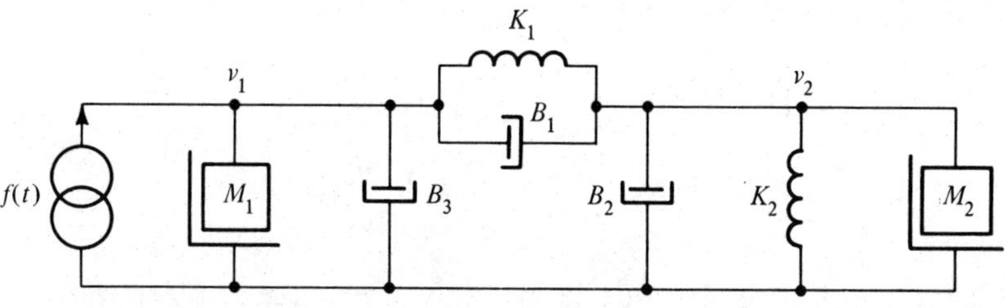

Figure 3.2-2 Mechanical system and equivalent mechanical network for
EXAMPLE 3.2-2

Figure (3.2-3b) shows the equivalent mechanical network for this
system. Application of D'Alembert's principle to the nodes of this mechanical system yields

$$-\tau_1(t) + J_1 \frac{d\omega_1(t)}{dt} + K\int_0^t [\omega_1(t) - \omega_2(t)]\, dt + \tau_K(0) + B_1\omega_1(t) = 0 \quad (1)$$

and

$$-\tau_2(t) + J_2 \frac{d\omega_2(t)}{dt} + K\int_0^t [\omega_2(t) - \omega_1(t)]\, dt - \tau_K(0) + B_2\omega_2(t) = 0 \quad (2)$$

We may take Laplace transforms of Eqs. (1) and (2) to obtain the
system input output transfer relations

$$(J_1 s^2 + B_1 s + K)W_1(s) - KW_2(s) = sT_1(s) + J_1 s\omega_1(0) + \tau_K(0)$$

$$-KW_1(s) + (J_2 s^2 + B_2 s + K)W_2(s) = sT_2(s) + J_2 s\omega_2(0) - \tau_K(0)$$

for this simple two coordinate rotational system.

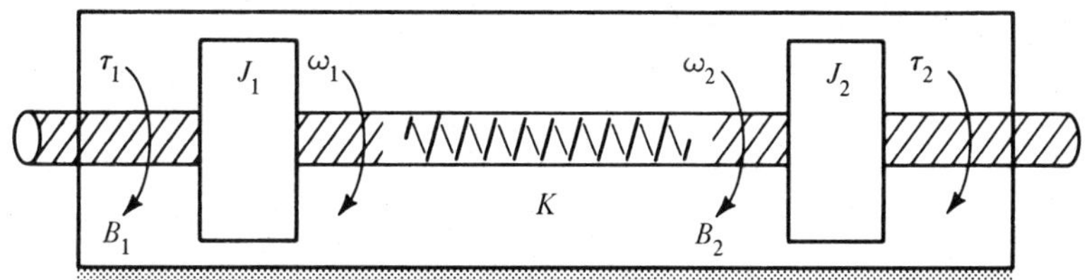

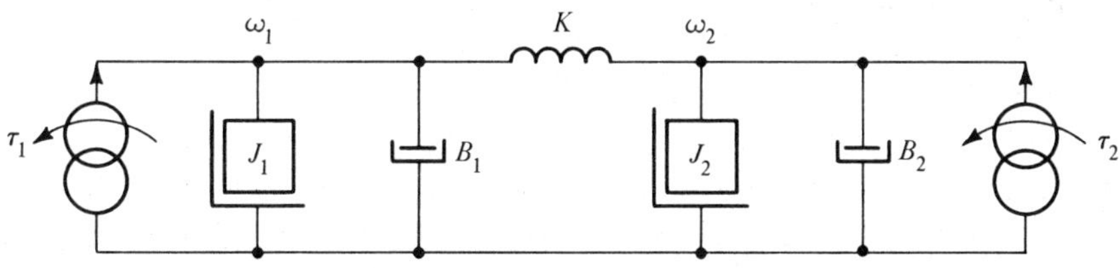

Figure 3.2-3   Mechanical system and equivalent mechanical network for EXAMPLE 3.2-3

We could extend our treatment of mechanical systems in several ways. We note that the circuit diagrams for electrical systems and mechanical systems are remarkably similar. In fact they are analog representations of one another. We note that a single loop initially unexcited series electric circuit driven by a voltage source is described by

$$L\frac{di(t)}{dt} + Ri(t) + \frac{1}{C}\int_0^t i(t)\,dt = v(t) \tag{3.2-1}$$

whereas a parallel initially unexcited electric circuit driven by a current source is described by

$$C\frac{dv(t)}{dt} + \frac{v(t)}{R} + \frac{1}{L}\int_0^t v(t)\,dt = i(t) \tag{3.2-2}$$

A "series" one coordinate initially unexcited translational mechanical network driven by a velocity source is described by

$$\frac{1}{K}\frac{df(t)}{dt} + \frac{f(t)}{B} + \frac{1}{M}\int_0^t f(t)\,dt = v(t) \qquad (3.2\text{-}3)$$

whereas a "series" initially unexcited one coordinate translational mechanical network driven by a force source is described by

$$M\frac{dv(t)}{dt} + Bv(t) + K\int_0^t v(t)\,dt = f(t) \qquad (3.2\text{-}4)$$

Single coordinate "series" rotational mechanical systems could be described by equations similar to Eqs. (3.2-3) and (3.2-4) except using rotational components.

As we easily see, all four of the foregoing equations are fundamentally the same. The electrical networks represented by Eqs. (3.2-1) and (3.2-2) are *dual* networks. If we replace a mechanical system described by Eq. (3.2-4) by the electrical system described by Eq. (3.2-2) such that current is analogous to force we have what is called a *direct analog* whereas if we replace the mechanical system described by Eq. (3.2-4) by the electrical system described by Eq. (3.2-1) such that force is analogous to voltage we have what is called an *indirect analog*. We could show that thermal and hydraulic and other non-electrical systems also have electrical analogs. It is of value to use electrical network analogs for linear mechanical systems since electric network theory has been developed to a very high degree. Since a thorough study of this topic would be somewhat peripheral to our primary goals we shall not explore this interesting topic here.

EXERCISE 3.2-1. Determine the differential equations which describe motion in the mechanical seismic system transducer model of Fig. (3.2-4). Be sure to draw a mechanical network. The source for this system is the horizontal ground displacement $x(t)$.

EXERCISE 3.2-2. Prepare a table of dual and analogous electrical and mechanical network elements.

EXERCISE 3.2-3. Determine the differential equations of motion describing the gear train system in Fig. (3.2-5). The number of teeth on

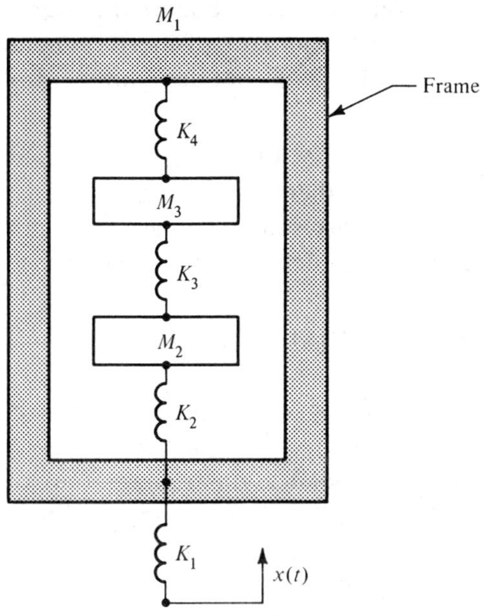

Figure 3.2-4 Model of seismic signal transducer

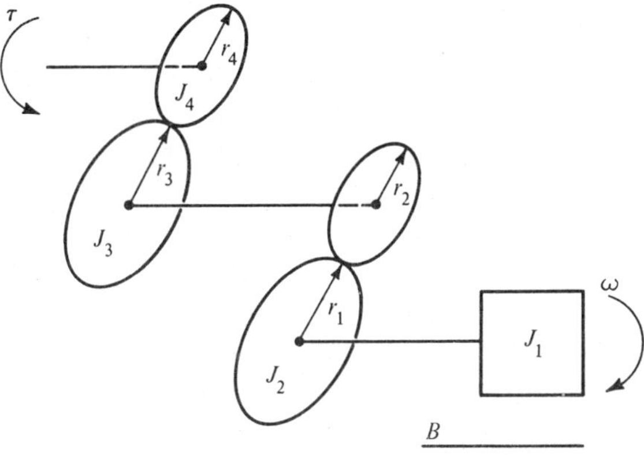

Figure 3.2-5 Rotational system for EXERCISE 3.2-3

each gear is proportional to the radius of the gear. Each gear and the associated rigid shaft has a moment of inertia. There is a single friction element associated with the load that has a moment of inertia J_1.

EXERCISE 3.2-4. Write the differential equation representing a mechanical system in which there is combined translation and rotation. Postulate an angular velocity source applied to a shaft with torsion that moves a conveyor belt on which a mass is located.

3.3 ELECTROMECHANICAL SYSTEMS

There are a great many electromechanical systems and elements in existence. Some of these such as microphones, speakers and strain gauges are primarily intended as instrumentation modulation devices in which a parameter in either the electrical or the mechanical portion of the system changes as a response to a change in conditions in the mechanical or electrical portion of the system. These electromechanical elements are particularly useful in control systems as sensing devices. Other electromechanical systems, such as motors, are primarily used as energy conversion devices and are thus useful as prime movers in control systems. This categorization of electromechanical systems into modulation devices for sensing and energy conversion devices for motion control is unfortunately slightly arbitrary in that many elements do act as energy convertors, the dynamic microphone for example, even though its primary purpose is to act as a sensor rather than a prime mover. Nevertheless, our categorization is useful.

One of the simplest modulation devices is the resistance divider or potentiometer shown schematically in Fig. (3.3-1a). A change in mechanical position produces a change in resistance measured from the electrical pickoff point v to ground and a change in the output voltage v. The resistance divider is a linear* device as long as the mechanical position is subject to the restriction $-\pi \leqslant \theta \leqslant \pi$. By grounding the center of the resistance divider we obtain an element with an output voltage symmetrical about 0. Also we may use a multiturn resistance divider. Elements such as this are used to convert mechanical shaft position into an electrical voltage and are, therefore, very useful as position sensors in rotational systems. In this function two resistance dividers are used, one to convert an input shaft position and the other

---

*Subject of course to granularity of the resistance element used to construct the resistance divider.

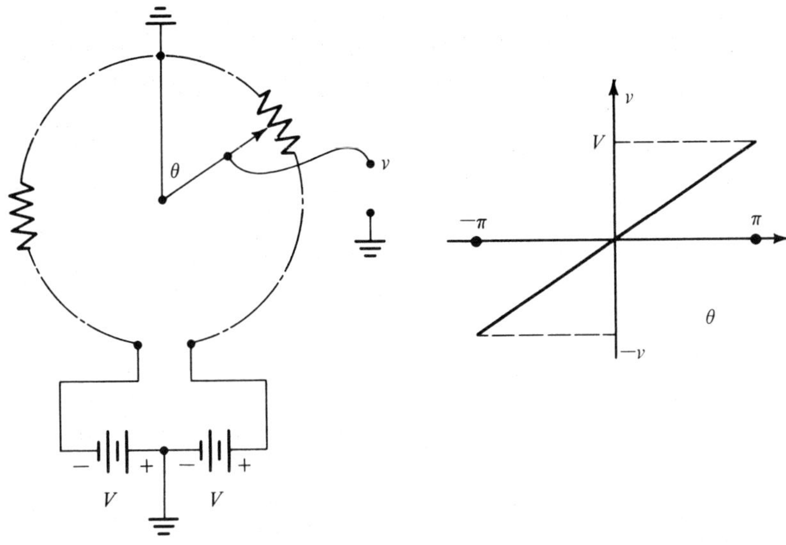

Figure 3.3-1 A simple electromechanical modulator—the resistance divider:
pictorial diagram and input output characteristics

to convert an output shaft position into voltages. These voltages are
subtracted to yield an error signal.

Somewhat related to the potential divider is the synchro. Although
there is no fundamental reason why a resistance divider cannot be used
with an alternating voltage supply rather than the direct voltage V, it
is generally more convenient to use a synchro since most ac control
systems are multiphase systems rather than single phase systems. A
synchro is basically a transformer-type device with a rotor applying a
single phase voltage to three phase transformer windings such that
output voltage is proportional to shaft position.

Tachometers are often used as sensors of shaft position also. When
an electric conductor moving with velocity dx/dt crosses a magnetic
field, there is induced in the conductor a voltage proportional to the
magnetic field intensity, the length of the conductor, and the velocity
with which it is moving. A tachometer is a rotational device, often with
a permanent magnet field for a direct current tachometer, whose input
output relation satisfies the equations $v(t) = K_T \omega(t)$, where v is the
output voltage generated by the tachometer, $\omega(t)$ is the input angular
velocity of the tachometer shaft, and K_T is a constant depending upon
the design parameters of the tachometer.

Direct current motors are often used as prime movers in control systems. Such motors may be series excited where the field winding is in series with the armature of the motor, shunt excited where the field winding is in parallel with the armature of the motor, field separately excited in which the field winding is excited from a constant current source, or field controlled or armature separately excited in which the armature is separately excited with a constant current. Only the latter two are of serious interest for precision linear control systems as the first two methods of excitation result in decidedly nonlinear characteristics.

Figure (3.3-2) shows a simplified schematic to represent a field separately excited motor. Two physical principles are needed in order to specify the dynamics of this motor. One of these relates to the force or torque exerted on the mechanical shaft due to the field flux and armature current. It can be shown that the torque induced on the shaft is linearly proportional to the armature current flowing and a constant of proportionality K_τ called the torque constant. The second one is the back electromotive force or back emf induced by the moving shaft. It turns out that this back emf is linearly proportional to the flux density induced by the field and the angular shaft velocity. Thus we write

$$\tau = K_\tau i_a \tag{3.3-1}$$

$$E = K_\omega \omega \tag{3.3-2}$$

The armature loop equation is

$$v(t) = Ri(t) + L\frac{di(t)}{dt} + K_\omega \omega(t) \tag{3.3-3}$$

and the torque equation is

$$\tau(t) = K_\tau i(t) = J\frac{d\omega(t)}{dt} + B\omega(t) \tag{3.3-4}$$

where the moment of inertia J includes the inertia of both the armature and the load, possibly reflected through a gear train. We can combine the foregoing two equations to give a single input output equation between input voltage and output angular velocity. This relation is

$$v(t) = \frac{LJ}{K_\tau} \frac{d^2\omega(t)}{dt} + \frac{1}{K_\tau}(BL + JR)\frac{d\omega(t)}{dt} + \left(K_\omega + \frac{BR}{K_\tau}\right)\omega(t) \quad (3.3\text{-}5)$$

and can be expressed as the transfer function, where $\omega = d\Theta/dt$,

$$\frac{\Theta(s)}{V(s)} = \frac{K_\tau}{JLs\left[s^2 + \left(\dfrac{B}{J} + \dfrac{R}{L}\right)s + \left(\dfrac{K_\tau K_\omega + BR}{JL}\right)\right]} \quad (3.3\text{-}6)$$

In many cases the armature inductance L will be very small and the armature current determined primarily by input voltage, armature resistance, and back emf. When we let $L = 0$, these equations simplify to

$$v(t) = \frac{JR}{K_\tau}\frac{d\omega(t)}{dt} + \left[K_\omega + \frac{BR}{K_\tau}\right]\omega(t) \quad (3.3\text{-}7)$$

and

$$\frac{\Theta(s)}{V(s)} = \frac{K_\tau}{JRs\left[s + \dfrac{K_\tau K_\omega + BR}{JR}\right]} \quad (3.3\text{-}8)$$

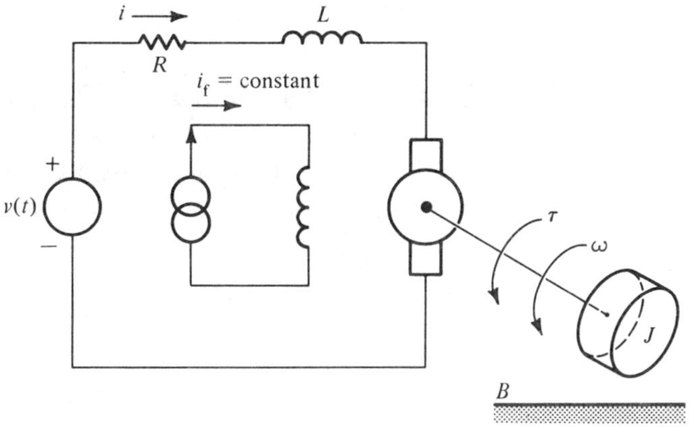

Figure 3.3-2 Field separately excited dc motor

Often we will find it convenient to write transfer functions in time constant form and we obtain

$$\frac{\Theta(s)}{V(s)} = \frac{K_\tau}{(K_\tau K_\omega + BR)s\left(\dfrac{JR}{K_\tau K_\omega + BR}s + 1\right)} \tag{3.3-9}$$

as the simplified transfer function for the armature controlled motor written in transfer function form.

A field separately excited dc motor requires a control voltage of sufficient power capacity to supply all the power needed for the armature circuit in the motor. This is a much greater amount of power than needed to supply the field. Thus there is motivation to consider a separately excited armature and control the field circuit as this would generally require lower power handling capability on the part of the amplifier driving the system. A simplified schematic of an armature separately excited dc motor is shown in Fig. (3.3-3). We can obtain the differential equations describing this electromechanical system in much the same way as we have obtained the relevant equations for the field separately excited dc motor. We have

$$v_f(t) = L_f \frac{di_f(t)}{dt} + R_f i_f \tag{3.3-10}$$

$$\tau(t) = K_\tau i_f = J \frac{d\omega(t)}{dt} + B\omega(t) \tag{3.3-11}$$

where we do not obtain the back emf term in the field voltage equation. The back emf appears in the armature circuit and we have supplied constant current to the armature. The single input output equation and transfer function describing this electromechanical system are

$$V_f(t) = \frac{L_f J}{K_\tau} \frac{d^2\omega(t)}{dt^2} + \left(\frac{R_f J}{K_\tau} + \frac{L_f B}{K_\tau}\right)\frac{d\omega(t)}{dt} + \frac{R_f B}{K_\tau}\omega(t) \tag{3.3-12}$$

$$\frac{\Theta(s)}{V_f(s)} = \frac{K_\tau}{L_f Js\left[s^2 + \dfrac{R_f J + L_f B}{L_f J}s + \dfrac{R_f B}{L_f J}\right]} \tag{3.3-13}$$

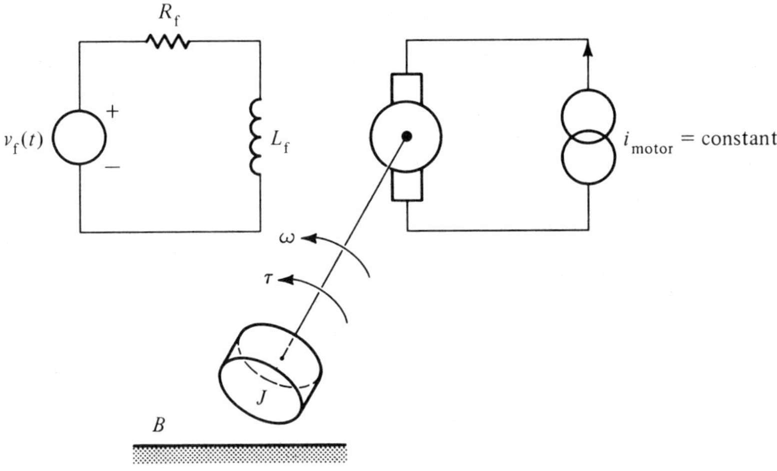

Figure 3.3-3   Armature separately excited dc motor

Often the armature separately excited motor is used in applications in which there is negligible friction on the output shaft. With $B = 0$, the transfer function of Eq. (3.3-12) becomes that of a double integrator system

$$\frac{\Theta(s)}{V_f(s)} = \frac{K_\tau}{L_f Js^2 \left(s + \dfrac{R_f}{L_f}\right)} = \frac{K_\tau}{R_f Js^2 \left(1 + \dfrac{L_f s}{R_f}\right)} \tag{3.3-14}$$

and when the field inductance L_f is negligible this becomes just that of a double integrator

$$\frac{\Theta(s)}{V_f(s)} = \frac{K_\tau}{R_f Js^2} \tag{3.3-15}$$

and this transfer function is often used to represent an armature separately excited dc motor.

EXERCISE 3.3-1.   Obtain the differential equations of the series connected self-excited dc motor. How can these be linearized?

EXERCISE 3.3-2. Obtain the differential equations of the shunt connected self-excited dc motor. How can these be linearized?

It is common to use two phase ac motors in control system design with low power requirements since the ac motor has no brushes and is easier to maintain than dc motors. Also ac amplifier design and balance problems are simpler than equivalent problems for dc amplifiers. Design of compensating networks is a bit more complex and often ac signals are demodulated, dc compensation networks used, and the resulting signal modulated again. A fixed or reference phase in a two phase ac control motor is supplied with a constant amplitude voltage at the prescribed ac frequency. An amplified error signal is applied to the other phase or control phase or field of the motor at a phase shifted 90 degrees relative to the reference phase. A simplified schematic diagram for a two phase motor is shown in Fig. (3.3-4) and a typically observed speed torque curve in Fig. (3.3-5). A speed torque curve of this form is also characteristic of the two dc motor configurations examined earlier. For small increments of angular shaft velocity ω, torque τ, and control field voltage v, we have the linearized relationship

$$\Delta\omega(t) = \frac{\partial\omega}{\partial v}\Delta v(t) + \frac{\partial\omega}{\partial\tau}\Delta\tau(t) \tag{3.3-16}$$

If the load contains friction and inertia only we have

$$\Delta\tau(t) = J\frac{\partial\Delta\omega(t)}{\partial t} + B\Delta\omega(t) \tag{3.3-17}$$

We obtain for the motor transfer function, upon taking the Laplace transform of the two foregoing equations

$$\frac{\Delta\omega(s)}{\Delta V(s)} = \frac{\dfrac{\partial\omega}{\partial v}}{1 - \dfrac{\partial\omega}{\partial\tau}[Js + B]} \tag{3.3-18}$$

The partial derivative $\partial\omega/\partial\tau$ is just the slope of the speed torque curve near the operating point and this is, from Fig. (3.3-5), always negative which insures stability of the motor. $\partial\omega/\partial v$ is positive as we see also from Fig. (3.3-5).

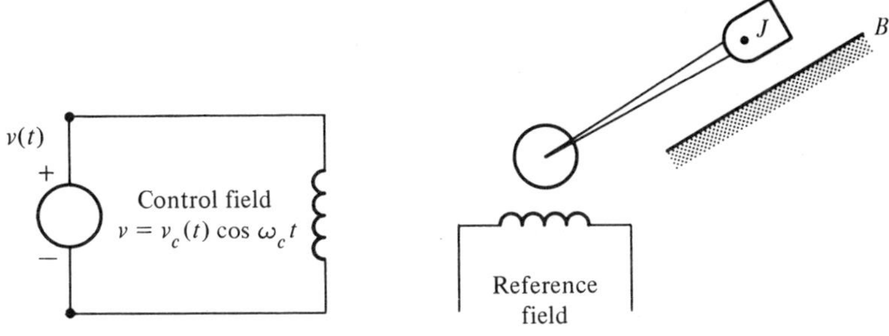

Figure 3.3-4   Two phase ac servomotor

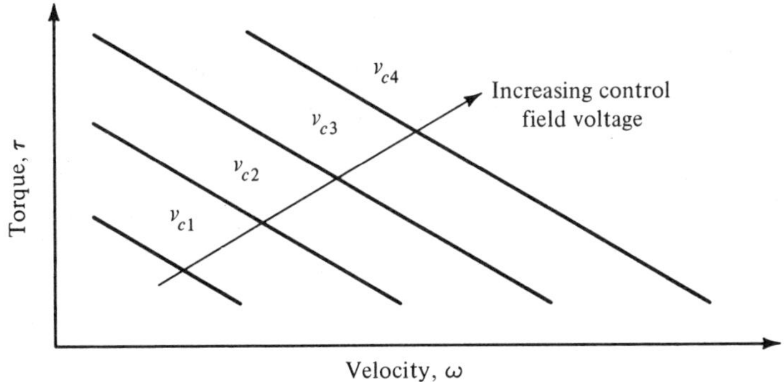

Figure 3.3-5   Linearized torque speed characteristics of two phase motor

EXERCISE 3.3-1.   Figure (3.3-6) illustrates a simplified schematic of a motor generator subsystem typical of that found in many positioning control systems. An advantage to this system is that an electronic amplifier need only have power capability needed for the field of the generator. Determine the differential equations which characterize this system assuming that the motor and generator have the characteristics $K_{\tau m}$, $K_{\omega m}$, $K_{\tau g}$, and $K_{\omega g}$. Answer

$$\frac{W(s)}{V(s)} = \frac{K_{\omega g}/K_{\omega m}}{(L_f s + R_f)\left[\dfrac{JL_a}{K_{\tau m}K_{\omega m}}s^2 + \dfrac{JR_a + BL_a}{K_{\omega m}K_{\tau m}}s + 1 + \dfrac{BR_a}{K_{\omega m}K_{\tau m}}\right]}$$

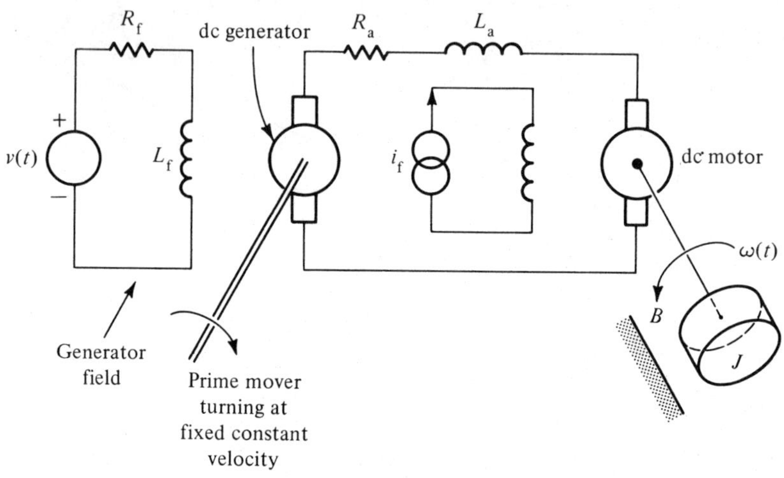

Figure 3.3-6 Motor generator subsystem

EXERCISE 3.3-2. Suppose that a separately excited constant field current dc motor is operating at a constant speed. The voltage input to the armature is suddenly switched to zero. Show that the effect of the motor is to stop movement more quickly than would otherwise occur.

3.4 THERMAL AND FLUID SYSTEMS

We continue our discussion of systems control elements in this section with a brief presentation of some simple thermal, or heat conduction, systems and fluid, or hydraulic, systems. These systems are inherently distributed spatially but most of them are such that the distribution of temperatures, velocities, pressures, reactivity, etc. are sufficiently uniform that lumped parameter models are possible.

Thermal energy will flow from one body to another if there is a temperature difference between the two bodies. The relationship expressing this flow is analogous to the flow of current in a resistor and is given by the empirical physical law

$$q(t) = \frac{T_1(t) - T_2(t)}{R}$$

(3.4-1)

where q is the heat flow rate or energy transfer rate from body 1 to body 2 (in energy units such as newton metres per second or calories per second), T_1 and T_2 are the temperatures of the two bodies, and R is the resistance to heat flow and depends on the geometries of the two bodies and the thermal conductivities of the bodies and any material separating them.

It also turns out that the rate of change of the temperature of a body is proportional to the net rate at which thermal energy is flowing into the body. This is a statement of the *first law of thermodynamics* which states that

$$\frac{dT(t)}{dt} = \frac{q(t)}{C} \tag{3.4-2}$$

where C is the thermal storage capacity of a given body. C is proportional to the mass of the thermal system M, and the specific heat K and is given by $C = MK$.

EXAMPLE 3.4-1. As a simple example of a thermal system let us consider the simple stirred tank thermal system of Fig. (3.4-1). A heater is immersed in the liquid and supplies energy at a rate of $q(t)$ calories per second. This energy raises $T(t)$, the temperature of the liquid, and supplies surface losses to the external boundary of the system assumed to be at fixed temperature T_f. The surface losses are assumed to vary linearly with the difference between the temperature of the liquid and the surrounding ambient temperature. Thus we have, from Eqs. (3.4-1) and (3.4-2)

$$q(t) = \frac{T(t) - T_f}{R} + C\frac{dT(t)}{dt}$$

which is the differential equation describing this simple system. The input output transfer relation for the system, assuming the temperature of the liquid is initially T_f, is

$$Q(s) = \left(\frac{1}{R} + sC\right)T(s) - \left(\frac{1}{R} + Cs\right)\frac{T_f}{s}$$

It is interesting to note that the $(1/R + Cs)$ term appears in both terms on the right-hand side of this equation. Thus the ambient temperature

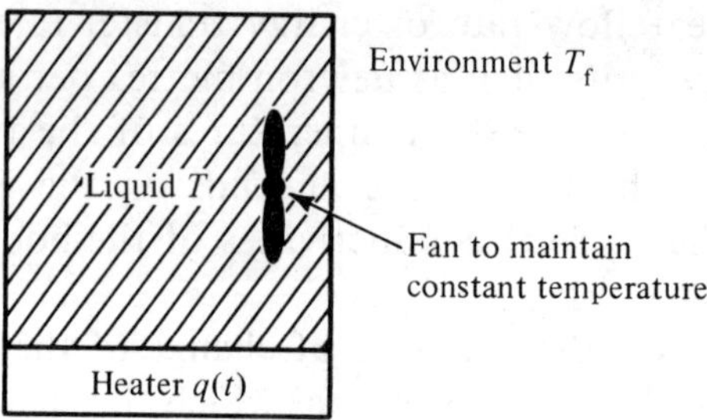

Figure 3.4-1   Simple stirred thermal system

is, in effect, a load on the system. If we define $\Delta T(t) = T - T_f$, then the latter two relations become

$$q(t) = \frac{\Delta T(t)}{R} + C\,\frac{d\Delta T(t)}{dt}$$

and

$$\frac{\Delta T(s)}{Q(s)} = \frac{1}{\dfrac{1}{R} + sC} = \frac{R}{RCs + 1}$$

EXAMPLE 3.4-2. We may extend the results of the previous example to yield a simple model of heat conduction from a rocket engine. Figure (3.4-2) represents a very simplified schematic diagram for this example. We assume that at the moment of rocket ignition the inside temperature T_1 rises to a constant value. The ambient temperature is constant at T_3 and the temperature in the rocket engine shell is T_2. There are resistances to heat flow R_1 and R_2 between the rocket chamber and the shell, and the shell and the external surroundings.

We obtain differential equations for our example by repeated application of Eqs. (3.4-1) and (3.4-2). We could draw a network representation of this system much as we have for the mechanical networks. This will be left as an exercise for the interested reader. We have for flow between the rocket engine chamber and the shell

$$q_{12}(t) = \frac{T_1 - T_2(t)}{R_1}$$

and for the thermal flow between the shell and the external environment

$$q_{23}(t) = \frac{T_2(t) - T_3}{R_2}$$

where R_1 and R_2 are the resistance to heat flow between chamber and shell and shell and environment. The net thermal energy flow across the shell is the difference between these two and it is this relationship we need to use to obtain the $q(t)$ for Eq. (3.4-2). Thus we have

$$\frac{q_{12}(t) - q_{23}(t)}{C} = \frac{dT_2(t)}{dt}$$

and so we obtain for the system differential equation

$$\frac{dT_2(t)}{dt} + \frac{1}{C}\left(\frac{1}{R_1} + \frac{1}{R_2}\right)T_2(t) = \frac{1}{C}\left(\frac{T_1}{R_1} + \frac{T_3}{R_2}\right)$$

which is our simple lumped model for temperature distribution in this rocket engine.

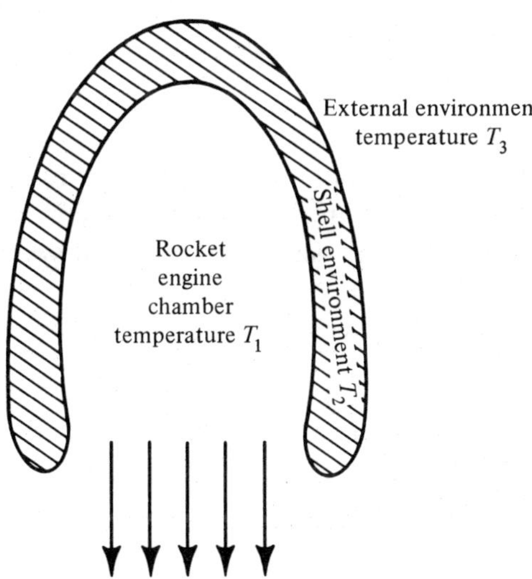

Figure 3.4-2 Simple model of rocket engine (chamber, shell, and environment)

EXERCISE 3.4-1. Determine "thermal network" models for EXAMPLES 3.4-1 and 3.4-2. Discuss direct and indirect analogies between thermal networks and electrical networks.

EXERCISE 3.4-2. Consider a motel consisting of three rooms in a row. Each room is cubic in shape. A heater delivers equal quantities of heat flow q_i, i = 1, 2, 3 to the rooms. The external temperature T_e is constant. The resistance of each internal wall is R_i and that of each external wall and floor is R_e. Derive differential equations for the temperature in each room. Assume that initially all rooms are at the same temperature.

There are numerous fluid systems of interest for control purposes. These vary from large water supply systems to rocket propulsion systems to hydraulic actuators used as control components. All of these can be analyzed using methods of fluid mechanics. We cannot attempt to present other than a very rapid brief excursion into this interesting area. We will be content to state three important equilibrium relations and consider two simple examples.

The *continuity equation*, a statement of the conservation of matter, must be satisfied by a fluid in an open system in which fluid may enter and leave. The continuity equation states that

$$\frac{dm(t)}{dt} = q_i - q_o \tag{3.4-3}$$

where $m(t)$ is the mass of fluid inside the imaginary envelope which constitutes the open system, and q_i and q_o are the mass inflow and outflow rates. The other two equilibrium relations are just consequences of Newton's second applied to translation mechanical systems

$$\sum_i f_i(t) = 0 \quad \text{(in any direction)} \tag{3.4-4}$$

and rotational mechanical systems

$$\sum_i \tau_i(t) = 0 \quad \text{(about any axis)} \tag{3.4-5}$$

where the f_i represent translational forces at any given node and the τ_i represent torques about any given node.

A given fluid system is analyzed by determining relationships between fluid density and pressure, between pressure and fluid flow rate, and between fluid flow rate and friction forces. The relationship between fluid density and pressure is generally assumed to be

$$p(t) = k\rho^{\gamma}(t) \tag{3.4-6}$$

for gases where $p(t)$ denotes pressure and $\rho(t)$ denotes density. k and γ are constants and are medium dependent. In air it turns out that $\gamma = 1.4$. For liquids we generally assume incompressability such that density and pressure are independent of one another. As is true with many fluid mechanics relationships, Eq. (3.4-6) is non-linear. If we assume that $p(t) = p_n(t) + \Delta p(t)$ and $\rho(t) = \rho_n(t) + \Delta\rho(t)$ and that at equilibrium $p_n(t) = k\rho_n^{\gamma}(t)$ then we have

$$\Delta p(t) = k\gamma\rho_n^{\gamma-1}(t)\Delta\rho(t) \tag{3.4-7}$$

to terms of first order and the relationship is linear in the perturbation terms $\Delta p(t)$ and $\Delta\rho(t)$.

Long tubes behave in a different fashion from short tubes as far as pressure and fluid flow rates are concerned. In a short tube there is assumed to be negligible fluid within the tube at a given time and the relationship between fluid flow and pressure drop is assumed given by

$$q(t) = \frac{1}{R}\left[p_i(t) - p_o(t)\right]^{\alpha}$$

where R is the resistance of the tube. For long tubes we must consider friction force on the fluid which will be proportional to fluid velocity according to

$$\text{force of friction} = cv^{\beta} = cv^{1/\alpha} \tag{3.4-8}$$

The values of γ, α, and β depend upon the types of fluid and its density and viscosity. Two simple examples will give an idea of the sort of dynamic equations we obtain for typical fluid systems.

EXAMPLE 3.4-3. In this example we consider a simplified representation of a water supply for a town. Figure (3.4-3) presents such a system. There is one very large reservoir such as might be formed from a dam located up mountain from a town. A long pipe carries water to a

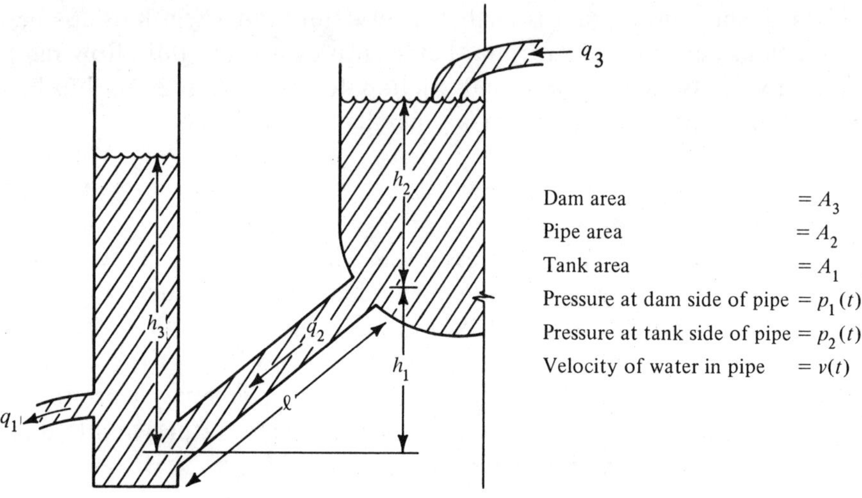

Figure 3.4-3 Simplified model of water supply system

storage tank reservoir. Citizens of the town use water from the storage tank at a mass flow rate $q_1(t)$. Water flows into the dam at a mass flow rate $q_3(t)$. Both dam and tank are so large that fluid flow rate within these is negligible. Also water is assumed incompressible.

There are a number of relations obtainable for this system from the fundamental equations (3.4-3) and (3.4-5). Water flows at a mass flow rate q_2 from the dam into the pipe and at a flow rate q_3 into the dam. If the horizontal area of the dam is A_3 and the height of the water is $h_1 + h_2$, then when we equate the flow rate in the dam to the net flow into the dam we have

$$\rho A_1 \frac{dh_2(t)}{dt} = q_3(t) - q_2(t) \tag{1}$$

assuming that the water level in the dam will always be above h_1. We have a similar relation expressing the net flow at the storage tank

$$\rho A_1 \frac{dh_3(t)}{dt} = q_2(t) - q_1(t) \tag{2}$$

where ρ is the density of the water and A_1 is the area of the storage tank.

We must now apply Newton's law, Eq. (3.4-4), to the long pipe. There are several forces involved. There are the forces due to the water pressure at the dam side of the pipe and the water pressure at the tank side. This force is just $p_1(t)A_2(t) - p_2(t)A_2(t)$. There is the force due to friction of the water running in the pipe. This force is given by Eq. (3.4-8) as $cv^\beta(t)$ where v is the water velocity in the pipe. The relation between this velocity and water flow is $q_2(t) = \rho A_2 v(t)$. There is also the force of the water accelerating in the pipe which is $mdv(t)/dt$ where m is the mass of the water in the pipe which is $\rho A_2 \ell$, the product of density of water, area of the pipe and length of the pipe. Finally there is the force due to gravity of the water in the pipe. This is mgh_1/ℓ, where h_1 is the height of the pipe, ℓ is the length of the pipe, and g is the gravitational constant. The h_1/ℓ term appears since the total force due to gravity is mg and we wish to find the component of this force directed along the axis of the pipe which is the sine of the angle θ between the pipe and horizontal. This sine θ is just h_1/ℓ. So we have for Newton's law applied to the pipe

$$p_1(t)A_2 - p_2(t)A_2 + \frac{mgh_1}{\ell} = \frac{cq_2^\beta(t)}{(\rho A_2)^\beta} + \frac{m}{\rho A_2}\frac{dq_2(t)}{dt} \tag{3}$$

The pressure in the two storage vessels at the exit and entry point to the pipe is proportional to the height of the water in the tanks above these points. We have

$$p_1(t) = \rho g h_2(t) \tag{4}$$

and

$$p_2(t) = \rho g h_3(t) \tag{5}$$

We substitute Eqs. (4) and (5) into Eq. (3). Then Eqs. (1)–(3) represent the final dynamic equations for this system. We note that these are nonlinear because of the presence of the β exponent in Eq. (3). As noted before, we can linearize equations such as these and use the linearized results as long as we are concerned primarily with small enough perturbations in the state variables. It turns out that the value of β in Eq. (3.4-8) is approximately 1 if the friction is predominantly viscous friction and the equations are then linear. Otherwise, we will need to use the linearized system relations

$$\frac{d\Delta h_3(t)}{dt} = \frac{1}{\rho A_1}\Delta q_2(t) - \frac{1}{\rho A_1}\Delta q_1(t) \tag{6}$$

$$\frac{d\Delta h_2(t)}{dt} = \frac{-1}{\rho A_3}\Delta q_2(t) + \frac{1}{\rho A_3}\Delta q_3(t) \tag{7}$$

$$\frac{d\Delta q_2(t)}{dt} = \frac{-c\beta q_2^{\beta-1}(t)}{\ell(\rho A_2)^\beta}\Delta q_2(t) - \frac{\rho g A_2}{\ell}\Delta h_3(t) + \frac{\rho g A_2}{\ell}\Delta h_2(t) \tag{8}$$

which is, of course, quite exact for the linear case and is a valid approximation for the nonlinear case. These differential equations (6)–(8) describe the system. These might comprise the fixed plant for a control problem in which we might wish to control the flow into the dam in order to maintain a desired h_3 to supply adequate water pressure to the town.

EXAMPLE 3.4-4. In this example we will consider a simple hydraulic valve with one master cylinder. A valve or actuator of this sort is often used to swivel antenna turrets, control surfaces on airplane wings, rudders, and rocket engines as well as other relatively large objects. Figure (3.4-4) shows a very simple representation of a hydraulic valve controlling the flaps of an airplane. A small valve shown at the top of the figure controls high pressure oil flowing into the large cylinder at the bottom of the figure and the high input pressure times piston area results in a very large torque to move the flap.

The rate of oil flow from the pump to the cylinder can be determined from Eq. (3.4-3) as

$$\frac{dm(t)}{dt} = A\rho\frac{dz(t)}{dt} = q(t) \tag{1}$$

where A is the cross-sectional area of the cylinder and ρ is the density of the oil. For the piston in the cylinder at the bottom of the figure, we have from Eq. (3.4-4)

$$p_3(t)A - p_4(t)A + f_f(t) = 0 \tag{2}$$

where $p_3(t)$ and $p_4(t)$ are the pressures on either side of the cylinder and $f_f(t)$ is the horizontal force due to flap reaction. This is a combined

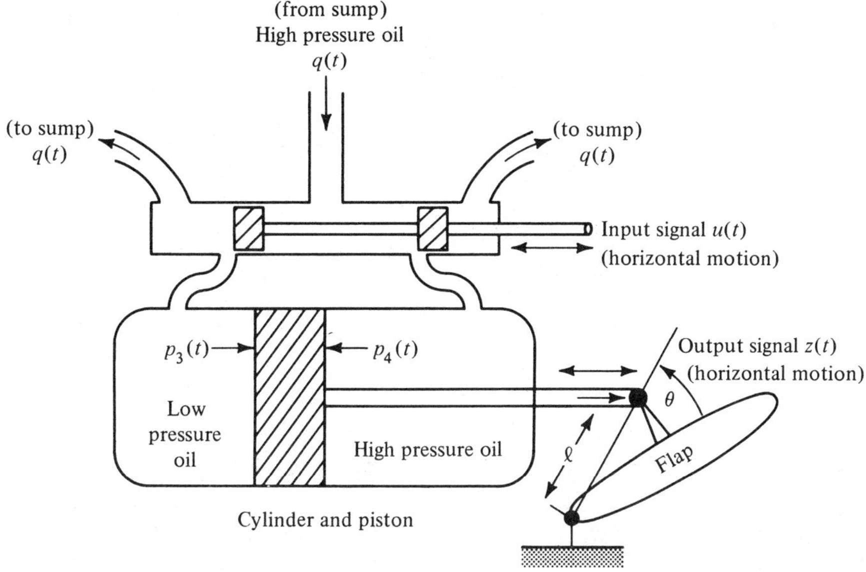

Figure 3.4-4 A simplified schematic of a hydraulic motor

translation rotation system and so we use the torque equation (3.4-5) to obtain

$$f_f(t)\ell - J\frac{d^2\theta(t)}{dt^2} = 0 \tag{3}$$

where J is the moment of inertia of the flap about the pivot point shown.

Equations (1)–(3) describe the hydraulic system when we use the fact that the mass flow into and out of the valve is proportional to the input control signal displacement $u(t)$. We might use

$$q(t) \doteq Cu(t) \tag{4}$$

to express the relationship between $u(t)$, the control input, which typically comes from a solenoid, and the mass flow rate $q(t)$. Also we have as the relationship between the output displacement $z(t)$ and the displacement θ

$$z(t) = \ell \sin\theta$$

We approximate this by

$$\Delta z(t) = \ell\, \Delta\theta \tag{5}$$

Combining Eqs. (1), (4), and (5), we have

$$\frac{d\Delta\theta(t)}{dt} = \frac{C}{A\rho\ell}\, u(t) \tag{6}$$

which indicates that an approximate transfer function for the hydraulic valve and actuator is

$$\frac{\Delta\Theta(s)}{\Delta U(s)} = \frac{C}{A\rho\ell s} \tag{7}$$

and the hydraulic motor functions as an integrator.

A more accurate representation of this system would not assume that the fluid is incompressible nor would it ignore fluid leakage or assume that mass flow rate is independent of pressure.* Thus we would not equate the mass flow rate in the valve to the mass flow rate in the cylinder. We have not used Eqs. (2) and (3) in our derivation of the transfer function of this simple hydraulic actuator. When we consider leakage and compressible fluids it turns out that these relations are needed and that the transfer function becomes

$$\frac{\Delta\Theta(s)}{\Delta U(s)} = \frac{K}{s(as^2 + bs + 1)} \tag{9}$$

where the a parameter vanishes if there is no compressibility of the fluid and where b vanishes if there is no leakage. With a small leakage the hydraulic system response can be underdamped but since fluids are not very compressible the resonant frequency would be very high. Presence of air in the fluid lines would make the total system very compressible and would reduce the resonant frequency dramatically and thus degrade system response significantly.

---

*We would assume that the mass flow rate is proportional to $-p_4$, the pressure differential in the valve, as well as displacement and linearize to obtain, instead of Eq. (4)

$$\Delta q(t) = C\Delta u(t) + D\Delta p(t) \tag{8}$$

EXERCISE 3.4-3. Show that inclusion of the differential pressure term of Eq. (8) in the solenoid characteristics leads to the need to include Eqs. (2) and (3) in the hydraulic valve dynamics and that the linearized transfer function is of the form

$$\frac{\Delta\Theta(s)}{U(s)} = \frac{K}{s(Ts + 1)}$$

What are K and T in terms of the physical parameters of the hydraulic system?

3.5 SUMMARY

In this chapter we have developed differential equations and transfer functions associated with a variety of electrical, mechanical, electromechanical, and thermal and fluid systems. We considered a variety of electrical networks that, as we shall see in later chapters, will be very useful as compensating networks in linear control systems. Because of the ubiquity of operational amplifiers in synthesizing control system compensating network transfer functions, as well as in modeling complete control systems for simulation purposes where operational amplifier configurations are used for an analog computer simulation, we have also discussed operational amplifier networks and linear analog simulation concepts.

Our examination of mechanical, electromechanical, and thermal and fluid systems demonstrated that the differential equations and transfer functions obtainable for these systems are much the same as those obtainable for electrical networks. As we have indicated there are direct and indirect analog and dual relationships between all of these networks.

At the expense of a very lengthy chapter we could now proceed to describe a large number of control systems plants, namely:

aeronautical

space

marine

pollution

biomedical

economic

nuclear

industrial

automotive

chemical

energy

natural resource

systems and determine differential equations and transfer functions for these systems. This would demonstrate that we can obtain differential equations and transfer functions that are, in mathematical or model form, quite similar although the fundamental physical (or social) process from which the systems arise may well be greatly different. Thus we see that the subject of automatic control has applicability to an enormous number of diverse subject areas. A person well trained in automatic control is prepared to undertake system studies in a diversity of interesting application areas once the requisite physical understanding of the particular applications area has been obtained. Often a systems control engineer will serve as a member of a team containing content specialists thoroughly familiar with a given application area. However, it will seldom, if ever, be true that a systems control engineer will be able to function well without at least some reasonable basic understanding, in fact often a very detailed one, of the physical problem under study.

In our next chapter we will study performance and performance specifications for typical low order linear control systems. Then we will turn our attention to a study of three design methods for linear control systems. On many occasions throughout our efforts we will refer to the material in this chapter as we indicate some simplified applications examples of real world control problems.

3.6 REFERENCES

The material in this chapter is both classic and basic to a full appreciation of control systems practice. There are many texts which discuss linear electric and mechanical networks. The classic reference

Gardner, M. F. and J. L. Barnes, *Transients in Linear Systems*, John Wiley and Sons, Inc., New York, 1942.

has much to recommend it. A very complete introductory treatment of electrical, mechanical and thermal and fluid networks is contained in the definitive and thorough text:

Cannon, R., *Dynamics of Physical Systems*, McGraw-Hill Book Co., New York 1967.

Detailed discussions of a number of useful systems control components is contained in

Ahrendt, W. R. and C. J. Savant, Jr., *Servomechanism Practice*, McGraw-Hill Book Co., Inc., New York, 1960.

Gibson, J. E. and F. B. Tuteur, *Control System Components*, McGraw-Hill Book Co., New York 1958.

as well as volume one and three in the three-volume Handbook

Grabbe, E. M., Ramo, S., and D. E. Wooldridge, *Handbook of Automation Computation and Control*, John Wiley and Sons, Inc., New York 1958 (Vol. I), 1961 (Vol. III).

and the later portions of

Gille, J. C., Pelegrin, M. J. and P. Decauline, *Feedback Control Systems*, McGraw-Hill Book Co., 1959.

An historical account of the development of feedback control including many of the earliest components and systems is contained in

Mayr, O., *The Origins of Feedback Control*, M.I.T. Press, 1970.

An outstanding treatment of analog computer simulation is contained in

Korn, G. A., and Korn, T. M., *Electronic Analog And Hybrid Computers*, Second Edition, McGraw-Hill, 1972.

3.7 PROBLEMS

1. Find the input output differential equation and the voltage transfer function of the electrical network represented by Fig. P3.7-1.

Transform these equations to an appropriate state variable representation and determine a matrix representation for this differential equation.

2. Repeat Problem 1 for the electrical network of Fig. P3.7-2.

3. Connect the networks of Problems 3.7-1 and 3.7-2 by joining the output of the first network to that of the second. Repeat Problem 3.7-1.

4. The circuit of Fig. P3.7-4 is in steady state at $t = 0$ when the open switch is closed. Find the output voltage as a function of time.

5. Suppose that the switch in Fig. P3.7-4 is initially closed and the circuit is in steady state. The switch opens at $t = 0$. Find the output voltage as a function of time.

6. Find an analog computer representation which will simulate the differential equation corresponding to Fig. P3.7-1.

7. Find an analog computer representation which will simulate the differential equation corresponding to Fig. P3.7-2.

8. Suppose that the analog computer configurations of Problems 3.7-6 and 3.7-7 are connected together with the analog computer output of Problem 3.7-6 being connected to the analog computer input of Problem 3.7-7. What is the new input output transfer function? Does this correspond to that obtained by connecting the electrical networks in Figs. P3.7-1 and P3.7-2 together?

9. What does the output transfer relationship become for an operational amplifier system in which the operational amplifier of Fig. 3.1-2 has an input impedance $Z_{in}(s)$ and output impedance $Z_{out}(s)$ such that the operational amplifier model is as indicated in Fig. P3.7-9?

10. Determine the differential equations which describe motion in the mechanical system of Fig. P3.7-10.

11. Draw a physical picture of at least two mechanical translational systems which can be represented by the output input differential equation

$$A \frac{d^2 z}{dt^2} + b \frac{dz}{dt} + cz = du(t)$$

12. Repeat Problem 3.7-11 for the differential equation

$$A \frac{d^2 z}{dt^2} + b \frac{dz}{dt} + cz = du + e \frac{du}{dt}$$

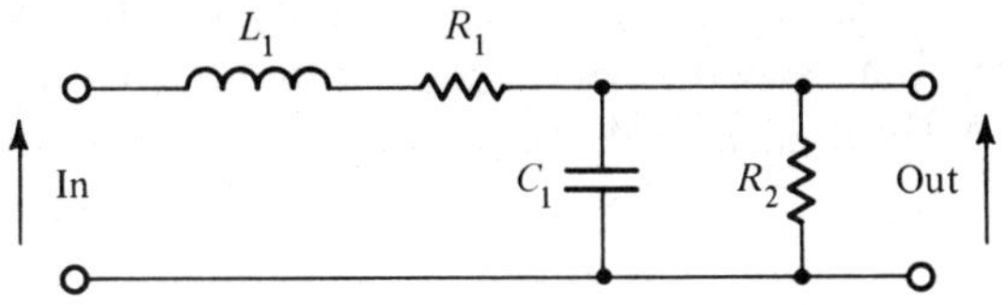

Figure P3.7-1 Network for Problem 3.7-1

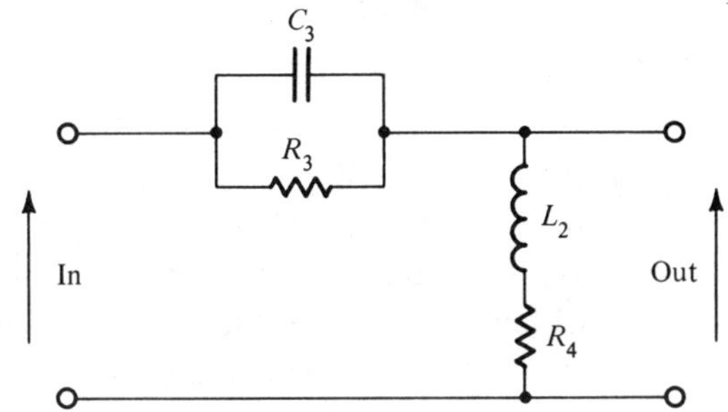

Figure P3.7-2 Network for Problem 3.7-2

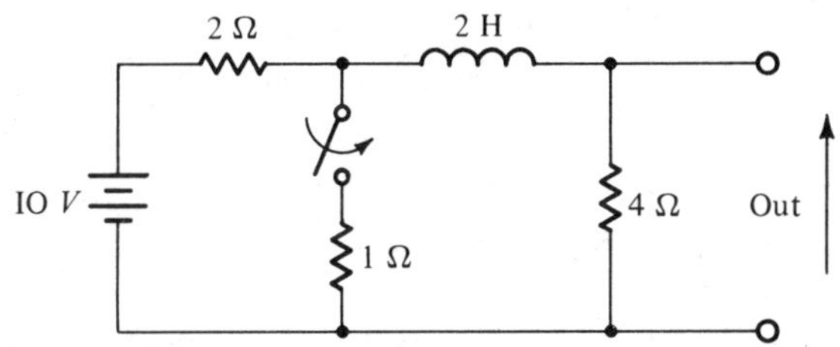

Figure P3.7-4 Network for Problems 3.7-4 and 3.7-5

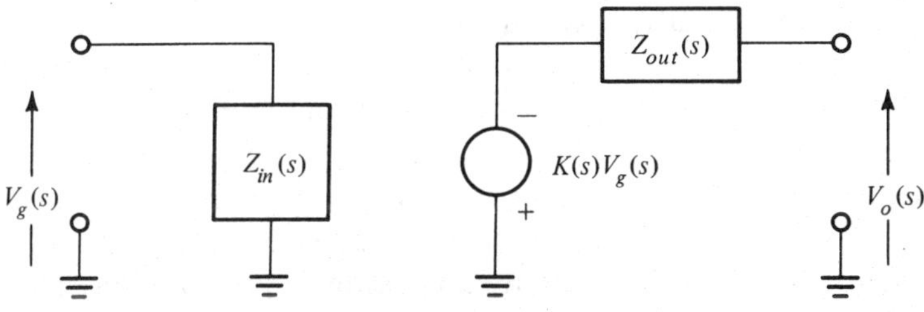

Figure P3.7-9 A more realistic model for the operational amplifier of Fig. (3.1-2)

13. A simplified clutch can be constructed by moving two right circular cylinder plates, with rotational moments of inertia J_1 and J_2, together. A fluid surrounds the clutch and the friction coefficient is B. The driving torque and the load torque are specified time functions. Determine a model and equations of motion for this fluid clutch.

14. Write the differential equation for the motion of a pendulum which consists of a point mass M suspended from a pivot point on a string of length L. The pendulum is offset at an angle θ with respect to vertical and released. Is this differential equation linear? How can it be linearized?

15. A simple electromagnetic loudspeaker or microphone can be created by attaching a cone to a coil of wire that is suspended in a magnetic field. As a speaker, current flowing through the coil moves the cones which vibrates the air to create sound. As a microphone, air pressure on the cone moves the coil through the magnetic field to induce an electro-motive force (voltage) in the coil. Determine a model for these systems and the system differential equation when the unit is utilized as a) a microphone, b) a speaker.

16. Construct a DELTA Chart illustrating a successful procedure for determination of the transfer function of a linear electrical network.

17. Construct a DELTA Chart illustrating a successful procedure to determine the transfer function of a linear mechanical circuit.

18. Construct a DELTA Chart illustrating a successful procedure to determine a direct and indirect electrical analog for a mechanical or thermal system.

19. A simple closed positioning system is shown in Fig. P3.7-19. The operational amplifiers are ideal. Determine a control system block diagram for this closed loop system. Be sure to include the various component transfer functions on your block diagrams. What is the input/output transfer function where the input is the angular shaft position $u = \theta_u$ and the output is the angular shaft position $z = \theta_z$? What is an appropriate state variable representation of the system? Which model is more appealing to you—the block diagram or the state variable model? Why?

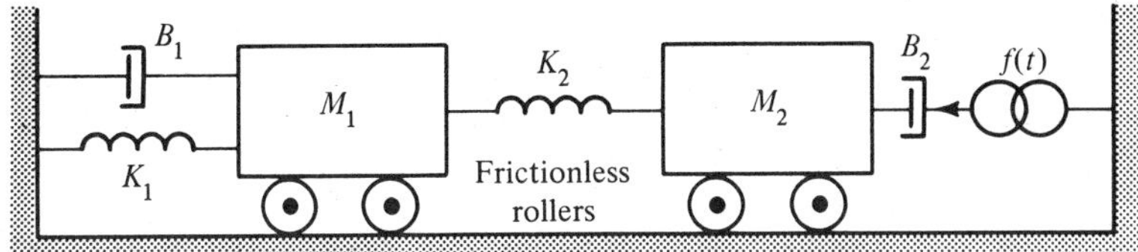

Figure P3.7-10 Mechanical system for Problem 3.7-10

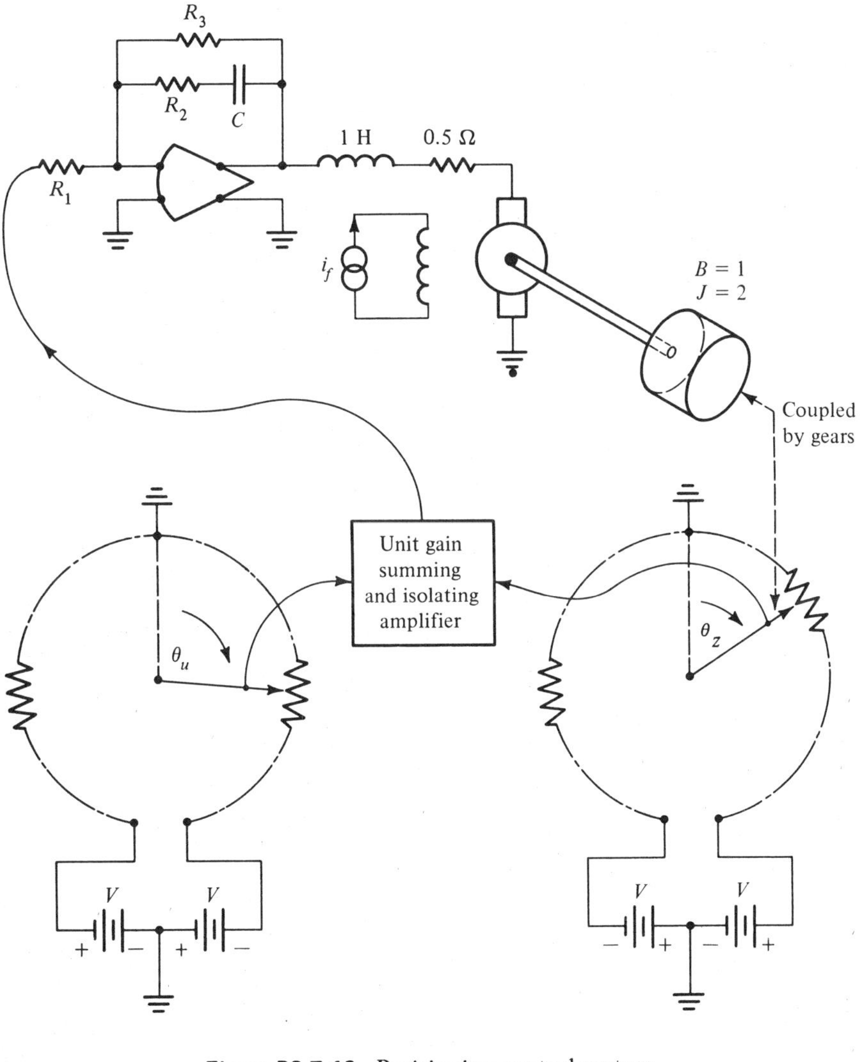

Figure P3.7-19 Positioning control system

4

SYSTEM RESPONSE STABILITY AND PERFORMANCE SPECIFICATIONS

As a preliminary to study of design methods for linear systems control we introduce a number of system concepts which yield information of value concerning the typical behavior of simple first and second order systems, requirements which must be imposed to insure system stability, and performance criteria and specifications. To these ends we discuss

a. time and frequency response of simple first and second order systems

b. system stability requirements

c. a variety of performance criteria and rule of thumb approximations for linear system design.

4.1 FIRST ORDER SYSTEMS

Virtually no realistic control system would be so simple that it could be characterized by a first order linear differential equation. Nevertheless we can learn quite a bit about typical system response by studying first order systems and attempting to characterize those performance characteristics which are applicable to higher order systems. We may characterize a first order system by the differential equation

$$\frac{dz}{dt} = K[u(t) - z(t)] \qquad (4.1\text{-}1)$$

or the equivalent equation in time constant form

$$T\frac{dz}{dt} = u(t) - z(t) \tag{4.1-2}$$

The unit impulse response for a system described by Eq. (4.1-2) is

$$z(t) = \frac{1}{T}e^{-t/T} \tag{4.1-3}$$

and the unit step response is

$$z(t) = 1 - e^{-t/T} \tag{4.1-4}$$

Figure 4.1-1 illustrates a time sketch of the unit impulse and step response given by Eq. (4.1-4). From this figure we can define the *rise time* response of this system. Actually several definitions of rise time are possible and have been used in practice. The time for the response to rise to 0.632, that is to $1 - e^{-1}$, of the final value is a very convenient rise time definition here. Thus the rise time definition here is just the system time constant. We could also use, for example, the time required for the response to reach 0.9502 of the final value (the zero to ninety-five percent rise time definition) in which case the rise time is $3T$. It might be preferable to simply call T the system time constant but even here we must be careful since this is but a first order system and we wish to adopt definitions that will be generally applicable to systems of any order.

The ramp $[u(t) = at]$ response of our simple first order system is

$$z(t) = a(t - T + Te^{-t/T}) \tag{4.1-5}$$

The *error* in the response, defined by

$$e(t) = u(t) - z(t) \tag{4.1-6}$$

is

$$e(t) = a(T - Te^{-t/T}) \tag{4.1-7}$$

The *steady state error* defined by

$$e_{ss} = \lim_{t \to \infty} e(t) \tag{4.1-8}$$

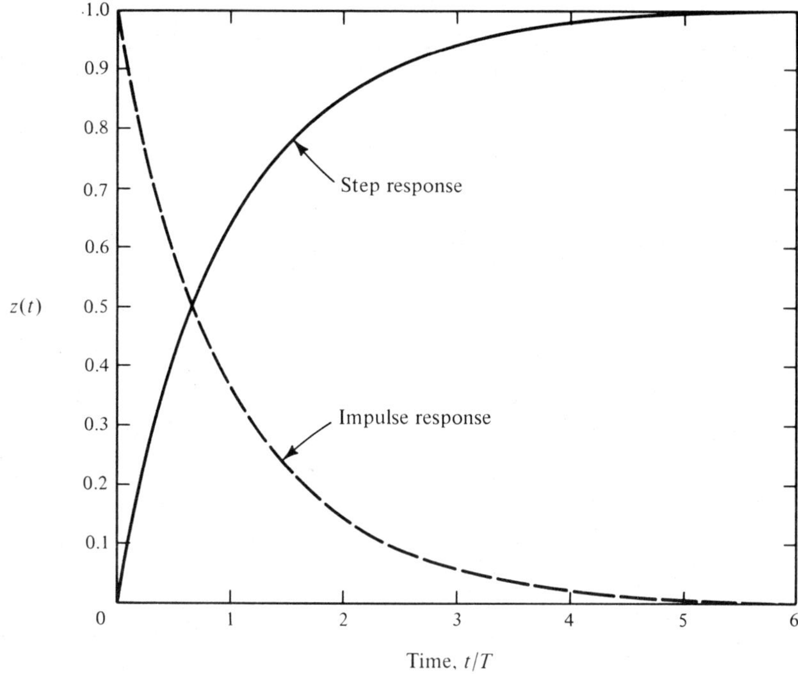

Figure 4.1-1 Unit impulse and step response of first order system

is just $e(t) = aT$ for this simple system with a ramp input. We see that this first order system will have a non-zero steady state error. This steady state error, sometimes called the velocity error, can be decreased by decreasing the system time constant or rise time T, which will also increase the speed of response of the system for a step input.

It is also desirable to determine the steady state response of the system of Eq. (4.1-1) to a sinusoidal input

$$u(t) = U \sin \omega t \tag{4.1-9}$$

applied at $t = 0$. It is a simple matter to show that the response of the system described by Eq. (4.1-2) to this input is

$$z(t) = \frac{UT\omega e^{-t/T}}{1 + T^2\omega^2} + \frac{U}{(1 + T^2\omega^2)^{1/2}} \sin(\omega t - \phi) \tag{4.1-10}$$

where

$$\phi = \tan^{-1} \omega T \tag{4.1-11}$$

The steady state response occurs when the exponential term decays to zero and this is given by

$$z(t) = |H(j\omega)| U \sin (\omega t + \beta) \tag{4.1-12}$$

where

$$|H(j\omega)| = \frac{1}{(1 + T^2 \omega^2)^{1/2}} \tag{4.1-13}$$

is the magnitude of the system transfer function $H(s)$ evaluated at $s = j\omega$ and

$$\beta = - \tan^{-1} \omega T \tag{4.1-14}$$

is the phase angle associated with the transfer function at frequency ω. It is very convenient to plot the magnitude of the transfer function, $|H(j\omega)|$, or system frequency response, on log log coordinate paper and the phase angle curve, or phase response, on semilog paper as illustrated in Figs. (4.1-2) and (4.1-3). Also shown in these figures are straight line approximations to the frequency and phase response. The frequency response straight line approximation has two asymptotes which intersect at the frequency $\omega = 1/T$ which is often called the *break point* or *break frequency*. At the break frequency the actual frequency response magnitude is 0.707 of the asymptotic value. The approximate curve is a better fit to the actual curve at all other frequencies. It will often be very convenient for us to replace an actual frequency response or gain magnitude curve

$$|H(s)| = |H(j\omega)| = \frac{1}{|1 + sT|} = \frac{1}{(1 + \omega^2 T^2)^{1/2}} \tag{4.1-15}$$

by either of its asymptotic values

$$|H(j\omega)| = \begin{cases} 1 & \omega T < 1 \\[2mm] \dfrac{1}{\omega T} & \omega T > 1 \end{cases} \tag{4.1-16}$$

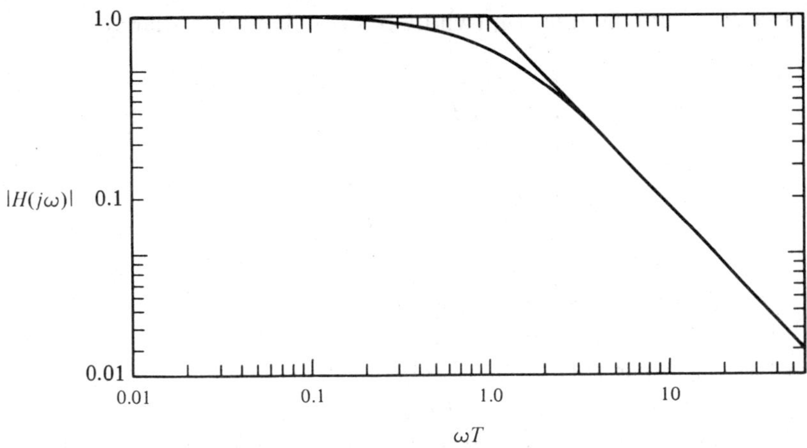

Figure 4.1-2 Magnitude of the frequency response curve for a first order system

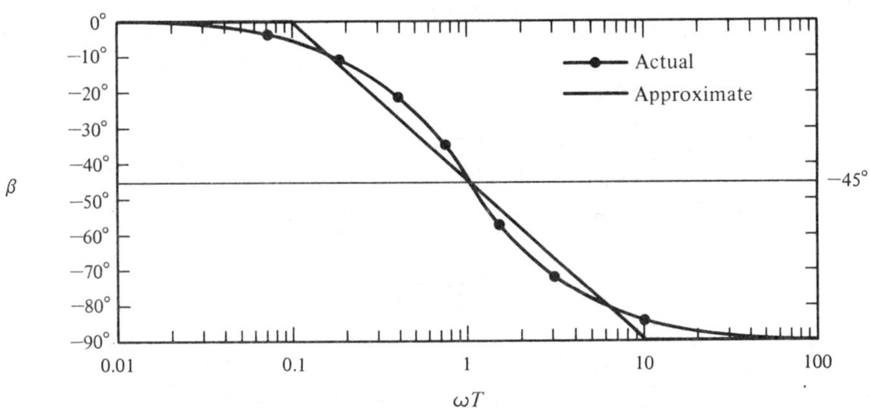

Figure 4.1-3 Curve for a first order system

and this will form the basis for a very important system control design approach that we will pay much attention to in our next chapter. In a similar way we will often find it convenient to use for the frequency response curve

$$\beta = -\tan^{-1} \omega T \qquad (4.1\text{-}17)$$

the analytical approximation (where β is obtained in radians)

$$\beta = \begin{cases} -\omega T & \omega < \dfrac{1}{T} \\[2ex] -\dfrac{\pi}{2} + \dfrac{1}{\omega T} & \omega > \dfrac{1}{T} \end{cases} \qquad (4.1\text{-}18)$$

which may be obtained by a Taylor series approximation to Eq. (4.1-17). The asymptotic curve shown in Fig. (4.1-3) is not obtained from Eq. (4.1-18) but is that curve obtained by assuming a straight line from $0°$ phase shift at $\omega T = 0.1$ to $-90°$ phase shift at $\omega T = 10$.

For some applications, a polar plot of the frequency response locus or transfer locus of the sinusoidal steady state response of this system is desirable. Figure (4.1-4) illustrates this for the first order system studied here.

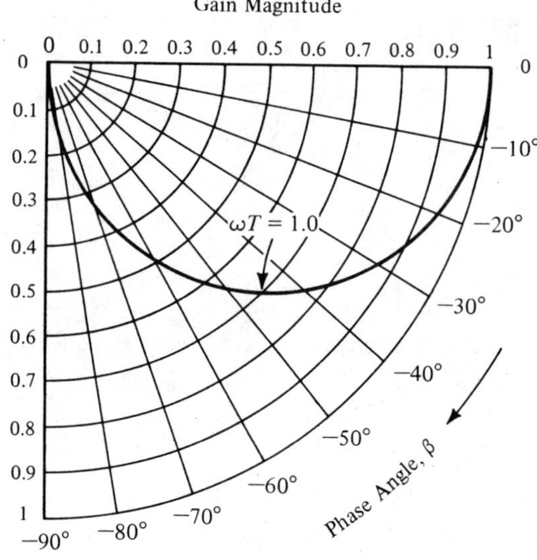

Figure 4.1-4   Polar plot of gain vs frequency curve

EXERCISE 4.1-1.   Derive Eqs. (4.1-3), (4.1-4), and (4.1-5).

EXERCISE 4.1-2.   Derive Eq. (4.1-10).

EXERCISE 4.1-3.　Derive Eq. (4.1-18) using a Taylor series approximation to Eq. (4.1-17).

4.2 SECOND ORDER SYSTEMS—no finite zeros

In this section we will consider the time and frequency response of
second order systems. First we will examine the case where there are no
finite zeros. We will consider the case where there is a single finite zero
in our next section. Since many performance criteria for high order systems are based upon concepts desired for second order systems, the
material in this section deserves rather careful and critical attention.

We first consider the input output differential equation

$$\frac{d^2 z}{dt^2} + 2\zeta\omega_n \frac{dz}{dt} + \omega_n^2 z(t) = \omega_n^2 u(t) \tag{4.2-1}$$

which can be represented by the transfer function

$$H(s) = \frac{Z(s)}{U(s)} = \frac{\omega_n^2}{s^2 + 2\zeta\omega_n s + \omega_n^2} \tag{4.2-2}$$

There are no finite zeroes for this transfer function and two finite poles
located at

$$s = -\zeta\omega_n \pm \omega_n\sqrt{\zeta^2 - 1} = \omega_n(-\zeta \pm j\sqrt{1 - \zeta^2}) \tag{4.2-3}$$

As the *damping ratio* ζ is varied the poles move about a circle in the s
plane as indicated in Fig. (4.2-1). We note that for $\zeta > 1$, the system
has two negative real roots. The unit impulse response is of the form

$$z(t) = \frac{\omega_n}{2\sqrt{\zeta^2 - 1}}$$

$$\exp\left[-(\zeta + \sqrt{\zeta^2 - 1})\omega_n t\right] - \exp\left[-(\zeta - \sqrt{\zeta^2 - 1})\omega_n t\right] \tag{4.2-4}$$

for this case. When $\zeta > 1$, the system is said to be *overdamped*.

For the very important *underdamped* case where $0 < \zeta < 1$, the
poles are complex conjugates as we see from Eq. (4.2-3). The unit impulse response is of the form

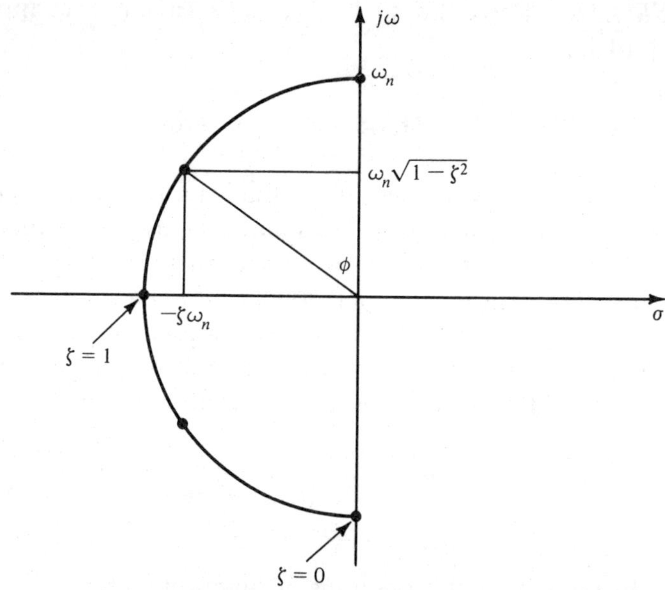

Figure 4.2-1   Poles and zeros of the transfer function of Eq. (4.2-2)

$$z(t) = \frac{\omega_n}{\sqrt{1-\zeta^2}}\, e^{-\zeta\omega_n t}\, \sin\cdot\omega_n\sqrt{1-\zeta^2}\, t \qquad (4.2\text{-}5)$$

which represents a sinusoidally damped or decaying exponential. The frequency of oscillation $\omega_d = \omega_n\sqrt{1-\zeta^2}$ is known as the *damped frequency*. For $-1 < \zeta < 0$, the response equation (4.2-5) is still valid but the response grows in time since the coefficient of the exponential in this equation is positive for positive time. In a similar way the response equation (4.2-4) is valid for $\zeta < -1$ and the response also grows as time increases. For each of these cases, that is whenever the damping ratio is less than zero, we say that the system is *unstable*. We will give more precise definitions of stability in a later section.

EXERCISE 4.2-1.   Verify that the unit impulse response of Eq. (4.2-2) is given by Eq. (4.2-4) or (4.2-5). Show that these two latter equations are equivalent and that both are valid regardless of the value of ζ. What does the response become for the case where $\zeta = 1$?

It is a straightforward matter to show that the unit step response of the second order system represented by Eq. (4.2-2) is

$$z(t) = 1 + \frac{1}{\sqrt{1-\zeta^2}}\, e^{-\zeta\omega_n t}\, \sin\left(\omega_n\sqrt{1-\zeta^2}\, t - \phi - \frac{\pi}{2}\right) \qquad (4.2\text{-}6)$$

where

$$\phi = \sin^{-1}\zeta$$

EXERCISE 4.2-2. What does Eq. (4.2-6) become for the overdamped case where $1 < \zeta$?

Figure (4.2-2) illustrates the unit step response of this second order system for various values of the damping ratio ζ. The time to the peak value of the output can easily be obtained by setting $dz(t)/dt = 0$ where $z(t)$ is the step response given in Eq. (4.2-6). Alternately, we can just find the time where the impulse response goes to zero. We obtain for the time to peak, t_p, the value of the peak output, z_p, and the peak overshoot, os, which is $z_p - 1$,

$$t_p = \frac{\pi}{\omega_n\sqrt{1-\zeta^2}} \qquad (4.2\text{-}7)$$

$$z_p = 1 + \exp\left(-\zeta\pi/\sqrt{1-\zeta^2}\right) = 1 + \text{os} \qquad (4.2\text{-}8)$$

The settling time t_s is often defined as the time when the response settles down to within 5% of its steady state value. This is, for all practical purposes, the time when $e^{-\zeta\omega_n t_s} = .05$ and this occurs when $\zeta\omega_n t_s = 3$ so we have

$$t_s = \frac{3}{\zeta\omega_n} \qquad (4.2\text{-}9)$$

It is interesting to note that the time to peak and the settling time vary inversely with the natural resonant frequency ω_n whereas the overshoot is independent of ω_n. Figure (4.2-3) illustrates design curves showing the variation of overshoot, time to peak, and settling time versus ζ and ω_n. Since $\zeta = 0$ is the limiting value for system stability and since the system becomes unstable for $\zeta < 0$, it is reasonable that the system overshoot is 1 at $\zeta = 0$ since this is the borderline stability case in which persistant sinusoidal oscillations occur, as we can easily see from Eq. (4.2-6).

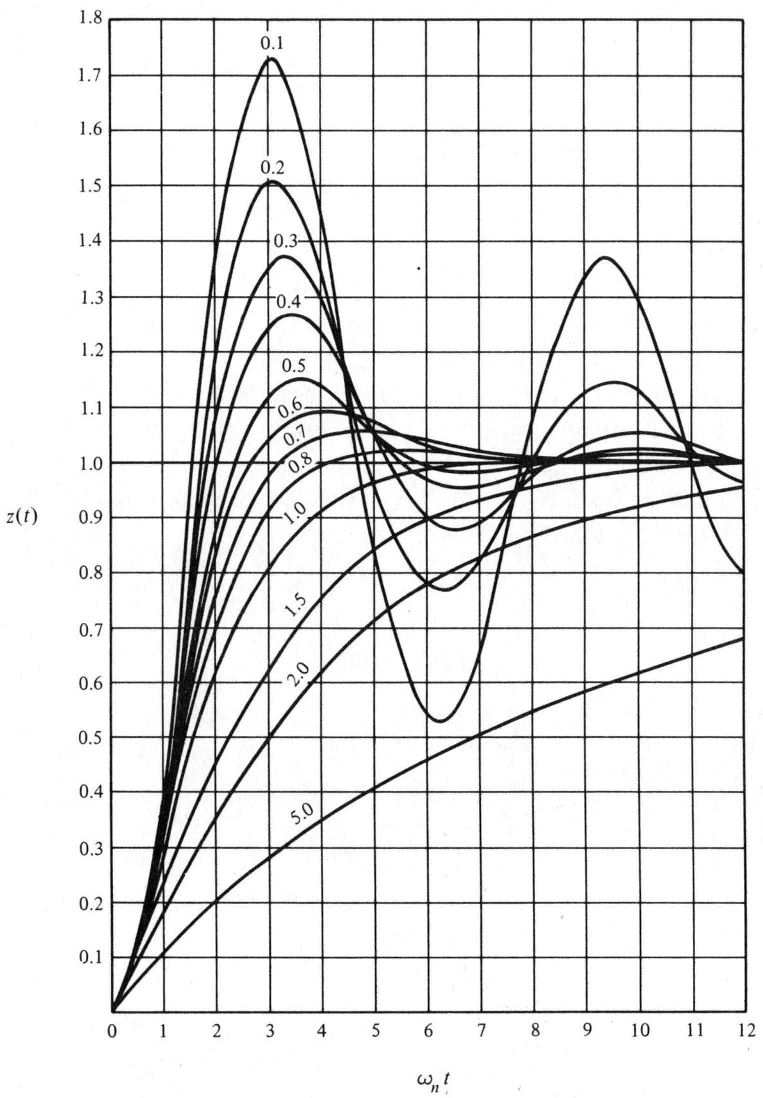

Figure 4.2-2 Unit step response of second order system for various damping ratios (ζ)

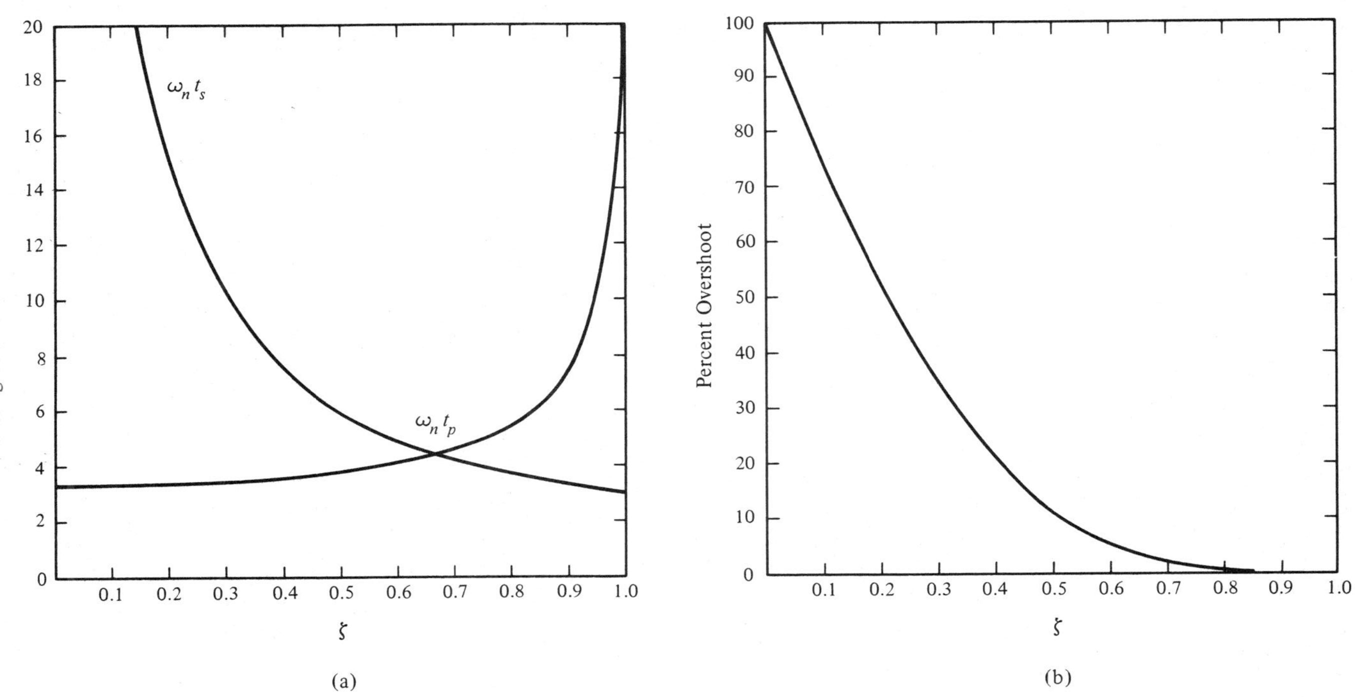

Figure 4.2-3 Settling time, time to peak, and percent overshoot for a second order system for various values of the damping ratio ζ

It is also of interest to determine the response of this second order system to a ramp input $u(t) = at$, for $t \geqslant 0$. We easily obtain

$$z(t) = at - \frac{2\zeta a}{\omega_n} + \frac{a}{\omega_n\sqrt{1 - \zeta^2}}\, e^{-\zeta\omega_n t} \sin\left[\omega_n\sqrt{1 - \zeta^2}\, t - \pi - 2\phi\right] \tag{4.2-10}$$

where

$$\phi = \sin^{-1} \zeta$$

This is the most convenient form of the response for the underdamped case where $0 \leqslant \zeta \leqslant 1$ but can be easily converted to a form convenient for the overdamped case. The last term in Eq. (4.2-10) decays to zero (for the stable case where $\zeta > 0$) and the steady state error becomes

$$e_{ss} = \lim_{t\to\infty} e(t) = \lim_{t\to\infty} [u(t) - z(t)] \tag{4.2-11}$$

$$= \frac{2\zeta a}{\omega_n}$$

We note that the steady state error for this system is a constant. It can be reduced by decreasing ζ but this makes the system less damped and the system transient response is poor for too small a value of ζ. We will later introduce a second order system with a finite zero such that the steady state error for a ramp input is zero. Prior to doing this we examine the frequency response characteristics of this system.

We obtain the frequency response curve for this second order system by evaluating the magnitude of the transfer function of Eq. (4.2-2) with $s = j\omega$. This yields

$$|H(j\omega)| = \frac{\omega_n^2}{\sqrt{(\omega_n^2 - \omega^2)^2 + (2\zeta\omega_n\omega)^2}} \tag{4.2-12}$$

The phase shift characteristic is obtained by evaluating the phase angle associated with the transfer function of Eq. (4.2-2) with $s = j\omega$. This yields

$$\beta(\omega) = -\tan^{-1} \frac{2\zeta\omega_n\omega}{\omega_n^2 - \omega^2} \tag{4.2-13}$$

Each of these expressions may be simplified by defining the normalized frequency $v = \omega/\omega_n$ such that we obtain

$$|H(jv)| = \frac{1}{\sqrt{(1 - v^2)^2 + (2\zeta v)^2}} \tag{4.2-14}$$

$$\beta(v) = -\tan^{-1} \frac{2\zeta v}{1 - v^2} \tag{4.2-15}$$

We may determine the *peak frequency* ω_p, or frequency where the magnitude of the transfer function is maximum, by setting the derivative of Eq. (4.2-14) with respect to v equal to zero and solving the resultant equation for $v = v_p$. We obtain after performing this operation

$$0 = 4v_p^3 - 4v_p + 8\zeta^2 v_p \tag{4.2-16}$$

which has the roots

$$v_p = 0, \quad \sqrt{1 - 2\zeta^2}$$

The first root of 0 is quite correct but this is a local maximum if the second root has a larger value. This will occur for $0 \leqslant \zeta \leqslant 0.707$ where the lower limit of zero is needed to insure system stability. Thus the normalized peak frequency is

$$v_p = \sqrt{1 - 2\zeta^2} \qquad 0 \leqslant \zeta \leqslant 0.707 \tag{4.2-17}$$

$$= 0 \qquad 0.707 \leqslant \zeta$$

The *peak value* or maximum value of the transfer function occurs at $v = v_p$ and this is, from Eqs. (4.2-14) and (4.2-17)

$$M_p = |H(jv)|\Big|_{v=v_p} = \frac{1}{2\zeta\sqrt{1 - \zeta^2}} \qquad 0 \leqslant \zeta \leqslant 0.707 \tag{4.2-18}$$

$$= 1 \qquad 0.707 \leqslant \zeta$$

Figure (4.2-4) illustrates the variation in the normalized peak frequency v_p (we note that $\omega_p = \omega_n v_p$) and M_p (called M peak) versus the damping ratio. These curves show that any value of ζ less than 0.25 will produce very large peaking in the system transfer function. This would be especially disadvantageous if there are noise components in the system at the peak frequency. Also as we have seen, the transient response shows much ringing for $\zeta \leqslant 0.25$.

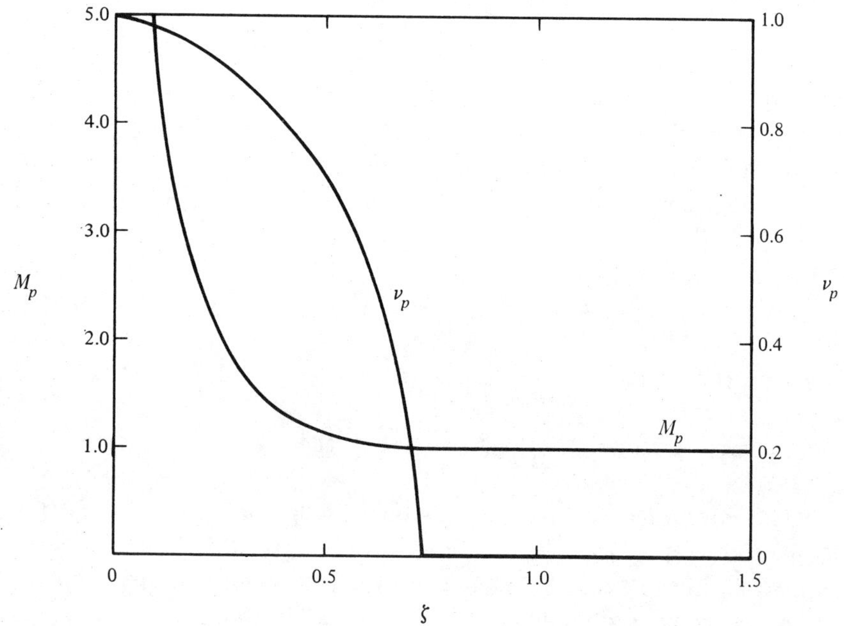

Figure 4.2-4 M_p and ν_p versus ζ for a second order system

We may also obtain frequency and phase shift curves for this second order system from Eqs. (4.2-14) and (4.2-13) and these are illustrated in Figs. (4.2-5) and (4.2-6). In order to get the asymptotic gain approximation for Eq. (4.2-14) we assume that for $\zeta < 1$, $\nu < 1$, we have $|H(j\nu)| = 1$ and for $\nu > 1$ we have $|H(j\nu)| = 1/\nu^2$ such that the asymptotic gain plot is as illustrated in the heavy dashed line of Fig. (4.2-5). We note that this is precisely the same asymptotic gain as if we had the transfer function

$$H'(j\nu) = \frac{1}{(j\nu + 1)^2} \tag{4.2-19}$$

In a similar way the asymptotic phase curve is obtained as if the transfer function were given by the foregoing transfer function with two real negative poles—regardless of the value of the damping ratio ζ. Alternately we might use the analytical approximation

$$\beta(v) = \begin{cases} -2v & v < 1 \\[2ex] -\pi + \dfrac{2}{v} & v > 1 \end{cases} \qquad (4.2\text{-}20)$$

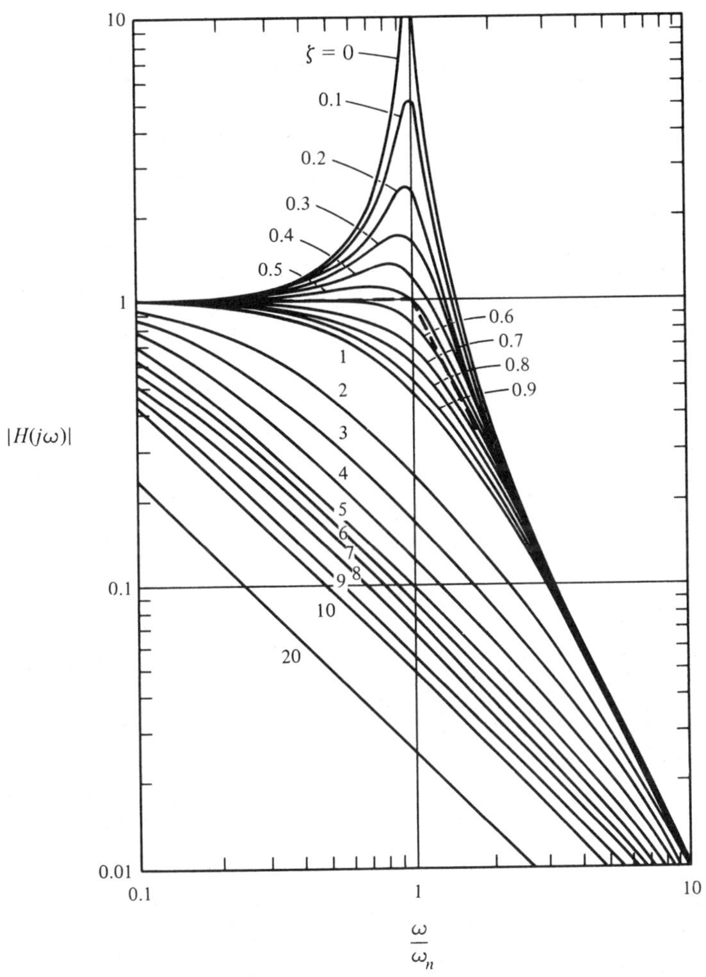

Figure 4.2-5 Normalized frequency response curve for second order system for various values of damping ratio (asymptotic approximation shown by dashed line)

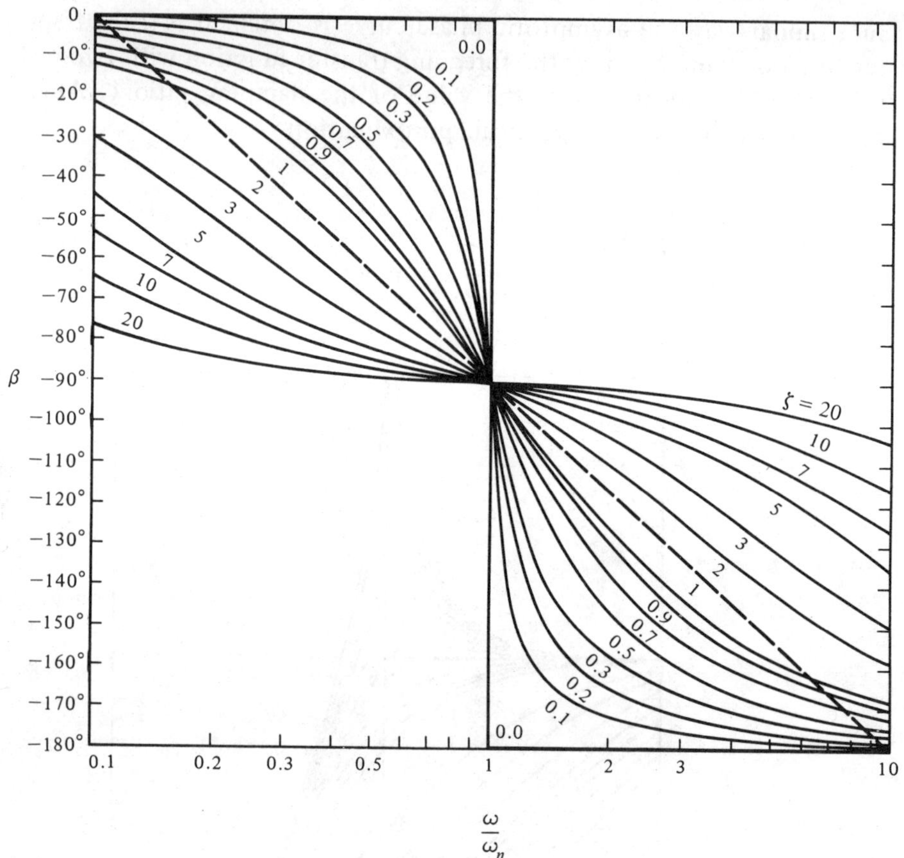

Figure 4.2-6 Phase shift curves vs frequency for various values of damping ratio ζ—
second order system (asymptotic approximation shown by dashed line)

Although these three approximations for the asymptotic gain, the
asymptotic phase and the analytical phase approximation may appear
poor, they are really quite good except at $\nu = 1$ for very small values
of ζ. Even for small ζ the approximations are quite good away from
$\nu = 1$. For values of $\zeta > 1$ the roots are real and we would use these
real distinct roots rather than assuming that both roots are at $j\nu = -1$
as in Eq. (4.2-19). Thus we would factor Eq. (4.2-2) as

$$H(s) = \frac{1}{(T_1 s + 1)(T_2 s + 1)} = \frac{1}{\left(\dfrac{s}{\omega_1} + 1\right)\left(\dfrac{s}{\omega_2} + 1\right)} \qquad (4.2\text{-}21)$$

where

$$T_1 T_2 = \omega_n^{-2}, \qquad T_1 + T_2 = 2\zeta\omega_n^{-1}$$

$$\omega_1 = T_1^{-1}, \qquad \omega_2 = T_2^{-1}$$

and use for the asymptotic approximation, where we assume $\omega_1 < \omega_2$,

$$|H(j\omega)| \approx \begin{cases} 1 & \omega < \omega_1 \\[2mm] \dfrac{\omega_1}{\omega} & \omega_1 < \omega < \omega_2 \\[2mm] \dfrac{\omega_1 \omega_2}{\omega^2} & \omega_2 < \omega \end{cases} \tag{4.2-22}$$

This asymptotic approximation will match quite well the actual frequency characteristics shown in Fig. (4.2-5) for the overdamped case ($\zeta > 1$). A similar approximation would be used for the phase shift characteristics. We shall explore the subject of asymptotic gain and phase approximations in much more detail in our subsequent efforts.

We can combine the frequency and phase response together on a polar plot which is called a *Nyquist Diagram*. This is just a graph of the locus of Eq. (4.2-2) which becomes, in normalized form

$$H(j\nu) = \frac{1}{(1 - \nu^2) + 2j\zeta\nu} \tag{4.2-23}$$

Figure (4.2-7) illustrates the loci corresponding to this equation. As we note from the figure the phase is always negative and decreases from a zero frequency value of 0 to $-180°$. The solid lines in this figure are for constant values of damping ratio ζ whereas the dashed lines are for constant normalized frequency ν. It is interesting and significant to note that at unity normalized frequency ($\nu = 1$) the phase shift is always $-90°$. For any positive value of ζ, the loci never reach the point where the gain is infinite. We will later show that this insures stability of this second order system.

EXERCISE 4.2-3. Determine asymptotic gain and phase shift versus frequency curves for values of $\zeta > 1$. Compare these with the actual gain and phase shift curves determined in this section.

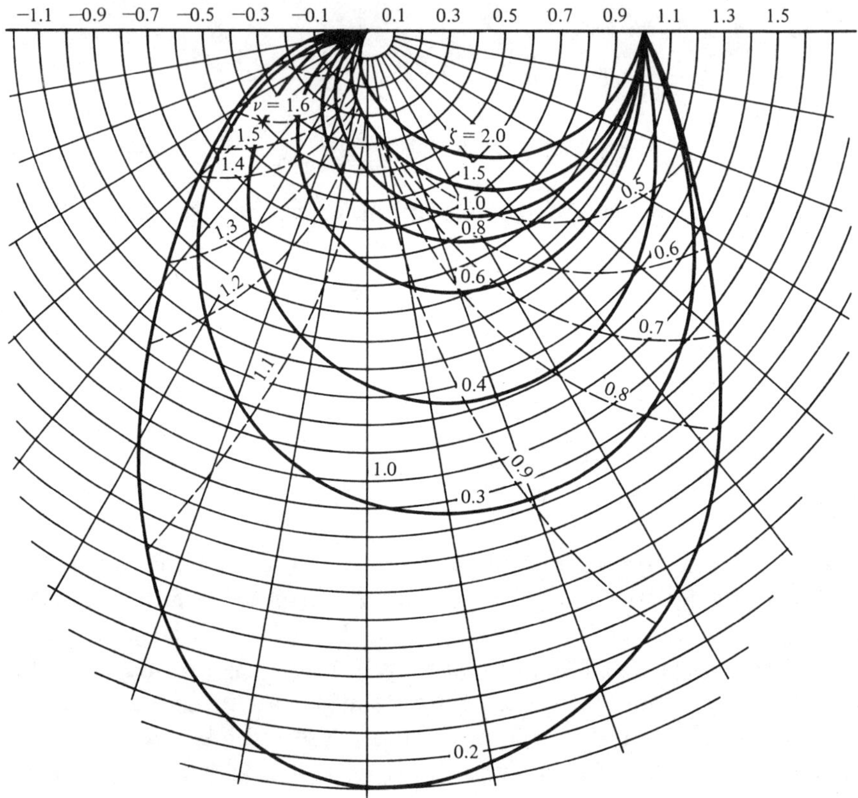

Figure 4.2-7 Polar or Nyquist plot of normalized second order system transfer

$$\text{function } H(j\nu) = \frac{1}{(1 - \nu^2) + 2j\zeta\nu}$$

EXERCISE 4.2-4. A second order unity feedback ratio control system has an open loop transfer function $G(s) = \dfrac{K}{s(s + 1)}$. Determine the nature of the solutions for this system in terms of K. Determine the nature of the unit step and frequency response curves for this system as a function of the gain K.

4.3 SECOND ORDER SYSTEMS—one finite zero

In examining the steady state error for a unit ramp or unit acceleration input we have just seen that a second order system with no finite zeros has a non-zero steady state error. In many cases this is highly desirable. We will show that the second system with a single finite zero that has a closed loop transfer function

$$H(s) = \frac{\omega_n^2 + 2\zeta\omega_n s}{\omega_n^2 + 2\zeta\omega_n s + s^2} \qquad (4.3\text{-}1)$$

will have zero steady state error for a unit ramp input. Such a system could be realized by a single loop input feedback ratio system with the open loop transfer function

$$G(s) = \frac{\omega_n^2 \left(1 + \dfrac{2\zeta}{\omega_n} s\right)}{s^2} \qquad (4.3\text{-}2)$$

We wish to examine some of the response performance properties of this system in this section.

The unit impulse response of this system can easily be determined as

$$z(t) = \frac{\omega_n}{\sqrt{1 - \zeta^2}} e^{-\zeta\omega_n t} \sin\left(\omega_n \sqrt{1 - \zeta^2}\, t + \phi\right)$$

$$(4.3\text{-}3)$$

$$\phi = \tan^{-1}\left[\frac{2\zeta\sqrt{1 - \zeta^2}}{1 - 2\zeta^2}\right]$$

The unit step response can be obtained either by integrating the unit impulse response or by use of Laplace transforms as

$$z(t) = 1 + \frac{1}{\sqrt{1 - \zeta^2}} e^{-\zeta\omega_n t} \sin\left(\omega_n \sqrt{1 - \zeta^2}\, t + \phi\right) \qquad (4.3\text{-}4)$$

$$\phi = \tan^{-1}\left[\frac{2\zeta\sqrt{1 - \zeta^2}}{1 - 2\zeta^2}\right] - \tan^{-1}\left[\frac{\sqrt{1 - \zeta^2}}{-\zeta}\right]$$

which is the appropriate form in the underdamped case where $\zeta < 1$.

EXERCISE 4.3-1. Show that the step response of this second order system with a finite zero can be obtained as an appropriately weighted sum of the step and impulse responses of the second order system with no finite zero.

EXERCISE 4.3-2. What is the appropriate form of Eq. (4.3-4) for the overdamped case?

The effect of this finite zero is to make the overshoot larger for a given damping ratio than would otherwise be the case. We can determine the time to the peak of the output response by taking the derivative of Eq. (4.3-4) and setting it equal to zero, or alternately simply equate the impulse response equal to zero, to obtain

$$t_p = \frac{\pi}{\omega_n\sqrt{1-\zeta^2}} - \frac{\tan^{-1}\left[2\zeta\sqrt{1-\zeta^2}/(1-2\zeta^2)\right]}{\omega_n\sqrt{1-\zeta^2}} \qquad (4.3\text{-}5)$$

We notice that the presence of the second term in the foregoing will decrease the time to the peak compared with the equivalent time to peak for a second order system with no finite zero as we see from Eq. (4.2-7). We can substitute Eq. (4.3-5) into Eq. (4.3-4) to find the peak value of the output response. Unfortunately this is a very cumbersome equation. For more complex systems expressions for such things as time to peak and percent overshoot become impossible to obtain analytically. For this reason it is very tempting to obtain rule of thumb approximations for higher order systems by considering similar low order systems. Figure (4.3-1) illustrates time to peak and percent overshoot as a function of damping ratio ζ for this system.

The unit ramp response may be obtained by evaluating the inverse transform of the product of the system transfer function given by Eq. (4.3-1) and $1/s^2$, the Laplace transform of the unit ramp. It is simpler, however, to evaluate the error for this particular system with a ramp input. We have

$$E(s) = U(s) - Z(s) = U(s)\left[1 - H(s)\right] = \frac{1}{\omega_n^2 + 2\zeta\omega_n s + s^2} \qquad (4.3\text{-}6)$$

and thus see that the transform of the error is just the transform of the response of the second order system, Eq. (4.2-2), to an impulse $u(t) = \delta(t)/\omega_n^2$. We have

$$e(t) = \frac{1}{\omega_n\sqrt{1-\zeta^2}}\, e^{-\zeta\omega_n t}\, \sin \omega_n\sqrt{1-\zeta^2}\; t \qquad (4.3\text{-}7)$$

and

$$z(t) = u(t) - e(t) = t - e(t) \qquad (4.3\text{-}8)$$

Figure (4.3-2) illustrates the error associated with this system in responding to a unit ramp input. Several conclusions are possible. The steady state error is zero as we have suggested. The response becomes quite oscillatory for values of ζ less than about 0.4. Increasing ω_n decreases the system error.

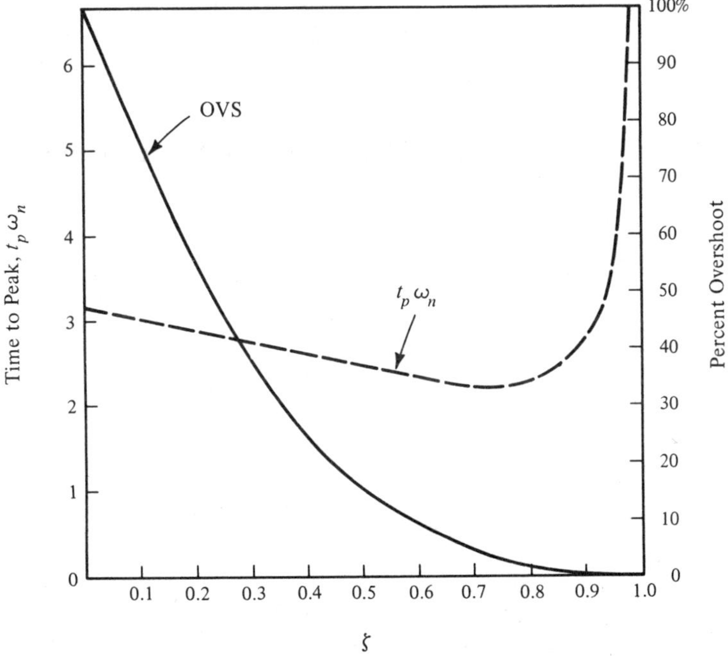

Figure 4.3-1 Time to peak and percent overshoot for second order system with finite zero

The frequency and phase response of this system is of interest. The asymptotic gain approximation to the system transfer function is exactly the same as if the transfer function were

$$H'(s) = \frac{1 + \dfrac{2\zeta s}{\omega_n}}{\left(1 + \dfrac{s}{\omega_n}\right)^2} \tag{4.3-9}$$

and this is, for $\zeta < 0.5$,

$$|H'(j\omega)| = \begin{cases} 1 & \omega < \omega_n \\[2ex] \dfrac{\omega_n}{\omega^2} & \omega_n < \omega < \dfrac{\omega_n}{2\zeta} \\[2ex] \dfrac{2\zeta\omega_n}{\omega} & \dfrac{\omega_n}{2\zeta} < \omega \end{cases} \tag{4.3-10}$$

whereas for $1 > \zeta > 0.5$ the gain magnitude frequency approximation is

$$|H'(j\omega)| = \begin{cases} 1 & \omega < \dfrac{\omega_n}{2\zeta} \\[2ex] \dfrac{2\zeta\omega}{\omega_n} & \dfrac{\omega_n}{2\zeta} < \omega < \omega_n \\[2ex] \dfrac{2\zeta\omega_n}{\omega} & \omega_n < \omega \end{cases} \tag{4.3-11}$$

When the damping ratio ζ is less than 0.5, the second order poles cause considerable peaking as we see from Figs. (4.2-4) or (4.2-5). An accurate representation of the frequency response curve should take this peaking ratio into account and this can be done using either of these two figures. When the damping ratio is between 0.5 and 1.0 there is either no or very little peaking and the approximation of Eq. (4.3-11) is quite a good one. For $\zeta > 1$, we have two real denominator roots. For ζ considerably greater than one, the poles are approximately at $s = -\omega_n/2, -2\zeta\omega_n$ such that the system transfer function is approximately

$$H'(s) = \frac{1 + \dfrac{2\zeta s}{\omega_n}}{\left[1 + \dfrac{2\zeta s}{\omega_n}\right]\left[1 + \dfrac{s}{2\zeta\omega_n}\right]}$$

and the frequency response is thus very nearly that of a first order system with transfer function

$$H'(s) = \frac{1}{1 + \dfrac{s}{2\zeta\omega_n}} \tag{4.3-12}$$

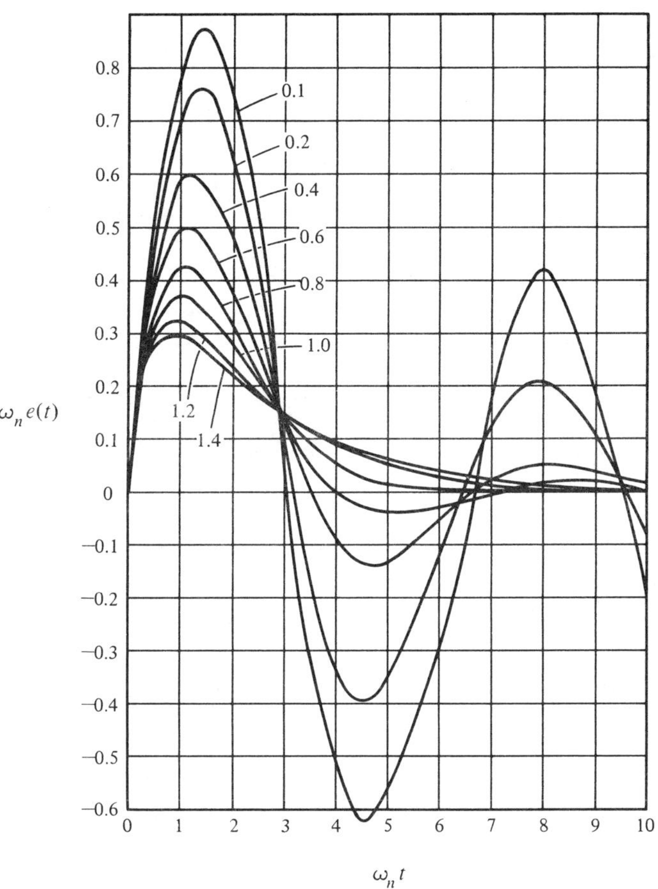

Figure 4.3-2 Error response of the second order—one finite zero system to a unit ramp input for various values of ζ

Figure (4.3-3) illustrates the asymptotic and actual gain versus frequency curves for the second order transfer function we have studied in this section. We shall find the ability to determine and use these asymptotic approximations invaluable for our efforts in the sequel.

EXERCISE 4.3-3.   Determine actual and asymptotic phase versus frequency curves for the transfer function of this section.

EXERCISE 4.3-4.   Determine a Nyquist plot for the transfer function of this section.

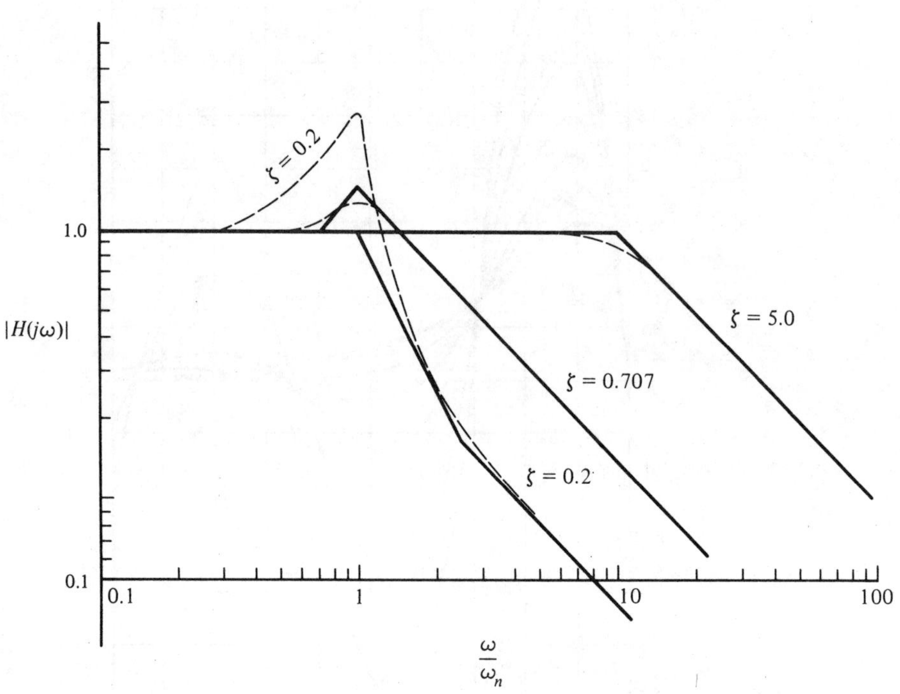

Figure 4.3-3   Asymptotic and actual gain magnitude vs frequency plot

4.4 STABILITY

If a control system is to be generally useful it must be *stable*. There are many definitions of stability available in the literature. In our earlier work we have seen that the transient response of a system is determined by the location of the poles of the system. If there is a pole in the right half of the s plane, then there will be a growing exponential function of time in the system transient response. For example a term $a(s - b)^{-1}$ in the partial fraction expansion of the system output will indicate an output component of the form ae^{bt}. Thus for a linear system we could say that a requirement for stability is that the system transfer function have no poles in the right half plane. Alternately we could write the state variable differential equation for a constant coefficient linear system as

$$\dot{\mathbf{x}} = \mathbf{A}\mathbf{x}(t) + \mathbf{b}u(t)$$

$$z(t) = \mathbf{c}\mathbf{x}(t) + du(t)$$

$$(4.4\text{-}1)$$

and write the system response for zero input $u(t)$ and a known but arbitrary initial condition as in Eq. (2.1-4) as

$$\mathbf{x}(t) = \boldsymbol{\phi}(t)\mathbf{x}(0) = e^{\mathbf{A}t}\mathbf{x}(0) \qquad (4.4\text{-}2)$$

where $\boldsymbol{\phi}(t)$ is the state transition matrix which satisfies

$$\frac{d\boldsymbol{\phi}(t)}{dt} = \mathbf{A}\boldsymbol{\phi}(t), \qquad \boldsymbol{\phi}(0) = \mathbf{I} \qquad (4.4\text{-}3)$$

In order for the system transient response to be bounded and ultimately decay to zero we must require that the sum of the square of the components in $\mathbf{x}(t)$ ultimately approach zero as time becomes infinitely large. Thus we require

$$\lim_{t \to \infty} |\,|\mathbf{x}(t)|\,|^2 = \lim_{t \to \infty} \mathbf{x}^T(t)\mathbf{x}(t) = 0 \qquad (4.4\text{-}4)$$

From Eq. (4.4-2) and the foregoing we have

$$\lim_{t \to \infty} \mathbf{x}^T(0)\boldsymbol{\phi}^T(t)\boldsymbol{\phi}(t)\mathbf{x}(0) = 0 \qquad (4.4\text{-}5)$$

We could examine conditions for $\phi(t)$ under which the foregoing limit will be satisfied. It turns out that we would require that each and every component of $\phi(t)$ approach zero for large time, or

$$\lim_{t \to \infty} \phi(t) = 0 \qquad (4.4\text{-}6)$$

If we take the Laplace transform of Eq. (4.4-3) we obtain

$$s\Phi(s) - \phi(0) = \mathbf{A}\Phi(s) \qquad (4.4\text{-}7)$$

or

$$\Phi(s) = (s\mathbf{I} - \mathbf{A})^{-1}\phi(0) \qquad (4.4\text{-}8)$$

The term $(s\mathbf{I} - \mathbf{A})^{-1}$ is very analogous, as we have seen in Chap. 2, to the system transfer function and so we conjecture again that the stability requirement is that the poles of the system transfer function be in the left half plane.

Other interpretations and stability definitions are also possible. We could, for example, require that a bounded input produce a bounded output. The convolution integral for a linear system, which we have obtained as Eq. (2.2-10), states that the system output is related to the input, for zero initial conditions, as

$$z(t) = \int_0^\infty h(\eta)u(t - \eta)\,d\eta \qquad (4.4\text{-}9)$$

The magnitude of this integral is

$$|z(t)| = \left| \int_0^\infty h(\eta)u(t - \eta)\,d\eta \right| \qquad (4.4\text{-}10)$$

We know that

$$\left| \int f(p)g(p)\,dp \right| \leqslant \int |f(p)|\,|g(p)|\,dp$$

since the integral of the absolute value of the terms in an integral cannot possibly be smaller than the magnitude of the integral of the terms. Application of the foregoing to Eq. (4.4-10) results in the bound

$$|z(t)| \leqslant \int_0^\infty |h(\eta)|\,|u(t - \eta)|\,d\eta \qquad (4.4\text{-}11)$$

We assume that the input is bounded in amplitude such that

$$|u(\eta)| \leqslant B_1 \tag{4.4-12}$$

so that Eq. (4.4-11) can be replaced by

$$|z(t)| \leqslant B_1 \int_0^\infty |h(\eta)| \, d\eta \tag{4.4-13}$$

Now if and only if the impulse response is bounded such that

$$\int_0^\infty |h(\eta)| \, d\eta \leqslant B_2 \tag{4.4-14}$$

will the system response

$$|z(t)| \leqslant B_1 B_2 \tag{4.4-15}$$

be also bounded.

We have just discussed four equivalent definitions of stability, namely

1. A linear system is stable if all poles of the (closed loop) transfer function are in the left half plane.

2. A linear system is stable if a bounded input produces a bounded output.

3. A linear system is stable if the system impulse response integral is bounded, i.e. $\int_0^\infty |h(\eta)| \, d\eta \leqslant B_2$.

4. A linear system is stable if all poles of the transfer matrix expression $(s\mathbf{I}-\mathbf{A})^{-1}$ are in the left half of the s plane.

Although we have not presented precise proofs of the equivalence of these definitions, we have given sufficient heuristic arguments to make this assertion most credible.

It is possible to present some very simple and at the same time very sophisticated stability concepts utilizing the second method of Lyapunov. Our preceding discussions are strictly applicable only to linear (constant coefficient) systems. Lyapunov stability concepts are also applicable to nonlinear systems. They are concerned with requirements that a system at some equilibrium condition $x(0)$ eventually returns to $x(0)$ for "slight" perturbations away from $x(0)$. While much of significance can be learned concerning system behavior through a study of Lyapunov stability, this excursion would require that we introduce some concepts which, while valuable, would necessarily take valuable time away from our study of linear systems control analysis and design procedures. Since a number of more advanced works present such discussions and since the second method approach is most valuable for nonlinear systems (where simpler approaches fail) and has not resulted in generally applicable and used linear design methods, we shall not discuss Lyapunov stability here.

The Routh Hurwitz Criterion

One approach to determination of system stability is to factor the poles of the system closed loop transfer function to see whether or not any lie in the right half plane. This approach is very useful for analysis, but not very useful for synthesis where we wish to determine parametric regions within which systems are stable. While there are a number of tests of a polynomial in s to determine the nature of the roots, the Routh Hurwitz test will be the only one described here because it is believed to be as easy to use and as generally applicable as any other available method.

The Routh Hurwitz criterion provides a simple method to determine whether the roots of a polynomial $P(s)$ are all in the right half plane. Three steps are needed to apply the Routh Hurwitz criterion and they are as follows:

1. Determine the closed loop system transfer function and the system characteristic equation.

2. Apply the Hurwitz criterion of necessary conditions to determine whether a polynomial has all roots in the left half plane.

3. Apply the Routh criterion to determine parametric requirements such that all poles of the system transfer function have positive real parts.

In order to determine the system closed loop transfer function and condition this for stability determination, we:

a) Write system input-output differential equations for appropriate subsystems.

b) Determine the system block diagram.

c) Take the Laplace transform of the system input-output differential equations for the subsystems to obtain the subsystem transfer functions.

d) Employ block diagram reduction techniques to obtain the overall system closed loop transfer function.

e) Prepare the system for examination by means of the Routh Hurwitz criterion by determining the characteristic equation of the system which is the denominator polynomial of the closed loop transfer function.

To employ the Hurwitz test we examine all the coefficients p_i of the characteristic equation polynomial $P(s)$ to see that:

a) All coefficients p_i are present for $i = 0, 1, 2, \ldots, n$ where the polynomial is of the order n.

b) All coefficients p_i are positive.

This test can be performed by physical inspection of the characteristic polynomial $P(s)$. Unfortunately, satisfaction of the Hurwitz test yields necessary but not sufficient conditions for a stable system. If a characteristic polynomial fails the test we can be quite sure that the system will be unstable. Thus we must modify the physical structure of the system, perhaps by inserting a compensating network, to insure that the Hurwitz test is satsified before proceeding further. Unfortunately many system characteristic equations may pass the Hurwitz test even though the system is unstable. To determine system stability after a characteristic equation passes the Hurwitz test we employ the Routh criterion. To do this we proceed as follows:

1. Obtain the characteristic polynomial describing pole loop locations for the closed loop system

$$P(s) = p_0 + p_1 s + p_2 s^2 + \ldots + p_n s^n = \sum_{i=0}^{n} p_i s^i$$

2. Form the first two Routh rows

$$
\begin{array}{cccccc}
s^n & p_n & p_{n-2} & p_{n-4} & p_{n-6} & \cdots \\
s^{n-1} & p_{n-1} & p_{n-3} & p_{n-5} & p_{n-7} & \cdots
\end{array}
$$

3. Determine the coefficients for a third Routh row

$$s^{n-2} \qquad a_1 \qquad a_2 \qquad a_3 \qquad \cdots$$

where the a_i coefficients are obtained by evaluating the determinant type cross multiplications from the first two Routh rows

$$a_1 = \frac{P_{n-1}P_{n-2} - P_n P_{n-3}}{P_{n-1}}, \qquad a_2 = \frac{P_{n-1}P_{n-4} - P_n P_{n-5}}{P_{n-1}}, \qquad \cdots$$

4. Augment the Routh rows with the new row just obtained

$$
\begin{array}{cccccc}
s^n & P_n & P_{n-2} & P_{n-4} & P_{n-6} & \cdots \\
s^{n-1} & P_{n-1} & P_{n-3} & P_{n-5} & P_{n-7} & \cdots \\
s^{n-2} & a_1 & a_2 & a_3 & & \cdots
\end{array}
$$

5. Compute the next Routh row where the row coefficients are obtained by evaluating determinant type cross multiplications from the last two Routh Rows

$$b_1 = \frac{a_1 P_{n-3} - a_2 P_{n-1}}{a_1}, \qquad b_2 = \frac{a_1 P_{n-5} - a_3 P_{n-1}}{a_1}, \qquad \cdots$$

6. Augment the Routh rows with the new row and continue until the s^0 row is found and the Routh array is completed, as

$$
\begin{array}{cccccc}
s^n & P_n & P_{n-2} & P_{n-4} & P_{n-6} & \cdots \\
s^{n-1} & P_{n-1} & P_{n-3} & P_{n-5} & P_{n-7} & \cdots \\
s^{n-2} & a_1 & a_2 & a_3 & a_4 & \cdots \\
s^{n-3} & b_1 & b_2 & b_3 & & \cdots \\
& & \vdots & & & \\
s^0 & n_1 & & & &
\end{array}
$$

7. After completion of the Routh array we examine terms in the first column of the array. The number of roots with positive real parts is simply equal to the number of changes in sign of terms in the first column.

This completes a description of the Routh Hurwitz stability test. Before we consider some examples, we note three special cases which occasionally occur in applying the criterion.

1. Often one row will contain very large numbers or very small numbers only. All entries in a row may be multiplied by any positive constant without altering the results obtained from the procedure.

2. If the first term in a row becomes zero, it should be assumed to be a small positive constant ϵ and the procedure continued.

3. Presence of roots equidistant and radially opposite from the origin will result in all terms in a computed row becoming zero. While this insures instability, it is helpful to continue the Routh Hurwitz test to determine the number of roots in the right half plane. To continue the test we:

   (a) form an auxiliary polynomial equation from the coefficients of the last non-vanishing row.

   (b) take the derivative of the auxiliary equation with respect to s.

   (c) replace the row with the coefficients of this derivative.

   (d) continue the test using the newly formed row.

Several examples should illustrate the Routh Hurwitz procedure for stability determination which is presented in DELTA chart form in Fig. (4.4-1).

EXAMPLE 4.4-1. As a simple example of the Routh Hurwitz test let us examine the polynomial

$$P(s) = (s + 1)\,(s + 2)\,(s + 1 + j2)\,(s + 1 - j2)\,(s - 1 + j3)\,(s - 1 - j3)$$

$$= s^6 + 4s^5 + 11s^4 + 53s^3 + 102s^2 + 150s + 100$$

With the roots given we know that the characteristic equation has roots

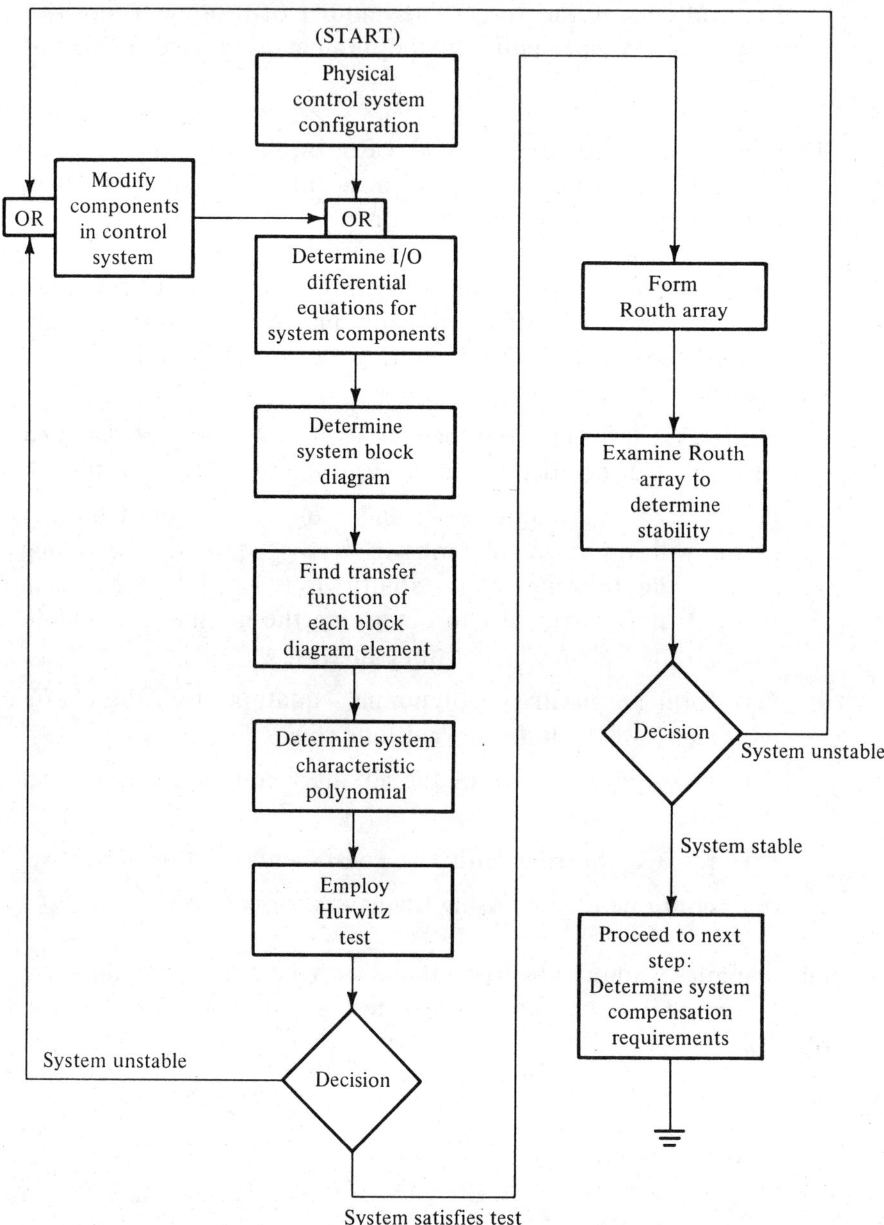

Figure 4.4-1 DELTA Chart for stability determination by the Routh Hurwitz criterion

in the right half plane. Let us not use this knowledge and employ the test.

The two requirements of the Hurwitz test are satisfied in that all polynomial coefficients are present and positive. We next obtain the first two Routh rows

$$s^6 \quad 1 \quad 11 \quad 102 \quad 100$$

$$s^5 \quad 4 \quad 53 \quad 150$$

From these two rows we obtain the next row

$$a_1 = \frac{4(11) - 1(53)}{4} = -2.25$$

$$a_2 = \frac{4(102) - 1(150)}{4} = 64.5$$

$$a_3 = \frac{4(100) - 1(0)}{4} = 100$$

and then have three Routh rows

$$s^6 \quad 1 \quad 11 \quad 102 \quad 100$$

$$s^5 \quad 4 \quad 53 \quad 150$$

$$s^4 \quad -2.25 \quad 64.5 \quad 100$$

We continue in this fashion until we obtain the complete Routh array with 6 rows

$$s^6 \quad 1 \quad 11 \quad 102 \quad 100$$

$$s^5 \quad 4 \quad 53 \quad 150$$

$$s^4 \quad -2.25 \quad 64.5 \quad 100$$

$$s^3 \quad 167.67 \quad 327.78$$

$$s^2 \quad 68.9 \quad 100$$

$$s^1 \qquad 84.43$$

$$s^0 \qquad 100$$

There are two changes in sign in the Routh array and so we conclude that there are two roots of this $P(s)$ in the right half plane. A system with these poles is unstable.

EXAMPLE 4.4-2.   We consider the polynomial

$$P(s) = (s + j1)\,(s - j1)\,(s + 1)^2 = s^4 + 2s^3 + 2s^2 + 2s + 1$$

The Hurwitz test is satisfied and so we form the Routh array

$$
\begin{array}{cccc}
s^4 & 1 & 2 & 1 \\
s^3 & 2 & 2 &
\end{array}
$$

The next row of the Routh array is

$$
\begin{array}{ccc}
s^2 & 1 & 1
\end{array}
$$

and the following row will consist of a single zero. Thus we find the auxiliary equation from the last non-vanishing row which is $s^2 + 1$, differentiate this with respect to s to obtain $2s$ and enter this as the s^1 row

$$
\begin{array}{cc}
s^1 & 2
\end{array}
$$

and continue the development of the Routh array. The complete Routh array is obtained as

$$
\begin{array}{cccc}
s^4 & 1 & 2 & 1 \\
s^3 & 2 & 2 & \\
s^2 & 1 & 1 & \\
s^1 & 2 & & \\
s^0 & 1 & &
\end{array}
$$

Since there are no changes in sign there are no roots in the right half plane. We note that poles on the $j\omega$ axis are not included in the right half plane insofar as the Routh Hurwitz test is concerned.

EXAMPLE 4.4-3. The stability examples we have considered so far are primarily analysis examples. The real use of the approach is in synthesis. Here we will determine the requirements on the gain K and time constant T for a unity ratio feedback system with open loop transfer function

$$G(s) = \frac{K(Ts^2 + 1)}{s^2(s + 1)}$$

such that the closed loop system is stable. The closed loop transfer function is

$$H(s) = \frac{G(s)}{1 + G(s)} = \frac{K(Ts^2 + 1)}{s^3 + s^2(1 + KT) + s + 1}$$

First we apply the Hurwitz test which requires that $K > 0$ and $1 + KT > 0$. Next we formulate the Routh array as

$$
\begin{array}{ccc}
s^3 & 1 & 1 \\
\\
s^2 & 1 + KT & K \\
\\
s^1 & \dfrac{1 + KT - K}{1 + KT} & \\
\\
s^0 & K &
\end{array}
$$

and see that, since we must require no changes of sign in the first column of this array, we require

$$1 + KT > 0 \qquad 1 + KT - K > 0 \qquad K > 0$$

The most restrictive of these requirements are $K > 0$ and $T > \dfrac{K - 1}{K}$.

Figure (4.4-2) illustrates the stability region in the KT plane for this example.

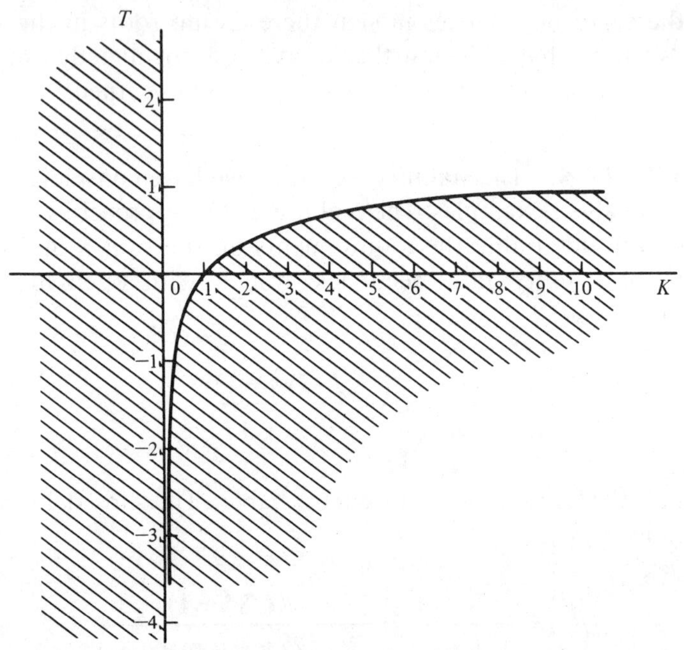

Figure 4.4-2 Stability boundary for EXAMPLE 4.4-3 —
shaded regions indicate instability

EXERCISE 4.4-1. Determine, using the Routh Hurwitz criterion, stability of the following polynomials which represent the denominator of a closed loop function:

(a) $P(s) = s^3 + 40s^2 + 36s + 800$

(b) $P(s) = s^4 + 4s^3 + 24s^2 + 64s + 128$

(c) $P(s) = s^4 + 2s^3 + 3s^2 + 8s + 8$

(d) $P(s) = s^5 + 2s^4 + 3s^3 + 7s^2 + 4s + 4$

(e) $P(s) = s^3 + 2s^2 + s + K$

EXERCISE 4.4-2. Under what conditions are closed loop unity ratio feedback systems with the following open loop transfer functions stable?

(a) $G(s) = \dfrac{K}{s(Ts + 1)}$

(b) $G(s) = \dfrac{K(T_1 s + 1)}{s^2(T_2 s + 1)}$

Use the Routh Hurwitz criterion to determine your solutions.

The Nyquist Criterion

In our subsection just completed we have illustrated use of the Routh Hurwitz criterion to determine system stability. This criterion determines whether or not the roots of a polynomial have positive real parts which indicate system instability. While this criterion is easy to apply it yields no information about relative stability. In a sense, the knowledge that a system is stable is not very valuable information for the systems control designer since much more than just stability is required in order for a system to have practical utility. The Nyquist criterion is one which not only yields information concerning stability— but relative stability as well. In fact many performance specifications of great usefulness for design purposes are defined using Nyquist diagrams.

We consider a single loop feedback system as shown in Fig. (4.4-3). The closed loop transfer function is

$$\frac{Z(s)}{U(s)} = H(s) = \frac{G_1(s)}{1 + G_1(s)G_2(s)} \qquad (4.4\text{-}16)$$

The closed loop poles are seen to be equal to the zeros of

$$D(s) = 1 + G_1(s)G_2(s) = 1 + G(s) \qquad (4.4\text{-}17)$$

In general, we can write $D(s)$ as a ratio of two polynomials

$$D(s) = \frac{A(s)}{B(s)} = \frac{(s + a_1)(s + a_2)(s + a_3)\ldots}{(s + b_1)(s + b_2)(s + b_3)\ldots} \qquad (4.4\text{-}18)$$

It is both interesting and important to note that the poles of $D(s)$, or roots of $B(s)$, are also the poles of $G(s) = G_1(s)G_2(s)$. However, the

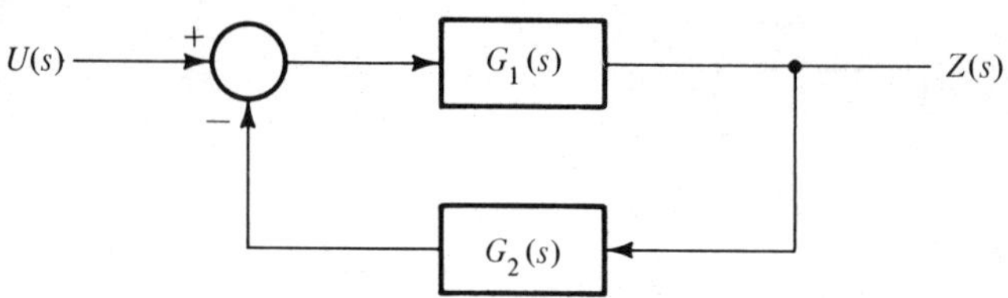

Figure 4.4-3 Single loop feedback system on which Nyquist criterion is based

zeros of $D(s)$, or roots of $A(s)$, are not at all the zeros of $G(s)$. In fact we must factor a polynomial, often of high order, in order to determine these roots. If the open loop system is stable, such that all poles of $G(s)$ are in the left half plane or $j\omega$ axis, then the poles of $D(s)$ are all in the left half plane at precisely the location of the poles of $G(s)$. We shall initially assume that this is the case and develop a method to determine the zeros of $1 + G(s) = D(s)$.

Let us assume that there are poles and zeros of $D(s)$ distributed in the s plane as shown in Fig. (4.4-4). The term $(s + a_1)$ is represented by the vector distance from the zero located at $-a_1$ to the point s in the complex plane. We now let point s move in some arbitrary but closed contour Γ as in Fig. (4.4-4). A value of the complex function $1 + G(s)$ is associated with every point on this closed contour and this value may be found simply by substituting the numerical value of s into $1 + G(s)$. The value of $D(s)$ may also be found by multiplying the distances from the point s to the zeros at $-a_i$ and dividing by the distances from s to the poles at $-b_i$. Suppose that the contour Γ is such that a point s on the contour moves in a clockwise path and encircles the zero at $-a_1$ but no other poles or zeros. The vector $s + a_1$ rotates through a net angle of -2π radians or $-360°$. If we plot the locus of points or values of $1 + G(s) = D(s)$ as s moves around closed contour Γ then we see that $D(s)$ moves in a closed contour Γ' which encircles the origin of the s plane, also in a clockwise direction as we see in the $D(s)$ plane of Fig. (4.4-5). The precise demography of these contours is not important now—only the direction and the fact that the contour Γ' will enclose the origin.

If the closed path Γ encloses the two zeros at $-a_1$ and $-a_2$ in a clockwise direction then the contour Γ' will enclose the origin twice in a clockwise direction. In general, the number of clockwise encirclements of Γ' about the origin will be equal to the number of zeros in the closed contour Γ which we move around in a clockwise direction.

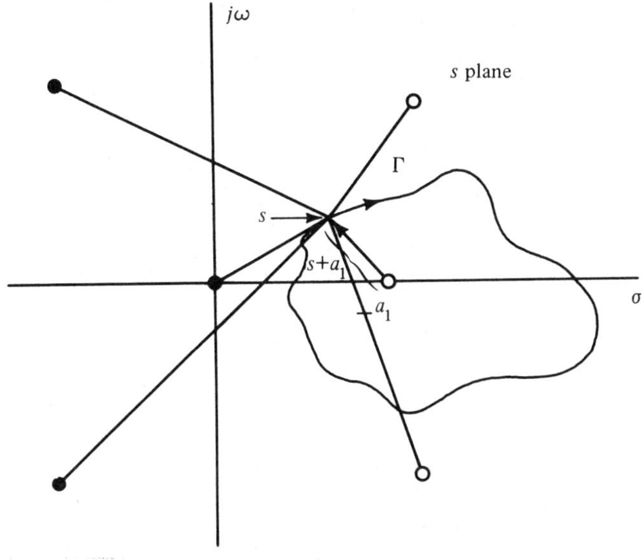

Figure 4.4-4 *s* plane and the closed contour Γ

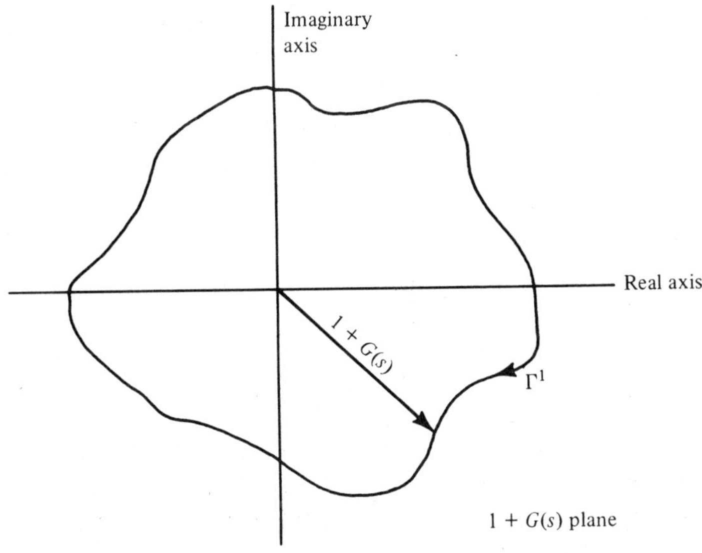

Figure 4.4-5 Polar plot of $1 + G(s)$

This is a consequence of a very famous principle in complex variable theory known as the principle of the argument and we have presented only a very heuristic argument concerning the validity of this principle here.

For stability determination we are only interested in determining whether or not any zeros of $D(s) = 1 + G(s)$ lie in the right half plane. Our system is assumed for the moment to be open loop stable so we know that there are no poles of $D(s)$ in the right half plane. Thus we could let Γ be the entire left half plane as shown in Fig. (4.4-6). If we plot the locus of points Γ' for points in s traversing Γ, the left half plane, in a clockwise direction then we can merely determine the number of clockwise encirclements Γ' makes around the origin and we immediately have the number of zeros of $D(s)$ in the right half plane. We assume that $G(s) = 0$ for $s = \infty \pm j\omega$, and this will always be true in practice for real systems, such that all of the infinite semicircle in the s plane of Fig. (4.4-4) maps into the point $1 + j0$ in the $D(s)$ plane of Fig. (4.4-5). We have thus obtained a polar plot, of the transfer function $1 + G(s)$ for $s = j\omega$. In actual practice we find it much more convenient to obtain a polar plot of the locus of $G(s)$ for $s = j\omega$. The only difference between the two curves so obtained is that the $G(s)$ locus is displaced one unit to the left of the $1 + G(s)$ locus. Since the critical occurrences in the $1 + G(s)$ locus are clockwise encirclements of the origin, the critical occurrence for the $G(s)$ locus are clockwise encirclements of the $-1 + j0$ point in the $G(s)$ plane. This is the critical point on a Nyquist diagram. We can now state the Nyquist criterion for the open loop stable case. An open loop stable system is closed loop stable if and only if the polar plot of $G(s)$ makes no encirclements of the point $-1 + j0$ as s varies from $-j\infty$ to $+j\infty$.

EXAMPLE (4.4-4).   We consider a single loop unity ratio feedback system with open loop transfer function

$$G(s) = \frac{K}{s(Ts + 1)} \tag{1}$$

We desire to determine the relation which must exist between K and T in order to insure system stability. We could either obtain a Nyquist plot of Eq. (1) and use $-1 + j0$ as the critical point or alternately obtain a Nyquist plot of

$$1 + G(s) = \frac{Ts^2 + s + K}{s(Ts + 1)} \tag{2}$$

We immediately see the great advantage to obtaining a Nyquist plot of Eq. (1) rather than Eq. (2). It is a much simpler thing to do. This is especially the case since we may use a Bode gain and phase diagram to construct a Nyquist diagram and it is relatively simple to obtain a Bode diagram for Eq. (1). The asymptotic gain approximation is

$$|G(j\omega)| = \begin{cases} \dfrac{K}{\omega} & \omega < \dfrac{1}{T} \\[2ex] \dfrac{K}{\omega^2 T} & \omega > \dfrac{1}{T} \end{cases}$$

and the asymptotic phase approximation is such that the phase shift is $-\pi/2$ until $\omega = 1/10T$. The phase shift is $-\pi$ above $\omega = 10/T$. A straight

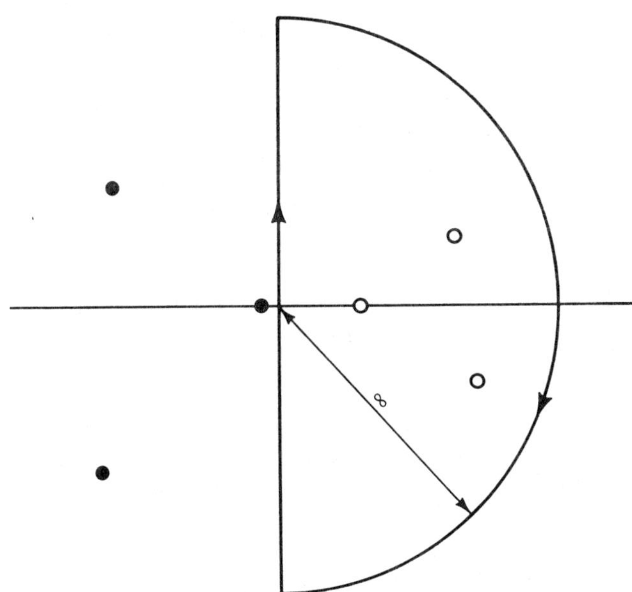

Figure 4.4-6 Infinite semicircular contour in the right half of the s plane—the Nyquist contour is the infinite semicircular contour in the left half of the s plane.

line connects the phase shift characteristics from $\omega = 1/10T$ to $\omega = 10/T$ with a $-3\pi/4$ phase shift at $\omega = 1/T$. Figures (4.4-7) and (4.4-8) illustrate the Bode and Nyquist diagrams for this example.

It is immediately apparent from the Nyquist plot that the polar plot of $G(s)$ cannot possibly encircle the $-1 + j0$ point and so this second order system is always stable. We can see from Fig. (4.4-8) that it is possible for the open loop gain to become arbitrarily close to the critical $-1 + j0$ point if the gain K is sufficiently large or if the time constant T is sufficiently small. Thus the response characteristics of the system could become deplorably poor—even though a system with the mathematical model described here cannot become unstable. As we shall soon see, an important performance criterion is phrased in terms of how close the Nyquist plot is to the critical point for a given system. In Fig. (4.4-8) the solid line indicates the polar plot corresponding to those values of Γ in the upper half of the right half plane. It is perhaps easier to visualize the polar plot if we let the portion of Γ near $s = 0$ be represented by $s = \epsilon e^{j\theta}$ where ϵ is an arbitrarily small number. To avoid the pole of $G(s)$ at $s = 0$ we must move slightly to the right of the origin as we go through the point $s = 0$ in the Γ plane of Fig. (4.4-6). The expression $G(s)$ is approximately K/s around $s = 0$ and we see that as we move around the pole at the origin the expression $G(s)$ becomes $K\epsilon^{-1}e^{-j\theta}$ and this produces the infinite semicircular curve sketched on the Nyquist plot. The dashed line in Fig. (4.4-8) shows the polar gain plot for negative frequency. Only in very rare cases is it ever necessary to sketch a Nyquist plot other than for the frequency range $0 < \omega < \infty$. We should also note that the Nyquist plot of Fig. (4.4-8) is not constructed to scale or it would be, of course, impossible to sketch the infinite semicircular part of the curve. It will often be true that much of value can be gained from a rough sketch of the Nyquist diagram, in fact essentially as much as from a detailed, drawn-to-scale, curve.

EXAMPLE 4.4-5. In this example we will consider the more complex unity ratio feedback system with transfer function

$$G(s) = \frac{100\left(1 + \dfrac{s}{10}\right)^2}{s\left(1 + \dfrac{s}{5}\right)^2\left(1 + \dfrac{s}{100}\right)^2} \tag{1}$$

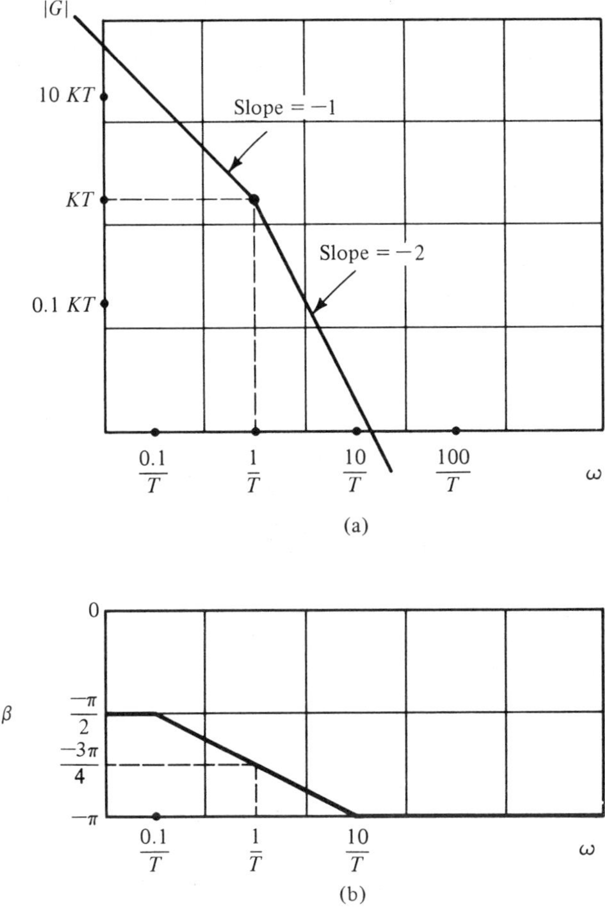

Figure 4.4-7 Bode gain (a) and phase (b) diagrams

The asymptotic Bode gain plot for this open loop transfer function is shown in Fig. (4.4-9). On this Bode plot are shown the slopes of the straight line segments on the log log coordinate paper which is by any criterion the coordinate system of choice for Bode amplitude plots. We could easily sketch in the actual Bode plot by noting that the approximate actual gain is less than that shown by a factor of 0.5 at $\omega = 10$ and lower than that shown by a factor of 0.5 at $\omega = 100$.

Over the frequency range $1 < \omega < 100$ the open loop gain varies from 100 to 0.3, over two and one-half orders of magnitude. There is

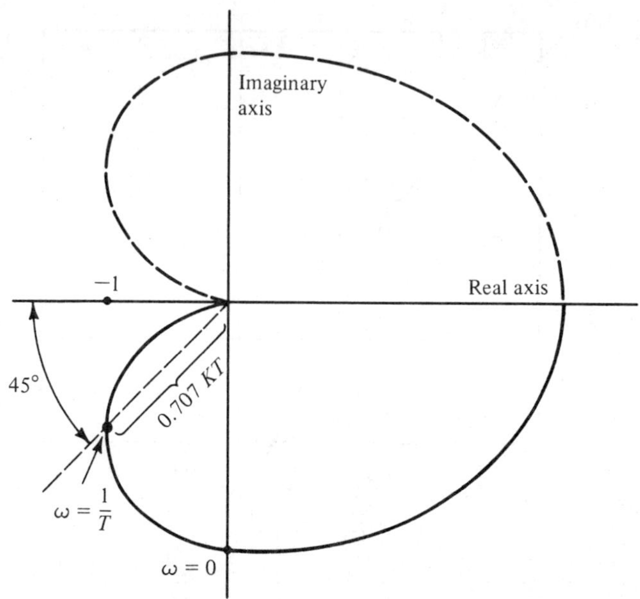

Figure 4.4-8   Nyquist plot for open loop system

essentially no way in which we can sketch a polar Nyquist plot over this frequency range on a scale that is meaningful in the sense of accurately and precisely displaying open loop behavior near the critical points where $|G| = 1$ and the angle of $G = -\pi$. Instead we often draw a Nyquist plot that is essentially constructed to scale in the vicinity of these critical points but which departs from scale elsewhere such that it is possible to sketch the complete Nyquist plot and still display accurate and meaningful information near the critical points.

From the open loop transfer function we can immediately write down the phase shifts characteristic

$$\beta(\omega) = -\frac{\pi}{2} - 2\tan^{-1}\frac{\omega}{5} + 2\tan^{-1}\frac{\omega}{10} - 2\tan^{-1}\left(\frac{\omega}{100}\right)$$

Using the arc tangent approximation we have

$$\beta(\omega) = -\frac{\pi}{2} + 2\left(-\frac{\omega}{5} + \frac{\omega}{10} - \frac{\omega}{100}\right), \qquad \omega < 5$$

$$= -\frac{\pi}{2} - 0.22\omega$$

$$\beta(\omega) = -\frac{\pi}{2} - 2\left(\frac{\pi}{2} - \frac{5}{\omega}\right) + 2\left(\frac{\omega}{10} - \frac{\omega}{100}\right), \qquad 5 < \omega < 10$$

$$= -\frac{3\omega}{2} + \frac{10}{\omega} + 0.018\omega$$

$$\beta(\omega) = -\frac{\pi}{2} - 2\left(\frac{\pi}{2} - \frac{5}{\omega}\right) + 2\left(\frac{\pi}{2} - \frac{10}{\omega}\right) - 2\left(\frac{\omega}{100}\right), \qquad 10 < \omega < 100$$

$$= -\frac{\pi}{2} - \frac{10}{\omega} - 0.02\omega$$

$$\beta(\omega) = -\frac{\pi}{2} - 2\left(\frac{\pi}{2} - \frac{5}{\omega}\right) + 2\left(\frac{\pi}{2} - \frac{10}{\omega}\right) - 2\left(\frac{\pi}{2} - \frac{100}{\omega}\right), \qquad 100 < \omega$$

$$= -\frac{3\pi}{2} + \frac{190}{\omega}$$

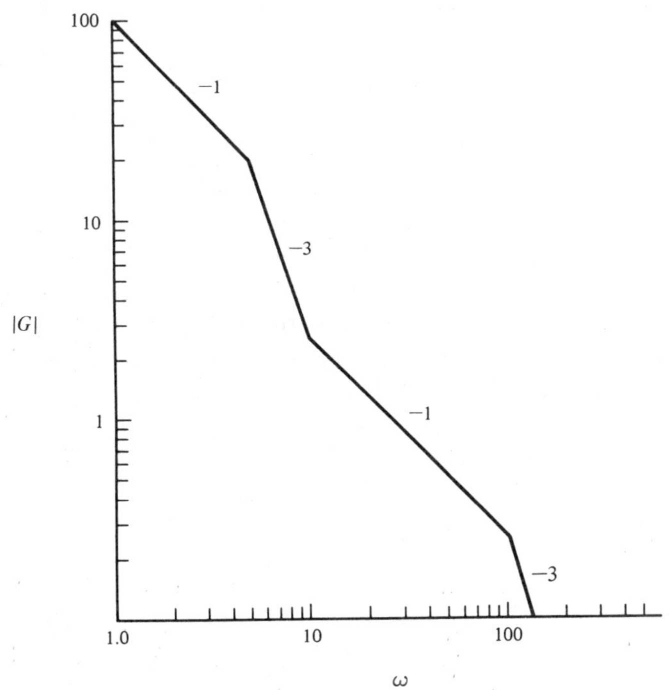

Figure 4.4-9　Bode amplitude plot for EXAMPLE 4.4-5

We could use these relations to determine phase shifts at appropriate frequencies and thus obtain information we need to sketch the Nyquist plot quite accurately if we desire. Of special interest is the phase shift when the gain is unity and this can be obtained from the approximation valid when $\omega = 25$ as

$$\beta_c = \beta(\omega_c) = \beta(25) = \frac{-\pi}{2} - \frac{10}{25} - .02(25) = 2.47 \text{ radians} = -147.6°$$

We use the symbol ω_c because the frequency where the open loop gain is one is normally called the *crossover frequency* and denoted by ω_c. The resulting phase shift is denoted β_c. The angle between the phase shift at crossover and the negative real axis is often called the *phase margin* and denoted PM. We have

$$PM = 180° + \beta_c = 38.4°$$

We will soon see that phase margin is a very important control system specification. We should also note that the gain and phase shift at $\omega = 25$ is obtained in a relatively straightforward way from the open loop gain function as

$$G(j25) = \frac{100(1 + j2.5)^2}{(j25)(1 + j5)^2(1 + j0.25)^2} = 1.05 \, \underline{/-139°}$$

These are very close to the values obtained from the asymptotic approximation. Certainly it is not difficult to substitute the value $\omega_c = 25$ into the exact $G(s)$ expression and calculate the gain and phase shift. Thus we might reasonably question the value of the approximations. The answer is that it would be very difficult to solve $|G(j\omega)| = 1$ using the exact $G(s)$ since this would result in a 5th order polynomial equation in ω. However, it is exceptionally easy to find the frequency where $|G| = 1$ on the Bode amplitude plot.

To solve for the frequency where the phase shift is $-180°$, we would need to solve the exact equation, from Eq. (1),

$$-\pi = 2 \tan^{-1} \frac{\omega}{10} - \frac{\pi}{2} - 2 \tan^{-1} \frac{\omega}{5} - 2 \tan^{-1} \frac{\omega}{100}$$

which is a difficult nonlinear transcendental equation. From the Bode amplitude plot we can guess that the $\beta = -\pi$ phase shift occurs for 10

$< \omega < 100$. Using the arc tangent approximation valid for this frequency range we have

$$-\pi = -\frac{\pi}{2} - \frac{10}{\omega} - 0.02\omega$$

and obtain as the solution

$$\omega = 71.6$$

This is a reasonably accurate value since the actual phase shift for $\omega = 71.6$ is obtained from Eq. (1) as $\beta(\omega = 71.6) = -170°$. The gain at $\omega = 71.6$ is obtained from the Bode plot as 0.35 and since this is considerably less than one the system is stable.

The Nyquist plot for this example is shown in Fig. (4.4-10). A system with this sort of Nyquist plot is called conditionally stable since the system could become unstable if the gain is reduced as it might well be in initial start up of the system. Normally we wish to avoid conditionally stable systems whenever possible.

Thus far we have not considered the possibility of an open loop unstable system in our development of the Nyquist criterion. If the $D(s)$ equation, Eq. (4.4-17), has one or more poles in the right half plane then these must be enclosed by the s plane contour Γ of Fig. (4.4-6). Suppose that there exists a single pole of $D(s)$ at $s = -b_1$ and no zeros in the right half plane. It is a straightforward task to show that when the contour Γ is such that a point s on the contour moves in a clockwise path and encircles the pole at $-b_1$ but no other poles or zeros then the vector $s + b_1$ rotates through a net angle of -2π radians or $-360°$. If we plot the locus of points or values of $1 + G(s) = D(s)$ as s moves around closed contour Γ we see that $D(s)$ moves in a closed contour Γ' which encircles the origin of the s plane in a counterclockwise direction. We can easily extend this to the case of multiple poles and poles and zeros in the right half plane of $D(s)$. The Nyquist criterion can now be stated in general: **If $G(s)$ is determined for s following an infinite semi-circular contour Γ, as in Fig. (4.4-6), then the polar plot of $G(s)$ encircles the critical point $-1 + j0$, in a clockwise direction, a number of times which is equal to the number of zeros minus the number of poles of $D(s) = 1 + G(s)$ which are in the right half s plane. For stability the Nyquist plot encircles, in a counterclockwise direction, the critical point $-1 + j0$ a number of times equal to the number of right half plane poles of $G(s)$.**

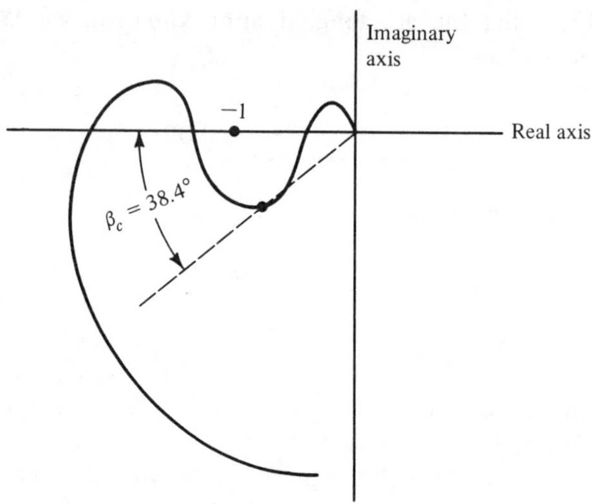

Figure 4.4-10   Nyquist plot of conditionally stable system

The DELTA chart of Fig. (4.4-11) provides a convenient and useful guide to stability determination by means of the Nyquist diagram.

EXAMPLE 4.4-6. We now consider the open loop unstable system with

$$G(s) = \frac{K(s + 1)}{s(s - 1)}$$

We can employ the Routh Hurwitz criterion to yield the range of K, $K > 1$, for which the system is stable. The asymptotic gain is just the gain magnitude for this example

$$|G(j\omega)| = \frac{K}{\omega}$$

The phase shift curve is given by

$$\beta(\omega) = \frac{-3\pi}{2} + 2 \tan^{-1} \omega$$

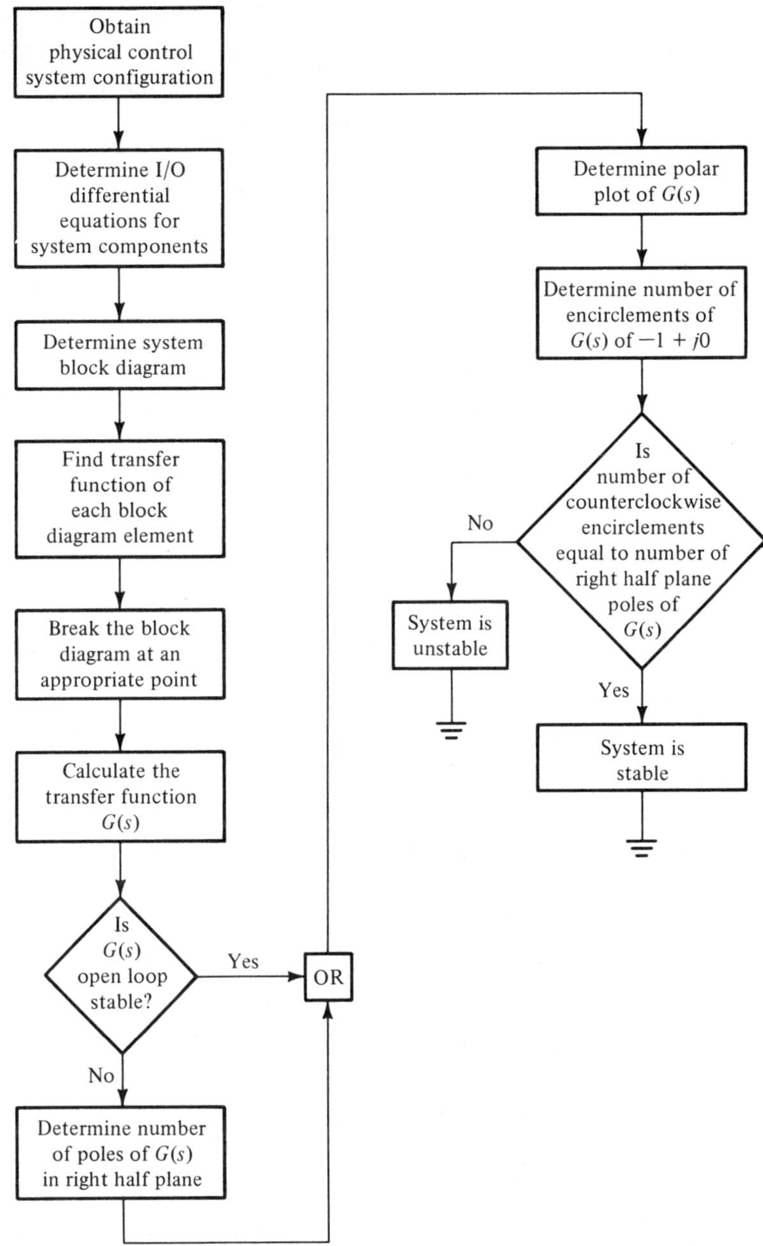

Figure 4.4-11 DELTA Chart for stability determination by Nyquist diagrams

The Nyquist plot for this system is shown in Fig. (4.4-12). From this Nyquist plot we see that there will be a counterclockwise encirclement of the critical point $-1 + j0$ if the gain K is sufficiently large. Beyond this critical value of $K(K > 1)$ the system will be stable. This is another conditionally stable system since it is unstable for small K. Conditional stability is a condition which we should normally try to avoid.

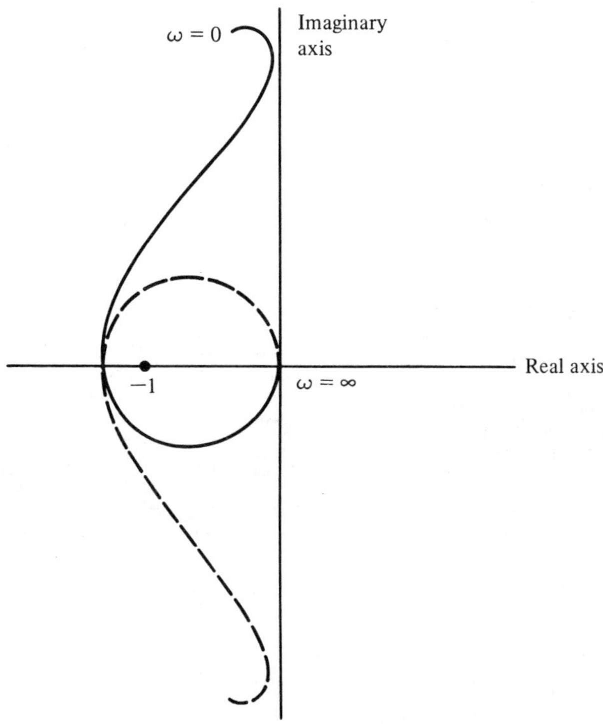

Figure 4.4-12 Nyquist diagram for EXAMPLE 4.4-6

EXAMPLE 4.4-7. The Laplace transform of a pure time delay is e^{-sT} where T is the amount of the delay. Many systems, especially in the process control industry, contain time delays. Often time delays are quite difficult to deal with since they have no finite poles or zeros; all zeros are at $s = \infty + j\omega$ whereas all poles are at $s = -\infty + j\omega$ for any ω. Often approximations such as

$$e^{-sT} \approx \frac{1 - sT}{1 + sT} \tag{1}$$

or

$$e^{-sT} \approx \frac{(1 - sT/n)^n}{(1 + sT/n)^n} \tag{2}$$

are used to approximate time delays for analysis purposes. Time delays present no special problem for us here and we consider the stability determination for the unity ratio closed loop system that has an open loop function

$$G(s) = \frac{Ke^{-sT}}{s}$$

Here we note that stability determination by means of the Routh Hurwitz test is not possible unless we use Eq. (1) or (2) or some similar relation to approximate the exact time delay.

The gain magnitude of this system is just

$$|G(j\omega)| = \frac{K}{\omega}$$

whereas the phase shift is

$$\beta(\omega) = \frac{-\pi}{2} - \omega T$$

The gain magnitude is one whenever $\omega_c = \omega = K$ and the phase shift is $\beta_c = \frac{-\pi}{2} - KT$. If this phase shift is less than $-\pi$ the system is unstable. Thus the system is stable whenever

$$\beta_c = \frac{-\pi}{2} - KT > -\pi$$

or whenever

$$K < \frac{\pi}{2T}$$

The interested reader should sketch the Nyquist plot for this example to verify that the above conclusion is indeed valid.

Some very important frequency domain specifications for linear systems control design are best defined by reference to a Nyquist plot. One of these is the phase margin concept. As we have indicated, **phase margin is the angle between the phase shift at crossover and the negative real axis.** Phase margin is a very important concept primarily because it is a single figure of merit that is easily calculated and typically gives a very good indication of the peaking that occurs in the closed loop system frequency response, the exact determination of which is often quite difficult.

The closed loop transfer function for a unity ratio system is related to the open loop transfer function $G(s)$ by

$$H(s) = \frac{G(s)}{1 + G(s)} \tag{4.4-19}$$

The magnitude of the closed loop transfer function

$$M(\omega) = |H(j\omega)| \tag{4.4-20}$$

is a very important performance criterion. Below crossover we have $|G(j\omega)| > 1$ and so we have for the approximate $M(\omega)$ curve

$$M(\omega) \approx 1 \qquad \omega < \omega_c$$

whereas for $\omega > \omega_c$ we have $|G(j\omega)| < 1$ so that

$$M(\omega) \approx |G(j\omega)| \qquad \omega > \omega_c$$

which is less than 1. Near the crossover frequency neither of these two approximations will generally yield good results.

We may determine a set of loci of constant $M(\omega)$ in the s plane used to represent a Nyquist plot such that the values of $M(\omega)$ for a given system may be determined directly from the Nyquist plot. The loci of constant $M(\omega)$ are easily determined. We let $G(j\omega) = x + jy$ such that from

$$M^2(\omega) = |H(j\omega)|^2 = H(j\omega)H(-j\omega)$$

we obtain

$$M^2 = \frac{x^2 + y^2}{(1 + x)^2 + y^2}$$

which can be rewritten as the equation for a circle with a center at $x = -\dfrac{M^2}{M^2 - 1}$, $y = 0$ and radius $= \left| \dfrac{M}{M^2 - 1} \right|$. This is

$$\left(x + \frac{M^2}{M^2 - 1} \right)^2 + y^2 = \left(\frac{M}{M^2 - 1} \right)^2 \qquad (4.4\text{-}21)$$

Figure (4.4-13) illustrates loci of constant M. Typical performance specifications for control systems require peak values of M, denoted M_p, in the range 1.2 to 1.4. In order to insure that $M_p < 1.4$ for example, we must restrict the Nyquist plot of $G(s)$ from entering the interior of the $M = 1.4$ circle. Normally trial and error approaches will be required to accomplish this.

We may determine an approximate value of phase shift at crossover which corresponds to a given M_p as follows. Figure (4.4-14) illustrates the $M = 1.4$ circle as well as the unit circle about the origin. It is reasonably clear from this figure that the phase margin will, for well-behaved systems, be approximately the radius of the M_p circle divided by the distance from the origin to the center of the M_p circle. Thus we have the approximation

$$\sin \text{PM} = \frac{\dfrac{M_p}{M_p^2 - 1}}{\dfrac{M_p^2}{M_p^2 - 1}} = \frac{1}{M_p} \qquad (4.4\text{-}22)$$

Our next chapter will be devoted exclusively to Bode diagram design techniques in which we may easily specify the phase shift at crossover β_c or the phase margin PM. For the vast majority of control systems we may rely upon the approximation of Eq. (4.4-22) such that we can convert specifications in terms of M_p into phase margin specifications.

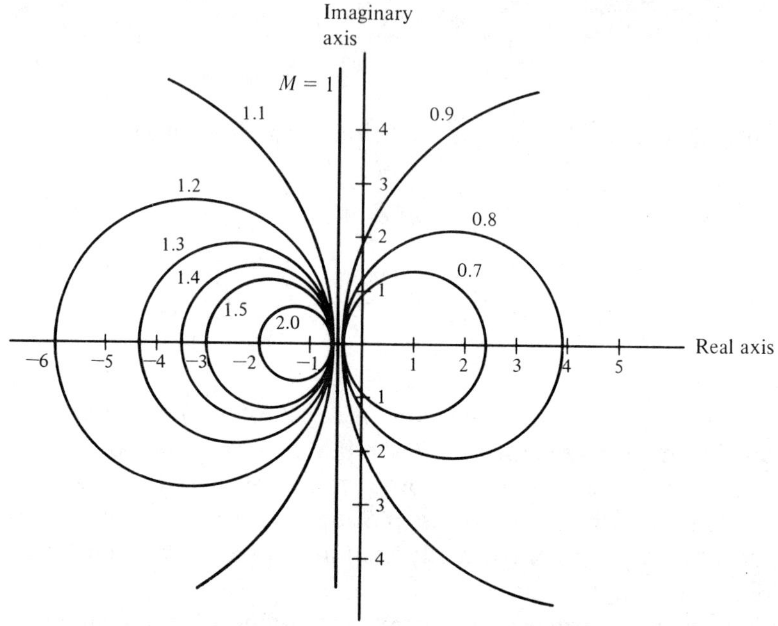

Figure 4.4-13 Loci for constant M circles

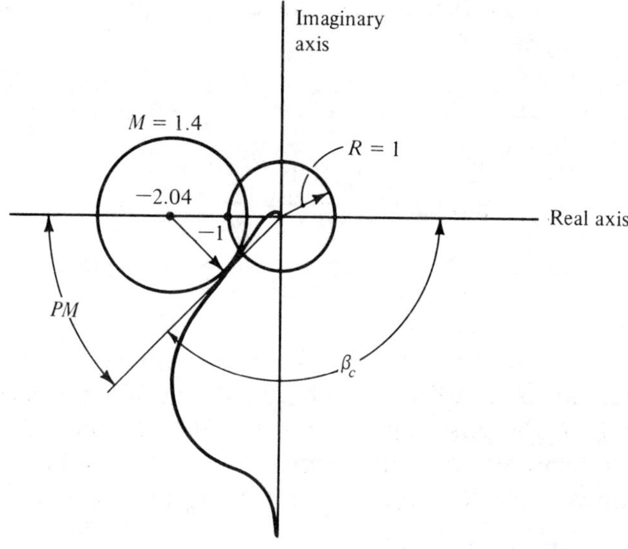

Figure 4.4-14 Typical plot of Nyquist diagram with $M_p = 1.4$

There are a number of other interesting topics concerning Nyquist diagrams that we could explore but will not do so here. For example, many systems control engineers will plot a Nyquist diagram using rectangular coordinates, typically log amplitude versus phase shift. The resulting chart, known as a Nichols chart, contains no more information than that shown on a Nyquist diagram although one can more easily visualize changes in M resulting from changes in gain than on a Nyquist diagram.

EXERCISE 4.4-3. Determine stability requirements of the following open loop transfer functions:

a) $\quad G(s) = \dfrac{K}{(s + 1)(s + 2)}$

b) $\quad G(s) = \dfrac{K}{s^2(1 + s)}$

c) $\quad G(s) = \dfrac{K(s - 1)}{s(s + 1)}$

d) $\quad G(s) = \dfrac{K(1 + Ts)}{s^2\left(1 + \dfrac{s}{10}\right)}$

e) $\quad G(s) = \dfrac{K(s + 1)}{s^2(s - 1)}$

f) $\quad G(s) = \dfrac{Ke^{-s}}{s(1 + s)}$

EXERCISE 4.4-4. What are appropriate gain (and time constants if appropriate) such that the systems of EXERCISE 4.4-3 have 45° phase margin? What is the resulting M_p? Use both the approximate relation as well as the exact value which you can get by sketching M circles on your Nyquist plots at the critical values of K for 45° phase margin.

4.5 ERROR COEFFICIENTS

For systems of higher order than the second it is usually not possible to determine relations between system parameters and system response that are sufficiently simple to be used for design purposes.

Consequently there is motivation to determine both approximate algorithms for system response and exact algorithms for limiting cases. In this section we will first examine static system accuracy or the error between the steady state desired output and the steady state actual output when the input is a polynomial of known order. Then we will investigate a series approximation to the dynamic accuracy of a system in responding to a given polynomial input of known order.

Static Error Coefficients

One of the simplest performance measures is the steady state error of a system in responding to a polynomial input of known order. We assume a unity ratio feedback control system with open loop transfer function $G(s)$. The closed loop transfer function is

$$\frac{Z(s)}{U(s)} = H(s) = \frac{G(s)}{1 + G(s)} \tag{4.5-1}$$

We define the error as the difference between input and output

$$E(s) = U(s) - Z(s) = U(s)[1 - H(s)] = \frac{U(s)}{1 + G(s)}$$

We can evaluate the infinite time error by use of the final value theorem*

$$e(t = \infty) = \lim_{s \to 0} sE(s) \tag{4.5-3}$$

Suppose that we consider a polynomial input

$$u(t) = \begin{cases} u_n(t) = \dfrac{t^n}{n!} & t \geq 0 \\[2ex] 0 & t < 0 \end{cases} \tag{4.5-4}$$

---

*The final value theorem is easily shown valid. First we establish the Laplace transform of $df(t)/dt$ using integration by parts

$$\mathcal{L}\,\frac{df(t)}{dt} = \int_0^\infty \frac{df(t)}{dt}\, e^{-st}\, dt = sF(s) - f(0)$$

then we simply let $s \to 0$ and evaluate the integral to obtain

$$f(t = \infty) = \lim_{s \to 0} sF(s)$$

The Laplace transform of this polynomial input is

$$U(s) = U_n(s) = \frac{1}{s^{n+1}} \qquad (4.5\text{-}5)$$

We consider a system with open loop transfer function

$$G(s) = \frac{K}{s^m} \frac{\displaystyle\prod_{i=1}^{\alpha}(1 + T_i s)}{\displaystyle\prod_{j=1}^{\beta}(1 + T_j s)} \qquad (4.5\text{-}6)$$

Thus we assume an open loop system with m poles at the origin, or m open loop integrations, and a number of other finite poles and zeros which may well appear in complex conjugate form but which cannot exist at the origin. We substitute Eq. (4.5-5) and (4.5-6) in Eq. (4.5-3) and obtain

$$e(t = \infty) = \lim_{s \to 0} \frac{1}{s^n} \frac{s^m \displaystyle\prod_{j=1}^{\beta}(1 + T_j s)}{s^m \displaystyle\prod_{j=1}^{\beta}(1 + T_j s) + K \displaystyle\prod_{i=1}^{\alpha}(1 + T_i s)}$$

Inspection of this equation shows that we will have either zero steady state error, finite steady state error, or infinite steady state error. In particular we have

$$e(t = \infty) = \begin{cases} 0 & m > n \\[2mm] \dfrac{1}{1 + K} & m = n = 0 \\[2mm] \dfrac{1}{K} & m = n > 0 \\[2mm] \infty & m < n \end{cases} \qquad (4.5\text{-}7)$$

It is common to use the words *type γ system* to denote a system with γ open loop integrations. Thus we can prepare TABLE 4.5-1 to indicate the steady state error for various types of systems and polynomial inputs. Unfortunately there is very limited information available from the static error coefficients since only one error constant is of significance. For example if we have a type one system or system with a single integration then we see that the steady state error is zero for any step input, a/K for a ramp input $u(t) = at$, and infinite for a parabola input. If we had a specification which indicated the maximum tolerable steady state error for a ramp input then we could determine a lower bound for the gain K. Thus we conclude that the static error coefficient method can yield important information, such as the lower bound on K, but that the information is limited since we have neither indication of the relative stability of a system nor indication of the way in which the steady state error is approached. The so-called "dynamic" error coefficients partially remove these difficulties.

Dynamic Error Coefficients

We now expand the error transfer function

$$\frac{E(s)}{U(s)} = \frac{1}{1 + G(s)} \tag{4.5-8}$$

in a Maclaurin series in s, or by simple division, to obtain

$$\frac{E(s)}{U(s)} = \frac{1}{K_0} + \frac{s}{K_1} + \frac{s^2}{K_2} + \ldots \tag{4.5-9}$$

The coefficients K_0, K_1, K_2, $\ldots$, are often called *dynamic error coefficients* although this is a misnomer since we have lost transient terms in obtaining the series of Eq. (4.5-9). A better expression might be particular error coefficients. We multiply both sides of Eq. (4.5-9) by $U(s)$ and formally take inverse transforms to obtain

$$e(t) = \frac{u(t)}{K_0} + \frac{\dot{u}(t)}{K_1} + \frac{\ddot{u}(t)}{K_2} + \ldots$$

$$= \sum_{i=0}^{\infty} \frac{1}{K_i} \frac{d^i u(t)}{dt^i} \tag{4.5-10}$$

The K_i in these expressions involve both the gain term K and the time constants in the open loop transfer function of Eq. (4.5-6). A simple example will illustrate the use of these error coefficients.

TABLE 4.5-1   TABLE OF STEADY STATE ERRORS FOR VARIOUS SYSTEM TYPES AND POLYNOMIAL INPUTS

	Type of System			
	0	*1*	*2*	*3*
Error for position input, $u(t) = a$	$\dfrac{a}{1+K}$	0	0	0
Error for velocity input, $u(t) = at$	∞	$\dfrac{a}{K}$	0	0
Error for acceleration input, $u(t) = \dfrac{at^2}{2}$	∞	∞	$\dfrac{a}{K}$	0
Error for jerk input, $u(t) = \dfrac{at^3}{6}$	∞	∞	∞	$\dfrac{a}{K}$

EXAMPLE 4.5-1. We consider a simple second order system with a single open loop integration

$$G(s) = \frac{K}{s(1 + Ts)} \tag{1}$$

The error transfer function is

$$\frac{E(s)}{U(s)} = \frac{1}{1 + G(s)} = \frac{s(1 + Ts)}{s(1 + Ts) + K} \tag{2}$$

which has the series expansion

$$\frac{E(s)}{U(s)} = \frac{1}{K}s + \left(\frac{KT - 1}{K^2}\right)s^2 + \left(\frac{1 - 2KT}{K^3}\right)s^3 + \ldots \tag{3}$$

The error series for this system is obtained by multiplying the foregoing equation by $U(s)$ and taking the inverse transform to obtain

$$e(t) = \frac{1}{K}\frac{du(t)}{dt} + \left(\frac{KT-1}{K^2}\right)\frac{d^2u(t)}{dt^2} + \left(\frac{1-2KT}{K^3}\right)\frac{d^3u(t)}{dt^3} + \ldots \quad (4)$$

For polynomial inputs of low order we get the same results as for the static error coefficients. However, for polynomials of order greater than that for which the static error is finite we actually obtain an approximate representation of the error as it increases in time. For example if $u(t) = \dfrac{at^2}{2}$ we obtain from Eq. (4)

$$e(t) = \frac{at}{K} + \frac{a(KT-1)}{K^2} \quad (5)$$

We may also obtain the "dynamic" error approximation for non-polynomial inputs. If $u(t) = \sin \omega t$, we have from Eq. (4)

$$e(t) = \frac{\omega}{K}\cos \omega t - \omega^2 \left(\frac{KT-1}{K^2}\right)\sin \omega t - \omega^3 \left(\frac{1-2KT}{K^3}\right)\cos \omega t + \ldots \quad (6)$$

This error series expression may well not converge. We see that unless $(\omega/K) < 1$ we are in trouble indeed if we wish to use Eq. (6) as an approximation to the system error. This is simply saying that the input frequency must be within the bound of frequencies the system will process with small error.

We see that the dynamic error series does yield more useful results than the static error coefficients. TABLE 4.5-2 presents the dynamic error coefficients for some typical systems.

EXERCISE 4.5-1. Derive the results in TABLE 4.5-1.

EXERCISE 4.5-2. Derive the results in TABLE 4.5-2.

EXERCISE 4.5-3. Determine the relationship between the non-zero finite static error coefficients for a type n system and the first non-vanishing dynamic error coefficient. (Answer: They are identical.)

TABLE 4.5-2 DYNAMIC ERROR COEFFICIENTS FOR FIRST, SECOND AND THIRD ORDER SYSTEMS

Transfer Function Open Loop $G(s)$	$\dfrac{1}{K_0}$	$\dfrac{1}{K_1}$	$\dfrac{1}{K_2}$	$\dfrac{1}{K_3}$
$\dfrac{K}{1+Ts}$	$\dfrac{1}{K+1}$	$\dfrac{TK}{(K+1)^2}$	$\dfrac{T^2K}{(K+1)^3}$	—
$\dfrac{K}{s(1+Ts)}$	0	$\dfrac{1}{K}$	$\dfrac{KT-1}{K^2}$	$\dfrac{1-2KT}{K^3}$
$\dfrac{K}{s(1+T_1 s)(1+T_2 s)}$	0	$\dfrac{1}{K}$	$\dfrac{K(T_1+T_2)-1}{K^2}$	$\dfrac{K^2-2K(T_1+T_2)+1}{K^3}$
$\dfrac{K(1+T_3 s)}{s(1+T_1 s)(1+T_2 s)}$	0	$\dfrac{1}{K}$	$\dfrac{(T_1+T_2)K-(1+KT_3)}{K^2}$	$\dfrac{(T_1 T_2-T_1 T_3-T_2 T_3)K^2-2K(T_1+T_2)+(1+KT_3)^2}{K^3}$
$\dfrac{K(1+T_2 s)}{s^2(1+\dot{T}_1 s)}$	0	0	$\dfrac{1}{K}$	$\dfrac{T_1-T_2}{K}$

4.6 PERFORMANCE SPECIFICATIONS
AND RULE OF THUMB CRITERIA

There are a great variety of performance specifications and rule of thumb criteria for linear control systems. We have discussed a number of these in our presentation thus far. Many of these performance criteria such as phase margin and M peak are somewhat artificial in that it would be highly unusual for the initial set of needs or other problem definition elements which lead to design of a control system to be expressed in these terms. However, even these artificial criteria can be approximately related to more realistic performance measures. There are some highly analytical performance measures more directly related to user performance requirements which can be, in theory, incorporated into systems control design using the technique of optimum systems control. However the analytical procedures of optimum systems control are more complex than those based on various approximate design approaches. Our efforts in the next two chapters will be concerned with two of the major approaches to trial and error or classical system control based on approximate performance specifications. Chapter 7 will be devoted to analytical or optimum systems control design techniques. There is no real dichotomy between any of these approaches. The modern systems control engineer uses those techniques which appear most promising in terms of problems resolution regardless of whether the design approaches are "classical" or "modern." In actual fact the "modern" techniques are based on mathematics that are of the same vintage as or even older than the "classical" techniques. Often a combination of several approaches are used on a given problem. While it is impossible to generalize typical uses of the various approaches it will often occur that the classical methods will be most useful for subsystem design such as instrument servomechanisms whereas the analytical or optimization based approaches will be most useful for large system design such as a guidance system for navigation or a large computer controlled process system. However we must emphasize that there will be many large scale problems where the classical trial and error approach will be eminently suitable.

In this section we will describe a number of useful performance specifications and rule of thumb design criteria for linear control systems. We have discussed many of these in our efforts thus far.

Crossover Frequency. The crossover frequency ω_c is the frequency at which the magnitude of the gain of a unity ratio feedback system with open loop transfer function $|G(j\omega_c)|$ is equal to one. The crossover frequency is easily determined from either a Bode or Nyquist plot of $G(j\omega)$.

Phase Shift at Crossover. The phase shift at crossover is the phase angle β_c associated with the frequency at which the magnitude of the open loop gain is equal to one. On a Nyquist diagram the phase shift at crossover $\beta_c = \beta(\omega_c)$ is the angle between the positive real axis and the point on the unit circle where $|G(j\omega_c)| = 1$. This phase shift may also be obtained directly from the Bode amplitude plot for a minimum phase system. Asymptotic phase shift curves may also be obtained in a straightforward way as we have seen.

Phase Margin. The phase margin PM of a system is defined as $\pi + \beta_c$. This is the angle on a Nyquist diagram between the $|G(j\omega_c)| = 1$ crossover point on the unit circle and the negative real axis. Phase margin is a simple easily computable performance measure that, with a very few exceptions, provides a very valuable indication of system relative stability. The phase margin criteria is especially appropriate for system design. Normally a phase margin of $40°$ or greater is acceptable for control systems.

Gain Margin. This is probably the simplest possible performance specification and can be calculated easily from the Routh Hurwitz test, a Bode diagram, or a Nyquist plot. It is the ratio of 1 to the actual gain magnitude $|G(j\omega)|$ at that frequency ω where the phase shift of $G(j\omega)$ is $-\pi$ radians or -180 degrees. A gain margin of less than 8 to 10 will generally be unsatisfactory but a gain margin greater than 8 to 10 is no assurance at all that the system response will be satisfactory. For a second order system with transfer function $G(s) = K/(s^2 + s)$, the gain margin is infinite regardless of K. Yet the system response will become very poor as K becomes much greater than one.

M **peak** (M_p). This frequency domain stability specification is just the magnitude of the peak value of the closed loop gain. M_p is the maximum of

$$|H(j\omega)| = \frac{|G(j\omega)|}{|1 + G(j\omega)|}$$

and can in theory be determined by setting the first derivative of $|H(j\omega)|$ with respect to ω equal to zero. This will be most difficult to accomplish in practice except for very simple systems. Circles of constant M [Eq. (4.4-21)] can be drawn on a Nyquist diagram, as we obtained in Fig. (4.4-13). It is considerably more difficult to design directly for a specified M_p than it is to design for a specified phase margin. Values of M_p less than 1.5 or perhaps 1.4 are generally regarded as acceptable. An approximate relation between phase margin and M_p which is very useful for design purposes was obtained in Eq. (4.4-22) as

$$\sin \text{PM} \approx \frac{1}{M_p}$$

Peak Frequency. The peak frequency ω_p is the frequency at which the M peak M_p occurs. It is simply the frequency which results in the maximum closed loop gain magnitude.

Oscillation Frequency. The frequency of transient damped sinusoidal oscillation ω_0 will generally be very close to the peak frequency ω_p. Also it turns out that a very useful rule of thumb relation between oscillation frequency and crossover frequency is $\omega_0 = \omega_p = 0.75\omega_c$.

EXAMPLE 4.6-1.   For a second order system with

$$G(s) = \frac{K}{s(Ts + 1)}$$

the crossover frequency ω_c occurs when the gain is unity and is therefore obtained, assuming $\omega_c > \frac{1}{T}$, from the asymptotic relation

$$\frac{K^2}{\omega_c^2(\omega_c^2 T^2)} = 1$$

as

$$\omega_c = \sqrt{K/T}$$

The closed loop transfer function for this system is

$$H(s) = \frac{G(s)}{1 + G(s)} = \frac{K}{Ts^2 + s + K}$$

which can be written in the standard form

$$H(s) = \frac{\omega_n^2}{s^2 + 2\zeta\omega_n s + \omega_n^2}$$

where

$$\omega_n = \sqrt{K/T}, \qquad \zeta = 1/(2\sqrt{KT})$$

Alternately we can write the closed loop transfer function in the form

$$H(s) = \frac{\alpha^2 + \omega_0^2}{(s + \alpha + j\omega_0)(s + \alpha - j\omega_0)} = \frac{\alpha^2 + \omega_0^2}{(s + \alpha)^2 + \omega_0^2}$$

where

$$\alpha = \zeta\omega_n, \qquad \omega_0 = \omega_n\sqrt{1 - \zeta^2}$$

or

$$\alpha = \frac{1}{2T}, \qquad \omega_0 = \sqrt{\frac{K}{T}}\sqrt{1 - \frac{1}{4KT}}$$

The frequency of the damped transient oscillations is ω_0. From our previous work we have already obtained the relationship between peak frequency and natural frequency ω_n. From Eq. (4.2-17) we have

$$\omega_p = \omega_n\sqrt{1 - 2\zeta^2}$$

If we plot the relations ω_0/ω_n and ω_p/ω_n versus ζ we see that the two curves yield essentially the same results for $0 < \zeta < 0.5$ which is the range for ζ where the frequency domain peaking may be significant.

If we divide the expressions for the oscillation frequency and the peak frequency we obtain

$$\frac{\omega_p}{\omega_c} = \sqrt{1 - \frac{1}{2KT}}$$

and as we might suspect, the ratio between these frequencies is a function of KT. We recognize that the damping ratio is determined by the

KT product as $\zeta = 1/(2\sqrt{KT})$. Values of ζ in the vicinity of approximately 0 to 0.5 are those where peaking is generally pronounced. This corresponds to $1 < KT < \infty$ and for this value of KT we have $0.71 < \omega_p/\omega_c < 1$. Thus it is not unreasonable to use the approximation $\omega_p \approx \omega_0 \approx 0.75\omega_c$. Also it is not unreasonable to use $\omega_n \approx \omega_p$ for this system if the parameters are adjusted such that $\zeta < 0.5$ so that there is peaking. The results of EXERCISES 4.6-1, 4.6-2 and 4.6-3 show that these results are even more valid for high order systems which can become unstable for sufficiently high open loop gain.

EXERCISE 4.6-1. Repeat the calculations of EXAMPLE 4.6-1 for the open loop transfer function $G(s) = \dfrac{K(Ts + 1)}{s^2}$ and show that the rule of thumb $\omega_0 = \omega_p = 0.75\omega_c$ is, again, a very reasonable one.

EXERCISE 4.6-2. Determine the M peak, phase margin, crossover frequency, oscillation frequency, peak frequency, and gain margin for the unity ratio feedback system with open loop transfer function

$$G(s) = \frac{4}{s\left(\dfrac{s}{10} + 1\right)^2}$$

Verify the approximate relations concerning performance specifications discussed thus far. (Suggestion—use a Nyquist diagram with the constant M circles superimposed.)

EXERCISE 4.6-3. Repeat EXERCISE 4.6-2 for

$$G(s) = \frac{5000(1 + 0.1s)^2}{s^3(1 + 0.004s)^2}$$

Damping Factor. The damping factor, or decrement factor as it is sometimes known, is the product of the damping ratio ζ and the natural frequency ω_n for a second order system. For a second order system there is an exponential decay term $e^{-\zeta\omega_n t}$ associated with the transient response and specification of the damping factor $\zeta\omega_n$ controls the envelope of the system transient response. The combination of the system damping factor and oscillation frequency ω_0 gives a rule of thumb guide concerning the number of oscillations in the system transient response. When $t\zeta\omega_n = 2$ the exponential e^{-2} has decayed to 13.5% of its initial value and in this time there will be $\dfrac{\omega_0}{\pi\zeta\omega_n}$ oscillations. For a second order system $\omega_0 = \sqrt{1 - \zeta^2}\,\omega_n$ and so we have about $\sqrt{1 - \zeta^2}/\pi\zeta$ discernible oscillations of the system transient response.

Time to Peak. For a second order system we can calculate the time to peak as in Eq. (4.2-7) and obtain $t_p = \pi/\omega_n\sqrt{1 - \zeta^2}$. Since an equivalent ω_n is not normally calculated for high order systems it is convenient to express this in terms of crossover frequency $\omega_c \approx \omega_0 = \omega_n\sqrt{1 - \zeta^2}$ as in EXAMPLE 4.6-1 such that we have the rule of thumb approximation for the system time to peak

$$t_p \approx \pi/\omega_c$$

Percentage Overshoot. We have previously obtained the percentage overshoot for simple second order systems with step function inputs. The peak response z_p for a two pole zero system with one open loop integration was shown to be given by Eq. (4.2-8) as

$$z_p = 1 + \exp\left(-\zeta\pi/\sqrt{1 - \zeta^2}\right)$$

Damping Ratio. The term ζ in the expression for the transfer function of the second order system

$$H(s) = \frac{\omega_n^2}{s^2 + 2\zeta\omega_n s + \omega_n^2}$$

is known as the damping ratio. Most systems are more complicated than a single second order system and so it is desirable to develop a rule of thumb approximation based on the damping ratio concept.

On a pole zero plot the damping ratio ζ is equal to $\sin\phi$ where ϕ is the angle between the vertical and the vector from the origin to the closed loop pole location on the locus of pole zero locations. Figure (4.2-1) illustrates this concept. There is a relationship between damping ratio and overshoot for a transient input but we must be careful to avoid the conclusion that no overshoot is possible unless $\zeta < 1$ for this is *not* correct. If there are no finite zeros then there will be no overshoot, or transient oscillations, if $\zeta > 0.707$ for all complex conjugate poles. Specification of a minimum allowable ζ will indeed control the transient type ringing oscillations but it will not necessarily control or eliminate the first overshoot. Damping ratios less than 0.5 are usually quite undesirable and use of the relation $\zeta = \sin\phi$ allows us to eliminate a region of the s plane as undesirable for closed loop pole locations. This will be of considerable value in our root locus efforts of Chap. 6.

A useful rule of thumb relationship between damping ratio and the closed loop gain at crossover $M_c = |H(j\omega_c)|$ can be obtained for the second order system of EXAMPLE 4.6-1. From the results of EXAMPLE 4.6-1 we have the asymptotic crossover frequency $\omega_c = \sqrt{K/T}$ and are thus able to evaluate the gain at crossover as

$$M_c = |H(j\omega_c)| = \sqrt{KT} = \frac{1}{2\zeta}$$

Thus we see that the damping ratio is approximately $1/(2M_c)$ where M_c is the closed loop system gain at the crossover frequency. For systems with low enough damping ratios to cause peaking in the amounts $1.2 < M_p < 1.6$, it also turns out that M_p is generally quite close to M_c and is actually given by $M_p = \dfrac{1}{2\zeta\sqrt{1-\zeta^2}}$. Thus we have the rule of thumb approximations

$$M_p \approx M_c \approx \frac{1}{2\zeta} \approx \frac{1}{\sin \text{PM}}$$

which relates the peak closed loop gain, the damping ratio, and the phase margin. Since we have rule of thumb relations between damping ratio ζ and other system specifications we may convert a system performance specified in terms of overshoot into phase margin or another criterion more appropriate for design purposes.

Final Value of Error. We have just completed a section concerning the final value of the system error for polynomial inputs. We can use this criterion to determine minimum gain values and the type of system required to yield specified maximum final steady state errors.

Rise Time. A number of different rise times have been defined for this response speed specification. The simplest definition for rise time is the reciprocal of the crossover frequency but this is perhaps best called the equivalent system time constant τ_{eq}.

$$\tau_{eq} = \frac{1}{\omega_c}$$

For a type one unity ratio system with $G(s) = K/s$, the crossover frequency is just $\omega_c = K$ and the exact time constant of the closed loop system which has a transfer function $H(s) = K/(s + K)$ is just $1/K$. For a first order system τ_{eq} is just the time required for the system response to a step input to reach 63.2% of its final value. If a high order system is reasonably well behaved $1/\omega_c$ is also a good rule of thumb indicator of the system *equivalent time constant*.

It is perhaps best to define rise time as the time required for the step response to rise from 0% to 90% of its final value. For the second order system response curve of Fig. (4.2-2), for say $\zeta = 0.7$, we see that the response reaches 0.9 at time $\omega_n t_r = 3$. The natural frequency ω_n is not very different from the crossover frequency for small ζ and so we have the rule of thumb approximation

$$t_r \approx \frac{3}{\omega_c}$$

Settling Time. We have previously defined the settling time for a second order system as the time required for the transient response to "settle" down within 5% of its final value. We derived the settling time $t_s = 3/\zeta\omega_n$ for a second order system in Eq. (4.2-9). When we use the approximate relations $\omega_0 = \omega_n\sqrt{1 - \zeta^2}$, we obtain

$$t_s = \frac{3\sqrt{1 - \zeta^2}}{\omega_0\zeta} \approx \frac{4\sqrt{1 - \zeta^2}}{\zeta\omega_c}$$

which we may use as a rule of thumb approximation for more complex systems.

Bandwidth. The bandwidth of a system is generally defined as the frequency at which the system frequency response is 0.707 of its value at zero frequency. For a first order system with open loop transfer function $G(s) = K/s$, the system bandwidth ω_b is just the crossover frequency $\omega_b = \omega_c = K$. For a second or higher order system the bandwidth is generally greater than the crossover frequency. From Fig. (4.2-5) and results determined in EXAMPLE 4.6-1 we see that a rule of thumb approximation for system bandwidth is

$$\omega_b \approx 1.55\omega_n \approx \frac{1.165\omega_c}{\sqrt{1-2\zeta^2}} \approx 1.165\omega_c$$

EXERCISE 4.6-4. Show that an exact expression for the bandwidth of a second order system with no finite zeros is given by

$$\omega_b = \omega_n \, [1 - 2\zeta^2 + (2 - 4\zeta^2 + 4\zeta^4)^{1/2}]^{1/2}$$

Integral of a Function of System Error. The performance criteria and rule of thumb approximations that we have obtained thus far are quite useful for trial and error design of linear systems. It is also possible for us to define a performance index such as the integral square

$$J_1 = \int_0^T e^2(t)\, dt$$

or the integral of time multiplied absolute value of error

$$J_2 = \int_0^T t|e(t)|\, dt$$

and adjust undetermined system parameters to seek a minimum of either of these performance indices. We will not develop this type of performance criteria here but merely wish to point out its existence. Chapter 7 will be devoted to a study of analytical design techniques which necessitate the use of performance measures such as these.

In this section we have considered a number of performance criteria and rule of thumb approximations. Both the performance criteria and rule of thumb approximations will be very useful to us in our next two chapters. We should emphasize that the performance criteria discussed here are perfectly general and applicable to linear systems of any order. The rule of thumb approximations are valid for step function inputs only in the case of the time domain criteria. Both time and frequency domain rule of thumb approximations are just approximations and assume reasonably behaved stable systems without excessive peaking (i.e. $M_p < 1.6$).

4.7 SUMMARY

In this chapter we have developed a number of system concepts that will be of considerable value in our study of linear systems control. A principal goal in our efforts was to develop a variety of performance criteria and rule of thumb approximations that will be useful in our future efforts. To do this we first looked in some detail at the performance of a simple first order system, a simple second order system with an open loop integration and no finite zeros, and a second order system with two open loop integrations and a single finite zero.

Following this we examined several classical methods for stability determination. As a byproduct of this we introduced some concepts concerning asymptotic gain and phase shift versus frequency plots (Bode plots) that will be of much value in actual system design. Finally, we examined static and dynamic error coefficients and presented a number of performance criteria and rule of thumb approximations that will allow us to convert user supplied specifications into specifications more suitable for actual design, the topic to which we now turn.

4.8 REFERENCES

There are a number of references to which the interested reader may turn for additional reading or supplementary material concerning the topics of this chapter.

A variety of linear systems texts discuss background material concerning the response of first and second order systems. The following references discuss not only response of simple systems but also present detailed tables or illustrations of the many interesting response-parameter variations and interactions.

Grabbe, E. M., Ramo, S., and Wooldridge, D. E., *Handbook of Automation Computation and Control*, Vol. I Control Fundamentals, John Wiley and Sons, Inc., New York, 1958.

Chestnut, H. and Mayer, R. W., *Servomechanisms and Regulating System Design*, Vol. I, John Wiley and Sons, Inc., New York, 1959.

Melsa, J. L. and Schultz, D. G., *Linear Control Systems*, McGraw-Hill Book Co., New York, 1969.

The above references also discuss the Routh Hurwitz and Nyquist stability criteria which we present here as well as performance measures and criteria. A very detailed discussion of stability criteria is contained in:

Kuo, B. C., *Automatic Control Systems*, Third Edition, Prentice-Hall, Inc., Englewood Cliffs, New Jersey, 1975.

For a simple discussion of Lyapunov stability, see

Sage, A. P. and White, C. C., III, *Optimum Systems Control*, 2nd Edition, Prentice-Hall, Inc., Englewood Cliffs, NJ, 1977.

4.9 PROBLEMS

1. Repeat the discussion of Sec. (4.2) for the unity ratio feedback system with transfer function $G(s) = \dfrac{K}{(T_1 s + 1)(T_2 s + 1)}$.

2. Repeat the discussion of Sec. (4.2) for the unity ratio feedback system with transfer function $G(s) = \dfrac{K(T_2 s + 1)}{s(T_1 s + 1)}$.

3. Determine stability requirements, using the Routh Hurwitz criterion, for the following $G(s)$ functions of a unity ratio feedback system

    (a) $G(s) = \dfrac{K(1 + T_1 s)^3}{s^3(1 + T_2 s)^3}$

    (b) $G(s) = \dfrac{K(s - 2)}{s(s + 4)}$

4. Determine a set of procedures which will enable you to discern whether a system with a major and a minor feedback loop is stable using the Nyquist criterion. Pattern your procedure to be especially useful for the system shown in Fig. P4.9-4. Construct a stability analysis guide in the form of a DELTA chart for this problem.

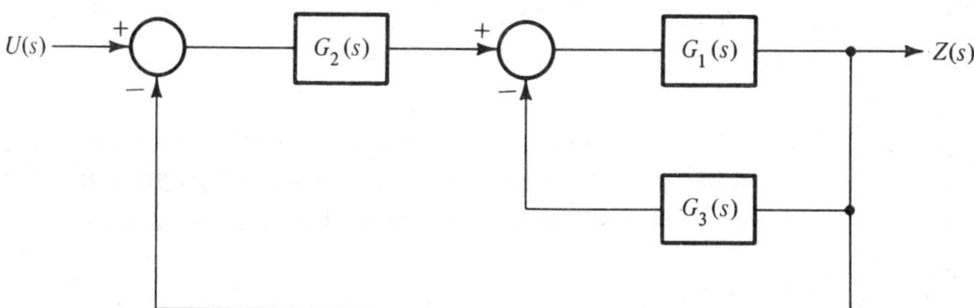

Figure P4.9-4   Block diagram for minor loop system within major loop

5. Use the procedure you have outlined in Problem 4.9-4 to determine stability of the system with

$$G_1(s) = \frac{1}{s^2(1 + 0.1s)}, \qquad G_3(s) = 1$$

and

(a) $G_2(s) = 10$

(b) $G_2(s) = 10(1 + s)$

check your results using the Routh Hurwitz criterion.

6. Reexamine Problem 4.9-4 from the viewpoints of

   (a) closing the minor loop first and then the major loop such that we analyze minor and major loops separately.

   (b) breaking both feedback loops simultaneously such as by breaking the loops at the output of subsystem $G_1(s)$.

   Contrast and compare these two points of view. Will they lead to similar Nyquist plots? to equivalent results?

7. Obtain results for Problem 4.9-6 using the transfer functions of Problem 4.9-5.

8. Determine stability, using the Nyquist diagram, for the unity ratio feedback system with

$$G(s) = \frac{100(1 + 0.1s)^2}{s^2(1 + 0.01s)^2}$$

   Obtain M_p and phase margin for this system. How good is the rule of thumb approximation $M_p = (\sin \mathrm{PM})^{-1}$ here? Determine an accurate Nyquist plot for this example with the M circles superimposed. How does this plot and $G(s)$ differ from other transfer functions considered in this chapter?

9. Determine the effect of a transport lag or time delay, which has a transfer function e^{-sT}, upon the static error coefficients.

10. Determine the dynamic error coefficients for the system of Fig. P4.9-4 with the transfer functions of Problem 4.9-5.

11. Accurately sketch Nyquist and Bode diagrams for the unity ratio feedback system with

$$G(s) = \frac{5}{s(1 + s)(1 + 0.1s)}$$

Determine all performance specifications and rule of thumb approximations discussed in Sec. (4.6).

12. The unity ratio feedback system with open loop transfer function

$$G(s) = \frac{\dfrac{3}{5}\left(\dfrac{7}{3}s + 1\right)}{s^2\left(\dfrac{s}{5} + 1\right)}$$

can be shown to have the closed loop transfer function

$$H(s) = \frac{\left(\dfrac{7}{3}s + 1\right)}{(s + 1)^2\left(\dfrac{s}{3} + 1\right)}$$

Determine and sketch accurately the step response of the system as well as a Bode and Nyquist diagram. Determine all performance specifications and rule of thumb approximations in Sec. (4.6) and compare with your determined step function response.

5

LINEAR SYSTEMS CONTROL DESIGN BY BODE DIAGRAMS

Our efforts thus far have been primarily concerned with analysis of linear control systems. We now turn our attention to system design. As much design involves trial and error repetition of analysis until a set of design specifications have been met, we will find our study of analysis methods to be most useful in the design process.

Figure (5.1) presents a DELTA chart of the design process and indicates the key role of analysis in linear systems control design. We will discuss one design method in this chapter—that based on Bode diagrams. We will discuss the use of both simple series equalizers and composite equalizers as well as the use of minor loop feedback in systems design. Our next chapter will discuss linear systems control design using the root locus approach and Chap. 7 will discuss analytical optimization techniques for the design of linear control systems. The DELTA chart of Fig. (5.1) is applicable to all three of the design methods presented in this text. There are five trial and error loops in the suggested design process. An experienced designer will often be able, primarily due to successful prior experience, to select a system structure and generic components such that specifications can be met with no or perhaps a very few iterations through the first iterative loop involving adjustment of equalizer or compensation parameters. Often an optimization approach can be used to select equalizer parameters to best meet specifications. If the optimization approach shows the specifications cannot be met, we are then assured that no equalizer of the form selected will meet specifications. The next design step, if needed, would consist of modification of the equalizer form or structure and repetition of the analysis process to determine equalizer parameter values to best meet specifications. If specifications still cannot be met, we will usually next modify generic fixed components used in the system such as selecting larger drive motors. The design process of iterative analysis is again repeated. If no reasonable fixed components can be obtained to meet specifications, then structural changes in the proposed system are next contemplated. If no structure can be found which allows satsifaction of specifications, the client must be requested to either relax the specifications or the project may be rejected as infeasible using present technology. As we might suspect, economics will play a dominant role in this design process. Changes made due to iteration in the inner loops of the DELTA chart of Fig. (5.1) normally involve little additional costs whereas those made due to iterations in the outer loops will often involve major cost changes.

5.1 THE BODE DIAGRAM

We have introduced the Bode diagram in our previous efforts. Since this diagram plays such a central role in our efforts in this chapter, we will now present a brief summary of the central features of this very useful diagram.

The steady state response of a stable linear constant coefficient system has particular significance as we know from many previous courses in electrical networks and circuits, and dynamics. We consider a stable linear system with transfer function

$$\frac{Z(s)}{U(s)} = H(s) \tag{5.1-1}$$

and assume a sinusoidal input $u(t) = \cos \omega t$ so that we have

$$Z(s) = \frac{sH(s)}{s^2 + \omega^2} \tag{5.1-2}$$

We expand this ratio using partial fractions to obtain

$$Z(s) = F(s) + \frac{a_1}{s + j\omega} + \frac{a_2}{s - j\omega} \tag{5.1-3}$$

Where $F(s)$ contains all the poles of $H(s)$ all of which lie in the left half plane since the system, represented by $H(s)$, is assumed stable. The coefficients a_1 and a_2 are determined as

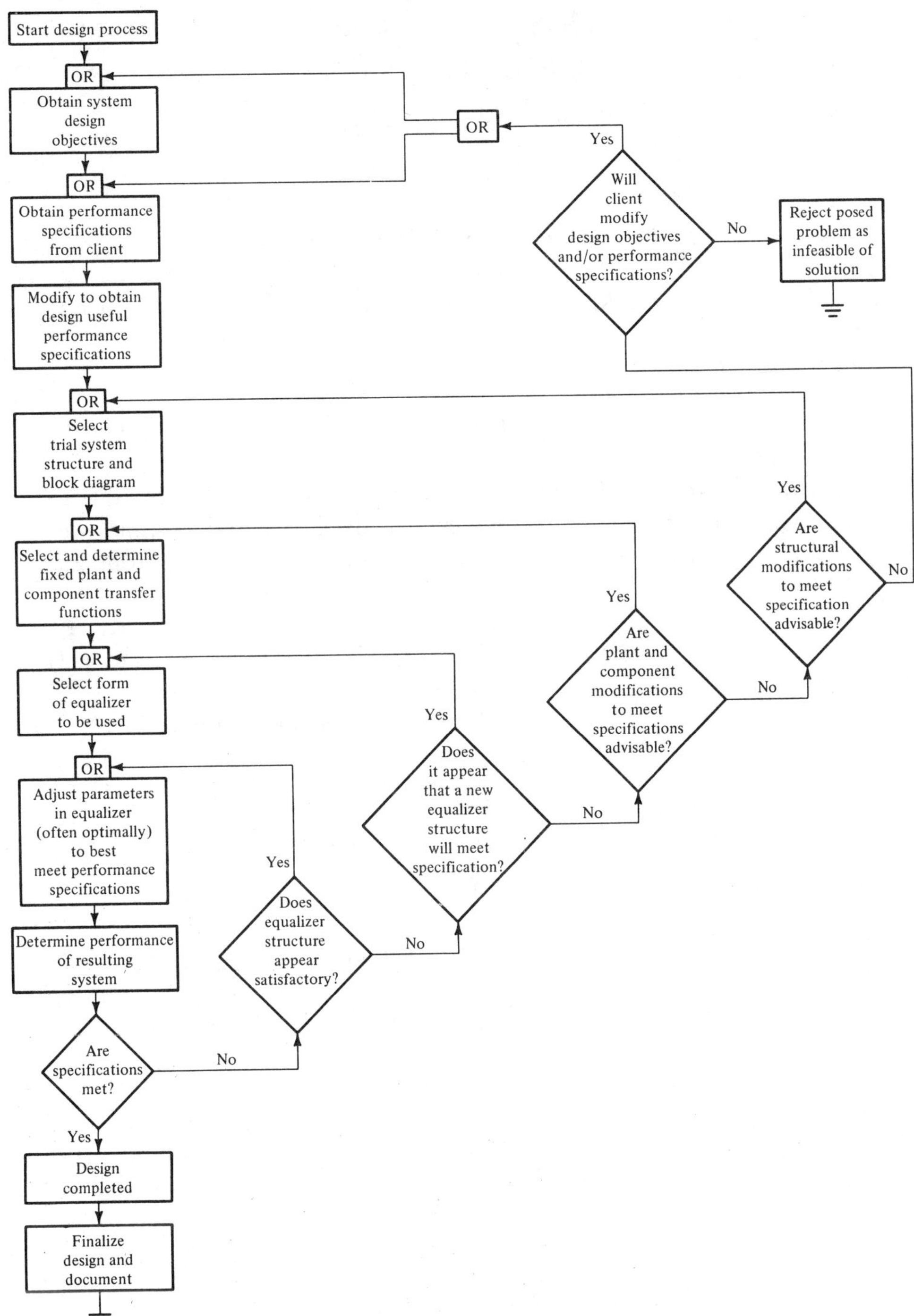

Figure 5.1   The linear systems control design process

$$a_1 = \frac{s\,H(s)}{s - j\omega}\bigg|_{s=-j\omega} = \frac{H(-j\omega)}{2}$$

$$a_2 = \frac{s\,H(s)}{s + j\omega}\bigg|_{s=j\omega} = \frac{H(j\omega)}{2}$$

$$(5.1\text{-}4)$$

We can represent the complex transfer function $H(j\omega)$ as

$$H(j\omega) = B(\omega) + jC(\omega) \tag{5.1-5}$$

Also we will have

$$H(-j\omega) = B(\omega) - jC(\omega) \tag{5.1-6}$$

The inverse transform of Eq. (5.1-3) will yield a transient term, due to the inverse transform of $F(s)$, which will decay to zero as time progresses. A steady state component will remain and this is, from Eq. (5.1-3)

$$z(t) = a_1\,e^{-j\omega t} + a_2\,e^{j\omega t} \tag{5.1-7}$$

We substitute Eqs. (5.1-4)–(5.1-6) into the foregoing and obtain

$$z(t) = B(\omega)\left[\frac{e^{j\omega t} + e^{-j\omega t}}{2}\right] - C(\omega)\left[\frac{e^{j\omega t} - e^{-j\omega t}}{2j}\right]$$

which becomes using the Euler identity*

$$z(t) = B(\omega)\cos \omega t - C(\omega)\sin \omega t \tag{5.1-8}$$

$$= [B^2(\omega) + C^2(\omega)]^{1/2}\cos(\omega t + \beta)$$

$$= |H(j\omega)|\cos(\omega t + \beta)$$

---

*The Euler identity $e^{jx} = \cos x + j \sin x$ is a very useful one and leads immediately to very helpful trigonometric identities such as

$$\sin x = \frac{e^{jx} - e^{-jx}}{2j}, \qquad \cos x = \frac{e^{jx} + e^{-jx}}{2}$$

$$\cos(x + y) = \cos x \cos y - \sin x \sin y$$

$$\sin(x + y) = \sin x \cos y + \cos x \sin y$$

where

$$\tan \beta(\omega) = C(\omega)/B(\omega) \qquad (5.1\text{-}9)$$

As we see, there is a very direct relationship between the transfer function of a linear constant coefficient system, the time response of a system to any known input, and the sinusoidal steady state response of the system. We can always determine any of these if we are given but one of them. This important conclusion justifies a design procedure for linear systems based only on sinusoidal steady state response as it is possible to determine transient response from steady state sinusoidal response.

The transfer function of Eq. (5.1-1) can be written as a ratio of zeros (numerator root locations in the s plane) and poles (denominator root locations in the s plane) as

$$H(s) = \frac{N(s)}{D(s)} = \frac{\displaystyle\prod_{i=1}^{m}\left(1 + \frac{s}{e_i}\right)K}{\displaystyle\prod_{i=1}^{n}\left(1 + \frac{s}{d_i}\right)} \qquad (5.1\text{-}10)$$

where some of the e_i and d_i may appear in complex conjugate pairs. It is easier to obtain asymptotic approximations to Eq. (5.1-10) for $s = j\omega$ than it is to obtain exact expressions. First we will replace all poles and zeros d_i and e_i with $|d_i|$ and $|e_i|$ such that the resulting expression has poles and zeros on the negative real axis. This is equivalent to regarding all d_i and e_i terms in Eq. (5.1-10) as being positive and real. Next we take the logarithm of the magnitude of Eq. (5.1-10) to obtain

$$\log |H(j\omega)| = \log K + \sum_{i=1}^{m} \log \left|1 + \frac{j\omega}{e_i}\right| - \sum_{i=1}^{n} \log \left|1 + \frac{j\omega}{d_i}\right| \qquad (5.1\text{-}11)$$

Now we consider a typical term in Eq. (5.1-11) and use either of two asymptotic approximations

$$\log \left|1 + \frac{j\omega}{p_i}\right| \cong \begin{cases} 0 & \omega < p_i \\[2mm] \log \dfrac{\omega}{p_i} & \omega > p_i \end{cases} \qquad (5.1\text{-}12)$$

and we see that this typical term will graph as a straight line (in logarithmic coordinates). Thus we see that the transfer function of Eq.

(5.1-11), or the original transfer function of Eq. (5.1-1) or (5.1-10) will also graph as a straight line if we use logarithmic coordinates and the asymptotic approximation of Eq. (5.1-12). We have previously obtained this result and have shown it to be a valid approximation at frequencies slightly distant from the break frequencies (pole and zero locations) and usually not a bad approximation even at the break frequencies. We have seen how we can correct the asymptotic plot to make it exact and have indicated that this will often not be necessary due to the generally small errors involved in the asymptotic gain plot.

We have introduced two approximations for the phase shift equation given by Eq. (5.1-9). From Eq. (5.1-10) we have

$$\beta = \sum_{i=1}^{m} \tan^{-1} \frac{\omega}{e_i} - \sum_{i=1}^{n} \tan^{-1} \frac{\omega}{d_i} \tag{5.1-13}$$

We can use the analytical arctangent approximation which yields for a typical term in Eq. (5.1-13)

$$\tan^{-1} \frac{\omega}{\alpha} \simeq \begin{cases} \dfrac{\omega}{\alpha} & \omega < \alpha \\[2ex] \dfrac{\pi}{2} - \dfrac{\alpha}{\omega} & \omega > \alpha \end{cases} \tag{5.1-14}$$

We have seen that errors using this approximation will generally be small. It is this approximation that we will use to develop our Bode diagram design procedure. Alternately we can draw straight line phase shift diagrams, on logarithmic frequency coordinates, by assuming for the typical term of Eq. (5.1-14) that

$$\tan^{-1} \frac{\omega}{\alpha} \cong \begin{cases} 0 & \omega < \alpha/a \\[1.5ex] \dfrac{\pi}{4} & \omega = \alpha \\[1.5ex] \dfrac{\pi}{2} & \omega > a\alpha \end{cases} \tag{5.1-15}$$

and drawing a straight line segment for $\dfrac{\alpha}{a} < \omega < a\alpha$. The value of a can be chosen according to several criteria. $a = 10$ is approximately the value of a which yields the best fit in a least square error sense whereas $a = 4.81$ is the value of a which results in the same slope for the actual arc tangent curve and the asymptotic and tangent curve at frequency α.

EXERCISE 5.1-1. Demonstrate the validity of Eq. (5.1-13).

EXERCISE 5.1-2. Show that $a = 4.81$ is indeed the value of a in Eq. (5.1-15) which results in $\dfrac{d\beta}{d(\ln \omega)}\Big|_{\omega=\alpha} = 1/2$ for both the actual phase shift curve and the straight line approximation.

EXERCISE 5.1-3. Determine the asymptotic gain relation and the asymptotic phase relations for the transfer functions given. Plot these and compare with exact values.

$$\text{(a)} \quad H(s) = \frac{(1 + 0.1s)}{(1 + 0.01s)(1 + 0.001s)}$$

$$\text{(b)} \quad H(s) = \frac{100}{s\left(1 + \dfrac{0.8s}{10} + \dfrac{s^2}{100}\right)}$$

5.2 BODE DIAGRAM DESIGN—SERIES EQUALIZERS

In this section we consider three types of series equalization:

1. Gain adjustment, normally attenuation by a constant at all frequencies.
2. Increasing the phase lead, or reducing the phase lag, at the crossover frequency by use of a phase lead network.
3. Attenuation of the gain at middle and high frequencies such that the crossover frequency will be decreased to a lower value where the phase lag is less, by use of a lag network.

In the following section we will first consider use of a composite lead lag network near crossover to attenuate gain only in order to reduce the crossover frequency to a value where the phase lag is less. Then we will consider more complex composite equalizers and state some general guidelines for Bode diagram design. We will use Bode diagram techniques to develop a design procedure for each of three types of series equalization in this section. We will generally develop these design approaches using specific examples first and will then state general features of a design procedure.

5.2-1 Gain Reduction

The majority of linear control systems can be made sufficiently stable merely by reduction of the open loop system gain to a sufficiently low value. This approach ignores all performance specifications however, except that of phase margin and is, therefore, usually not a satisfactory approach. It is a very simple one, however, and serves to illustrate the approach to be taken in more complex cases.

EXAMPLE 5.2-1. We consider the system illustrated in Fig. (5.2-1) which has an open loop transfer function

$$G(s) = \frac{10,000}{s\left(1 + \dfrac{s}{100}\right)^2} \tag{1}$$

From Fig. (5.2-1), or use of the asymptotic gain approximation for $\omega_c > 100$

$$|G(\omega)| \cong \frac{10^8}{\omega^3} \tag{2}$$

we have the crossover frequency

$$\omega_c \cong 10^{8/3} = 215$$

The phase shift at the crossover frequency, obtained as

$$\beta_c = \frac{-\pi}{2} - 2 \tan^{-1} \frac{\omega_c}{100} \simeq \frac{-\pi}{2} - 2\left(\frac{\pi}{2} - \frac{100}{\omega_c}\right) \tag{3}$$

$$= -3.78 \text{ radians} = -216.70 \text{ degrees}$$

indicates that the closed loop system is unstable. The phase shift at $\omega = 100$ is $-\pi$ radians or $-180°$ so that any crossover frequency greater than 100 radians per second is bound to result in an unstable system.

Suppose that we desire a phase margin of $45° = \pi/4$ radians. The crossover frequency must be less than 100 radians per second such that the use of the appropriate arctangent approximation yields

$$\beta = \frac{-\pi}{2} - 2 \tan^{-1} \frac{\omega}{100} = \frac{-\pi}{2} - \frac{2\omega}{100}$$

For a 45° phase margin we have

$$\beta_c = -\frac{3\pi}{4} = \frac{-\pi}{2} - \frac{2\omega_c}{100}$$

and find the crossover frequency is $\omega_c = 39.27$. The asymptotic gain expression for $G(s)$ which is valid for $\omega < 100$ is given by

$$|G(j\omega)| = \frac{10,000\,\alpha}{\omega}$$

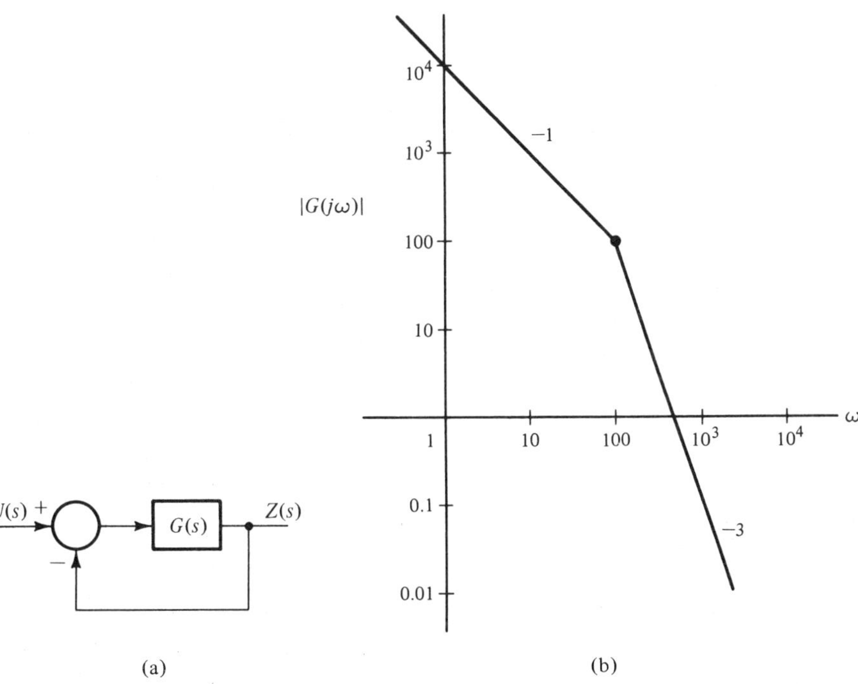

Figure 5.2-1 Block diagram and open loop Bode diagram for system with open loop transfer function

$$G(s) = \frac{10,000}{s\left(1 + \dfrac{s}{100}\right)^2}$$

where α represents the attenuation to be inserted to give a crossover frequency of 39.27 radians per second. We have

$$1 = \frac{10{,}000\,\alpha}{39.27}$$

and so see that we must introduce the rather large attenuation of 254.7 such that $\alpha = 0.00392$. In this example we have no real control over the closed loop system bandwidth or the crossover frequency and the system response may well be sluggish.

The following steps constitute an appropriate Bode diagram design procedure for compensation by gain adjustment:

1. Determine the required phase margin PM and the corresponding phase shift $\beta_c = -\pi + \text{PM}$.

2. Determine the frequency ω_c at which the phase shift is such as to yield the phase shift at crossover required to give the desired PM.

3. Adjust the gain such that the actual crossover frequency occurs at the value computed in step 2.

EXERCISE 5.2-1. Find the value of the gain K which will result in a $45°$ phase margin for the single loop system with open loop transfer function

$$G(s) = \frac{K}{s\left(1 + \dfrac{s}{10}\right)\left(1 + \dfrac{s}{100}\right)}$$

5.2-2 Phase Lead Compensation

In compensation using a phase lead network we increase the phase lead at the crossover frequency such that we meet a performance specification concerning phase shift. A phase lead compensating network transfer function is

$$G_c(s) = \frac{1 + s/\omega_1}{1 + s/\omega_2} \qquad \omega_1 < \omega_2 \qquad\qquad (5.2\text{-}1)$$

Figure (5.2-2) illustrates the gain versus frequency and phase versus frequency curves for a simple lead network with the transfer function of Eq. (5.2-1). The maximum phase lead obtainable from a phase lead network depends upon the ratio ω_2/ω_1 used in designing the network. From the expression for the phase shift of the transfer function of Eq. (5.2-1)

$$\beta = \tan^{-1}\frac{\omega}{\omega_1} - \tan^{-1}\frac{\omega}{\omega_2} \tag{5.2-2}$$

we see that maximum phase lead* occurs at

$$\left.\frac{d\beta}{d\omega}\right|_{\omega=\omega_m} = 0 = \frac{1}{\omega_1\left[1+\left(\dfrac{\omega}{\omega_1}\right)^2\right]} - \frac{1}{\omega_2\left[1+\left(\dfrac{\omega}{\omega_2}\right)^2\right]}$$

or

$$\omega_m = \sqrt{\omega_1\omega_2} \tag{5.2-3}$$

which is at the center of the two break frequencies for the lead network on a Bode log asymptotic gain plot. It is interesting to note that this is exactly the same frequency that we would obtain using the arctangent approximation with the assumption that $\omega_1 < \omega < \omega_2$. The maximum value of the phase shift obtained at $\omega = \omega_m$ is

$$\beta_m(\omega_m) = \tan^{-1}\sqrt{\omega_2/\omega_1} - \tan^{-1}\sqrt{\omega_1/\omega_2} \tag{5.2-4}$$

$$= \frac{\pi}{2} - 2\tan^{-1}\sqrt{\omega_1/\omega_2} = -\frac{\pi}{2} + 2\tan^{-1}\sqrt{\omega_2/\omega_1}$$

*Recall that if

$$x = \tan^{-1}\frac{\omega}{a}$$

then

$$\frac{dx}{d\omega} = \frac{1/a}{1+\dfrac{\omega^2}{a^2}}$$

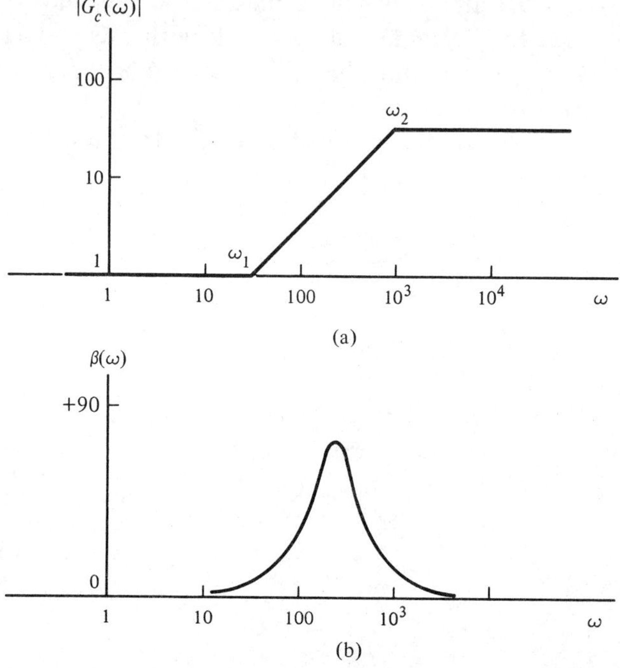

Figure 5.2-2 Gain and phase shift curves for a simple lead network

which can be approximated in more useable form using the arctangent approximation as

$$\beta_m(\omega_m) \simeq \frac{\pi}{2} - 2\sqrt{\omega_1/\omega_2} \qquad (5.2\text{-}5)$$

The gain magnitude of the lead network at the frequency of maximum phase shift can be computed from

$$|G_c(\omega_m)| = \sqrt{\frac{1 + \omega_m^2/\omega_1^2}{1 + \omega_m^2/\omega_2^2}} = \sqrt{\omega_2/\omega_1} \qquad (5.2\text{-}6)$$

or from the asymptotic gain approximation

$$|G_c(\omega_m)| \simeq \omega_m/\omega_1 = \sqrt{\omega_2/\omega_1} \qquad (5.2\text{-}7)$$

It will sometimes occur that about 45 degrees or $\pi/4$ radians of phase lead must be inserted by a lead network. This yields a ratio $\omega_1/\omega_2 =$

$(0.1542) = (\pi/8)^2$ and corresponds to a gain $|G(j\omega_m)| = 8/\pi = 2.55$. Thus we see that the crossover frequency will generally be raised by using a lead network since the gain at the frequency of maximum phase shift is greater than 1. This must be considered in our design procedure if we are designing the system for a specified closed loop bandwidth or open loop crossover frequency. Usually rule of thumb approximations are used to do this as exact approaches are generally cumbersome. Figure (5.2-3) presents a graph of gain magnitude obtained at the frequency maximum phase shift and the amount of the maximum phase shift for the simple lead network of Eq. (5.2-1).

There are many ways of realizing a lead network. All methods require the use of an active element since the gain of the lead network at high frequencies is greater than one. A simple electrical network realization is that shown in Fig. (5.2-4).

It appears simplest to indicate our suggested design procedure using lead networks by means of a simple example.

EXAMPLE 5.2-2. Suppose that we specify a type 2 system with a crossover frequency of 100 radians per second and a phase margin of $\pi/4$ radians. We assume that we have an open loop system with transfer function

$$G_f(s) = \frac{10^4}{s^2} \tag{1}$$

which is a type 2 system, such that there will be zero steady state error for velocity input $u(t) = t$, with a crossover frequency of 100 radians per second. The phase margin is zero for the uncompensated system and we attempt compensation by means of a lead network.

We first design the system without regard for a (slight) increase in crossover frequency. We assume that we wish to retain the acceleration error coefficient K_a of 10^4. For a variety of reasons it is usually desirable to obtain maximum phase shift at the crossover frequency. In so doing we will use the lead network with the smallest ω_2/ω_1 ratio and therefore the smallest additional gain will be required. Also we could obtain a greater phase margin by (slight) gain adjustment if we do not have maximum phase shift at crossover. Usually it is desirable that we design a system for the maximum phase margin possible under the constraints of the system configuration assumed. Thus we will use as an additional design objective the use of a lead network such that the phase shift at crossover is the maximum possible.

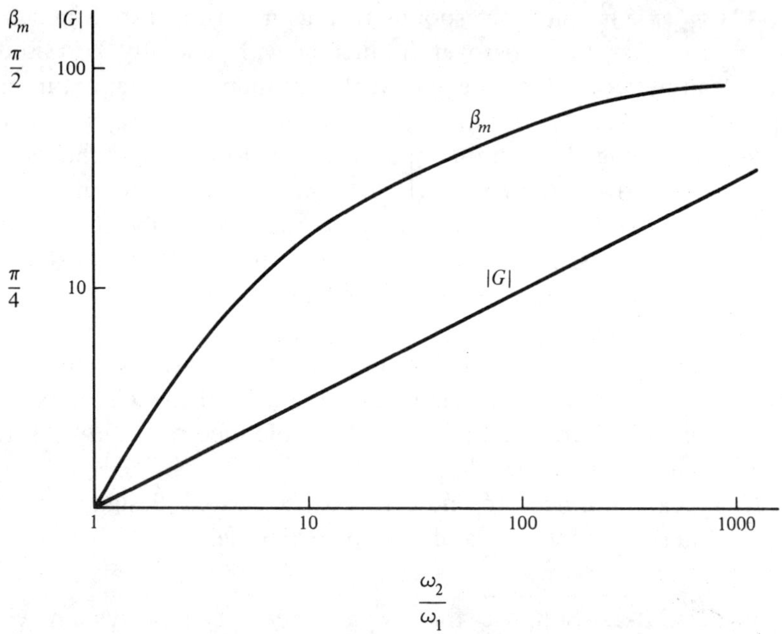

Figure 5.2-3   Design chart showing gain ratio increases and phase shift (lead) as functions of ω_2/ω_1 for simple lead network

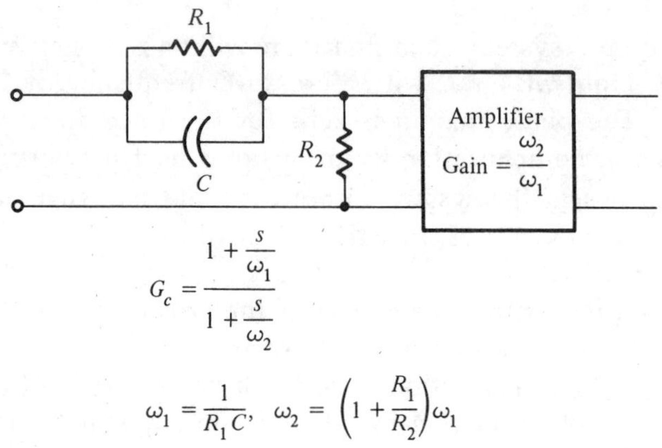

Figure 5.2-4   A simple electrical lead network

The asymptotic gain diagram for this example is as shown in Fig. (5.2-5). We have an open loop transfer function

$$G(s) = \frac{10^4}{s^2} \frac{1 + \dfrac{s}{\omega_1}}{1 + \dfrac{s}{\omega_2}} \tag{2}$$

and wish to select the break frequencies ω_1 and ω_2 such that the phase shift at crossover is maximum and, further, we want this maximum phase shift to be $-3\pi/4$ such that we have a $\pi/4$ radian or $45°$ phase margin.

Since the crossover frequency is such that $\omega_1 < \omega_c < \omega_2$ we have for the arc tangent approximation to the phase shift in the vicinity of crossover

$$\beta(\omega) = -\pi + \tan^{-1}\frac{\omega}{\omega_1} - \tan^{-1}\frac{\omega}{\omega_2} \tag{3}$$

$$\cong \frac{-\pi}{2} - \frac{\omega_1}{\omega} - \frac{\omega}{\omega_2}$$

To maximize the phase shift at crossover we set

$$\frac{d\beta(\omega)}{d\omega}\bigg|_{\omega=\omega_c} = 0$$

and obtain

$$\omega_c = \sqrt{\omega_1 \omega_2} \tag{4}$$

The crossover frequency is thus half way between the two break frequencies ω_1 and ω_2 on a logarithmic frequency coordinate. The phase shift at this optimum value of crossover frequency becomes

$$\beta_c = \beta(\omega_c) = \frac{-\pi}{2} - 2\sqrt{\omega_1/\omega_2}$$

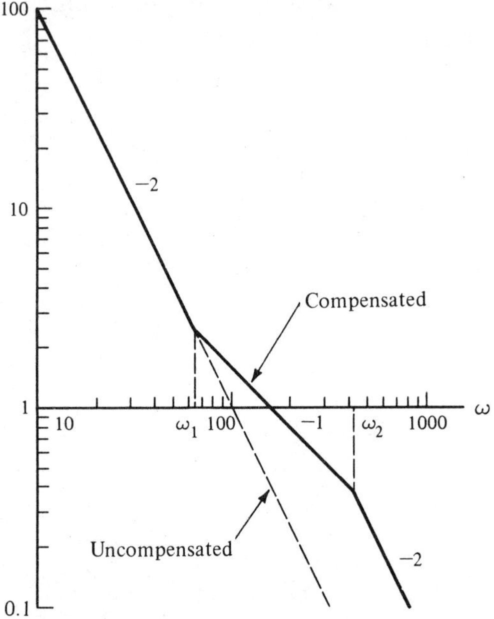

Figure 5.2-5 Bode diagram of uncompensated and compensated system

For a phase margin of $-3\pi/4$ we have

$$\frac{-3\pi}{4} = -\frac{\pi}{2} - 2\sqrt{\omega_1/\omega_2}$$

and

$$\frac{\omega_1}{\omega_2} = 0.1542 \tag{5}$$

We see that we have need for a lead network with a gain of $\omega_2/\omega_1 = 6.485$. The gain at the crossover frequency is 1 and from the asymptotic approximation to Eq. (2) for $\omega_1 < \omega < \omega_2$ we have

$$|G(j\omega)| = \frac{10^4}{\omega\omega_1}$$

$$|G(j\omega_c)| = 1 = \frac{10^4}{\omega_c\omega_1} \tag{6}$$

Solving Eqs. (4), (5), and (6) for the lead network equalizer parameters we obtain

$$\omega_c = 159.58 \qquad \omega_1 = 62.66 \qquad \omega_2 = 406.37$$

We note that we have increased the crossover frequency to 159.58 by the design procedure we have employed here. This will result in a system with the required degree of stability but with a faster speed of response, due to the crossover frequency, which is higher than that specified. Figure (5.2-5) illustrates the open loop Bode diagram for this system which has the transfer function

$$G(s) = \frac{10^4}{s^2} \; \frac{1 + \dfrac{s}{62.66}}{1 + \dfrac{s}{406.37}} \tag{7}$$

We should note that the use of two decimal digit accuracy for the lead network equalizer is not warranted. The phase margin criterion is an approximate one only and the original transfer function for the fixed plant

$$G_f(s) = \frac{10^4}{s^2}$$

is probably not known with two decimal digit accuracy. Thus it is entirely reasonable to write the compensating network transfer function as

$$G_c(s) = \frac{1 + \dfrac{s}{60}}{1 + \dfrac{s}{400}}$$

such that the overall open loop transfer function becomes

$$G(s) = G_f(s)G_c(s) = \frac{10^4}{s^2} \; \frac{1 + \dfrac{s}{60}}{1 + \dfrac{s}{400}}$$

If we wish to retain the $\omega_c = 100$ specification and are unconcerned with retaining the acceleration error coefficient of 10^4, there are several modified approaches which could be taken. We could note from Fig. (5.2-3) or Eq. (5.2-6) that a gain of $(6.485)^{1/2}$ occurs at the crossover frequency of a lead network compensating network with a $\omega_2/\omega_1 = 6.485$. This is the break point frequency ratio to yield a phase lead of $\beta_m = \pi/4$, the amount required here to obtain a phase margin of $\pi/4$. Thus we should reduce the gain of the open loop system by $(6.485)^{1/2}$, so that we have

$$G(s) = G_f(s)G_c(s) = \frac{10^4}{2.547\,s^2}\,\frac{1 + \dfrac{s}{\omega_1}}{1 + \dfrac{s}{\omega_2}} \tag{8}$$

To get the $\omega_c = 100$ crossover frequency we use $|G(j\omega_c)| = 1$ and obtain $\omega_1 = 39.27$ from the foregoing. This completes the equalizer design as we obtain for the open loop transfer function

$$G(s) = \frac{3927}{s^2}\,\frac{1 + \dfrac{s}{39.27}}{1 + \dfrac{s}{254.67}} \tag{9}$$

Another procedure which works quite satisfactorily consists of specifying an open loop transfer function of the required form which has been adjusted such that: the crossover frequency occurs at 100, crossover frequency occurs midway between the two break frequencies, and the ratio of the break frequencies is such that the required phase margin is obtained. Figure (5.2-6) illustrates the Bode diagram for this procedure. The asymptotic gain near the crossover frequency is obtained from

$$G(s) = \frac{K}{s^2}\,\frac{1 + s/\omega_1}{1 + s/\omega_2}$$

as

$$|G(j\omega)| = \frac{100}{\omega^2} = \frac{K}{\omega_1\omega} \tag{10}$$

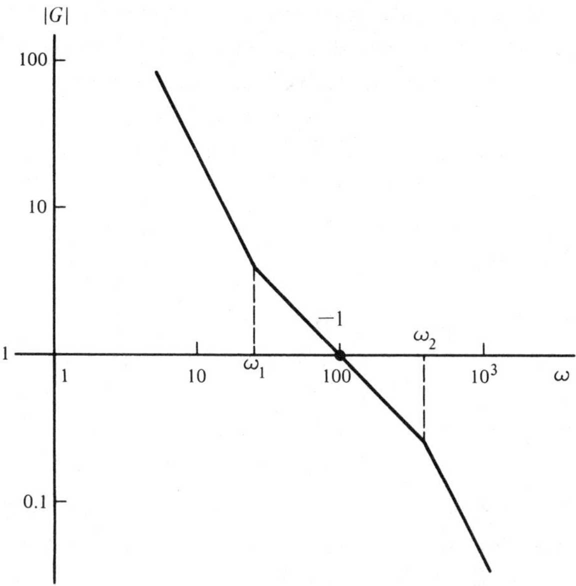

Figure 5.2-6 Partial Bode diagram of compensated system

and the phase shift equation is obtained from the arctangent approximation as

$$\beta(\omega) \cong -\frac{\pi}{2} - \frac{\omega_1}{\omega} - \frac{\omega}{\omega_2}$$

To maximize the phase shift at crossover we have, from $\partial\beta(\omega_c)/\partial\omega_c = 0$,

$$\omega_c = \sqrt{\omega_1 \omega_2} \tag{11}$$

so we have by combining the foregoing two relations,

$$\beta(\omega_c) = \frac{-\pi}{2} - 2\sqrt{\omega_1/\omega_2} = \frac{-3\pi}{4} \tag{12}$$

We obtain from solution of Eqs. (10) through (12)

$$K = 3927 \qquad \omega_c = 100 \qquad \omega_1 = 39.27 \qquad \omega_2 = 254.67$$

and the design is again completed.

It may appear that we have obtained two different lead network parameter designs for the same problem specifications but this is not the case. Our first set of lead network equalizer break frequencies was obtained for the specification that the acceleration error coefficient remain 10^4 whereas the second design was conducted using the specification that the crossover frequency remain at 100. The first design resulted in a crossover frequency of 159.58 radians per second whereas the second resulted in an acceleration error coefficient of 3926.19. We cannot say which design is "best" since we really did not state any design objectives or specifications. The first system has a larger bandwidth and acceleration error coefficient, but this does not necessarily mean that it is the best system since we have, in this example, no standards against which to make such a decision.

EXAMPLE 5.2-3. In this example we will introduce performance specifications on the closed loop system and design a lead network in order to meet these specifications. We assume the fixed uncompensated open loop system has the transfer function

$$G_f(s) = \frac{1000}{s(1 + s/100)} \tag{1}$$

and that we desire a closed loop system that meets the following design specifications:

1. The velocity error constant (K_v) is 1000.
2. Sinusoidal inputs at any frequency up to 10 radians per second must be reproduced with an error no greater than 2%.
3. The phase margin is 45°.

First we will see what changes need to be made to satisfy specifications 1 and 2. The Bode diagram for the fixed plant is as shown in Fig. (5.2-7). The present velocity error coefficient is 1000 and we do not need to change the low frequency gain. In order to have an error transfer function

$$\frac{E(s)}{U(s)} = \frac{1}{1 + G_f(s)G_c(s)} \tag{2}$$

with a magnitude less than 0.02 for all frequencies less than 10 radians per second we see that the magnitude of the open loop transfer func-

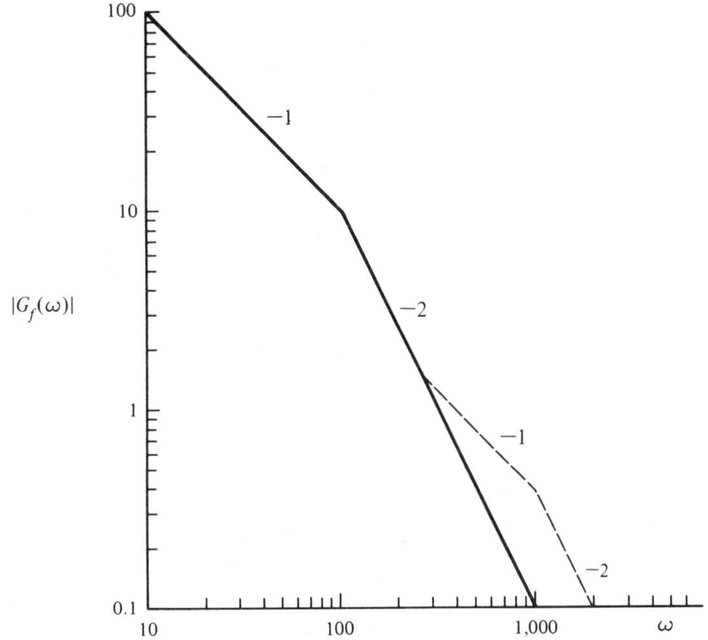

Figure 5.2-7 Bode diagram of open loop system for EXAMPLE 5.2-3

tion $|G_f(s)G_c(s)|$ must be greater than 50 for all frequencies less than 10. We obtain this number by noting that where $|G_f(s)G_c(s)| > 1$ we can drop the one term in Eq. (2) to obtain the approximation

$$\frac{E(s)}{U(s)} \cong \frac{1}{G_f(s)G_c(s)} \tag{3}$$

for all frequencies where the gain is high and the error low. Use of Eq. (3) and the sinusoidal error criterion immediately leads to the open loop gain requirement that the gain for all frequencies less than 10 be greater than 50. We note that the open loop gain at a frequency of 10 radians per second is now 100 and thus can not be reduced except by a very small factor. So we must use a network which does not decrease the low frequency gain and this leads us to consider a lead network as a compensatory network. Some sort of equalization must be used since the phase margin for the uncompensated system is only 17.88°. No high frequency specification is given which would require restricting the gain at high frequencies. Also the system is of relatively low order and we do not get major increases in phase shift for the uncompensated system at high frequencies. These two reasons further justify our choice of a lead network.

The crossover frequency for the uncompensated system occurs at 316.23 radians per second and the phase shift at this frequency is $-162.12°$. To achieve a phase margin of $45°$ at this frequency we would have to add about $27.12°$ of phase lead. Of course the crossover frequency will be increased by use of the lead network. We assume that the first break of the lead network equalizer occurs at a frequency greater than 100 radians per second such that the general shape of the compensated system open loop gain curve is as shown by the dotted line in Fig. (5.2-7). We desire to determine the two break frequencies ω_1 and ω_2 such that we obtain a $45°$ phase margin and the ratio ω_2/ω_1 is the minimum possible in order that we may use the smallest amount of gain to realize the lead network.

Several design approaches are possible. Most of these are very similar. One popular approach starts with the realization that $27.12°$ of additional phase lead would be needed if the crossover frequency were not increased. The crossover frequency will be increased however and there will be additional phase lag due to the pole at $s = -100$. Thus we might attempt a design to yield say $30°$ of phase shift due to the lead network. The maximum phase shift occurs as we have seen at the geometric mean of the break frequencies ω_1 and ω_2, so we use

$$\omega_c = \sqrt{\omega_1 \omega_2} \tag{4}$$

The transfer function of the compensated system is

$$G_f(s)G_c(s) = \frac{10^3}{s(1 + s/100)} \cdot \frac{1 + s/\omega_1}{1 + s/\omega_2} \tag{5}$$

Since the crossover occurs on the second -1 slope we have in the vicinity of crossover

$$|G_f(\omega)G_c(\omega)| \cong \frac{10^5}{\omega_1 \omega} \tag{6}$$

and

$$1 = \frac{10^5}{\omega_1 \omega_c} \tag{7}$$

We wish to obtain $30° = \pi/6$ radians phase lead from the equalizer network

$$G_c(s) = \frac{1 + s/\omega_1}{1 + s/\omega_2} \tag{8}$$

at the crossover frequency. This phase shift may be determined from Eq. (5.2-4) as

$$\beta(\omega_c) = \frac{\pi}{2} - \tan^{-1} \sqrt{\omega_1/\omega_2} \tag{9}$$

so we see that for $\beta(\omega_c) = \pi/6$ we have

$$\frac{\omega_2}{\omega_1} = 3.0 \tag{10}$$

Alternately we could have used the arctangent approximation of Eq. (5.2-5) to obtain $\omega_2/\omega_1 = 3.65$. Simultaneous solution of Eqs. (4), (7), and (10) leads to the specifications

$$\omega_1 = 240.28 \qquad \omega_2 = 720.84 \qquad \omega_c = 416.18$$

Use of the ratio $\omega_2/\omega_1 = 3.65$ obtained from the arctangent approximation leads to the specifications

$$\omega_1 = 228.78 \qquad \omega_2 = 835.06 \qquad \omega_c = 437.09$$

Since the crossover frequency was not known prior to making our assumption that $30°$ additional phase lead was needed, we should now check to see that the design does lead to a phase margin that meets specifications. Using the parameters obtained from the ratio $\omega_2/\omega_1 = 3.0$ we obtain a phase margin of $43.51°$ whereas using the parameters obtained from the ratio $\omega_2/\omega_1 = 3.65$ results in a system phase margin of $47.63°$. Curiously enough, the design using the arctangent approximation meets and exceeds the phase margin specifications whereas that obtained using the exact relationship of Eq. (5.2-5) fails, although only by a slight amount, to meet specifications.

A problem with this particular design approach, which is a common approach, is that we had to guess the additional amount of phase shift needed by the lead network due to an unknown increase in the crossover frequency that results from using the lead network. We now present a slightly different and more exact approach in which we design directly for the required phase margin. As we will see, the exact equations for the phase shift are transcendental and not easily solved. Thus we must use the arctangent approximation to easily obtain the design parameters. However, the arctangent approximation is entirely consistent, in terms of accuracy, with the asymptotic gain approximation. Since the method we have just presented uses the asymptotic gain approximation (and in fact use also of the arctangent approximation for this example leads to a better design!) we believe the design method to be presented, which designs directly in a specified phase margin, is slightly preferable to the method just presented. Both are, however, quite acceptable, especially since the entire Bode diagram design philosophy as illustrated in Fig. 5-1 is a trial and error approach and the phase margin criterion is an approximate specification of system performance for all but second order systems. Further, the fixed plant transfer function parameter may be known only to within $\pm 10°$ and thus the approximations we advocate are entirely reasonable.

In the direct approach to design for a specified phase margin we assume a single lead network equalizer such that the open loop system transfer function of Eq. (5) results. We will illustrate this approach to design by the following steps:

1. We find an equation for the gain at the crossover frequency in terms of the compensated open loop system break frequencies.

2. We find an equation for the phase shift at crossover.

3. We find the relationship between equalizer parameters and crossover frequency such that the phase shift at crossover is the maximum possible and a minimum of additional gain is needed.

4. We determine all parameter specifications to meet the phase margin specifications.

5. We check to see that all design specifications have been met. If they have not, we iterate the design process.

We assume that crossover occurs at a frequency $\omega_1 < \omega_c < \omega_2$ such that the asymptotic gain magnitude from Eq. (5) is

$$|G_f(\omega)G_c(\omega)| \cong \frac{10^5}{\omega_1 \omega}$$

The gain is unity at crossover so

$$\omega_1 \omega_c = 10^5 \tag{11}$$

The phase shift is determined from Eq. (5) as

$$\beta(\omega) = \frac{-\pi}{2} - \tan^{-1}\frac{\omega}{100} + \tan^{-1}\frac{\omega}{\omega_1} - \tan^{-1}\frac{\omega}{\omega_2} \tag{12}$$

Within the frequency range $\omega_1 < \omega < \omega_2$ use of the arctangent approximation gives

$$\beta(\omega) \cong \frac{-\pi}{2} + \frac{100}{\omega} - \frac{\omega_1}{\omega} - \frac{\omega}{\omega_2} \tag{13}$$

To accomplish step 3 of our design procedure we set the first derivative of β with respect to ω equal to zero at the crossover frequency and obtain

$$\left.\frac{\partial \beta(\omega)}{\partial \omega}\right|_{\omega=\omega_c} = 0 = \frac{-100}{\omega_c^2} + \frac{\omega_1}{\omega_c^2} - \frac{1}{\omega_2}$$

so that

$$\omega_c = \sqrt{\omega_2(\omega_1 - 100)} \tag{14}$$

Simultaneous solution of Eqs. (11), (13), and (14), with $\beta(\omega_c) = -3\pi/4$ yields a quadratic relationship

$$\omega_c^2 + \frac{800}{\pi}\omega_c - \frac{8 \times 10^5}{\pi} = 0$$

which has the solutions $\omega_c = 393.12$ and $\omega_c = -647.77$. Clearly the frequency -647.77 is meaningless and so we accept 393.12 as the solution. The other parameters are easily obtained and the complete design specification is

$$\omega_1 = 254.38 \qquad \omega_2 = 1001.06 \qquad \omega_c = 393.12$$

We check the phase margin and obtain a phase margin, using Eq. (12), of 49.93°. Thus we see that each of the two designs leads to essentially equivalent results.

In the preceding example we have illustrated two approaches to the Bode diagram design of a simple lead network. It is convenient and helpful to illustrate these two approaches by means of DELTA Charts. Figures (5.2-8) and (5.2-9) illustrate these two methods of lead network design by means of Bode diagrams.

EXAMPLE 5.2-4. There are a number of fixed plants which cannot be compensated by means of a simple lead network. We consider the fixed plant

$$G_f(s) = \frac{100^2 e^{-.01s}}{s^2}$$

and impose the requirement that the velocity constant remain at 10^4 and the phase margin be 45°. The uncompensated crossover frequency is 100 radians per second and the uncompensated closed loop system is unstable. At the crossover frequency of 100 the phase shift is -4.14 radians or $-237.30°$. We would need to add $102.30°$ phase lead at crossover even if the crossover frequency would not increase due to use of the lead network. This amount of phase lead cannot be obtained from a single stage lead network.

As the preceding example has illustrated, there are some fixed plants which cannot be compensated to yield a given set of performance specifications using a simple lead network. Also, the additional high frequency gain inherent in any lead network will amplify any high frequency system noise that may be present and the equalizer or compensating network cannot be realized using only passive elements. We will now turn our attention to other forms of series equalization.

EXERCISE 5.2-2. A unity ratio linear control system has the fixed plant transfer function

$$G_f(s) = \frac{10000}{s(1 + s/10)}$$

Compensate the system using a lead network equalizer to achieve a 45° phase margin and a velocity error coefficient of 10^3. Use the two approaches suggested and contrast and compare the results obtained.

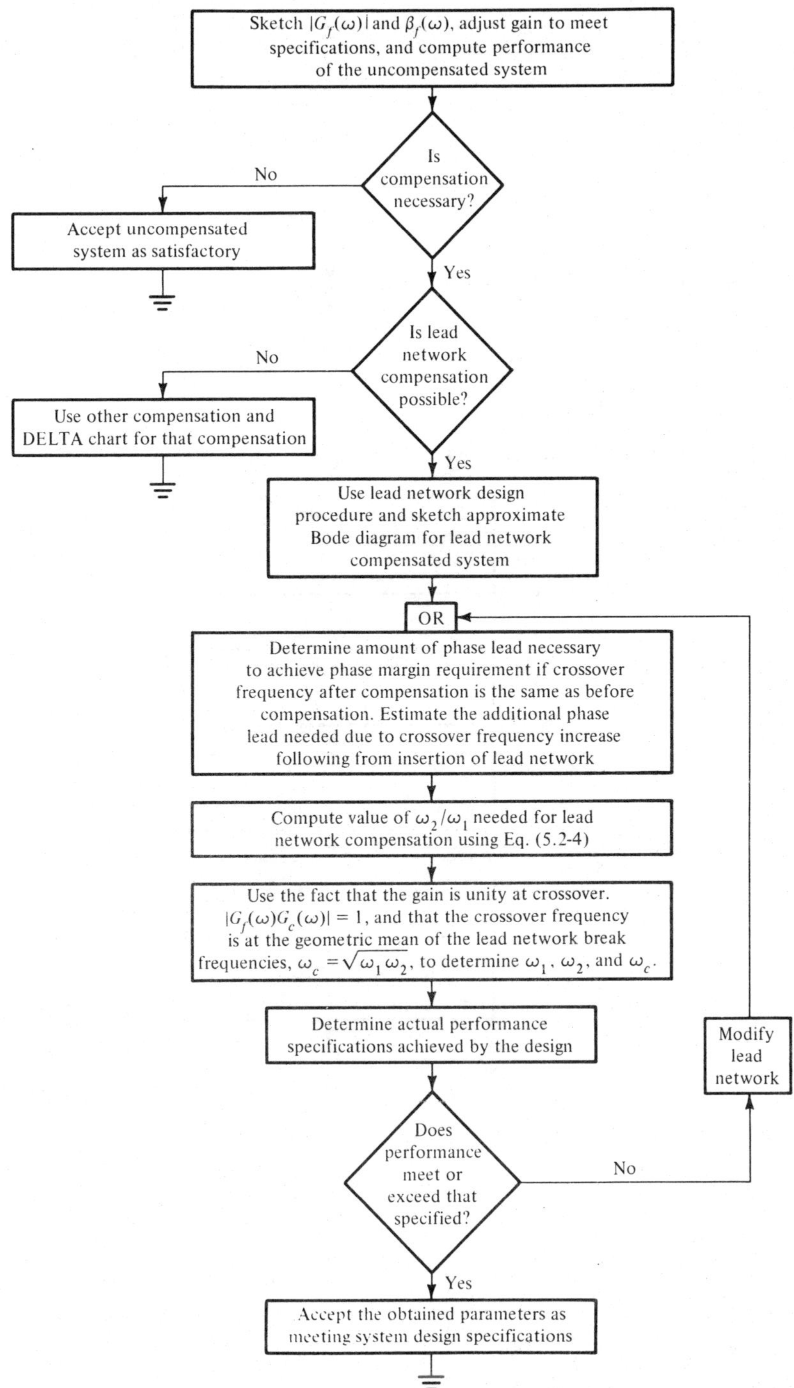

Figure 5.2-8 DELTA Chart of lead compensation design procedure using estimate of phase lag due to increased crossover frequency

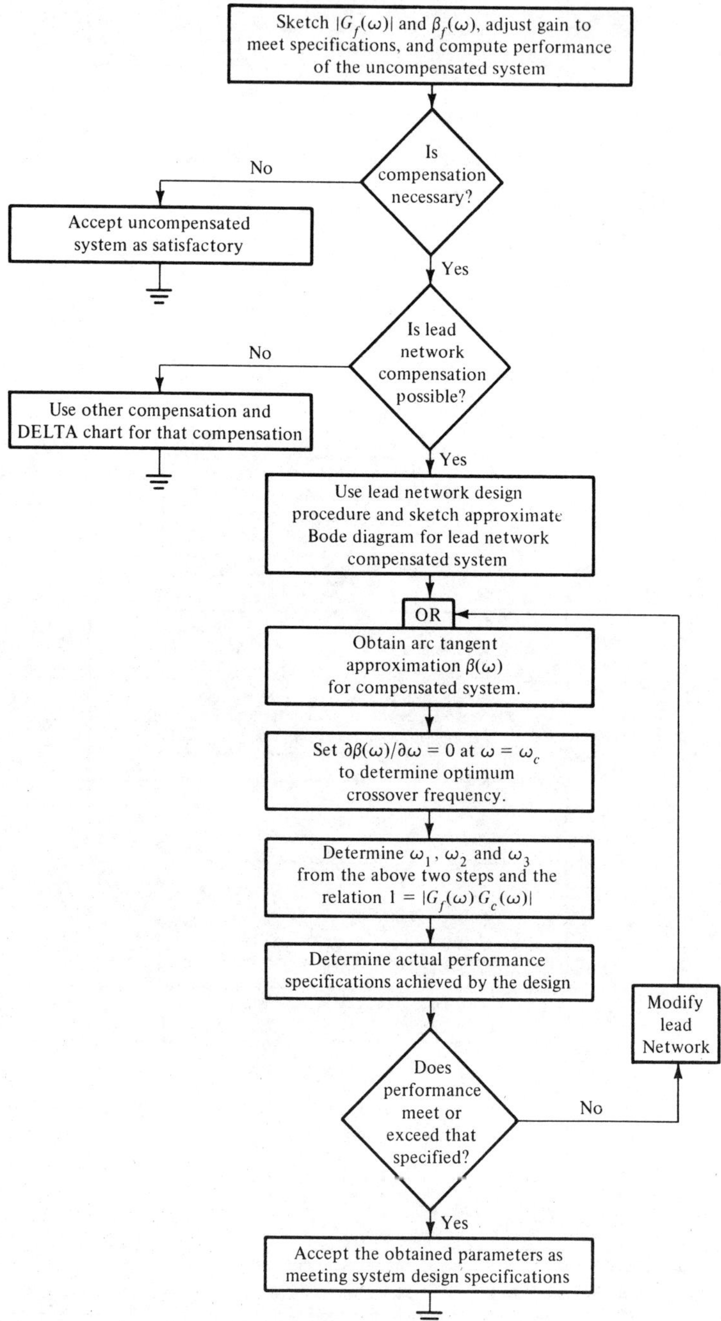

Figure 5.2-9 DELTA Chart for lead network design to meet actual phase shift at crossover specification and obtain optimum crossover frequency

EXERCISE 5.2-3. A unity ratio linear control system is compensated with a 10:1 break frequency ratio lead network such that the open loop transfer function becomes

$$G_f(s)G_c(s) = \frac{10000}{s(1 + s/10)} \frac{1 + s/\omega_1}{1 + s/10\omega_1}$$

Find the value of ω_1 which maximizes the phase margin. What is this maximum phase margin? What is the gain margin for this system? the M_p?

EXERCISE 5.2-4. Use a lead network to design a system with fixed plant

$$G_f(s) = \frac{500}{s(1 + s/10)}$$

for a phase margin of $45°$ and a velocity error coefficient of 5000.

EXERCISE 5.2-5. Can a single stage lead network be found to design a system with fixed plant

$$G_f(s) = \frac{100}{s(1 + s)^2}$$

such that the phase margin is $45°$ and the velocity error coefficient is 100? If so, determine the lead network transfer function.

EXERCISE 5.2-6. Repeat EXERCISE 5.2-5 for the fixed plant transfer function

$$G_f(s) = \frac{100}{s^2(1 + s/100)}$$

5.2-3 Phase Lag Compensation

In phase lag compensation we reduce the gain at low frequencies such that crossover occurs before the phase lag has had a chance to become intolerably large. A simple single stage phase lag compensating network transfer function is

$$G_c(s) = \frac{1 + s/\omega_2}{1 + s/\omega_1} \qquad \omega_1 < \omega_2 \tag{5.2-8}$$

Figure (5.2-10) illustrates the gain and phase versus frequency curves for a simple lag network with the transfer function of Eq. (5.2-8). The maximum phase lag obtainable from a phase lag network depends upon the ratio ω_2/ω_1 used in designing the network. From the expression for the phase shift of the transfer function of Eq. (5.2-8)

$$\beta = \tan^{-1} \frac{\omega}{\omega_2} - \tan^{-1} \frac{\omega}{\omega_1} \tag{5.2-9}$$

we see that maximum phase lag occurs at the frequency ω where $\partial\beta/\partial\omega = 0$. We have

$$\omega_m = \sqrt{\omega_1 \omega_2} \tag{5.2-10}$$

which is at the center of the two break frequencies for the lag network on a Bode diagram log log asymptotic gain plot.

The maximum value of the phase lag obtained at $\omega = \omega_m$ is

$$\beta_m(\omega_m) = \frac{\pi}{2} - 2 \tan^{-1} \sqrt{\omega_2/\omega_1}$$

$$= \frac{-\pi}{2} + 2 \tan^{-1} \sqrt{\omega_1/\omega_2} \tag{5.2-11}$$

which can be approximated in a more useable form (using the arctangent approximation) as

$$\beta_m(\omega_m) \cong -\frac{\pi}{2} + 2\sqrt{\omega_1/\omega_2} \tag{5.2-12}$$

The attenuation of the lag network at the frequency of minimum phase shift, or maximum phase lag, is obtained from the asymptotic approximation for the gain at crossover as

$$|G_c(\omega_m)| = \sqrt{\omega_1/\omega_2} \tag{5.2-13}$$

Figure (5.2-11) presents a curve of attenuation magnitude obtainable at the frequency of maximum phase lag and the amount of the phase lag for various ratios ω_2/ω_1 for the simple lag network of Eq. (5.2-8).

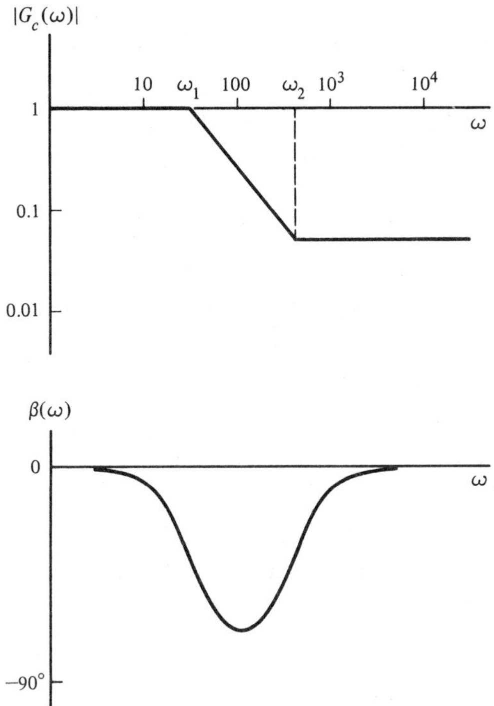

Figure 5.2-10 Gain and phase shift curves for a simple lag network

There are many ways of realizing a lag network. Since the network only attenuates at some frequencies (It never has a gain greater than 1 at any frequency.), it can be realized with passive components only. Figure (5.2-12) presents an electrical realization of the simple lag network.

It appears best, just as in our discussion of series equalization by means of lead network compensation, to present the approach for lag network design using specific examples first and then state some general rules.

EXAMPLE 5.2-5. In this example we consider the same fixed plant

$$G_f(s) = \frac{1000}{s(1 + s/100)} \tag{1}$$

as in EXAMPLE 5.2-3. As design specifications we assume

1. The velocity error constant K_v is 1000.
2. The phase margin is $45°$.

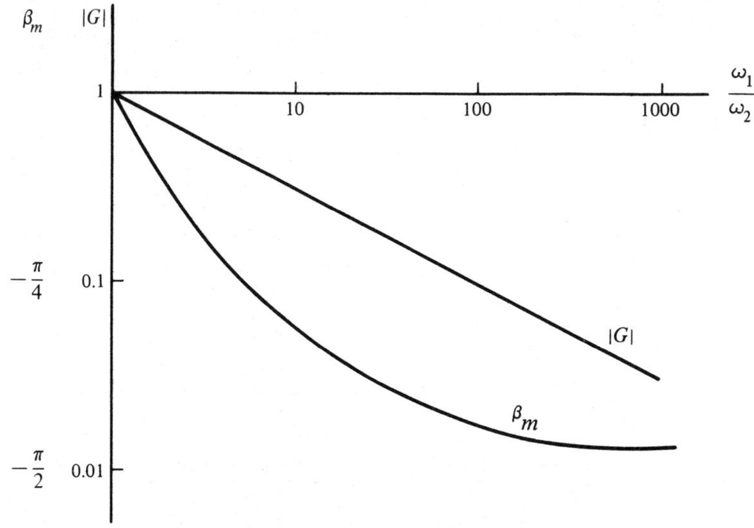

Figure 5.2-11 Design chart showing gain ratio increases and phase shift as functions of ω_1/ω_2 for a simple lag network

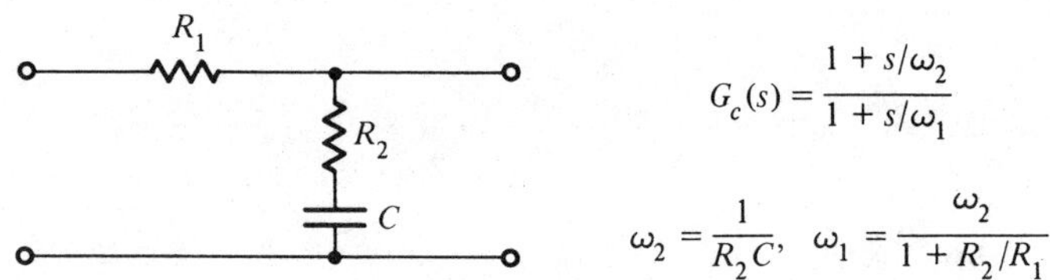

$$G_c(s) = \frac{1 + s/\omega_2}{1 + s/\omega_1}$$

$$\omega_2 = \frac{1}{R_2 C}, \quad \omega_1 = \frac{\omega_2}{1 + R_2/R_1}$$

Figure 5.2-12 A simple electrical lag network

3. Sinusoidal inputs for any frequency greater than 1000 radians per second must be attenuated by a factor of at least 10.

We note that specifications 1 and 2 are the same as in EXAMPLE 5.2-3. The third specification can be converted into an open loop gain approximation from an approximation

$$\frac{Z(s)}{U(s)} = \frac{G_f(s)G_c(s)}{1 + G_f(s)G_c(s)} \cong G_f(s)G_c(s) \tag{2}$$

which is valid for $|G_f(\omega)G_c(\omega)| \ll 1$. Thus we see that specification 3 can be met by restricting $|G_f(\omega)G_c(\omega)| < 0.1$ for $\omega > 1000$. From the

Bode diagram of the fixed plant shown in Fig. (5.2-13) we see that a lead network cannot be used. Thus we attempt design by means of a lag network.

As in the lead network case, there are several slightly different design approaches we could use. One popular approach assumes that the compensation is first achieved using a constant attenuation at all frequencies. The crossover frequency that would result in a 45° phase margin is then determined. Clearly this frequency is $\omega_c = 100$ radians per second. We calculate the ratio of the original crossover frequency and the crossover frequency which, under gain reduction, results in a 45° phase margin. We assume that we need a lag network equalizer with an attenuation ratio equal to this ratio of crossover frequencies or

$$\frac{\omega_2}{\omega_1} = \frac{316.22}{100} = 3.16$$

A little thought indicates that we will need an attenuation greater than this because the lag network will introduce not only attenuation but phase lag as well.* Thus we would need a larger attenuation ratio such that the phase margin, assuming gain reduction only, would be larger than our previously calculated value by 10° to 15°. Once we have decided upon an attenuation ratio, we could locate the actual crossover frequency at the geometric mean of the break frequencies $\omega_c = \sqrt{\omega_1 \omega_2}$, and then use the fact that the magnitude of the gain is unity, $|G_f(\omega_c)G_c(\omega_c)| = 1$, to solve for all equalizer parameters.

Even though the maximum phase lag of the lag network occurs at the geometric mean of the lag network break frequencies it is not true that we necessarily obtain maximum phase margin by setting $\omega_c = \sqrt{\omega_1 \omega_2}$. A somewhat better approach would be to write the arctangent approximation for the phase shift in the vicinity of crossover and obtain maximum phase margin at crossover by setting the derivative of this phase margin expression with respect to the crossover frequency equal to zero in order to determine the optimum crossover frequency.

A third method of design can be obtained based on the concept that we desire the lag network to act as an attenuator only at the crossover frequency. If the largest break frequency ω_2 is small with respect to the crossover frequency, this will be essentially the case. Thus we

---

*Actually, since reduction of the gain by this amount puts the crossover frequency at $\omega = 100$, the asymptotic gain approximation is in modest error and we actually need an attenuation quite a bit greater than 3.16—as we should see.

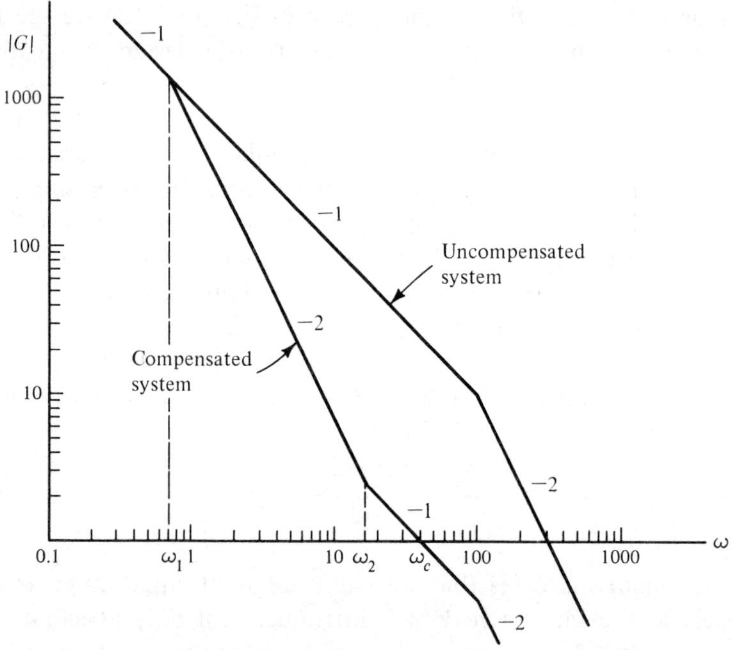

Figure 5.2-13 $G_f(s)$ and $G_f(s)G_c(s)$ for lag network design EXAMPLE (5.2-5)

might set $\omega_c/\omega_2 = 6$ or 10 and use the equation for the phase shift at crossover equal to that specified, and the fact that the gain at crossover is unity to obtain three equations to solve for the ω_1, ω_2, and ω_c— parameters necessary to complete the design.

Each of these three methods can lead to satisfactory results as the results of the exercises at the conclusion of this subsection suggest. However, each approach is based upon an assumption used only to simplify the design procedure. The first and second approaches assume a lag network without phase shift in order to compute an approximate value of the lag network attenuation ratio ω_2/ω_1. The third approach uses this same philosophy of a phaseless lag network and selects the ratio of crossover frequency to largest lag network break frequency such that the assumption of an essentially phaseless lag network is approximately correct. Since the design approach is an iterative one, any of these three approaches should converge to an acceptable design. Often a number of iterations will be needed and we prefer a design method, which we now introduce, that will allow us to select the best lag network in a single iteration for most cases.

We develop a fourth method here that does not employ intuitive guesses concerning the attenuation ratio of the lag network. The approach to determine the frequencies ω_1, ω_2, and ω_c is based upon satisfying three equations or requirements:

1. The magnitude of the open loop gain at the crossover frequency is unity.

2. The phase shift at the crossover frequency is that specified.

3. The phase shift at crossover (phase margin) is the minimum (maximum) possible for a minimum attenuation lag network.

In addition, we set the constant gain such that the static error coefficient specification is satisfied. This approach is no more difficult than the other three approaches and, because it leads to an optimum result in terms of a minimum attenuation lag network, this is the approach we will develop fully here and use in our subsequent efforts.

As we have indicated, we use three simple relations to determine ω_1, ω_2, and ω_c, parameters that we need to determine to satify design specifications. We assume that crossover occurs on the -1 slope as shown in Fig. (5.2-13). The open loop compensated system transfer function is

$$G_f(s)G_c(s) = \frac{1000(1 + s/\omega_2)}{s(1 + s/100)(1 + s/\omega_1)} \tag{3}$$

The asymptotic gain at crossover is

$$1 = \frac{10^3\,\omega_1}{\omega_c\omega_2} \tag{4}$$

The phase shift in the vicinity of crossover is

$$\beta(\omega) = -\frac{\pi}{2} - \left(\frac{\pi}{2} - \frac{\omega_1}{\omega_c}\right) - \frac{\omega_c}{100} + \left(\frac{\pi}{2} - \frac{\omega_2}{\omega_c}\right)$$

$$= -\frac{\pi}{2} - \frac{\omega_c}{100} + \frac{\omega_1 - \omega_2}{\omega_c} \tag{5}$$

In order to obtain the minimum attenuation lag network which meets specifications we set

$$\frac{\partial \beta(\omega)}{\partial \omega}\bigg|_{\omega=\omega_c} = 0 = -\frac{1}{100} + \frac{\omega_2 - \omega_1}{\omega_c^2}$$

to obtain the expression

$$\omega_c = \sqrt{100(\omega_2 - \omega_1)} \tag{6}$$

Equation (4) is responsive to our first requirement for this design procedure whereas Eq. (6) is responsive to the third. We satisfy the second requirement by setting the phase shift $\beta(\omega)$ at $\omega = \omega_c$ equal to $-3\pi/4$ and obtain, from Eq. (5), an equation responsive to the second requirement

$$-\frac{3\pi}{4} = -\frac{\pi}{2} - \frac{\omega_c}{100} + \frac{\omega_1 - \omega_2}{\omega_c} \tag{7}$$

Simultaneous solution of Eqs. (4), (6), and (7) leads to

$$\omega_1 = 0.63 \qquad \omega_2 = 16.04 \qquad \omega_c = 39.27$$

as the parameters for the lag network and the crossover frequency. Figure (5.2-13) illustrates the Bode diagram of the compensated open loop system.

Two of the break frequencies in the fixed plant, namely those at $\omega = 100$ and $\omega = 16.04$ are reasonably close to the crossover frequency. Thus it is desirable to use actual gains and phase shifts to compute the system phase margin rather than the asymptotic values. We obtain from Eq. (3) actual gain and phase shift of 1.0056 and $-132.73°$ and so we see that crossover occurs around $\omega_c = 39.27$ and the actual phase margin is $47.27°$. This is so close to the desired value that we accept this design.

Figure (5.2-14) presents a DELTA chart illustrating the design procedure presented here for lag network design. There are several minor modifications which could be made to this chart to accommodate any of the other three design methods discussed.

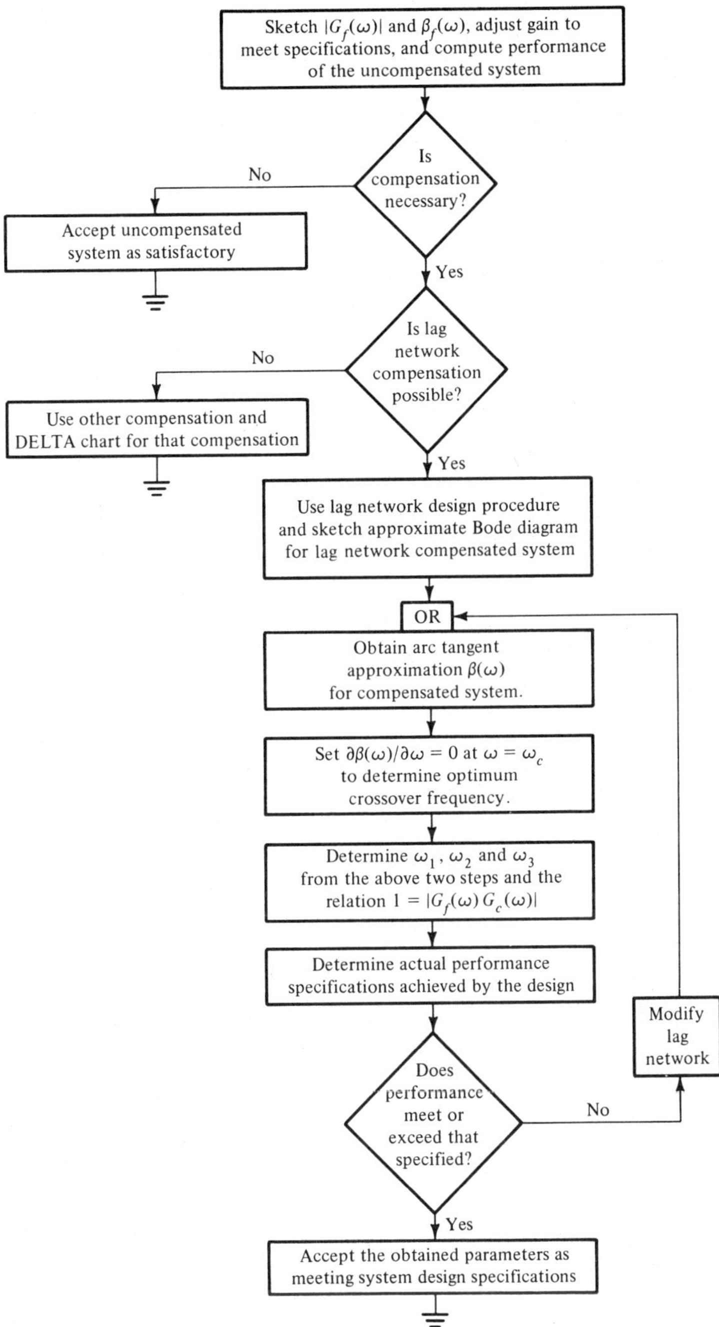

Figure 5.2-14 DELTA Chart for lag network design to meet actual phase margin specifications with a minimum attenuation lag network

All of our examples thus far have involved relatively low order systems. To indicate that the design procedure presented is not restricted to low order systems we consider two additional examples of high order systems.

EXAMPLE 5.2-6.   In this example we consider the fixed plant

$$G_f(s) = \frac{100}{s(1 + s/100)^{11}}$$

which will result in an unstable system if a unity feedback ratio loop is closed with no loop compensation. We leave the velocity error coefficient unchanged and insert a lag network to obtain a 45° phase margin. The open loop compensated system transfer function becomes

$$G_f(s)G_c(s) = \frac{100\left(1 + \dfrac{s}{\omega_2}\right)}{s\left(1 + \dfrac{s}{100}\right)^{11}\left(1 + \dfrac{s}{\omega_1}\right)} \tag{1}$$

and is illustrated in Fig. (5.2-15). Using the three relations for our suggested design procedure we obtain:

1. Gain at crossover equals 1

$$1 = \frac{10^2\,\omega_1}{\omega_c \omega_2} \tag{2}$$

2. Phase shift at crossover equals $-3\pi/4$

$$-\frac{3\pi}{4} = -\frac{\pi}{2} - 11\left(\frac{\omega_c}{100}\right) - \left(\frac{\pi}{2} - \frac{\omega_1}{\omega_c}\right) + \frac{\pi}{2} - \frac{\omega_2}{\omega_c} \tag{3}$$

3. $\left.\dfrac{\partial \beta(\omega)}{\partial \omega}\right|_{\omega=\omega_c} = 0$, to insure minimum attenuation lag network

$$\omega_c^2 = \frac{100}{11}(\omega_2 - \omega_1) \tag{4}$$

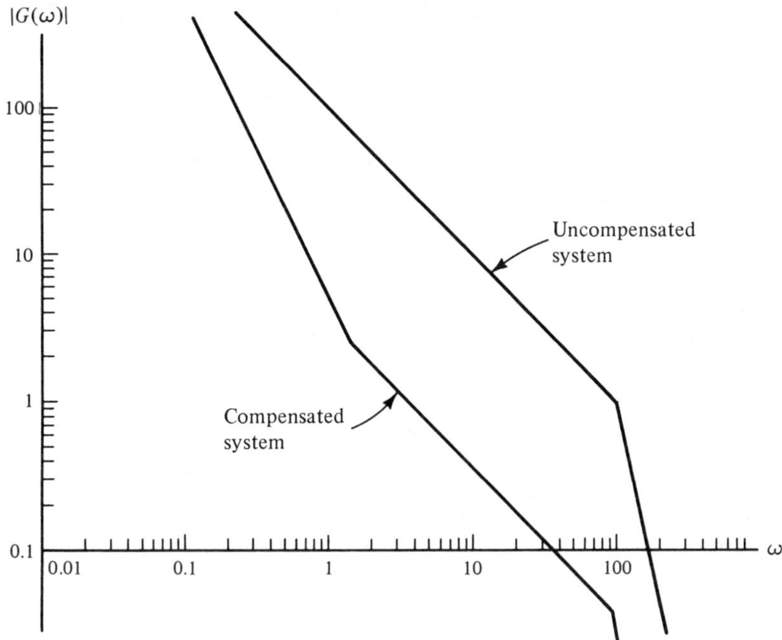

Figure 5.2-15   Bode diagram design using a lag network

We solve Eqs. (2), (3), and (4) and obtain

$$\omega_1 = 0.052 \qquad \omega_2 = 1.45 \qquad \omega_c = 3.57$$

We note that this is another fixed plant that cannot be compensated using a single stage lead network. The crossover frequency has been reduced considerably from the uncompensated case and the crossover frequency is much closer to ω_2 than it is to the 11th order pole at $s = -100$. This is reasonable since we must reduce the gain quite a bit below unity before $\omega = 100$, because much phase lag will occur around this frequency. Figure (5.2-15) illustrates the final design which meets design specifications.

EXAMPLE 5.2-7.   Our final example of simple lag network design involves the non-minimum phase time delay system with fixed plant transfer function

$$G_f(s) = \frac{100}{s} e^{-0.1s} \tag{1}$$

which will be unstable if a unity feedback ratio loop is closed. This is another example of a transfer function for a system that cannot be compensated using a lead network. A simple lag network will allow stabilization, however. We will design such a network such that the velocity error coefficient is 100 and the phase margin is $45°$.

The open loop compensated transfer function is, with lag network compensation,

$$G_f(s)G_c(s) = \frac{100(1 + s/\omega_2)}{s(1 + s/\omega_1)} e^{-0.1s} \tag{2}$$

The same three-part design procedure used in the previous two examples will be used here. In order that the gain be unity at the crossover frequency we have

$$1 = \frac{100\omega_1}{\omega_c \omega_2} \tag{3}$$

The phase shift in the vicinity of crossover is

$$\beta(\omega) = \frac{-\pi}{2} - \left(\frac{\pi}{2} - \frac{\omega_1}{\omega}\right) + \left(\frac{\pi}{2} - \frac{\omega_2}{\omega}\right) - 0.1\omega \tag{4}$$

To obtain the minimum attenuation lag network we set

$$\left. \frac{\partial \beta(\omega)}{\partial \omega} \right|_{\omega = \omega_c} = 0$$

and obtain

$$\omega_c^2 = 10(\omega_2 - \omega_1) \tag{5}$$

We set the phase shift at crossover equal to $-3\pi/4$ and obtain, from Eq. (4)

$$\frac{-\pi}{4} = \frac{\omega_1 - \omega_2}{\omega_c} - 0.1\omega_c \tag{6}$$

Simultaneous solution of Eqs. (3), (5), and (6) leads to

$$\omega_1 = 0.063 \qquad \omega_2 = 1.61 \qquad \omega_c = 3.93$$

This particular open loop system corresponds to a difference differential equation representing a process with a pure time delay. The transfer functions we are working with here are not ratios of rational polynomials in s. Thus such operations as inverse Laplace transforms to recover time functions are difficult to obtain. Yet the Bode diagram design to a specified phase margin is straightforward.

As we have indicated, the object of lag network design is to reduce the gain at frequencies lower than the original crossover frequency in order to reduce the open loop gain to unity before the phase shift becomes so excessive that the system phase margin is too small. A disadvantage of lag network compensation is that the attenuation introduced reduces the crossover frequency and makes the system slower in terms of its transient response. Of course, this would be advantageous if high frequency noise is present and we wish to reduce its effect. The lag network is an entirely passive device and thus is more economical to instrument than the lead network.

In lead network compensation we actually insert phase lead in the vicinity of the crossover frequency in order to increase the phase margin. Thus we realize a specified phase margin without lowering the medium frequency system gain. We see that the disadvantages of the lag network are the advantages of the lead network and the advantages of the lag network are the disadvantages of the lead network.

We can attempt to combine the lag network with the lead network into an all-passive structure called a lag lead network. Generally we obtain better results than we can achieve using either a lead or a lag network. We will consider design using lag lead networks in our next section as well as more complex composite equalization networks.

EXERCISE 5.2-7. A unity ratio linear control system has the fixed plant

$$G_f(s) = \frac{10,000}{s(1 + s/10)}$$

Compensate the system using a lag network equalizer to achieve a 45° phase margin and a velocity error coefficient of 10^3. Compare your results with the results of EXAMPLE 5.2-2.

EXERCISE 5.2-8.   A unity ratio linear control system is compensated with a 10:1 attenuation ratio lag network such that the open loop transfer function becomes

$$G_f(s)G_c(s) = \frac{10,000}{s(1 + s/10)} \; \frac{1 + s/10\omega_1}{1 + s/\omega_1}$$

Find the value of ω_1 which maximizes the phase margin. What is the phase margin?

EXERCISE 5.2-9.   Can a single-stage lag network be found such as to design a system with fixed plant

$$G_f(s) = \frac{100}{s(1 + s)^2}$$

for a 45° phase margin? What is the transfer function of the compensating network?

EXERCISE 5.2-10.   Repeat EXERCISE 5.2-9 for the fixed plant transfer function

$$G_f(s) = \frac{100}{s^2(1 + s/100)}$$

5.3 COMPOSITE EQUALIZERS

In our section just completed we examined the simplest forms of series equalization, namely gain adjustment, lead network compensation, and lag network compensation. In this section we will consider more complex design examples in which composite equalizers will be used for series compensation. The same design principles used earlier in this chapter will be used here as well.

5.3-1 Lag Lead Network Design

The prime purpose of a lead network is to add phase lead near the crossover frequency in order to increase the phase margin. Accompanied with this phase lead is a gain increase that will increase the crossover frequency. This will sometimes cause difficulties if there is much phase lag in the uncompensated system at high frequencies. As we have seen, there may be situations where use of a phase lead network is not possible due to too many high frequency poles.

The basic idea behind lag network design is to reduce the gain at "middle" frequencies such as to reduce the crossover frequency to a lower value than for the uncompensated system. If the phase lag is less at this lower frequency, then the phase margin will be increased by use of the lag network. We have seen that it is not possible to use a lag network in situations where there is not a frequency where an acceptable phase margin would exist if this frequency were the crossover frequency. Even if use of a lag network is possible, the significantly reduced crossover frequency resulting from its use may make the system so slow and sluggish in response to an input that system performance is unacceptable even though the relative stability of the system is acceptable.

Examination of these characteristics or attributes of lead network and lag network compensation suggests that it might be possible to combine the two approaches in order to hopefully achieve the desirable features of each approach. Thus we will attempt to provide attenuation below the crossover frequency to decrease the phase lag at crossover and phase lead closer to the crossover frequency in order to increase the phase lead of the uncompensated system at the crossover frequency.

The transfer function of the basic lag lead network is

$$G_c(s) = \frac{(1 + s/\omega_2)(1 + s/\omega_3)}{(1 + s/\omega_1)(1 + s/\omega_4)} \qquad (5.3\text{-}1)$$

where

$$\omega_4 > \omega_3 > \omega_2 > \omega_1$$

Often it is desirable that $\omega_2 \omega_3 = \omega_1 \omega_4$ such that the high frequency gain of the equalizer is unity. It is generally not desirable that $\omega_1 \omega_4 >$

$\omega_2\omega_3$ as this indicates a high frequency gain greater than one and this will require an active network, or gain, and a passive equalizer. It is a fact that we should always be able to realize a linear minimum phase network using passive components only if the network has a rational transfer function* with a gain magnitude that is no greater than one at any (real) frequency. We will not offer a proof of this assertion here.

Figure (5.3-1) illustrates the gain magnitude and phase shift curves for a single stage lag lead network equalizer with the transfer function of Eq. (5.3-1). Figure (5.3-2) illustrates an electrical network realization of a passive lag lead network equalizer. Parameter matching can be used to determine the electrical network parameters which yield a specified transfer function. Because the relationships between the break frequencies and the equalizer component values are complex it may be desirable, particularly in preliminary instrumentation of the control system, to use analog computer programming techniques (such as those discussed briefly in Chap. 2) to construct the equalizer.

EXAMPLE 5.3-1. As a simple example to illustrate lag lead series equalizer design, let us consider the fixed plant

$$G_f(s) = \frac{1000}{s(1 + s/100)} \tag{1}$$

with design specifications:

1. $K_v = 1000$.

2. Phase margin $= 45°$.

3. Sinusoidal inputs at frequencies up to 10 radians per second must be passed with less than 2% error.

4. Sinusoidal inputs at frequencies greater than 1000 radians per second must be attenuated at the output by a factor of at least 10.

The first specification is satisfied by the open loop system without compensation. The second specification is not satisfied by the uncompensated system and we attempt series equalization. To determine the

---

*A rational transfer function $H(s)$ is one of the form $H(s) = A(s)/B(s)$ where $A(s)$ and $B(s)$ are rational polynomials in s, that is, polynomials of the form $a_0 + a_1 s + a_2 s^2 + \ldots$.

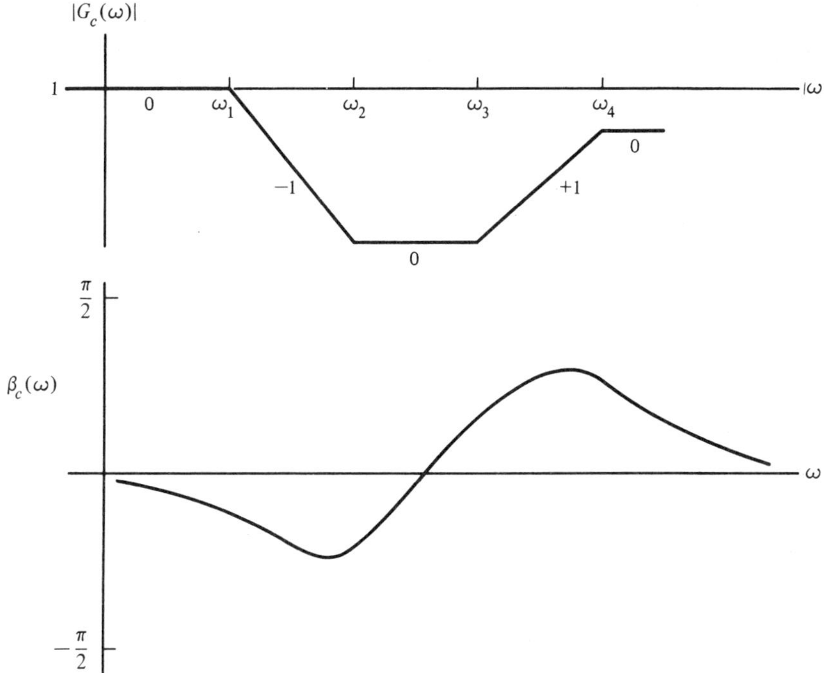

Figure 5.3-1 Gain and phase shift curves for a typical lag lead network

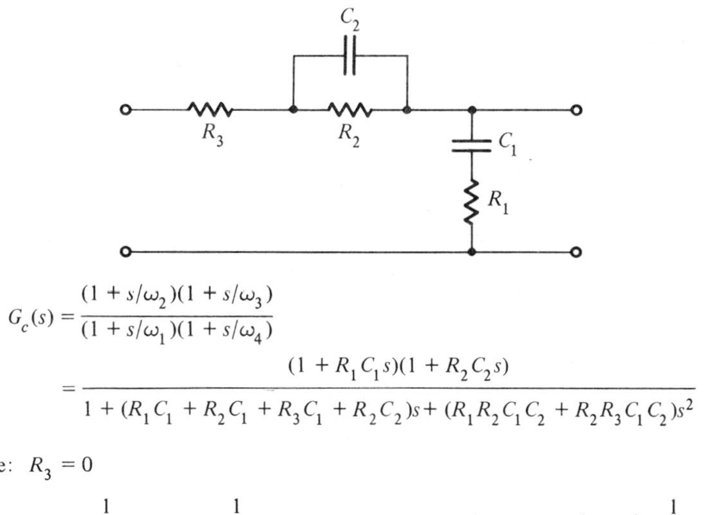

$$G_c(s) = \frac{(1 + s/\omega_2)(1 + s/\omega_3)}{(1 + s/\omega_1)(1 + s/\omega_4)}$$

$$= \frac{(1 + R_1 C_1 s)(1 + R_2 C_2 s)}{1 + (R_1 C_1 + R_2 C_1 + R_3 C_1 + R_2 C_2)s + (R_1 R_2 C_1 C_2 + R_2 R_3 C_1 C_2)s^2}$$

Special case: $R_3 = 0$

$$\omega_2 = \frac{1}{R_1 C_1}, \quad \omega_3 = \frac{1}{R_2 C_2}, \quad \omega_1 \omega_4 = \omega_2 \omega_3, \quad \omega_1 + \omega_4 = \omega_2 + \omega_3 + \frac{1}{R_1 C_2}$$

Figure 5.3-2 Electrical lag lead network and design parameter relations

type of compensation needed we examine specifications three and four. The error transfer function for $|G_f(\omega)G_c(\omega)| > 1$ is

$$\frac{E(s)}{U(s)} = \frac{1}{1 + G_f(s)G_c(s)} \cong \frac{1}{G_f(s)G_c(s)}$$

so we see that we must require in order to satisfy specification 3

$$|G_f(\omega)G_c(\omega)| \geqslant 50, \quad \text{all } \omega < 10$$

Simple lag network compensation will not allow us to meet this specification as we saw in the previous section. To meet specification 4 we use the approximation for the closed loop transfer function

$$\frac{Z(s)}{U(s)} = \frac{G_f(s)G_c(s)}{1 + G_f(s)G_c(s)} \cong G_f(s)G_c(s)$$

which is valid for $|G_f(\omega)G_c(\omega)| < 1$. Thus we see that we must require

$$|G_f(\omega)G_c(\omega)| < 0.10, \quad \text{all } \omega > 1000$$

in order to satisfy specification 4. A simple lead network design will not allow us to satisfy this specification as we saw in the previous section. Thus we turn our attention to use of a lag lead network.

Figure (5.3-3) illustrates the Bode amplitude diagram for the uncompensated system, requirements for the compensated system to satisfy specifications 3 and 4, and a suggested equalized system. We will now determine the parameters for this equalized system and will then state a suggested design procedure for lag lead network design.

The compensated system transfer function is

$$G_c(s)G_f(s) = \frac{1000(1 + s/\omega_2)}{s(1 + s/\omega_1)(1 + s/\omega_4)} \tag{2}$$

which requires the lag lead network compensation

$$G_c(s) = \frac{(1 + s/\omega_2)(1 + s/100)}{(1 + s/\omega_1)(1 + s/\omega_4)} \tag{3}$$

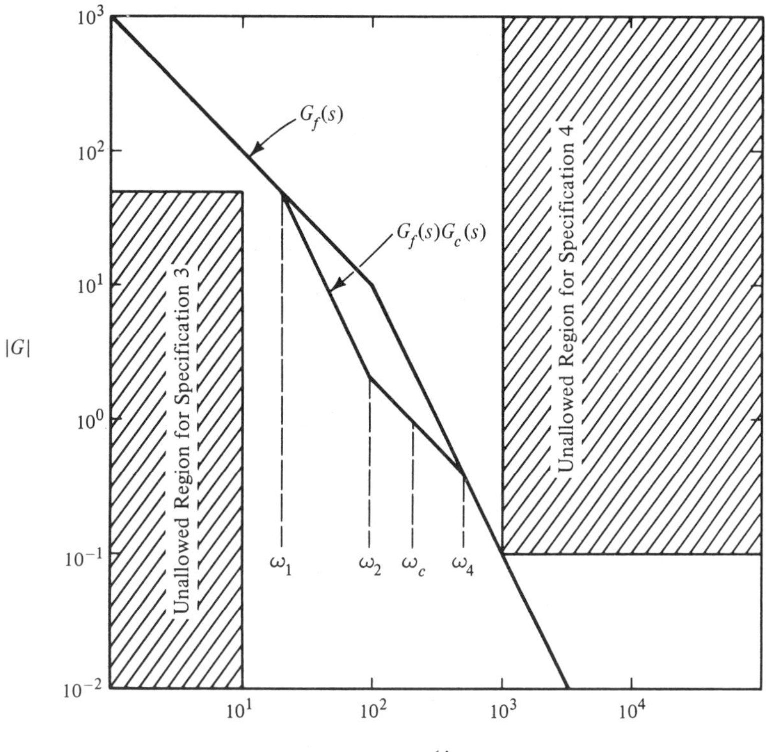

Figure 5.3-3 Bode amplitude diagram for EXAMPLE 5.3-1

We have four parameters to determine, ω_1, ω_2, ω_4, and ω_c. Thus we need four equations. The four equations are:

1. The magnitude of the gain at the crossover frequency is one.

2. The phase shift at the crossover frequency is $-3\pi/4$ radians (the phase margin is $45°$).

3. ١The phase shift versus frequency curve is flat at crossover such that we get maximum phase shift at crossover.

4. The high frequency gain of the lag lead network equalizer is one.

Requirements 1 and 2 are necessary to satisfy system specifications. Requirement 3 insures optimum placement of the compensating network break frequencies such that we put the maximum phase lead frequency of the compensated system at the crossover frequency. This results in the minimum reduction in crossover frequency possible. Requirement 4 also keeps the closed loop system bandwidth as large as possible with a passive equalizer. If the high frequency gain had to be reduced to satisfy specifications, we would use an equation to insure this in place of requirement 4.

Use of Eq. (1), assuming that $\omega_2 < \omega_c < \omega_4$, leads to

$$1 = \frac{10^3 \omega_1}{\omega_c \omega_2} \tag{4}$$

The arctangent approximation for the phase shift for $\omega_2 < \omega < \omega_4$ is

$$\beta(\omega) \cong -\frac{\pi}{2} - \left(\frac{\pi}{2} - \frac{\omega_1}{\omega}\right) + \left(\frac{\pi}{2} - \frac{\omega_2}{\omega_1}\right) - \frac{\omega}{\omega_4} \tag{5}$$

To satisfy requirement 2 we have

$$\beta(\omega_c) = -\frac{3\pi}{4} = \frac{-\pi}{2} + \frac{\omega_1 - \omega_2}{\omega_c} - \frac{\omega_c}{\omega_4} \tag{6}$$

To satisfy requirement 3 we set

$$\frac{\partial \beta(\omega)}{\partial \omega}\bigg|_{\omega = \omega_c} = 0$$

and obtain

$$\omega_c^2 = \omega_4 (\omega_2 - \omega_1) \tag{7}$$

Finally we satisfy requirement 4 by

$$\omega_1 \omega_4 = 100 \omega_2 \tag{8}$$

Simultaneous solution of Eqs. (4), (6), (7), and (8) leads to the parameters that satisfy system specifications. We combine Eqs. (6) and (7) to obtain

$$\frac{\omega_c}{\omega_4} = \frac{\pi}{8} \tag{9}$$

Combining Eqs. (4) and (9) leads to

$$\frac{\pi}{8}\,\omega_2\,\omega_4 = 10^3\,\omega_1 \tag{10}$$

Combining Eqs. (8) and (10) leads to

$$\omega_4 = 504.63$$

and we then obtain $\omega_c = 198.17$ from Eq. (9). Combining Eqs. (7) and (8) leads to

$$\omega_1 = \frac{100\omega_c^2}{\omega_4(\omega_4 - 100)} = 19.23$$

and use of Eq. (4) gives us $\omega_2 = 97.05$. This completes the determination of the compensating network transfer function. The details of this compensation have been illustrated here both to guide the reader in the generally simple task of analytic solution of the suggested design approach as well as to illustrate that the computational effort involved is not great and, in fact, is often much less than the effort involved in iterative procedures which involve guessing, perhaps in an intelligent fashion, various compensating network transfer functions and then evaluating the phase margin for the resulting compensated system.

It is now a relatively straightforward matter to determine that all system specifications have been satisfied. At the calculated crossover frequency of 198.17 the actual open loop gain magnitude is 1.03 and the phase shift is $-132.67°$, which yields a phase margin of $47.33°$. So we see that we have indeed satisfied all specifications.

EXAMPLE 5.3-2. We assume that specification 4 of the previous example has been changed such that it becomes

    4. Sinusoidal inputs at frequencies greater than 1000 radians per second must be attenuated at the output by a factor of at least 25.

The fourth of our network parameter determining requirements must now be changed since we must require $|G_f(\omega)G_c(\omega)| < 0.04$ in order to satisfy the new specification 4. We thus have as the new requirement 4:

4.   The compensating network must be designed such that

$$G_f(10^3)G_c(10^3) = 0.04$$

Figure (5.3-4) illustrates the Bode amplitude diagram for this example. Requirement 4 is satisfied by noting that the uncompensated system gain at $\omega = 10^3$ is 0.1. We must reduce this to 0.04 and so we require that the high frequency gain of the lag lead network of Eq. (3) of the foregoing example is

$$G_c(\omega > 1000) = \frac{\omega_1 \omega_4}{100\omega_2} = \frac{0.04}{0.1} = 0.4 \tag{1}$$

where we assume that $\omega_4 < 1000$.

Simultaneous solution of Eq. (4), (6), and (7) of EXAMPLE 5.3-1 and Eq. (1) of this example is a simple task and leads to the design parameters

$$\omega_1 = 7.05 \qquad \omega_2 = 56.27 \qquad \omega_c = 125.33 \qquad \omega_4 = 319.15$$

which are such that the specifications for this example are met.

It is of interest to briefly compare the results of this example with the results of EXAMPLE 5.3-1. Each results in a phase margin of $45°$. The crossover frequency for this example is 125.33 whereas in EXAMPLE 5.3-1 it was 198.17. The closed loop response for this example will be a bit sluggish compared with that of EXAMPLE 5.3-1, but that is the price we must pay for the specification which required greater input output attenuation at high frequencies. From inspection of Fig. (5.3-3) or (5.3-4), we see that it would be possible to set specifications on low frequency error performance and high frequency noise rejection such that there would be no feasible region left and a design to meet specifications would be impossible.

Figure (5.3-5) presents a DELTA Chart which we may use for lag lead network design. We see that this DELTA Chart has much in common with the charts and design procedures for lead network and lag

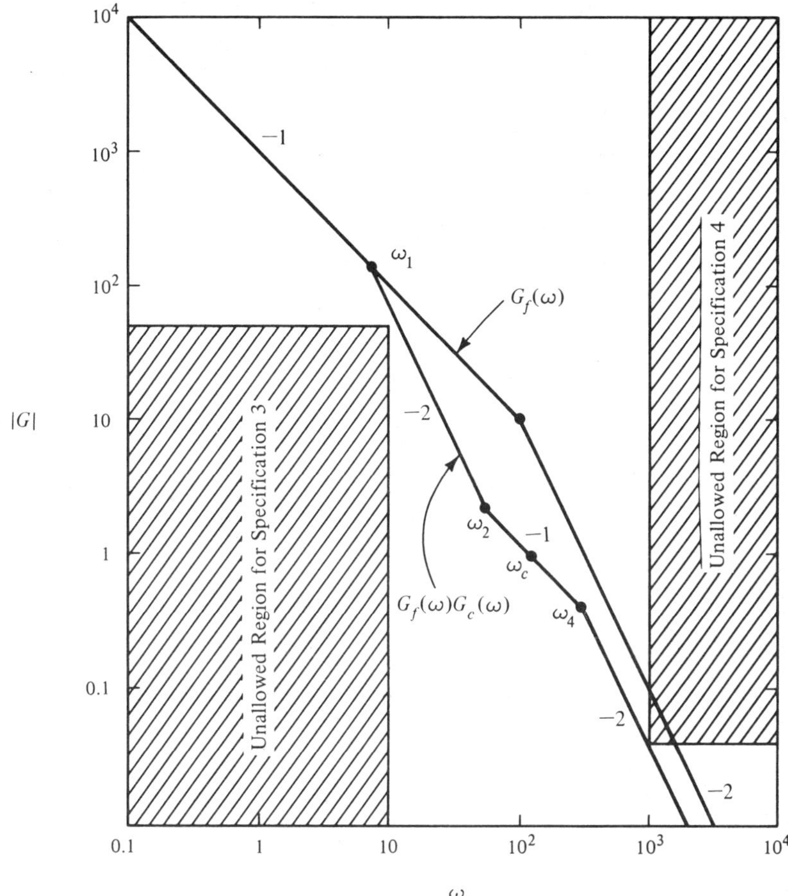

Figure 5.3-4   Bode amplitude diagram for EXAMPLE 5.3-2

network design, and that each of these approaches first involves determining or obtaining a set of desired specifications for the control system. Next, the form of a trial compensating network and the number of break frequencies in the network are selected. We must then obtain a number of equations, equal to the number of network break frequencies plus one. One of these equations shows that the gain magnitude is one at the crossover frequency. The second equation will be an equation for the phase shift at crossover. It is generally desirable that there be at least two unspecified compensating network break frequencies such that we may use a third equation, the optimality of the phase shift at crossover equation, in which we set $\partial\beta/\partial\omega|_{\omega=\omega_c} = 0$. If other equations are needed we obtain these from system specifications.

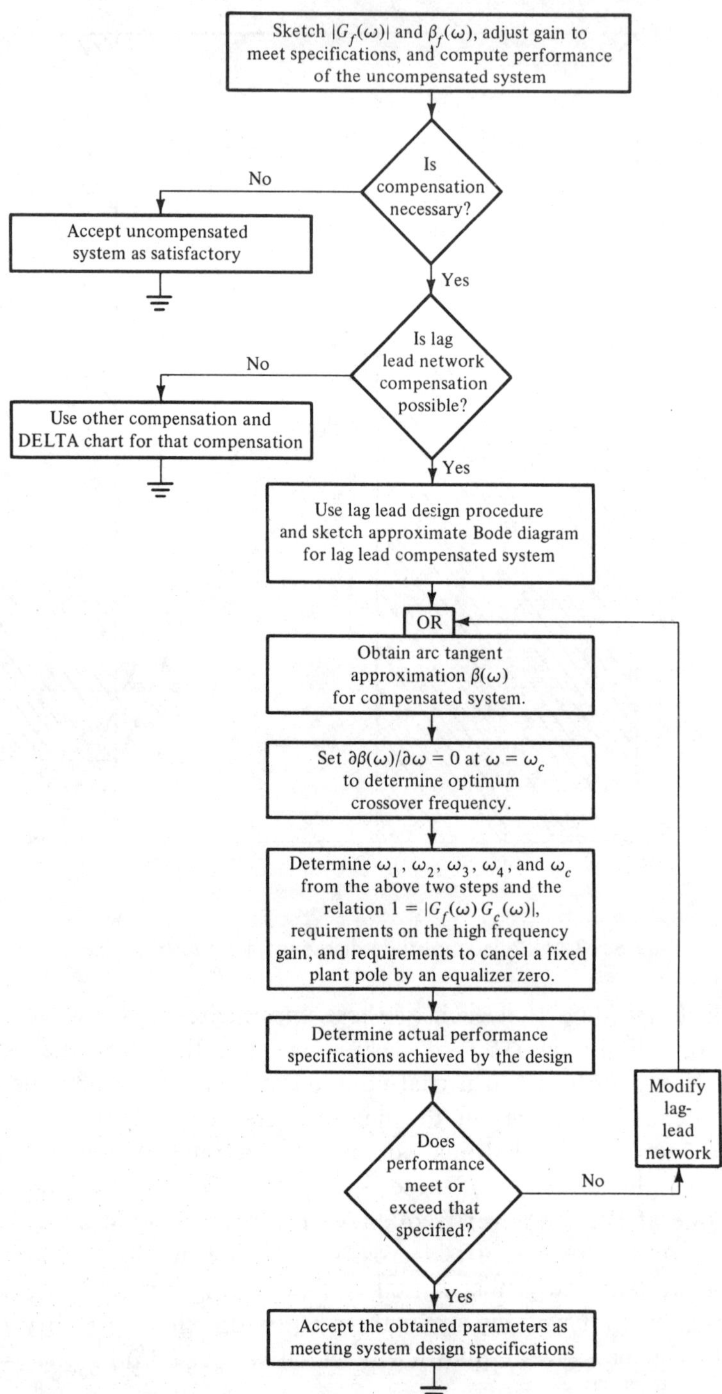

Figure 5.3-5 DELTA Chart for lag lead network design

5.3-2 General Bode Diagram Design

Figure (5.3-6) presents a DELTA Chart of a general design procedure for Bode diagram design. As we will see in the next section, a minor modification of this DELTA Chart can be used to accomplish design using minor loop feedback or a combination of minor loop and series equalizations. These detailed DELTA Charts for Bode diagram design are, of course, part of the overall design procedure of Fig. (5.1).

EXERCISE 5.3-1. Determine lag lead network compensation for the fixed plant transfer function

$$G_f(s) = \frac{100}{s(1 + s/10)}$$

in order to meet the following specifications

1. $K_v = 100$
2. Phase Margin $= 60°$
3. Sinusoidal inputs for frequencies of less than 1 radian per second must be reproduced with less than 1% error
4. The input output attenuation for frequencies greater than 1000 radians per second must be at least 10.

EXERCISE 5.3-2. Determine which of the following open loop transfer functions can be compensated for a 45° phase margin by

1. gain adjustment.
2. lead network.
3. lag network compensation.
4. lag lead network compensation.

In each case sketch a Bode amplitude diagram of the compensated network transfer function

a. $G_f(s) = \dfrac{100}{s^3}$

b. $G_f(s) = \dfrac{100}{s^2(1 + s)}$

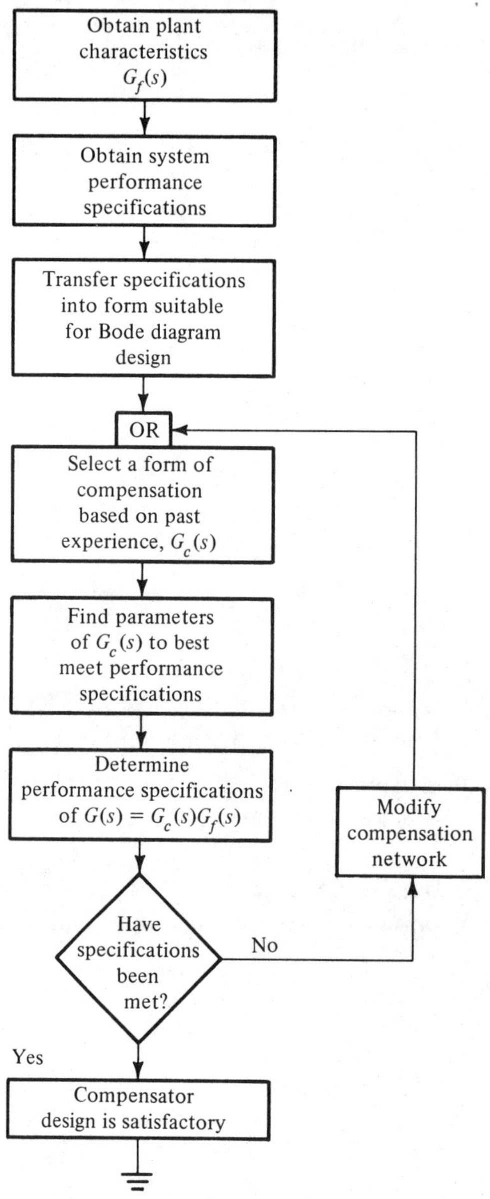

Figure 5.3-6   DELTA Chart of general Bode diagram design procedure

c. $G_f(s) = \dfrac{100(1-s)}{s(1+s)}$

d. $G_f(s) = \dfrac{1000}{(1+s/1000)^4}$

EXERCISE 5.3-3. Determine lag lead network compensation for the fixed plant transfer function

$$G_f(s) = \frac{1000}{s(1 + s/100)(1 + s/1000)^2}$$

with design specifications

    a. $K_v = 1000$

    b. Phase Margin $= 45°$

    c. Sinusoidal inputs at frequencies up to 10 radians per second must be passed with less than 2% error.

    d. Sinusoidal inputs at frequencies greater than 1000 radians per second must be attenuated by a factor of at least 20.

It is not difficult to think of a number of realistic fixed plant closed loop specification combinations such that compensation by one of the four basic compensation networks given here is not possible. Thus far we have considered:

    1. An attenuation network

    2. A lead network

    3. A lag network

    4. A lag lead network

Obviously, we can combine any of the basic networks suggested above and, as we have seen, a lag lead network is a combination of a lag network and a lead network. We will consider some examples as indicative of the general procedures to be followed.

EXAMPLE 5.3-3. Suppose that we have a double integrator fixed plant with

$$G_f(s) = \frac{100}{s^2} \tag{1}$$

Suppose further that performance specifications are:

    1. $K_a = 100$

    2. Phase Margin $= 45°$

3. Sinusoidal inputs at frequencies greater than 40 radians per second must be attenuated by a factor of at least 10.

Figure (5.3-7) illustrates the gain magnitude for the open loop system and the forbidden high frequency region. It is clear that we cannot use a lead network for compensation since the gain at high frequencies cannot be raised. A lag network will not work for there is no frequency where the fixed plant has a phase shift of less than 180° phase lag. A lag lead network might work and we try a special form of lag lead network with

$$G_c(s) = \frac{(1 + s/\omega_2)^2}{(1 + s/\omega_1)(1 + s/\omega_4)} \tag{2}$$

and will restrict the high frequency gain to one by requiring

$$\omega_1 \omega_4 = \omega_2^2 \tag{3}$$

In order to restrict the gain at crossover to one we require

$$1 = \frac{100\omega_1}{\omega_c \omega_2^2} \tag{4}$$

The phase shift in the vicinity of crossover is

$$\beta(\omega) = -\pi - \left(\frac{\pi}{2} - \frac{\omega_1}{\omega_c}\right) + 2\left(\frac{\pi}{2} - \frac{\omega_2}{\omega_c}\right) - \frac{\omega_c}{\omega_4} \tag{5}$$

In order to obtain maximum crossover frequency we set

$$\left.\frac{\partial \beta(\omega)}{\partial \omega}\right|_{\omega=\omega_c} = 0$$

and thus have

$$\omega_c^2 = \omega_4(2\omega_2 - \omega_1) \tag{6}$$

To obtain a 45° phase margin we set $\beta(\omega_c) = -3\pi/4$. Eq. (5) becomes

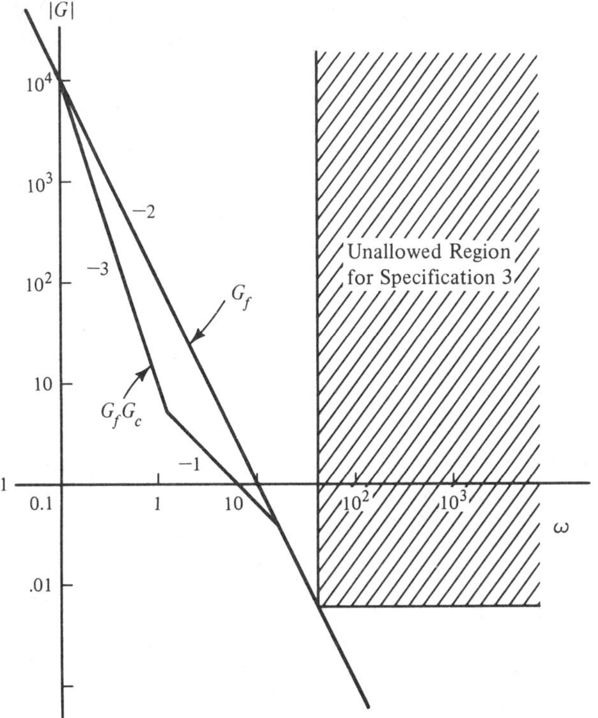

Figure 5.3-7 Bode diagram for EXAMPLE 5.3-3

$$\frac{-\pi}{4} = \frac{\omega_1 - 2\omega_2}{\omega_c} - \frac{\omega_c}{\omega_4} \tag{7}$$

Simultaneous solution of Eqs. (3), (4), (6), and (7) leads to the compensating network break frequencies. Combination of Eqs. (6) and (7) gives

$$\omega_c = \frac{\pi}{8}\,\omega_4 \tag{8}$$

Substitution of Eq. (8) into Eq. (4) yields

$$\frac{\pi}{8}\,\omega_4\,\omega_2^2 = 100\omega_1 \tag{9}$$

Combination of Eqs. (3) and (9) leads to $\omega_4 = 15.96$. We then obtain $\omega_c = 6.27$ from Eq. (8). Combination of Eqs. (3) and (6) to eliminate ω_1 leads to

$$\omega_2^2 - 2\omega_2\omega_4 + \omega_c^2 = 0 \qquad (10)$$

We may easily solve this quadratic equation for ω_2 and obtain

$$\omega_2 = \omega_4 \pm \omega_4 \sqrt{1 - (\omega_c^2/\omega_4^2)}$$

Using the known values of ω_4 and ω_c we obtain $\omega_2 = 1.28$ and $\omega_2 = 30.64$. We must reject the $\omega_2 = 30.64$ solution since we have assumed that $\omega_2 < \omega_c$ and $\omega_c = 6.27$. Finally we obtain $\omega_1 = 0.103$ from Eq. (4) and this completes our design. The compensated open loop system is shown in Fig. (5.3-7).

All of the examples we have considered in this chapter lead to the conclusion that satisfactory linear systems control design is such that the crossover frequency occurs on a -1 slope gain curve. In the vicinity of crossover we may approximate any minimum phase transfer function, with crossover on a -1 slope, by

$$G(s) = G_f(s)G_c(s) = \frac{\omega_c\,\omega_1^{n-1}\,(1 + s/\omega_1)^{n-1}}{s^n(1 + s/\omega_2)^{m-1}} \qquad \omega_1 < \omega_c < \omega_2 \quad (5.3\text{-}2)$$

where ω_1 is the break frequency just prior to crossover and ω_2 is the break frequency just after crossover. It is easy to verify that, in Eq. (5.3-2), we have $|G(j\omega_c)| = 1$ if $\omega_1 < \omega_c < \omega_2$. Figure (5.3-8) illustrates this rather general approximation to a compensated system Bode diagram in the vicinity of the crossover frequency. We will conclude this section by determining some general design requirements for a system with the transfer function of Eq. (5.3-2) and Fig. (5.3-8).

There are three unknown frequencies in Eq. (5.3-2). Thus we need three requirements or equations in order to determine design parameters. We will use the same three requirements used thus far in all of our efforts in this chapter, namely:

1. The gain at crossover is one.

2. The phase margin is $45°$ or any other specified value.

3. The phase margin at crossover is the maximum possible for a given ω_2/ω_1 ratio.

We see that the first requirement, that the gain is one at the crossover frequency, is satisfied by Eq. (5.3-2) if crossover occurs on the -1 slope portion of the gain curve as assumed in Fig. (5.3-8). We use the arctangent approximation to obtain the phase shift in the vicinity of crossover as

$$\beta(\omega) = -\frac{n\pi}{2} + (n-1)\left(\frac{\pi}{2} - \frac{\omega_1}{\omega_c}\right) - (m-1)\frac{\omega_c}{\omega_2} \qquad (5.3\text{-}3)$$

To satisfy requirement 3 we set

$$\frac{\partial\beta(\omega)}{\partial\omega}\bigg|_{\omega=\omega_c} = 0 = \frac{(n-1)\omega_1}{\omega_c^2} - \frac{m-1}{\omega_2}$$

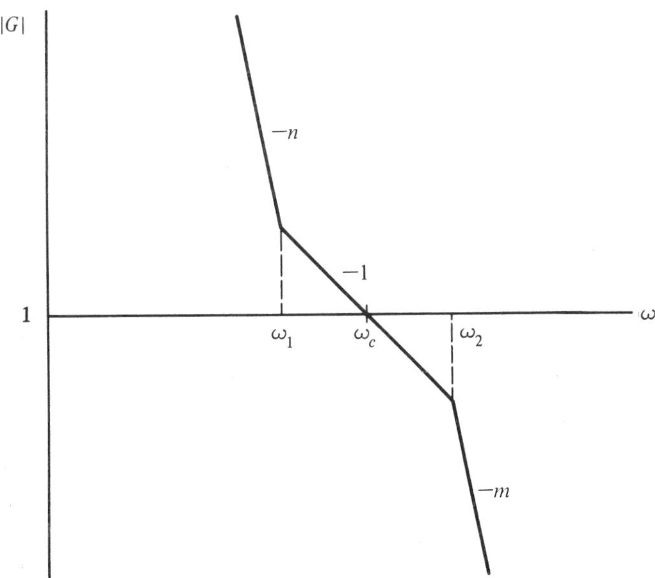

Figure 5.3-8 General compensated system Bode diagram in the vicinity of

and obtain

$$\omega_c^2 = \frac{n-1}{m-1}\,\omega_1\omega_2 \tag{5.3-4}$$

as the optimum setting for the crossover frequency.

Substitution of the "optimum" frequency given by Eq. (5.3-4) into Eq. (5.3-3) yields

$$\beta(\omega_c) = \frac{-\pi}{2} - 2\sqrt{(m-1)(n-1)}\,\sqrt{\omega_1/\omega_2} \tag{5.3-5}$$

We desire a specific phase margin here, $\beta(\omega_c) = -3\pi/4$, and so the equalizer break frequency locations are specified by Eq. (5.3-4) and

$$\sqrt{(m-1)(n-1)}\,\sqrt{\omega_1/\omega_2} = \frac{\pi}{8} \tag{5.3-6}$$

There is a single parameter here which is unspecified and an additional equation must be found in any specific application. Alternately we could simply assume a nominal crossover frequency of unity or simply normalize frequencies ω_1 and ω_2 in terms of the crossover frequency by use of the normalized frequencies

$$\omega_1 = W_1\omega_c \qquad \omega_2 = W_2\omega_c$$

such that Eqs. (5.3-4) and (5.3-6) become

$$1 = \frac{n-1}{m-1}\,W_1 W_2 \tag{5.3-7}$$

and

$$\sqrt{(m-1)(n-1)}\,\sqrt{W_1/W_2} = \frac{\pi}{8} \tag{5.3-8}$$

Solution of these two equations yields

$$W_1 = \frac{\omega_1}{\omega_c} = \frac{\pi}{(n-1)8}$$

$$\tag{5.3-9}$$

$$W_2 = \frac{\omega_2}{\omega_c} = \frac{8(m-1)}{\pi}$$

It is a relatively simple matter to show that for a specified phase margin (PM), expressed in radians, the foregoing relations become

$$W_1 = \frac{\omega_1}{\omega_c} = \frac{PM}{2(n-1)}$$

$$(5.3\text{-}10)$$

$$W_2 = \frac{\omega_2}{\omega_c} = \frac{2(m-1)}{PM}$$

It is helpful to develop a set of design charts based on these relations. Either the design charts shown in Fig. (5.3-9), and TABLE 5.3-1 or Eq. (5.3-9) or Eq. (5.3-10) if the phase margin is other than $\pi/4$, may be used as aids to design. They indicate how we will have to structure the system in the vicinity of crossover in order to obtain a specified phase

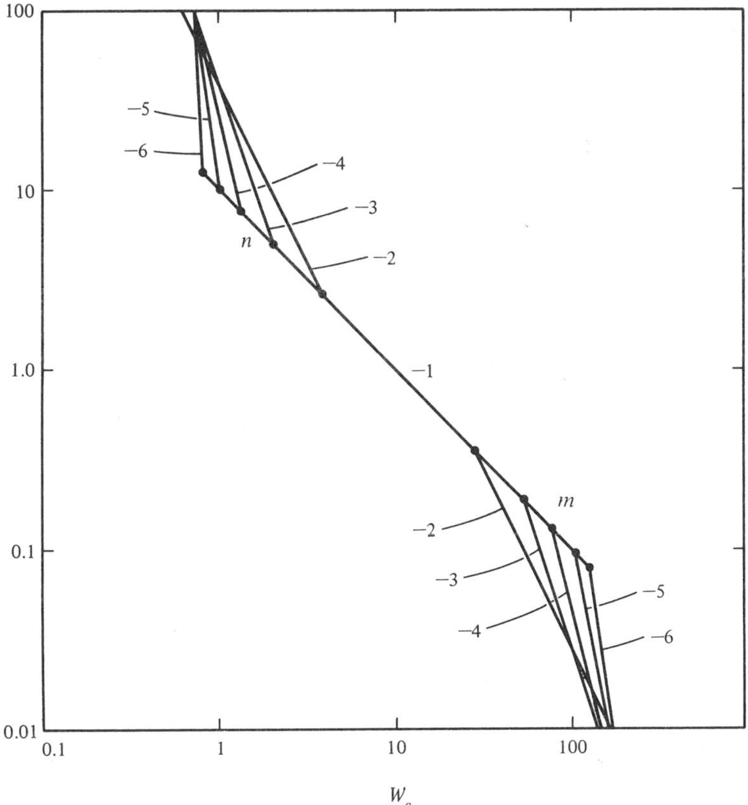

Figure 5.3-9   Design chart for a 45° phase margin

TABLE 5.3-1   DESIGN TABLE FOR 45° PHASE MARGIN WITH CROSSOVER BEHAVIOR OF EQ. (5.3-2)

M	W_2	N	W_1
1	$\emptyset$	1	$\emptyset$
2	2.55	2	0.39
3	5.09	3	0.20
4	7.64	4	0.13
5	10.19	5	0.10
6	12.73	6	0.08

margin. As we would suspect the length of the -1 slope depends upon the number of break frequencies above and below crossover and increases as the number of zeros and poles increases.

EXAMPLE 5.3-4. To illustrate use of this design chart approach to Bode diagram design for a specified phase margin we reconsider EXAMPLE 5.3-3. We pick a compensating network structure that is the same as in EXAMPLE 5.3-3 and so have the compensated network

$$G_f(s)G_c(s) = \frac{100(1 + s/\omega_2)^2}{s^2(1 + s/\omega_1)(1 + s/\omega_4)} \tag{1}$$

In the vicinity of crossover we assume that the pole at $s = -\omega_1$ is replaced by an integrator by the approximation

$$\frac{1}{1 + s/\omega_1} \cong \frac{\omega_1}{s}$$

such that the compensated open loop system becomes

$$G_f(s)G_c(s) \cong \frac{100\omega_1(1 + s/\omega_2)^2}{s^3(1 + s/\omega_4)} \tag{2}$$

The normalized break frequencies can be determined from Eq. (5.3-9) or TABLE 5.3-1 as

$$W_1 = 0.20 \qquad W_2 = 2.55$$

The actual crossover frequency can be determined from Eq. (2) as the frequency where the gain is one,

$$\omega_c = \frac{100\omega_1}{\omega_2^2} \tag{3}$$

The pole location after crossover is denoted $-\omega_4$ in Eq. (2) and it is this pole location which, when normalized, becomes W_2. Thus we have

$$\omega_4 = \omega_c W_2 = \frac{255\omega_1}{\omega_2^2} \tag{4}$$

In a similar fashion the zero location before crossover is denoted $-\omega_2$ in Eq. (2). It is this zero location which, when normalized, becomes W_1. Thus we have

$$\omega_2 = \omega_c W_1 = \frac{20\omega_1}{\omega_2^2} \tag{5}$$

or

$$\omega_2 = 2.71\omega_1^{1/3} \tag{6}$$

thus

$$\omega_4 = \frac{255\omega_1}{\omega_2^2} = 34.6\omega_1^{1/3} \tag{7}$$

The transfer function of Eq. (1) becomes

$$G_f(s)G_c(s) = \frac{100(1 + s/2.71\omega_1^{1/3})^2}{s^2(1 + s/\omega_1)(1 + s/34.6\omega_1^{1/3})} \tag{8}$$

and crossover occurs at

$$\omega_c = \frac{100\omega_1}{\omega_2^2} = 13.57\omega_1^{1/3} \tag{9}$$

We have now gone as far as we can using the design charts. The system has been optimally designed in terms of having

1. Gain at crossover equals 1

2. Phase shift at crossover $= \dfrac{-3\pi}{4}$

3. Maximum phase margin

for the approximate system represented by Eq. (2). One parameter remains to be chosen: ω_1. This should be chosen to meet the specification that the open loop compensated system behaves as the fixed plant at high frequencies. At frequencies above ω_4, we may approximate Eq. (8) as

$$\left| G_f(\omega)G_c(\omega) \right| \cong \frac{100(34.6)\omega_1^{4/3}}{\omega^2 (2.71\omega_1^{1/3})^2}$$

$$\cong \frac{454.21\omega_1^{2/3}}{\omega^2}$$

and in order for this to behave as $100/\omega^2$ we must have

$$\omega_1 = 0.10$$

We then obtain, from Eqs. (6), (9), and (7)

$$\omega_2 = 1.27 \qquad \omega_c = 6.20 \qquad \omega_4 = 16.06$$

The system designed using the design chart approach suggested here is almost identical to the results of EXAMPLE 5.3-3 which were

$$\omega_1 = 0.10 \qquad \omega_2 = 1.28 \qquad \omega_c = 6.27 \qquad \omega_4 = 15.96$$

and the only reason for any difference at all is that the pole at $s = -\omega_1$ in EXAMPLE 5.3-3 is assumed to occur at the origin in this design chart approach which only considers the most significant break frequencies which occur just above and just below crossover.

Figure (5.3-9) presents a DELTA Chart illustrating a suggested Bode diagram design procedure based upon the design chart we have been using here which considers only these break frequencies immediately above and below crossover and which approximates all others. Break frequencies far below crossover are approximated by integrations or differentiations (poles or zeros at $s = 0$) and break frequencies far above crossover are ignored.

EXAMPLE 5.3-5. In our concluding example of this section we will reconsider EXAMPLE 5.3-2 using the approach suggested here. As suggested by the DELTA Chart of Fig. (5.3-10) we rewrite the compensated system transfer function, which is Eq. (2) of EXAMPLE (5.3-1),

$$G_c(s)G_f(s) = \frac{1000(1 + s/\omega_2)}{s(1 + s/\omega_1)(1 + s/\omega_4)} \tag{1}$$

by considering only these break frequencies just above and just below crossover. We obtain

$$G_c(s)G_f(s) = \frac{1000\omega_1(1 + s/\omega_2)}{s^2(1 + s/\omega_4)} \tag{2}$$

and immediately recognize that this transfer function fits the requirements for our design approach where $m = n = 2$. From TABLE 5.3-1 we have $W_1 = 0.39$ and $W_2 = 2.55$. The crossover frequency is obtained from Eq. (2) as that frequency where the gain is one, or

$$\omega_c = \frac{1000\omega_1}{\omega_2}$$

We employ the normalization equations and find that, for the poles at $-\omega_2$ and $-\omega_4$ in Eq. (2)

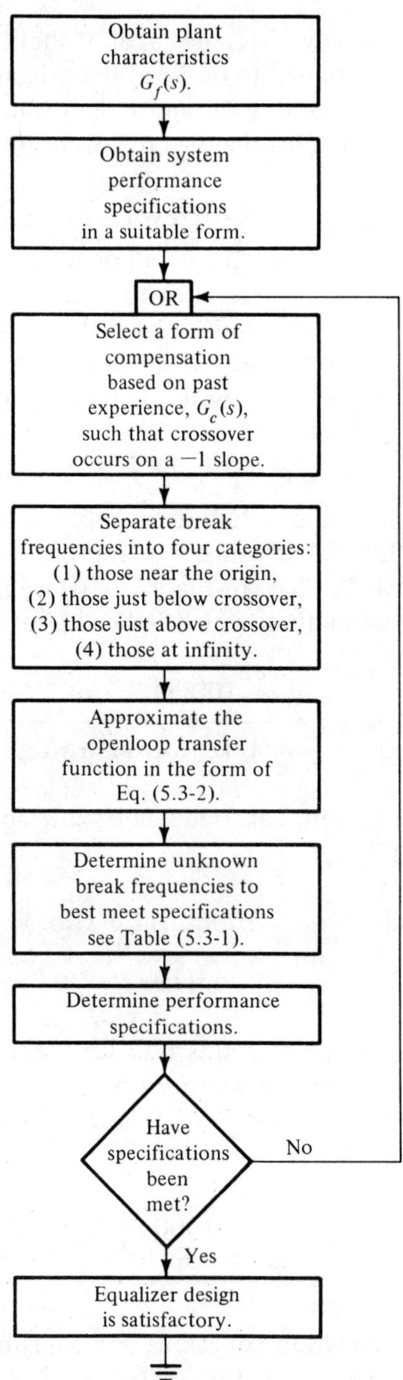

Figure 5.3-10 DELTA Chart of approximate crossover behavior design procedure

$$\omega_2 = W_1 \omega_c = \frac{390\omega_1}{\omega_2}$$

$$\omega_4 = W_2 \omega_c = \frac{2550\omega_1}{\omega_2}$$

which we rewrite as

$$\omega_2 = 19.75\omega_1^{1/2}$$

$$\omega_4 = 129.11\omega_1^{1/2}$$

Again there is a single unspecified parameter, in this case ω_1. We have utilized our normalized design approach and now must obtain the parameter ω_1 to satisfy specific system requirements. For EXAMPLE 5.3-2 specification 4 required that the input output attenuation be greater than 25 for frequencies greater than 1000 radians per second. This requires that the open loop gain be less than 0.04 for frequencies greater than 1000. Letting the gain magnitude of Eq. (1) equal to 0.04 at $\omega = 1000$ yields $\omega_1 = 6.12$. The system is completely specified by the parameters

$$\omega_1 = 6.12 \qquad \omega_2 = 48.85 \qquad \omega_c = 125.28 \qquad \omega_4 = 319.37$$

Again we see that the results obtained from using this approach are very close to those obtained by our earlier more exact approach. The small differences are due to considering only those break frequencies just above and just below crossover in the design chart approach.

EXERCISE 5.3-4. Rework EXAMPLE 5.3-1 using the design chart approach we have presented here.

EXERCISE 5.3-5. The fixed plant of a control system has the transfer function

$$G_f(s) = \frac{500}{s^2(1 + s/10)}$$

Determine a maximum crossover frequency system including the compensating network and associated break frequencies to meet the following specifications

a. $K_a = 500$

b. Phase Margin $= 45°$

c. Sinusoidal inputs for frequencies greater than 20 radians per second must be attenuated by a factor of at least 10.

EXERCISE 5.3-6. Rework the lead network design EXAMPLE 5.2-3 using the design chart approach suggested here.

EXERCISE 5.3-7. Rework the lag network design EXAMPLE 5.2-4 using the design chart approach suggested here.

EXERCISE 5.3-8. Consider lead network equalization of the fixed plant $G_f(s) = \dfrac{1000}{s^3}$ to produce a $45°$ phase margin. Design a minimum gain two stage lead network $G_c(s) = \dfrac{(1 + s/\omega_1)^2}{(1 + s/\omega_2)^2}$ which meets specifications. Use the approach suggested in Sec. 5.3-2 as well as the design chart approach suggested here. Contrast and compare the results of the two approaches.

5.4 MINOR LOOP DESIGN

In our efforts thus far in this chapter we have assumed that compensating networks would be placed in series with the fixed plant and then a unity feedback ratio loop closed around these elements to yield the closed loop system. In many applications it may be physically convenient, perhaps because of instrumentation considerations, to use one or more minor loops in order to obtain a desired compensation of a fixed plant transfer function.

For a single input single output linear system there are no theoretical advantages whatever to any minor loop compensation to series compensation as the same closed loop transfer function can be realized by all procedures. However when there are multiple inputs or outputs then there may be considerable advantages to minor loop design as contrasted to series compensation design. Multiple inputs often occur when there is a single signal input and one or more noise or disturbance inputs present and a task of the system is to pass the signal inputs and reject the noise inputs. Also there may be saturation type nonlinearities present and we may be concerned not only with the primary system output but also with keeping the output at the saturation point within bounds such that the system remains linear. Thus there are reasons why minor loop design may be preferable to series equalization.

We have discussed block diagram reduction in Chap. 2. It is desirable here to review some concepts that will be of value for our discussion of minor loop design. Figure (5.4-1) illustrates a relatively general linear control system with a single minor loop. This block diagram could represent many simple control systems. $G_1(s)$ could represent a discriminator and series compensation, $G_2(s)$ an amplifier and that part of a motor transfer function excluding the final integration to convert velocity to position. $G_3(s)$ would then represent an integrator. $G_4(s)$ would represent a minor loop compensation transfer function, such as that of a tachometer.

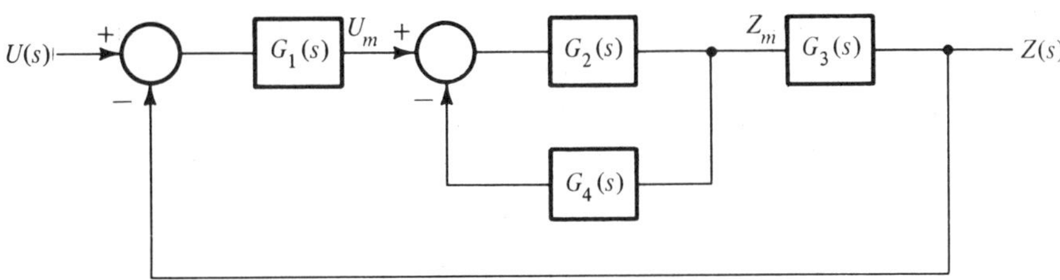

Figure 5.4-1 Block diagram of unity feedback ratio system with a single minor loop

The closed loop transfer function for this system is

$$\frac{Z(s)}{U(s)} = H(s) = \frac{G_1(s)G_2(s)G_3(s)}{1 + G_2(s)G_4(s) + G_1(s)G_2(s)G_3(s)} \tag{5.4-1}$$

It is convenient to define several other transfer functions from Fig. (5.4-1). First there is the minor loop gain

$$G_m(s) = G_2(s)G_4(s) \tag{5.4-2}$$

which is just the loop gain of the minor loop only as shown in Fig. (5.4-2). The minor loop has the transfer function

$$\frac{Z_m(s)}{U_m(s)} = H_m(s) = \frac{G_2(s)}{1 + G_2(s)G_4(s)} = \frac{G_2(s)}{1 + G_m(s)} \tag{5.4-3}$$

There will usually be a range or ranges of frequency for which the minor loop gain magnitude is much less than one and we then have

$$\frac{Z_m(s)}{U_m(s)} = H_m(s) \cong G_2(s) \qquad |G_m(\omega)| \ll 1 \qquad (5.4\text{-}4)$$

There will also generally be ranges of frequency for which the minor loop gain magnitude is much greater than one and we then have

$$\frac{Z_m(s)}{U_m(s)} = H_m(s) \cong \frac{1}{G_4(s)} \qquad |G_m(\omega)| \gg 1 \qquad (5.4\text{-}5)$$

We will use these two relations to considerably simplify our approach to the minor loop design problem.

We will use two major loop gain functions. First we will consider the major loop gain with the minor loop compensating network removed such that $G_4(s) = 0$. This represents the standard situation we have examined in the last section. This uncompensated major loop transfer function is

$$G_{Mu}(s) = G_1(s)G_2(s)G_3(s) \qquad (5.4\text{-}6)$$

With the minor loop compensation inserted, the major loop gain, the input output transfer function with the unity ratio feedback open, is

$$G_{Mc}(s) = \frac{G_1(s)G_2(s)G_3(s)}{1 + G_m(s)} \qquad (5.4\text{-}7)$$

The complete closed loop transfer function is expressable as

$$\frac{Z(s)}{U(s)} = H(s) = \frac{G_{Mc}(s)}{1 + G_{Mc}(s)} \qquad (5.4\text{-}8)$$

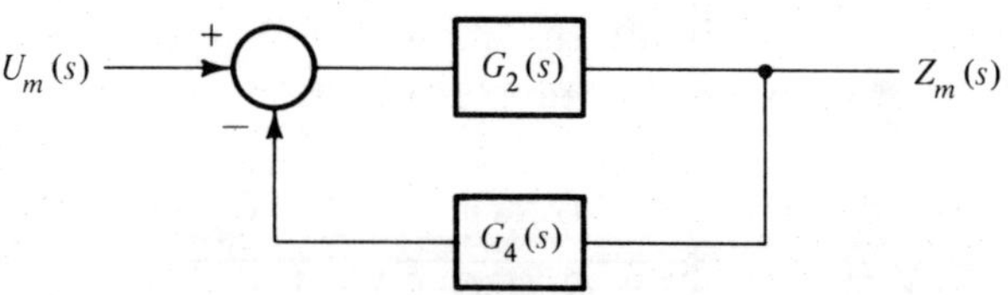

Figure 5.4-2   Block diagram of minor loop only

   A particularly useful relationship may be obtained by combining Eqs. (5.4-6) and (5.4-7) as

$$G_{Mc}(s) = \frac{G_{Mu}(s)}{1 + G_m(s)} \tag{5.4-9}$$

We may give this equation a particularly simple interpretation. For frequencies where the minor loop gain $G_m(s)$ is low the minor loop closed major loop transfer function, $G_{Mc}(s)$, is approximately that of the minor loop open major loop transfer function $G_{Mu}(s)$ in that

$$G_{Mc}(s) \cong G_{Mu}(s) \qquad |G_m(\omega)| \ll 1 \tag{5.4-10}$$

For frequencies where the minor loop gain $G_m(s)$ is high the minor loop closed major loop transfer function is just

$$G_{Mc}(s) \cong \frac{G_{Mu}(s)}{G_m(s)} \qquad |G_m(\omega)| \gg 1 \tag{5.4-11}$$

This has an especially simple interpretation on the log-log plots we use for Bode diagrams for we may simply subtract the minor loop gain, $G_m(s)$, from the minor loop open major loop gain, $G_{Mu}(s)$, to obtain the compensated system gain as transfer function $G_{Mc}(s)$.

EXAMPLE 5.4-1. To illustrate some of the essentials of minor loop analysis, which form the key to minor loop design, let us consider a simple minor loop analysis problem in which

$$G_1(s) = K_1 \qquad\qquad\qquad G_3(s) = 1/s$$

$$G_2(s) = \frac{K_2}{(1 + s/\omega_1)(1 + s/\omega_2)} \qquad G_4(s) = \frac{s}{1 + s/\omega_3}$$

The minor loop gain is, from Eq. (5.4-2)

$$G_m(s) = \frac{K_2 s}{(1 + s/\omega_1)(1 + s/\omega_2)(1 + s/\omega_3)} \tag{1}$$

and has the Bode amplitude diagram of Fig. (5.4-3) for $\omega_1 > \omega_2 > \omega_3$.

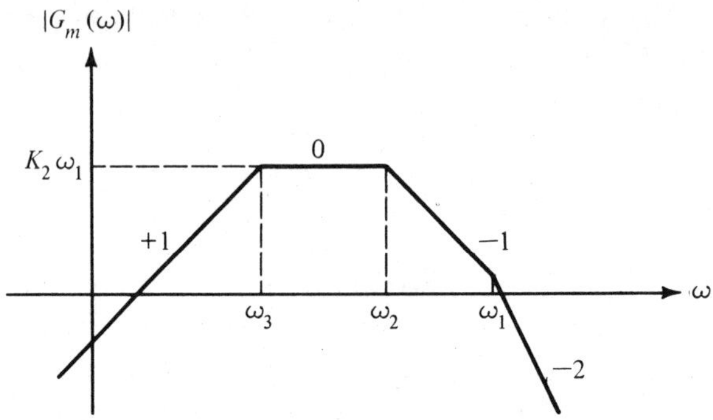

Figure 5.4-3   Minor loop gain

The minor loop open major loop transfer function is

$$G_{\mathrm{Mu}}(s) = \frac{K_1 K_2}{s(1 + s/\omega_1)(1 + s/\omega_2)} \tag{2}$$

and is shown in Fig. (5.4-4). With the minor loop closed the major loop transfer function is

$$\tag{3}$$

$$G_{\mathrm{Mc}}(s) = \frac{K_1 K_2 (1 + s/\omega_3)}{s\left[1 + s\left(\dfrac{1}{\omega_1} + \dfrac{1}{\omega_2} + \dfrac{1}{\omega_3} + K_2\right) + s^2\left(\dfrac{1}{\omega_1 \omega_2} + \dfrac{1}{\omega_2 \omega_3} + \dfrac{1}{\omega_1 \omega_3}\right) + s^3 \dfrac{1}{\omega_1 \omega_2 \omega_3}\right]}$$

and we see that this could become a rather difficult function to work with in a system of higher order. Even in this simple system we would have to factor a cubic equation in order to determine the pole locations for $G_{\mathrm{Mc}}(s)$. More importantly this would cause considerable insight into system behavior to be lost. Thus we use the minor loop gain approximations of Eqs. (5.4-4) and (5.4-5).

When $|G_{\mathrm{m}}(\omega)| \ll 1$ we have

$$G_{\mathrm{Mc}}(s) \cong G_1(s)G_2(s)G_3(s) = \frac{K_1 K_2}{s(1 + s/\omega_1)(1 + s/\omega_2)} \tag{4}$$

and when $|G_{\mathrm{m}}(\omega)| \gg 1$ we have

$$G_{\mathrm{Mc}}(s) \cong \frac{G_1(s)G_3(s)}{G_4(s)} = \frac{K_1(1 + s/\omega_3)}{s^2} \tag{5}$$

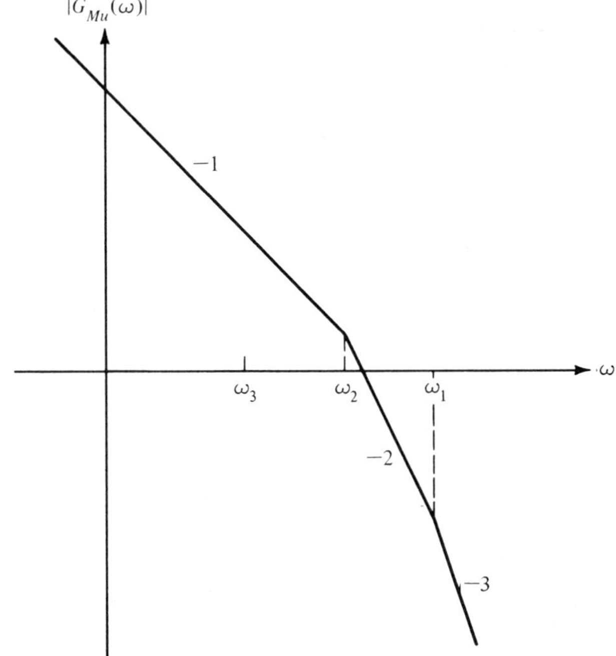

Figure 5.4-4   Major loop gain with minor loop open

Inspection of Fig. (5.4-3) indicates that at low and high frequencies Eq. (4) is the appropriate equation since the minor loop gain of Fig. (5.4-3) is less than 1. In the middle frequency range Eq. (5) is the appropriate equation. Figure (5.4-5) represents those portions of the Bode diagram which we can infer from Eqs. (4) and (5).

When we recall the basic procedure we used to construct an asymptotic Bode diagram we realize that all we have to do to determine the asymptotic gain curve for $G_{Mc}(s)$ is to extend the straight asymptotic lines in Fig. (5.4-5) until they intersect. This is precisely the same as equating Eqs. (4) and (5) to determine the frequency intersection. We have at the low frequency intersection

$$\frac{K_1 K_2}{\omega_{low}} = \frac{K_1}{\omega^2_{low}}$$

and obtain

$$\omega_{low} = \frac{1}{K_2} \tag{6}$$

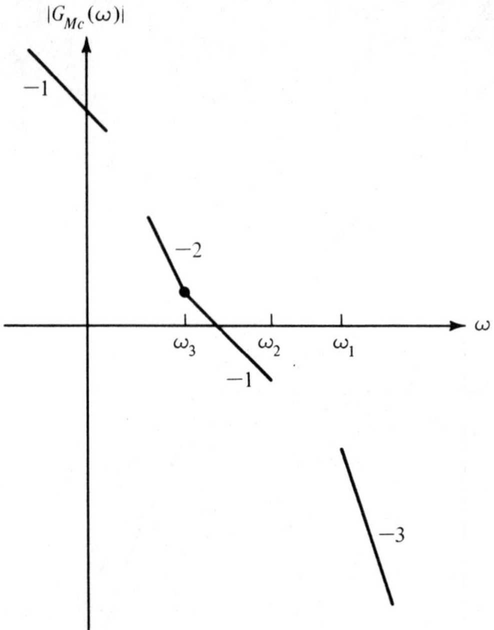

Figure 5.4-5　Partial construction of $G_{Mc}(s)$

For the high frequency intersection we have

$$\frac{K_1 K_2 \omega_1 \omega_2}{\omega_{high}^3} = \frac{K_1}{\omega_3 \omega_{high}}$$

and so

$$\omega_{high} = \sqrt{K_2 \omega_1 \omega_2 \omega_3} \tag{7}$$

Figure (5.4-6) indicates the Bode diagram for $G_{Mc}(s)$. It is this Bode diagram which we should examine to determine system phase margin.

The procedure we have used here is certainly an approximate one although it is a very realistic one both from the point of view of accuracy and general design purposes. For example if the original parameters are $K_1 = 10^5$, $K_2 = 10^{-1}$, $\omega_1 = 10^4$, $\omega_2 = 10^3$, and $\omega_3 = 10^2$ then the minor loop open major loop transfer function is as shown in Fig. (5.4-7) and given by

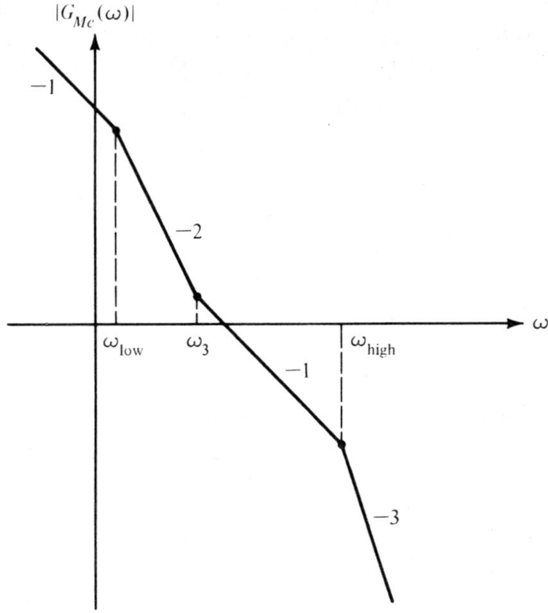

Figure 5.4-6 Bode diagram for $G_{Mc}(s)$

$$G_{Mu}(s) = \frac{10^4}{s(1 + s/10^4)(1 + s/10^3)}$$

The minor loop transfer function is

$$G_m(s) = \frac{10^{-1}s}{(1 + s/10^2)(1 + s/10^3)(1 + s/10^4)}$$

and this is also shown in Fig. (5.4-7). For frequencies where $|G_m(\omega)|$ $\ll 1$ the minor loop closed major loop transfer function $G_{Mc}(s)$ is just $G_{Mu}(s)$, the minor loop open major loop transfer function. For frequencies where $|G_m(\omega)| \gg 1$ the minor loop closed major loop transfer function is approximately

$$G_{Mc}(s) = \frac{10^5(1 + s/10^2)}{s^2} \qquad |G_m(\omega)| \gg 1$$

Both of these approximations are indicated by dotted lines on Fig. (5.2-7). We extend these asymptotic straight lines and see that they intersect at $\omega_{low} = 10$ and $\omega_{high} = 10^4$. Thus the (approximate) minor loop closed major loop transfer function is

$$G_{Mc}(s) = \frac{10^4(1 + s/10^2)}{s(1 + s/10)(1 + s/10^4)^2}$$

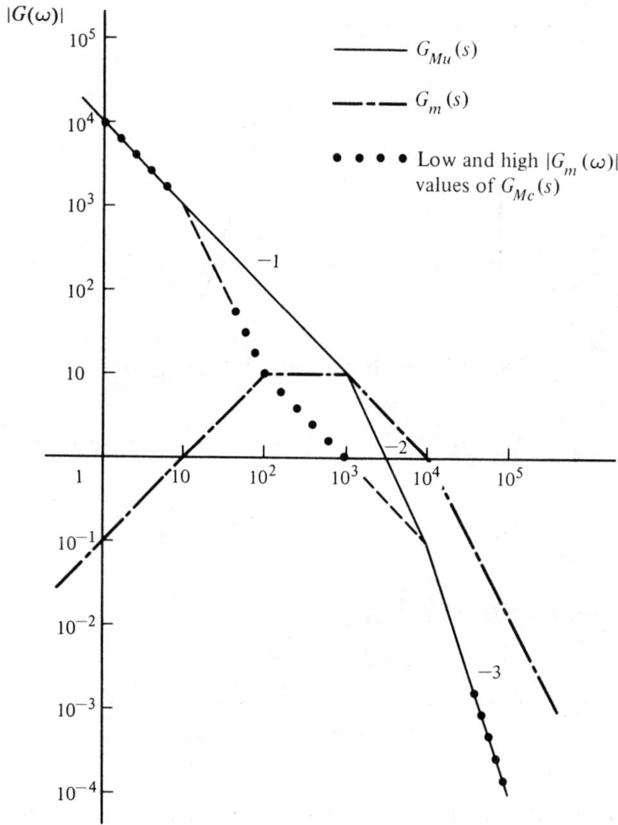

Figure 5.4-7 Minor loop Bode diagram analysis EXAMPLE (5.4-1)

If the $G_{Mu}(s)$ represented a fixed plant transfer function and $G_{Mc}(s)$ the transfer function of the compensated system we see that we have increased the phase margin by minor loop compensation.

EXERCISE 5.4-1. Use Eq. (3) of EXAMPLE 5.4-1 to find the exact pole zero locations and transfer function $G_{Mc}(s)$ and compare with the results of the example given here.

We may easily convert our analysis procedure for systems with minor loop feedback to a design or synthesis procedure. The basic procedure is in fact rather simple. We first determine the characteristics or transfer function of the fixed plant. From the system specifications and the fixed plant transfer function we determine the major loop gain $G_{Mc}(s)$ which will meet specifications and which meets certain guidelines such that minor loop compensation is possible. We extract any part of the compensation $G_{Mc}(s)/G_f(s)$ which we might wish to realize using series compensation and then design a minor loop equalizer to realize the overall $G_{Mc}(s)$. Let us examine two examples and then state a general design procedure in the form of a DELTA Chart.

EXAMPLE 5.4-2. We will reexamine EXAMPLE 5.3-3 in which the fixed plant had the transfer function

$$G_f(s) = \frac{100}{s^2} \tag{1}$$

and the compensated major loop transfer function was

$$G_{Mc}(s) = \frac{100(1 + s/1.28)^2}{s^2(1 + s/0.103)(1 + s/15.96)} \tag{2}$$

In EXAMPLE 5.3-3 we realized the compensated major loop by means of a lag lead network series compensation

$$G_c(s) = \frac{(1 + s/1.28)^2}{(1 + s/0.103)(1 + s/15.96)} \tag{3}$$

The fixed plant and compensated system Bode diagrams are shown in Fig. (5.4-8) and these diagrams are independent of the particular form of compensation selected.

We shall assume that all compensation is to be inserted in a minor loop. Figure (5.4-9) illustrates a block diagram of the assumed system. For this example the minor loop open major loop transfer function is

$$G_{\mathrm{Mu}}(s) = G_{\mathrm{f}}(s) = \frac{100}{s^2} \tag{4}$$

We can determine the minor loop compensation by comparison of $G_{\mathrm{Mu}}(s)$ and $G_{\mathrm{Mc}}(s)$. At low frequencies and at high frequencies the compensated system $G_{\mathrm{Mc}}(s)$ is the same as the fixed plant. Thus at low and high frequencies, the minor loop gain $G_{\mathrm{m}}(s)$ must be less than one.

At frequencies above 0.103 radians per second the compensated system gain is less than the fixed plant gain. The difference is due to a slope of one and so we see that the minor loop Bode diagram must go through the point where the gain is one and the frequency 0.103 with a $+1$ slope. Only in this way will the compensated system transfer function deviate from the fixed plant transfer function above $\omega = 0.103$. Let us again justify this assertion.

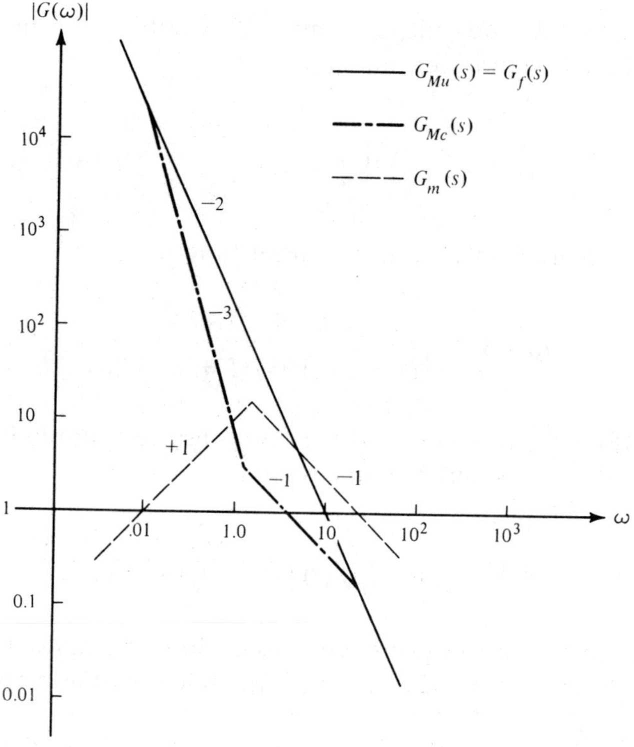

Figure 5.4-8   Bode diagrams for minor loop design EXAMPLE 5.4-2

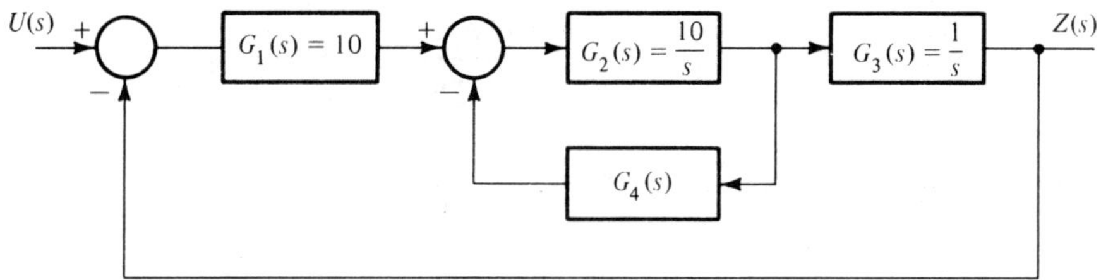

Figure 5.4-9 Block diagram for minor loop EXAMPLE 5.4-2

We may rewrite Eq. (5.4-7) as before as

$$G_{\text{Mc}}(s) = \frac{G_{\text{Mu}}(s)}{1 + G_{\text{m}}(s)} \tag{5}$$

and so we see that the minor loop transfer function may be obtained by subtracting $G_{\text{Mu}}(s)$ from $G_{\text{Mc}}(s)$ for these frequencies where the gain $|G_{\text{m}}(\omega)|$ is greater than one. For frequencies where $|G_{\text{m}}(\omega)|$ is less than one we have $G_{\text{Mc}}(s) = G_{\text{Mu}}(s)$. Thus we see that our assertion is indeed correct.

We subtract $G_{\text{Mu}}(s)$ from $G_{\text{Mc}}(s)$ on the Bode diagram of Fig. (5.4-8) for $0.103 < \omega < 15.96$. We note that the $G_{\text{m}}(s)$ so obtained has a gain magnitude of one at $\omega = 0.103$ and $\omega = 15.96$. We simply extend the curves for $G_{\text{m}}(s)$ for frequencies less than $\omega = 0.103$ and for frequencies greater than 15.96 to obtain the $G_{\text{m}}(s)$ shown in Fig. (5.4-8).

We have just obtained the minor loop transfer function

$$G_{\text{m}}(s) = \frac{s}{0.103(1 + s/1.28)^2} \tag{6}$$

This minor loop transfer function is independent of the particular forward paths about which minor loop feedback is taken. We may select the minor loop pickoff point from instrumentation considerations, output disturbance rejection considerations, ease of realization considerations, and other factors.

With the particular forward transfer function, $G_2(s) = \dfrac{10}{s}$, shown in Fig. (5.4-9), we obtain

$$G_4(s) = \frac{0.97s^2}{(1 + s/1.28)^2} \tag{7}$$

which we may realize in a variety of ways.

This completes our minor loop design for this example. We should generally verify that the minor loop is stable and has a phase margin of at least $10°$ to $15°$ or serious practical difficulties may occur if the major feedback loop is open circuited. This poses no problems for this example since minor loop crossover occurs on a $+1$ and -1 slope. The minor loop phase margin is about $90°$. For a more complex example there might well be a problem and minor loop phase margins should be calculated. If they are not sufficiently large or if the minor loop is unstable, addition or removal of forward transfer elements in the minor loop is called for, or a combination of series compensation and minor loop compensation might be tried.

To be sure, we do not obtain precisely the $G_{Mc}(s)$ of Eq. (2) with the $G_{Mu}(s) = G_f(s)$ of Eqs. (2) and (4), and the $G_m(s)$ of Eq. (6). Use of Eq. (5) leads to

$$G_{Mc}(s) = \frac{100(1 + s/1.28)^2}{s^2(1 + s/0.0892)(1 + s/18.38)} \tag{8}$$

which is very close to that obtained from EXAMPLE 5.3-3 and given by Eq. (2) of this example.

As we have mentioned, Eq. (5.4-9) and the approximations to this relation given by Eqs. (5.4-10) and (5.4-11) are the key relations for minor loop design. These relations indicate that some forms of series compensation yield a given major loop transfer function $G_{Mc}(s)$ that will not be appropriate for realization by minor loop compensation. In particular a lead network series compensation cannot be realized by means of equivalent minor loop compensation. The gain of the fixed plant $G_{Mu}(s)$ is raised at high frequencies due to the use of a lead network compensation. Equations (5.4-10) and (5.4-11) indicate that $G_{Mc}(s)$ can only be lowered by use of a minor loop gain $G_m(s)$.

A lag network series compensating network will result in a reduction in the fixed plant gain $|G_{Mu}(\omega)|$ at all high frequencies. This can only be achieved if the minor loop transfer gain $G_m(s)$ is constant for high frequencies. In some cases this may be achievable but often it will not be. Thus while it is possible to realize the equivalent of lag network series equalization by means of a minor loop equalizer it appears generally best to attempt to use minor loop equalization for systems where the low and high frequency behavior of $G_{Mu}(s)$ (or $G_f(s)$) and $G_{Mc}(s)$ are the same and where the gain magnitude of the compensated system $|G_{Mc}(s)|$ is at no frequency any greater than the gain magnitude of the fixed plant $|G_f(s)|$ or minor loop open major loop transfer function $|G_{Mu}(s)|$. Thus we see that lag lead network series equalization is an ideal type of equalization to realize by means of equivalent minor loop equalization.

Figure (5.4-10) represents a DELTA Chart of a suggested design procedure for minor loop compensator design. Let us consider an example using the suggested procedure. Following this we will demonstrate some of the many advantages of minor loop design.

EXAMPLE 5.4-3. We will reexamine EXAMPLE 5.3-2 with the intent of obtaining minor loop equalization for this problem. The fixed plant transfer function for this example was

$$G_f(s) = G_{Mu}(s) = \frac{1000}{s(1 + s/100)}$$

and series equalization led to the major loop transfer function

$$G_{Mc}(s) = \frac{1000(1 + s/97.05)}{s(1 + s/19.23)(1 + s/504.63)}$$

Figure (5.4-11) presents pertinent Bode diagrams for this example. We subtract $G_{Mu}(s)$ from $G_{Mc}(s)$ for all frequencies where these two gains differ. Then we extend the asymptotes at low and high frequencies in order to obtain the minor loop gain

$$G_m(s) = \frac{s}{19.23(1 + s/97.05)(1 + s/100)}$$

which has a very acceptable phase margin. We elect to partition the system for minor loop feedback purposes as in Fig. (5.4-12). Since

$$G_2(s) = \frac{10}{1 + s/100}$$

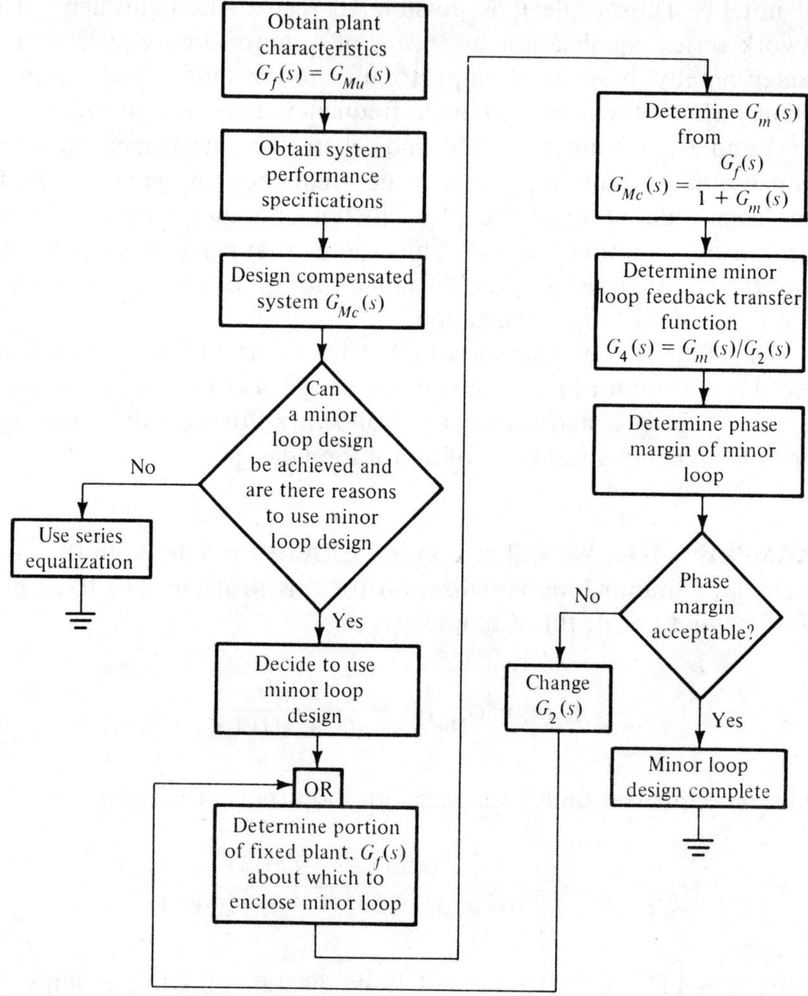

Figure 5.4-10   DELTA Chart of minor loop design process for block diagram of Fig. (5.4-1)

we have for the minor feedback loop transfer function

$$G_4(s) = \frac{s}{192.3(1 + s/97.05)}$$

which can be realized with any of several simple passive physical networks. Thus our minor loop design is complete.

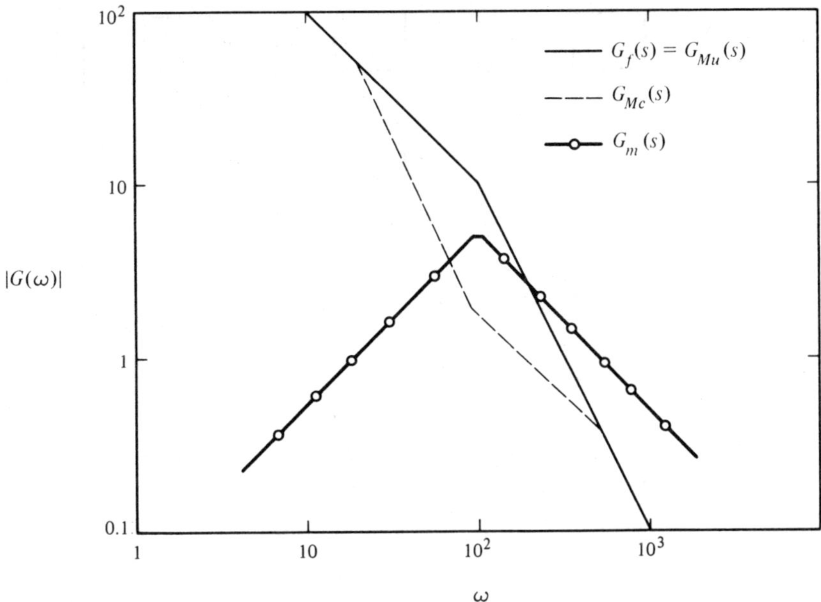

Figure 5.4-11   Bode diagrams for minor loop design EXAMPLE 5.4-3

EXERCISE 5.4-2.   Synthesize a minor loop compensation for EXAM-PLE 5.3-3.

In our examples we have enclosed only a small number of poles with our minor loop and minor loop stability did not pose a problem. If we attempt to enclose a large number of poles, stability of the minor loop will pose a problem. Three alternatives suggest themselves. One of these, that of enclosing fewer poles by the minor loop, has already been suggested. A combination of series compensation and minor loop equalization may be attempted as may the use of more than one minor loop. We will leave development of design criteria for these cases to the interested reader.

In our work thus far we have assumed that parameters were constant and known. Such is seldom the case and we must naturally be concerned with the effects of parameter variations, disturbances, and nonlinearities upon system performance. Suppose for example that we design a system with a certain gain assumed as K_1. If the system operates open loop and the gain K_1 is in cascade or series with the other

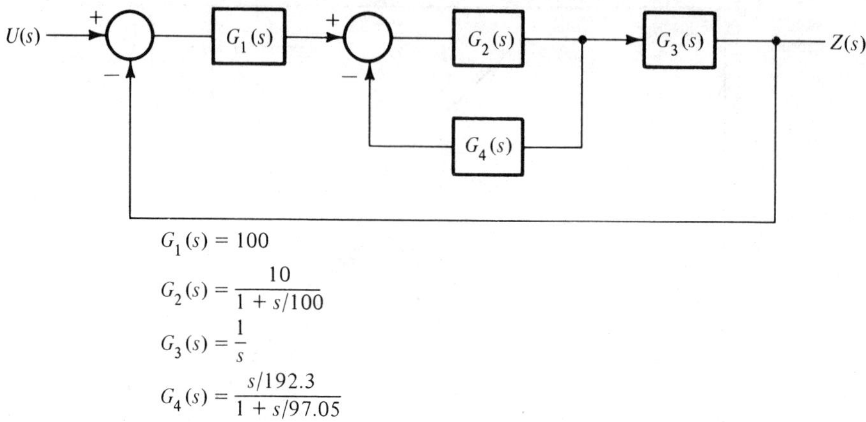

$$G_1(s) = 100$$

$$G_2(s) = \frac{10}{1 + s/100}$$

$$G_3(s) = \frac{1}{s}$$

$$G_4(s) = \frac{s/192.3}{1 + s/97.05}$$

Figure 5.4-12   Block diagram of control system elements and transfer functions for EXAMPLE 5.4-3

input output components, then the overall transfer function changes by precisely the same factor as K_1 changes. But if we have an amplifier with unity ratio feedback around a gain K_1, the situation is much different. The closed loop gain would nominally be $K_1/(1 + K_1)$ and a change to $2K_1$ would give a closed loop gain $2K_1/(1 + 2K_1)$. If K_1 is large, say 10^3, then the new gain is 0.99950025 which is a percentage change of less than 0.05% for a change in gain of 100%.

Let us expand upon this sensitivity concept. First we consider non-dynamic systems with feedback with open loop transfer function G_1 and feedback transfer function G_2. The loop gain is $G_1 G_2$ and, if we assume negative feedback, the closed loop transfer function is

$$H = \frac{G_1}{1 + G_1 G_2} \tag{5.4-12}$$

We consider that G_1 is a parameter that may vary. The change in G_1 is denoted ΔG_1 and the relative change is $\Delta G_1/G_1$. The corresponding change in H is ΔH and the relative change $\Delta H/H$. A sensitivity function $S^H_{G_1}$ is then defined as

$$S^H_{G_1} = \frac{\dfrac{\Delta H}{H}}{\dfrac{\Delta G_1}{G_1}} = \frac{\dfrac{\Delta H}{\Delta G_1}}{\dfrac{H}{G_1}} \cong \frac{\partial H}{\partial G_1} \frac{G_1}{H} \tag{5.4-13}$$

and we obtain

$$S_{G_1}^{H} = \frac{1}{1 + G_1 G_2} \tag{5.4-14}$$

which shows that the sensitivity function can be made as small as we desire by increasing the "return difference" $1 + G_1 G_2$. The sensitivity function depends upon which parameter we are changing and where that parameter is located. The sensitivity of Eq. (5.4-12) to a variation in the feedback parameter G_2 is

$$S_{G_2}^{H} = \frac{\partial H}{\partial G_2} \frac{G_2}{H} = -G_2 H \tag{5.4-15}$$

and so the system is quite sensitive to changes in G_2.

We can formally extend this sensitivity concept to dynamic systems by considering a closed loop transfer function

$$H(s) = \frac{G_1(s)}{1 + G_1(s)G_2(s)} \tag{5.4-16}$$

and show that the sensitivity defined by

$$S_{G_1}^{H}(s) = \frac{\partial H(s)}{\partial G_1(s)} \frac{G_1(s)}{H(s)} \tag{5.4-17}$$

is given by

$$S_{G_1}^{H}(s) = \frac{1}{1 + G_1(s)G_2(s)} \tag{5.4-18}$$

and note that the effect of the feedback control system in reducing sensitivity is achieved for those frequencies where $|1 + G_1(\omega)G_2(\omega)|$ is large. We can easily determine this frequency range by examining $|G_1(\omega)G_2(\omega)|$ on a Bode or Nyquist plot.

Usually we are interested in sensitivity due to changes in individual parameters of a transfer function and not just changes due to changing the entire transfer function. For example the unity ratio system with the open loop gain of that in EXAMPLE 5.4-3 is such that the phase margin decreases from $45°$ to $22°$ if the gain changes from 10^3 to 10^4. If the minor loop gain $G_m(s)$ changes by a factor of 10 then the minor loop becomes effective at an earlier frequency, about 0.2 radians per second, and continues to exert control until a later frequency, about 5500 radians per second. We recall from Eqs. (5.4-5) and (5.4-7) that

$$G_{Mc}(s) \cong \frac{G_1(s)G_3(s)}{G_4(s)} \qquad |G_2(\omega)G_4(\omega)| \gg 1 \qquad (5.4\text{-}19)$$

and so we see that in the important region near crossover the major loop gain $G_{Mc}(s)$ is nearly unaffected by this change in gain since the changing gain element has a minor loop around it.

EXERCISE 5.4-3.   Determine $G_{Mc}(s)$ in EXAMPLE 5.4-3 for the case where the gain in $G_2(s)$ is increased by a factor of 10. Determine the system phase margin.

Another advantage of minor loop feedback occurs when there are output disturbances such as those due to load disturbances caused by wind gusts on an antenna, etc. We consider the case illustrated in Fig. (5.4-13). For convenience we will assume that $G_3(s) = 1$ in our other minor loop block diagrams. The response due to $D(s)$ alone is

$$\frac{Z(s)}{D(s)} = \frac{1}{1 + G_2(s)G_4(s) + G_1(s)G_2(s)} \qquad (5.4\text{-}20)$$

When we use the relation for the minor loop gain

$$G_m(s) = G_2(s)G_4(s)$$

and the major loop gain

$$G_{Mc}(s) = \frac{G_1(s)G_2(s)}{1 + G_2(s)G_4(s)}$$

we can rewrite Eq. (5.4-20) as

$$\frac{Z(s)}{D(s)} = \frac{1}{[1 + G_m(s)][1 + G_{Mc}(s)]} \qquad (5.4\text{-}21)$$

Using the relation between the major loop gain corrected transfer function and the uncorrected function obtained as Eq. (5.4-9) we have

$$\frac{Z(s)}{D(s)} \cong \frac{1}{\left[1 + G_m(s)\right]\left[1 + \dfrac{G_{Mu}(s)}{1 + G_m(s)}\right]} \qquad (5.4\text{-}22)$$

Over the range of frequency where $|G_{Mc}(\omega)| \gg 1$ the foregoing relation may be approximated as

$$\frac{Z(s)}{D(s)} \cong \frac{1}{G_{Mu}(s)} \qquad |G_{Mc}(\omega)| \gg 1 \qquad (5.4\text{-}23)$$

Thus over the frequency range where the corrected loop gain is large the attenuation of a load disturbance is proportional to the uncorrected loop gain. This is generally larger over a wider frequency range than the corrected loop gain $G_{Mc}(\omega)$ which is what the attenuation would be if series compensation were used.

Over the range of frequencies where the minor loop gain is large but the corrected loop gain small, $|G_m(\omega)| > 1$ and $|G_{Mc}(\omega)| < 1$, Eq. (5.4-22) becomes approximately

$$\frac{Z(s)}{D(s)} \cong \frac{1}{G_m(s)} \qquad (5.4\text{-}24)$$

and the output disturbance is attenuated by the minor loop gain rather than unattenuated as would be the case if series compensation had been used.

At frequencies where both the minor loop gain transfer and the major loop gain is small we have

$$\frac{Z(s)}{D(s)} \cong 1 \qquad (5.4\text{-}25)$$

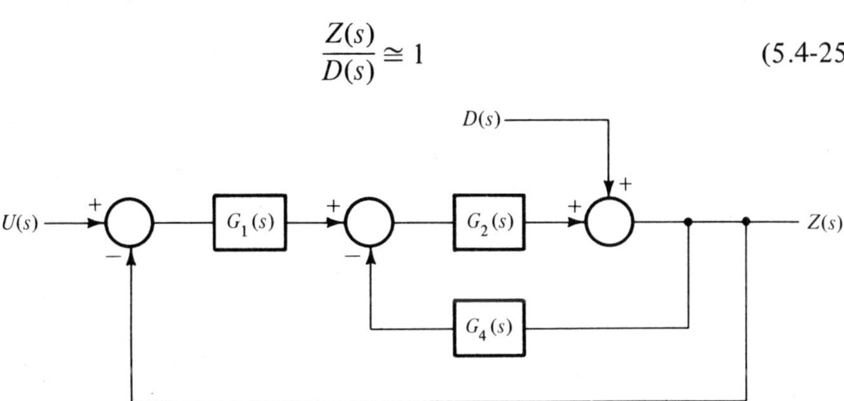

Figure 5.4-13 Block diagram of minor loop feedback system with load disturbance

and over this range of frequencies neither minor loop compensation nor series equalization is useful in reducing the effect of a load disturbance. We therefore see that there are a number of advantages to minor loop compensation as compared to series equalization.

EXERCISE 5.4-3. An uncompensated fixed plant

$$G_f(s) = \frac{39(1 + s/0.375)}{s^2(1 + s/35)^3}$$

can be made to have a $45°$ phase margin by series or minor loop compensation of the system to yield

$$G_{Mc}(s) = \frac{39(1 + s/3.9)}{s^2(1 + s/76.4)^3}$$

Discuss the use of series and minor loop compensation to achieve the specified degree of stability. In particular determine the minor loop compensation and stability characteristics of the minor loop and the disturbance response of the two systems.

5.5 SUMMARY

In this chapter, which is in many respects the focal chapter for our efforts in this text, we have examined the subject of linear system compensation by means of Bode diagrams. Our approach has been entirely in the frequency domain and has relied heavily upon performance criteria concepts we discussed in Chap. 4. We have discussed a variety of compensation networks including:

1. Gain attenuation
2. Lead networks
3. Lag networks
4. Lag-lead networks and Composite equalizers
5. Minor loop feedback

In our next chapter we will discuss the root locus method of linear systems control analysis and design. This method together with the Bode diagram approach forms a most useful and most used approach for the design of linear systems.

5.6 REFERENCES

The Bode diagram frequency domain approach to linear systems control design has been proven in extensive use for many years. Essentially every beginning control systems text discusses this method although perhaps not in quite the same way presented here nor, in many recent texts, with the same emphasis. Since this method is the primary design tool of many practitioners it is an approach that should not be slighted. Although there are many textbooks that discuss Bode diagram design techniques, a truly outstanding account is contained in:

Bower, J. L. and Schultheiss, P. M., *Introduction to the Design of Servo-mechanisms*, John Wiley and Sons, New York, N. Y., 1958.

5.7 PROBLEMS

1. Obtain a DELTA Chart for design for a specified phase margin by gain adjustment.

2. Obtain a DELTA Chart for design for maximum phase margin using a lead network for the case where the break frequency ratio ω_2/ω_1 of the lead network is fixed.

3. For the fixed plant and compensating network

$$G_f(s) = \frac{100}{s^2} \qquad G_c(s) = \frac{1 + s/\omega_1}{1 + s/10\omega_1}$$

what is the value of ω_1 which yields maximum phase margin and what is that phase margin?

4. For the fixed plant and compensating network

$$G_f(s) = \frac{100}{s^2} \qquad G_c(s) = \frac{1 + s/\omega_1}{1 + s/\alpha\omega_1}$$

find the maximum phase margin as a function of α and plot a design curve of this relationship.

5. For the fixed plant and compensating network

$$G_f(s) = \frac{100}{s(1 + s/10)} \qquad G_c(s) = \frac{1 + s/10\omega_1}{1 + s/\omega_1}$$

find the optimum value of ω_1 which yields maximum phase margin. What is this phase margin?

6. For the fixed plant and compensating network

$$G_f(s) = \frac{100}{s(1 + s/10)} \qquad G_c(s) = \frac{1 + s/\alpha\omega_1}{1 + s/\omega_1}$$

find the maximum phase margin as a function of α. Consider both $0 < \alpha < 1$ and $\alpha > 1$ and plot design curves for this example.

7. The fixed plant of a linear control system has the transfer function

$$G_f(s) = \frac{100s}{(1 + s/10)^3}$$

Discuss use of and design a minimum gain lead network compensator which will yield a $45°$ phase margin.

8. The open loop transfer function of a unity feedback ratio system is

$$G(s) = \frac{K}{s(1 + 1.4s + s^2)}$$

Find the gain margin, phase margin, and M_p as a function of K.

9. The fixed plant transfer function for a linear control system representative of an amplidyne designed to position the cutting head of a machine tool system is

$$G_f(s) = \frac{K}{s(1 + s/\omega_f)(1 + s/\omega_q)(1 + s/\omega_m)}$$

where

K = system velocity constant K_ν

ω_m = motor load break frequency = 10 rad/sec

ω_q = quadrature axis break frequency = 50 rad/sec

ω_f = control field break frequency = 1000 rad/sec

The system must have a phase margin of $45°$

   a. What value of K will yield a $45°$ phase margin? What is the crossover frequency at this value of K?

   b. In an attempt to increase the crossover frequency, a 5 to 1 lead network of the form

$$G_c(s) = \frac{1 + s/\omega_1}{1 + s/5\omega_1}$$

is inserted in series with $G_f(s)$. What are the values of K and ω_1 which yield maximum crossover frequency and $45°$ phase margin? What is this crossover frequency?

10. A unity feedback ratio system has the transfer function

$$G_f(s) = \frac{100e^{-0.025s}}{s(1 + s)}$$

The specifications for the system are:

   a. $K_v = 100$

   b. phase margin $= 45°$

   c. for sinusoidal inputs up to 0.2 radians per second the reproduction error must not exceed 1%.

Design a series compensation network to meet these specifications.

11. The open loop transfer function of the fixed plant of a unity feedback ratio system is

$$G_f(s) = \frac{1000}{s(1 + s/100)^4}$$

Design a maximum crossover frequency lag lead network which has a phase margin of $45°$. Design the compensating network using the two methods presented in Sec. 5.3 and compare the results.

12. Determine design charts and a DELTA Chart for an appropriate design procedure to yield a $45°$ phase margin for an open loop system of the form

$$G_f(s)G_c(s) = \left[\frac{\omega_c \omega_1^{n-1}(1 + s/\omega_1)^{n-1}}{s^n(1 + s/\omega_2)^{m-1}} \right] e^{-sT}$$

which is precisely the same as the transfer function we assumed in Sec. 5.3 plus a time delay nonminimum phase component.

13. Determine a minor loop realization of the compensated system of Problem 11.

14. We desire a compensated open loop system with the transfer function

$$G_{Mc}(s) = \frac{1000(1 + s/50)}{s(1 + s/5)(1 + s/300)}$$

The fixed plant has the transfer function

$$G_f(s) = \frac{1000}{s(1 + s/30)}$$

Two suggested methods of realization of the system are shown in Fig. P5.14. A disturbance of frequency 50 radians per second is inserted as shown. Discuss the performance of the two systems in terms of noise rejection.

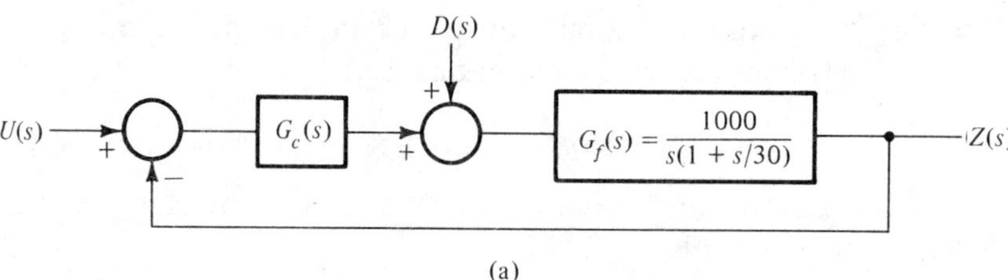

(a)

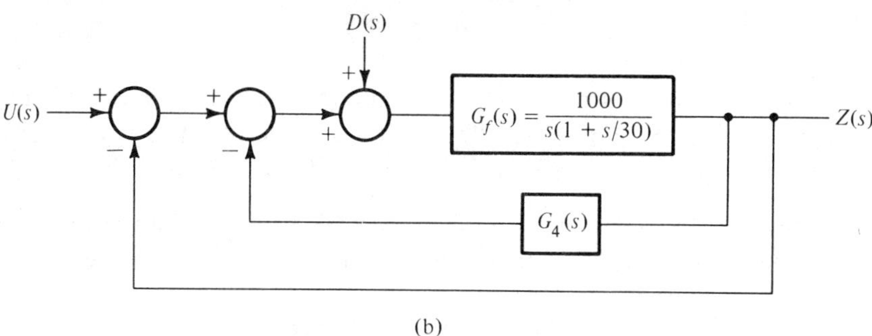

(b)

Figure P5.14   Block diagrams for
(a) Series compensation and
(b) Minor loop compensation

15. We have indicated that it is not possible to obtain the equivalent of series lead network compensation by means of minor loop equalization. Show that this can be done by means of a combination of series and minor loop equalization. In particular show that it is possible to compensate a fixed plant

$$G_f(s) = \frac{39}{s^2}$$

such as to yield a compensated open loop system

$$G_{Mc}(s) = \frac{39(1 + s/3.9)}{s^2(1 + s/25.5)}$$

by use of a gain $G_1(s) = 6.54$ and an appropriate minor loop transfer function. All transfer functions refer to the standard minor loop block diagram of Fig. (5.4-1). Find an appropriate minor loop gain $G_m(s)$ and minor loop feedback element $G_4(s)$. Compare the output disturbance response of your minor loop system to that for a series compensated system.

16. Determine suitable minor loop compensation $G_4(s)$ for the fixed plant

$$G_f(s) = \frac{10(1 - s/10)}{s(1 + s/10)}$$

such as to yield a 45° phase margin.

17. Determine suitable minor loop compensation $G_4(s)$ for the fixed plant

$$G_f(s) = \frac{10}{s(1 - s/10)}$$

Be sure to design a system such that the minor loop is stable. A system phase margin of 45° is desirable.

6

THE ROOT LOCUS METHOD

In the Bode diagram design approach of our previous chapter we assume that the complex variable s is replaced by the imaginary variable $s = j\omega$ where ω represents the real frequency variable. We developed very useful design procedures to determine open loop pole and zero locations to insure a specified degree of stability and smoothness of the compensated systems closed loop response. However the precise location of the closed loop system poles are not specified or controlled in the Bode diagram design process.

In this chapter we will present a discussion of the root locus method, originated by W. R. Evans in 1948. This method provides us with a useful tool for obtaining the roots of the numerator and denominator polynomial of a closed loop system. These roots, the zero and pole locations for a closed loop system, provide an indication of stability as well as transient and steady state system behavior. Basically we consider a unity feedback ratio system with open loop transfer function $G(s)$. The closed loop transfer function is $Z(s)/U(s) = G(s)/1 + G(s)$. We see that the closed loop zeros are the zeros of $G(s)$ and the closed loop poles are the zeros of $1 + G(s)$. We could obtain the closed loop zero and pole locations by using a numerical analysis root finding algorithm. If we represent $G(s)$ by the ratio of two rational polynomials in s, $N(s)/D(s)$, then the roots of $N(s)$ are the zero locations, and the roots of $N(s) + D(s)$ the pole locations. However, such a procedure would not lead to anywhere near as useful design information as obtainable from use of the root locus method in that very little insight into sensitivity questions, relative stability questions, and the effects of different forms of compensating networks would result from use of root finding numerical analysis algorithms. Of course these algorithms would be very useful as part of a collection of algorithms with which to determine the root locus.

In this chapter we will first provide some motivation for the root locus method by considering three simple analysis examples. Then we will introduce the root locus procedure which is a procedure that assumes that a trial complex number s_T is a root of the numerator of the characteristic equation $1 + G(s)$. A test is devised to see whether the trial root s_T can be a root of this characteristic equation and if the result of the test is favorable the loop-gain which would make this trial root s_T an actual root is found. A graph of the sequence of complex numbers which could be possible roots of $1 + G(s)$ is known as a root locus graph or root locus plot associated with this graph and points representing possible roots of the characteristic equation are the values of the loop-gain which make these roots possible. After developing rules for the root locus design procedure we will illustrate a useful DELTA Chart for analysis and design using the root locus. We will also examine some extensions of the root locus to include a discussion of root locus diagrams for changing time contents, root locus methods for nonrational transfer functions, and the root locus method for multiple loop systems.

6.1 NEED FOR THE ROOT-LOCUS: SOME SIMPLE ILLUSTRATIONS

It is helpful, prior to examining root locus construction rules in detail, to consider some simple examples which illustrate the need for the root locus procedure and interconnections between the root locus, our Bode diagram frequency domain design approach, and system performance. We will consider three very simple examples that have easily obtained analytical solutions to motivate our discussions to follow. All of these examples will assume a unity feedback ratio system such as illustrated in Fig. (6.1-1). All parameters for these examples will be assumed real and positive.

EXAMPLE 6.1-1. We assume that the open loop transfer function is that of a simple integrator with gain

$$G(s) = \frac{K}{s}$$

The phase margin of this system is $90°$ and is independent of the gain K. The gain margin is infinite which indicates that the closed loop system is always stable. The closed loop transfer function is

$$\frac{Z(s)}{U(s)} = H(s) = \frac{K}{s + K}$$

and so we see that there is a single closed loop pole at $s = -K$ and no finite zeros. We may illustrate this as in Fig. (6.1-2) which represents a root locus diagram for this very simple system. The closed loop frequency response of the system is

$$\frac{Z(\omega)}{U(\omega)} = H(w) = \frac{K}{\sqrt{\omega^2 + K^2}}$$

and has a simple single break from 0 to a -1 slope occurring at $\omega = K$. The unit step response of the system is

$$z(t) = 1 - e^{-tK}$$

and we see that increasing K increases the speed of response and decreases the rise time. All of these relations are quite consistent. The system phase margin is always large. The root locus diagram never shows any poles located off of the negative real axis. The closed loop frequency response is very smooth and the transient response is just that of exponentially decreasing error.

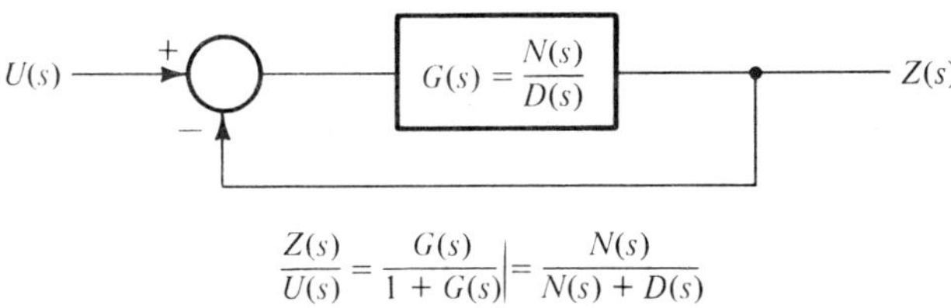

$$\frac{Z(s)}{U(s)} = \frac{G(s)}{1 + G(s)} = \frac{N(s)}{N(s) + D(s)}$$

Figure 6.1-1   Unity ratio feedback system

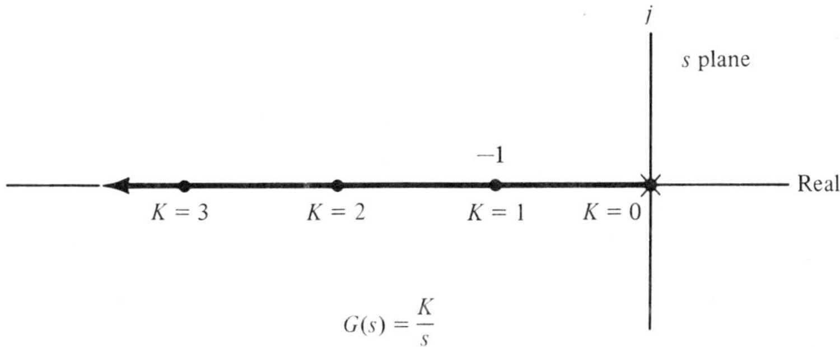

$$G(s) = \frac{K}{s}$$

Figure 6.1-2   Root locus for the simple system of EXAMPLE 6.1-1

EXAMPLE 6.1-2. As a slightly more complex system we consider the second order type one system with fixed plant transfer function

$$G(s) = \frac{K_v}{s(1 + s/\omega_1)} \tag{1}$$

The velocity error coefficient of this system is K_v. The gain margin is infinite for K_v and ω_1 greater than zero so the system is always stable. The exact crossover frequency is obtainable from

$$\frac{\omega_c^4}{\omega_1^2} + \omega_c^2 = K_v^2 \tag{2}$$

or

$$\omega_c = 0.707\omega_1 \left[(1 + 4K_v^2/\omega_1^2)^{1/2} - 1 \right]^{1/2} \tag{3}$$

and the exact phase shift at crossover is given by

$$\beta(\omega_c) = \frac{-\pi}{2} - \tan^{-1}\left(\frac{\omega_c}{\omega_1}\right) \tag{4}$$

As we know from our discussions in Chaps. 4 and 5, the asymptotic approximations are much easier to use than the exact expressions of Eqs. (2) and (4) and these are for $K_v < \omega_1$

$$\omega_c \cong K_v \qquad\qquad K_v < \omega_1$$

$$\beta(\omega_c) \cong \frac{-\pi}{2} - \frac{\omega_c}{\omega_1}$$

or for $K_v > \omega_1$

$$\omega_c \cong (K_v \omega_1)^{1/2}$$

$$\beta(\omega_c) \cong -\pi + \frac{\omega_1}{\omega_c}$$

Figure (6.1-3) illustrates the phase margin, $\text{PM} = \pi + \beta(\omega_c)$, and crossover frequency variation with the gain K_v.* We see that the system phase margin starts at $90°$ for very small K_v and decreases to zero as the gain K_v is increased. As with the many other examples we have considered the asymptotic approximations are very good ones.

We will now determine the closed loop system roots. The closed loop transfer function is

$$\frac{Z(s)}{U(s)} = H(s) = \frac{G(s)}{1 + G(s)} = \frac{K_v \omega_1}{s^2 + \omega_1 s + K_v \omega_1} \tag{5}$$

There are no finite closed loop zeros. There are closed loop poles at

$$s^2 + \omega_1 s + K_v \omega_1 = 0 \tag{6}$$

or

$$s = -\frac{\omega_1}{2} \pm \frac{\omega_1}{2}\left(1 - \frac{4K_v}{\omega_1}\right)^{1/2}$$

Figure (6.1-4) illustrates the root locus diagram for this simple second order example. We can visualize and even obtain the transient response of this system directly from this root locus plot. The "standard" transfer function for a second order system was written in Chap. 4 as

$$\frac{Z(s)}{U(s)} = H(s) = \frac{\omega_n^2}{s^2 + 2\zeta\omega_n s + \omega_n^2} \tag{7}$$

and we see that we can easily obtain the damping ratio and natural frequency

$$\omega_n = (K_v \omega_1)^{1/2}$$

$$\zeta = \frac{1}{2}(\omega_1 / K_v)^{1/2}$$

---

*The loop gain of this system is normally called $-K_v$ although some authors refer to it as $+K_v$. The minus sign occurs, of course, because of the subtraction element in Fig. (6.1-1).

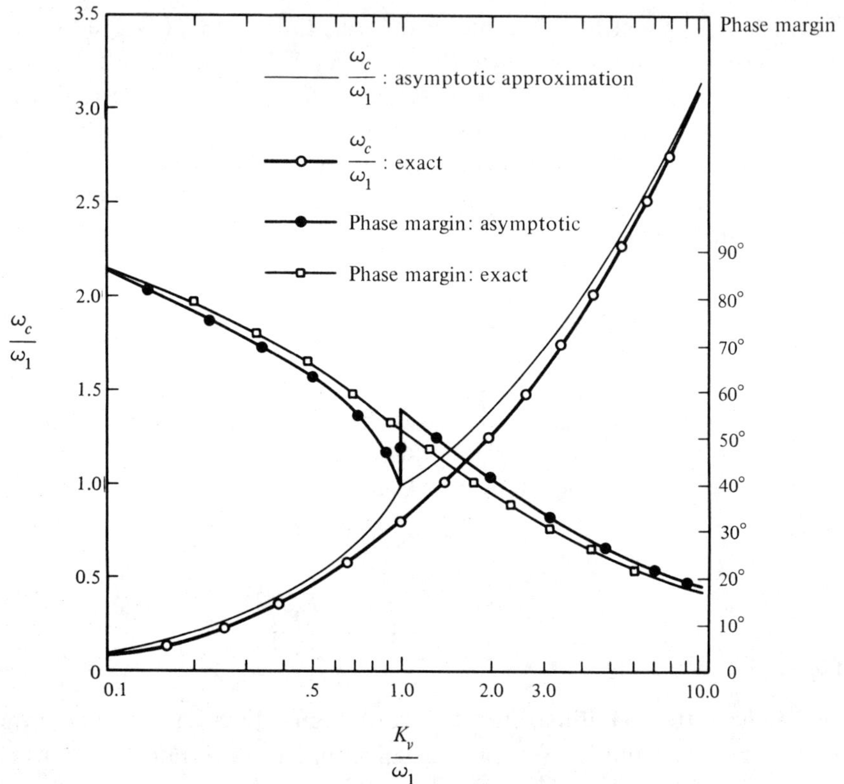

Figure 6.1-3 Frequency domain specifications for EXAMPLE 6.1-2

directly from this root locus diagram. The pole locations for Eqs. (5) and (7) are not quite the same since the root locus for Eq. (5), which is shown in Fig. (6.1-4), is that for varying the loop gain, $-K_v$, in the open loop system

$$G(s) = \frac{K_v}{s(1 + s/\omega_1)} \tag{1}$$

where ω_1 is presumably a fixed unalterable constant. An equivalent open loop transfer function for Eq. (7) is

$$G(s) = \frac{\omega_n^2}{s(s + 2\zeta\omega_n)} = \frac{\dfrac{\omega_n}{2\zeta}}{s(1 + s/2\zeta\omega_n)} \tag{8}$$

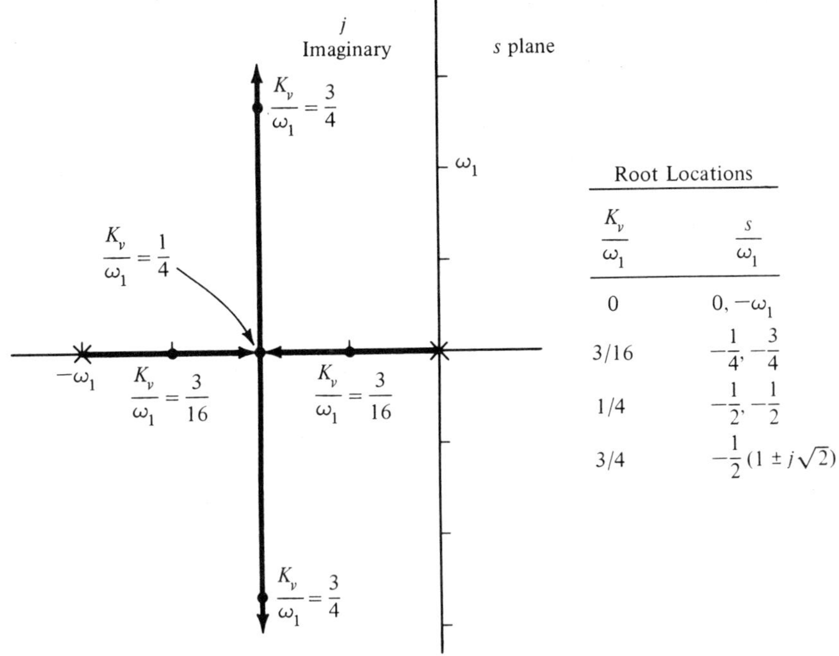

Root Locations

$\dfrac{K_v}{\omega_1}$	$\dfrac{s}{\omega_1}$
0	$0, -\omega_1$
3/16	$-\dfrac{1}{4}, -\dfrac{3}{4}$
1/4	$-\dfrac{1}{2}, -\dfrac{1}{2}$
3/4	$-\dfrac{1}{2}(1 \pm j\sqrt{2})$

Figure 6.1-4   Root locus diagram for EXAMPLE 6.1-2

The usual locus of Eq. (7) is obtained for varying the damping ratio ζ and we see from Eq. (8) that varying the damping ratio varies both the velocity coefficient $K_v = \omega_n/2\zeta$ in Eq. (8) as well as the break frequency $\omega_1 = 2\zeta\omega_n$. Thus we should indeed expect the two loci to be different.

An acceptable design for the system of Eq. (8) might be the one where the damping ratio ζ is 0.707. This corresponds to a velocity error coefficient

$$\omega_1/K_v = 4\zeta^2 = 2.0$$

which yields a phase margin of approximately 63° and a crossover frequency, ω_c/ω_1 of approximately 0.5 as we see from Fig. (6.1-3). We see that our most natural performance specifications will be in terms such as damping ratio, resonant frequencies, etc. for the closed loop system rather than the specifications most natural for design using Bode diagrams.

EXAMPLE 6.1-3. As our final example in this section illustrating need for and uses of the root locus procedure, let us consider the open loop system

$$G(s) = \frac{K_v(1 - s/\omega_1)}{s(1 + s/\omega_1)} \tag{1}$$

which we recognize as a nonminimum phase system with a crossover frequency

$$\omega_c = K_v \tag{2}$$

and an exact phase shift at crossover

$$\beta(\omega_c) = -\frac{\pi}{2} - 2\,\tan^{-1}\frac{\omega_c}{\omega_1} \tag{3}$$

which can be approximated by the arctangent asymptotic approximations

$$\beta(\omega_c) = \begin{cases} -\dfrac{\pi}{2} - \dfrac{2\omega_c}{\omega_1} & K_v < \omega_1 \\[2ex] -\dfrac{3\pi}{2} + \dfrac{2\omega_1}{\omega_c} & K_v > \omega_1 \end{cases}$$

We see that this system can become unstable for the phase margin goes to zero when $K_v = \omega_1$. For a 45° phase margin we would set $\beta(\omega_c) = -\dfrac{3\pi}{4}$ to obtain $K_v/\omega_1 = \pi/8 = 0.39$ from the arctangent approximation and $K_v/\omega_1 = 0.41$ from the exact phase shift equation (3).

   To obtain the root locus diagram in a brute force fashion we find the closed loop transfer function

$$H(s) = \frac{G(s)}{1 + G(s)} = \frac{K_v(1 - s/\omega_1)}{s(1 + s/\omega_1) + K_v(1 - s/\omega_1)}$$

and rewrite this in a more standard fashion as

$$H(s) = \frac{K_v(\omega_1 - s)}{s^2 + s(\omega_1 - K_v) + K_v\omega_1} \tag{4}$$

There is a single closed loop zero located in the right half plane of the s plane at $+\omega_1$. There are two closed loop poles located at

$$s^2 + s(\omega_1 - K_v) + K_v\omega_1 = 0 \tag{5}$$

Thus we have real poles in the s plane when $(\omega_1 - K_v)^2 > 4K_v\omega_1$. Thus there exist real poles for $K_v < 0.17\omega_1$ and $K_v > 5.83\omega_1$. For the case where the roots are complex and given by

$$s = \frac{-(\omega_1 - K_v) \pm j(6\omega_1 K_v - \omega_1^2 - K_v^2)^{1/2}}{2}$$

we see that the real axis displacement is

$$\sigma = -\left(\frac{\omega_1 - K_v}{2}\right) \tag{7}$$

and the imaginary axis displacement is

$$\omega = \pm\frac{1}{2}(6\omega_1 K_v - \omega_1^2 - K_v^2)^{1/2} \tag{8}$$

The equation for a circle of radius r centered at point (x_0, y_0) in the (x, y) plane is

$$(x - x_0)^2 + (y - y_0)^2 = r^2$$

So we note that the behavior of the root locus, for $0.17\omega_1 < K_v < 5.83\omega_1$ is that of a circle with center at $\sigma = \omega_1$ and radius $1.415\omega_1$. This is so since

$$(\sigma - \omega_1)^2 + \omega^2 = (1.415\omega_1)^2$$

where σ and ω are given by Eqs. (7) and (8).

The roots of Eq. (6) are equal when $K_v = 0.17\omega_1$ where the double roots are located at $s = -0.415\omega_1$, and when $K_v = 5.83\omega_1$ where the double roots are located at $s = 2.415\omega_1$.

Figure (6.1-5) illustrates the root locus for this simple example. We must be a bit careful in interpreting the $\zeta = 0.707$ case for this system since it is not a standard two pole no finite zero system. The $\zeta = 0.707$ case corresponds to $K_v = 0.27\omega_1$ and this yields a crossover frequency of $0.27\omega_1$ and a phase margin of $60°$ which indicates that the relative stability of the system is quite good. We could tolerate a smaller value of ζ for this system and still have a reasonable phase margin.

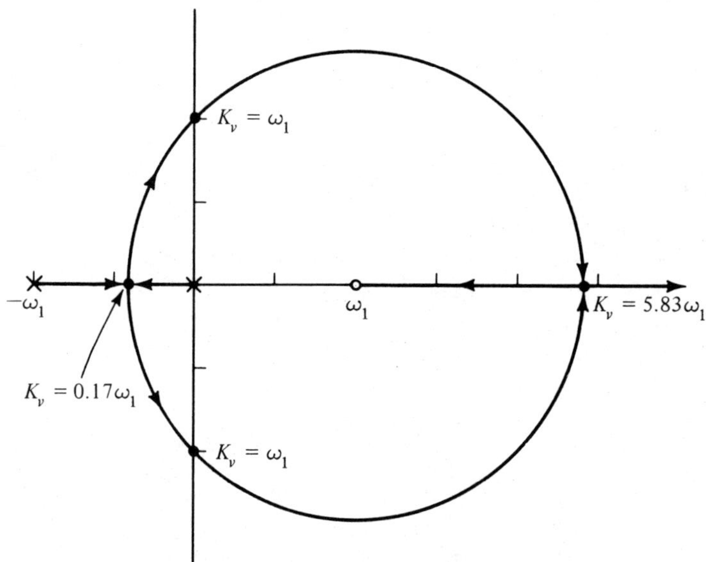

Figure 6.1-5 Root locus for $G(s) = \dfrac{K_v(1 + s/\omega_1)}{s(1 + s/\omega_1)}$

We see that this brute force technique of determining the root locus will pose considerable numerical problems if we deal with systems of order higher than the second. Thus we develop a set of rules that will aid in constructing the root locus. These examples furnish a guide to the points on the root locus that are critical in determining it. We have seen, for example, that we desire to be able to determine such items as: the number of branches on the root locus, starting points and terminal points for the root locus, conditions under which the locus will exist on the real axis, angles and points of departure of complex conjugate poles from the real axis, angles and points of arrival of complex conjugate poles on the real axis, large gain behavior of the root locus, and points of intersection with the $j\omega$ (imaginary) axis. We now turn our attention to the development of some general rules for construction of root locus diagrams.

EXERCISE 6.1-1.   Determine the root locus diagram for the open loop system

$$G(s) = \frac{\omega_1^2 T}{s(1 + sT)}$$

where ω_1 is assumed fixed and T is the variable parameter. Follow the protocol established in EXAMPLE 6.1-2 and contrast and compare your results with the results of that example.

EXERCISE 6.1-2.   Repeat the calculations of EXAMPLE 6.1-3 for the open loop transfer function

$$G(s) = \frac{K_a(1 + s/\omega_1)}{s^2}$$

Determine the root locus diagram for this system.

EXERCISE 6.1-3.   Repeat the calculations of EXAMPLE 6.1-2 for negative K.

6.2 RULES FOR CONSTRUCTING ROOT LOCI

Prior to establishing rules we may use in constructing root locus diagrams let us examine some fundamental relations which determine the locus of roots of a linear control system. We again assume the block diagram of Fig. (6.1-1) and have for the closed loop transfer function

$$\frac{Z(s)}{U(s)} = H(s) = \frac{G(s)}{1 + G(s)} = \frac{N(s)}{N(s) + D(s)} \tag{6.2-1}$$

where $N(s)$ is the numerator polynomial and $D(s)$ the denominator polynomial of $G(s)$.

We obtain the system characteristic equation by setting the numerator of $1 + G(s)$ equal to zero. We assume that the open loop transfer function is a ratio of rational polynomials in s of the form

$$G(s) = \frac{N(s)}{D(s)} = \frac{K(s^n + a_1 s^{n-1} + a_2 s^{n-2} + \ldots + a_n)}{s^d + b_1 s^{d-1} + b_2 s^{d-2} + \ldots + b_{d-1} s + b_d} \qquad (6.2\text{-}2)$$

thus we obtain

$$N(s) + D(s) = s^d + b_1 s^{d-1} + \ldots + b_d +$$

$$K(s^n + a_1 s^{n-1} + \ldots + a_n) = 0 \qquad (6.2\text{-}3)$$

as the system characteristic equation. Setting Eq. (6.2-3) equal to zero is clearly the same as setting the characteristic equation equal to zero or

$$1 + G(s) = 0 \qquad (6.2\text{-}4)$$

This will occur only if

$$G(s) = -1 \qquad (6.2\text{-}5)$$

The basic root locus will be determined by changing the negative loop gain, or K, in Eq. (6.2-2). Thus it is convenient to define a normalized transfer function with K removed

$$G_n(s) = \frac{G(s)}{K} = \frac{s^n + a_1 s^{n-1} + \ldots + a_n}{s^d + b_1 s^{d-1} + \ldots + b_d} \qquad (6.2\text{-}6)$$

such that the condition of Eq. (6.2-5) becomes

$$G_n(s) = \frac{-1}{K} \qquad (6.2\text{-}7)$$

This equality may be satisfied, in general, in several ways. First of all we must have

$$|G_n(s)| = \frac{1}{|K|}, \qquad s = \sigma + j\omega \qquad (6.2\text{-}8)$$

and then we must also have for positive K

$$\underline{/G_n(s)} = (2k + 1)\pi \qquad K > 0 \qquad (6.2\text{-}9)$$

where $\underline{/}$ denotes angle of. For negative K we have

$$\underline{/G_n(s)} = 2k\pi \qquad K < 0 \qquad (6.2\text{-}10)$$

In our previous section we considered only positive gains but here we will consider negative gains as well as we develop a general theory for the root locus. We will develop a root locus by finding these values of s for which Eqs. (6.2-9) or (6.2-10) are satisfied. This can be done independent of K since K does not enter either of these two equations. Then we will find the values of K which satisfy Eq. (6.2-8).

We assume that the open loop transfer function $G(s)$ is available in or can be put in standard factored form

$$G(s) = \frac{K(s + n_1)(s + n_2) \ldots (s + n_n)}{(s + d_1)(s + d_2) \ldots (s + d_d)} = KG_n(s) \qquad (6.2\text{-}11)$$

The angle criterion of Eq. (6.2-9) becomes, upon using Eq. (6.2-11),

$$\underline{/G_n(s)} = \sum_{i=1}^{n} \underline{/s + n_i} - \sum_{i=1}^{d} \underline{/s + d_i} = (2k + 1)\pi \qquad (6.2\text{-}12)$$

for $0 < K < \infty$. For negative K we use Eq. (6.2-10) to obtain

$$\underline{/G_n(s)} = \sum_{i=1}^{n} \underline{/s + n_i} - \sum_{i=1}^{d} \underline{/s + d_i} = 2k\pi \qquad (6.2\text{-}13)$$

Each of these equations may be used in order to construct the complete root locus. Often only positive gains are of significance and, in this case, we need only find the values of $s = \sigma + j\omega$ which satisfy Eq. (6.2-12). We must solve these equations for all integer values of k. Let us illustrate some of the properties of the relations with an example and then state a number of rules for constructing a root locus.

EXAMPLE 6.2-1. We consider the closed loop system with open loop transfer function

$$G(s) = \frac{K}{s(s+1)(s+3)} \qquad K > 0 \qquad\qquad (1)$$

To satisfy the angle criteria of Eq. (6.2-9) we must have, where $s = \sigma + j\omega$

$$-\tan^{-1}\frac{\omega}{\sigma} - \tan^{-1}\frac{\omega}{\sigma+1} - \tan^{-1}\frac{\omega}{\sigma+3} = (2k+1)\pi \qquad (2)$$

This is a difficult equation to solve and we will soon develop procedures to extract maximum information from this equation. Suppose we restrict trial values of s to those near the real axis. For $0 > \sigma > -1$ we have for small ω

$$\tan^{-1}\frac{\omega}{\sigma} \cong \pi + \frac{\omega}{\sigma}$$

$$\tan^{-1}\frac{\omega}{\sigma+1} \cong \frac{\omega}{\sigma+1}$$

$$\tan^{-1}\frac{\omega}{\sigma+3} \cong \frac{\omega}{\sigma+3}$$

and Eq. (2) becomes

$$-\pi - \frac{\omega}{\sigma} - \frac{\omega}{\sigma+1} - \frac{\omega}{\sigma+3} = (2k+1)\pi \qquad (3)$$

On the real axis where $\omega = 0$ we see that the foregoing equation is satisfied for $k = -1$. For $-3 < \sigma < -1$ we have for small ω

$$\tan^{-1}\frac{\omega}{\sigma} \cong \pi + \frac{\omega}{\sigma}$$

$$\tan^{-1}\frac{\omega}{\sigma+1} \cong \pi + \frac{\omega}{\sigma+1}$$

$$\tan^{-1}\frac{\omega}{\sigma+3} \cong \frac{\omega}{\sigma+3}$$

and Eq. (2) becomes

$$-2\pi - \frac{\omega}{\sigma} - \frac{\omega}{\sigma+1} - \frac{\omega}{\sigma+3} \cong (2k+1)\pi \tag{4}$$

On the real axis where $\omega = 0$ we see that this equation is nowhere satisfied. Thus we see that there is not a valid solution for the loci of the roots on the real axis interval $-3 < \sigma < -1$. For $-3 > \sigma$ the arctangent approximations to Eq. (2) become

$$\tan^{-1} \frac{\omega}{\sigma} \cong \pi + \frac{\omega}{\sigma}$$

$$\tan^{-1} \frac{\omega}{\sigma+1} \cong \pi + \frac{\omega}{\sigma+1}$$

$$\tan^{-1} \frac{\omega}{\sigma+3} \cong \pi + \frac{\omega}{\sigma+3}$$

and Eq. (2) becomes

$$-3\pi - \frac{\omega}{\sigma} - \frac{\omega}{\sigma+1} - \frac{\omega}{\sigma+3} \cong (2k+1)\pi \tag{5}$$

On the real axis where $-3 > \sigma$ we see that we can have valid closed loop poles for $k = -2$. In a similar way we can show that Eq. (2) has no solution for $\sigma > 0$. Thus we have determined possible root locus solutions on the real axis. These are illustrated in Fig. (6.2-1).

It is intuitively reasonable, given Fig. (6.2-1), that the poles at $s = 0$ and $s = -1$ move towards one another and intersect somewhere on the real axis with $-1 < \sigma < 0$. The pole at $s = -3$ must move towards the $-\infty$ point on the real axis.

When the poles at $s = 0$ and $s = -1$ intersect on the real axis they will "break away" from the real axis and become complex conjugate poles. We can easily calculate the break away point. Suppose that we move a very small distance ω from the real axis with $-1 < \sigma < 0$. Equation (6.2-9) must still be valid and for small ω the arctangent approximation of Eq. (3) is also a valid approximation. This equation is satisfied for $k = -1$ and it then becomes

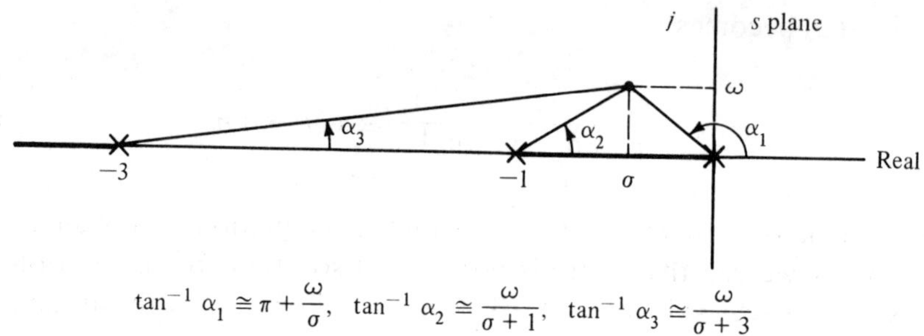

$$\tan^{-1}\alpha_1 \cong \pi + \frac{\omega}{\sigma}, \quad \tan^{-1}\alpha_2 \cong \frac{\omega}{\sigma+1}, \quad \tan^{-1}\alpha_3 \cong \frac{\omega}{\sigma+3}$$

Figure 6.2-1 Real axis root locus for $G(s) = \dfrac{K}{s(s+1)(s+3)}$ showing permissible behavior on the real axis and arctangent approximation valid for $-1 < \sigma < 0$

$$-\frac{\omega}{\sigma} - \frac{\omega}{\sigma+1} - \frac{\omega}{\sigma+3} = 0 \tag{6}$$

We can cancel the ω term in this equation and solve for σ to obtain $\sigma = -2.22$ and $\sigma = -0.45$. The $\sigma = -2.22$ root is outside the interval $-1 < \sigma < 0$ where the arctangent approximation is valid and so it is not a valid solution. Thus we have found a break away point at $\sigma = -0.45$.

The behavior of the locus near $|s| = \infty$ is also of interest. Near $|s| = \infty$ we may approximate Eq. (1) by

$$G(s) \cong \frac{K}{s^3} \tag{7}$$

and so the characteristic equation becomes

$$1 + G(s) \cong 0 = \frac{s^3 + K}{s^3} \tag{8}$$

Thus we see that there will be zeros of the characteristic equation at

$$s^3 + K = 0 \tag{9}$$

or at

$$s = -K^{1/3}, \qquad K^{1/3}\,e^{j60°}, \qquad K^{1/3}\,e^{-j60°}$$

We could use Eqs. (6.2-8) and (6.2-9) to obtain this asymptotic large gain behavior also. We see that one pole approaches $-\infty$ for large K whereas the other two poles move out at an angle of $\pm 60°$ from the origin.

We will not expect these asymptotes to intersect at the origin but instead meet at some point on the real axis. We can find this point of asymptote intersection by examining the closed loop transfer function which becomes

$$H(s) = \frac{K}{s^3 + 4s^2 + 3s + K} \tag{10}$$

The coefficient +4 represents the negative of the sum of the root locations as is known from elementary algebra. There are three poles and no zeros associated with this transfer function. Thus the "average pole" location or asymptote meeting point is $-4/3 = -1.33$.

There are a number of other information items we could obtain concerning the behavior of the root locus without actually factoring Eq. (10). In particular we can obtain the gain directly from our sketch of the root locus in Fig. (6.2-2) by use of Eq. (6.2-8). All we need to use is a scale to measure distance magnitudes on the root locus plot and then multiply and divide these values according to Eq. (6.2-8).

Now that we have calculated the root locus for a simple third order example using the fundamental constructs and relations for root behavior let us turn to a more systematic presentation of the root locus method.

EXERCISE 6.2-1. Determine the phase margin as a function of the gain K for EXAMPLE 6.2-1 and compare the analysis of this system using the root locus method to that obtained using Bode diagrams. In particular determine stability bounds for the system using the Bode diagram and the gain and phase margin which results in or from $\zeta = 0.707$ on the root locus diagram.

There are a number of ways of stating formal rules or theorems useful for root locus construction and, as we might suspect, all of these are very closely related. We will present fourteen aids to root locus construction here and will provide a simple proof and example calculation for each whenever the rule or its application is not intentively trivial.

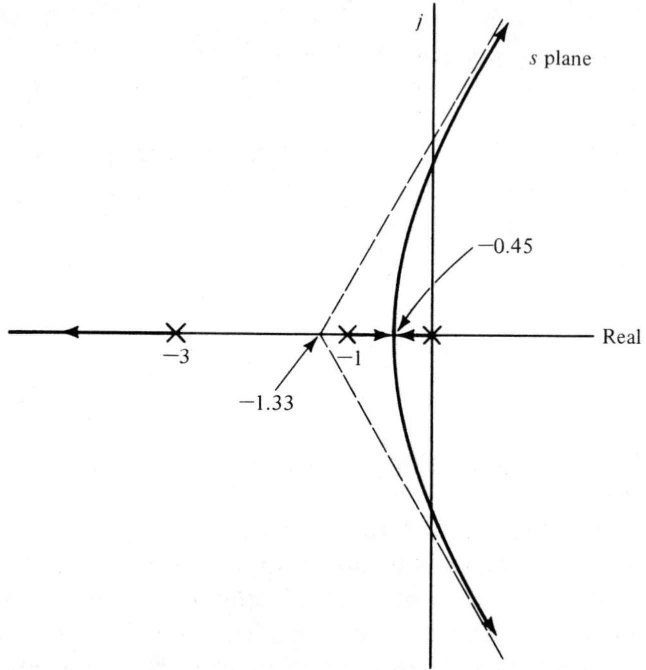

Figure 6.2-2   Root locus for $G(s) = \dfrac{K}{s(s+1)(s+3)}$

1. *Number of Branches of the Root Locus.* The total numbers of branches of the root locus will be equal to the number of poles in the open loop transfer function $G(s)$.

2. *Closed loop zeros are the same as open loop zeros.*

3. *Starting points for the Root Locus $(K = 0)$.* The starting points for the root locus at $K = 0$ are at the poles of the open loop transfer function $G(s)$.

These rules are easily established and appear almost trivial. From Eqs. (6.2-6) and (6.2-11) we have

$$G_n(s) = \frac{G(s)}{K} = \frac{\displaystyle\prod_{i=1}^{n}(s + n_i)}{\displaystyle\prod_{i=1}^{d}(s + d_i)} \qquad\qquad (6.2\text{-}14)$$

as the open loop transfer function. A closed loop root occurs when the characteristic equation $1 + G(s)$ is equal to zero or where $G(s) = -1$. As K approaches zero the only way for $G_n(s)$ in Eq. (6.2-14) to approach ∞ such that $G(s) = -1$ is for $G_n(s)$ to become infinity. This occurs at the poles of $G(s)$. The zeros of the closed loop transfer function $H(s) = G(s)/1 + G(s)$ are the zeros of $G(s)$ and these do not change as K changes. The poles of the closed loop transfer function are the zeros of $1 + G(s)$ which are found by equating $1 + G(s) = 0$. For $K = 0$ we have just determined that these occur at the poles of $G(s)$.

4. *Ending Points for the Root Locus* $(K = \infty)$. The branches of pole movement on the root locus will terminate at open loop zeros at infinite gain.

When $G(s) = -1$ to satisfy the characteristic equation and $K = \infty$ we see that $G_n(s) = 0$ in Eq. (6.2-14). This will occur at the zeros of $G_n(s)$ and so we see that one closed loop pole will necessarily terminate at each open loop zero. In the usual control system we observe that there are more poles than zeros. Thus we realistically must question what happens to the excess poles that do not have finite zeros at which to terminate. The answer is that we can always insert infinite zeros of the form $1 + s(\sigma_\infty + j\omega_\infty)$ where $|\sigma_\infty + j\omega_\infty| = \infty$ and not change the transfer function at all. This provides zeros at infinity, properly located, and excess open loop poles will terminate at these zeros at infinity.

5. *Symmetry of the Root Loci.* The root loci must be symmetrical with respect to the s plane real axis.

This rule is essentially self evident if we accept the fact that the roots of a polynomial with real coefficients will be real and/or occur in complex conjugate pairs. We can also show that the root locus will also be symmetrical with respect to the axis of symmetry of the zeros and poles of the open loop transfer function.

EXAMPLE 6.2-2. To illustrate this rule we consider the open loop transfer function of EXAMPLE 6.2-1. The poles in that example are symmetrical to the $s = -4/3$ axis. The poles are of course always symmetrical about the real line or real axis which is the line $s = j0$. Inspection of Fig. (6.2-2) illustrates the symmetry about the real axis, $s = j0$, and the $s = -4/3$ axis.

6. *Real Axis Root Loci.* A point on the real axis will be a point on the root locus (for $K > 0$) if there are an odd number of open loop zeros and poles on the real axis to the right of this point.

Use of the phase angle criterion of Eq. (6.2-9) easily verifies this rule. We do not need to be concerned with zeros and poles off of the real axis since these must occur in complex conjugate pairs and thus there must always be an even number of zeros and poles off of the real axis and they contribute a phase shift of either 0 or 360°.

EXAMPLE 6.2-3. We consider real axis behavior of EXAMPLE 6.2-1. Inspection of Fig. (6.2-1) indicates that the angle criterion will be satisfied only if

$$\alpha_1 + \alpha_2 + \alpha_3 = k\pi + \pi$$

and we immediately obtained the regions sketched in this figure as valid regions for real axis closed loop poles.

For $K < 0$ we use the angle criterion of Eq. (6.2-9) and so we see that rule 6 is reversed in that a point on the real axis will be a point on the root locus (for $K < 0$) if there are an even number of open loop zeros and poles on the real axis to the right of this point. Thus, regions on the real axis which are not allowable loci for $K > 0$ are allowable for $K < 0$.

7. *Asymptotic Behavior of Loci for Large Gain (or large s).* For large values of the gain K, the loci of the system closed loop poles are asymptotic to straight lines given by the angle of the asymptotes

$$\alpha_{as} = \frac{(1 + 2k)180°}{n - d} \tag{6.2-15}$$

where $k = 0, 1, 2, \ldots, |d - n| - 1$ and where d is the number of open loop poles and n is the number of open loop zeros.

We may easily verify the correctness of this rule by examining the transfer function of Eq. (6.2-2) and noting that as the gain increases the s

values of the loci will have to increase also or it will not be possible to maintain the characteristic equation $G(s) = -1$.

Thus evaluation of the asymptotic behavior of the loci for large K is the same as investigating this behavior for large s. For large s we may approximate Eq. (6.2-2) by

$$G(s) = \frac{Ks^n}{s^d} \tag{6.2-16}$$

and we seek the conditions under which the magnitude and angle characteristic equations are satisfied. For $K > 0$ the angle condition becomes

$$\underline{\big|s^{n-d}} = (1 + 2k)\pi$$

taking the $(n - d)$th root of the above we obtain

$$\alpha_{as} \overset{\Delta}{=} \underline{\big|s} = \frac{(1 + 2k)\pi}{n - d} = \frac{(1 + 2k)180°}{n - d}$$

which is the desired relation for the angle of the asymptotes.

EXAMPLE 6.2-4. We will use rule 7 to compute the angle of the asymptotes for EXAMPLE 6.2-1. We have simply, since $n = 0$, $d = 3$

$$\alpha_{as} = \frac{(1 + 2k)180°}{-3}$$

and obtain $-60°$ for $k = 0$, $-180°$ for $k = 1$ and $-300°$. The values are in complete agreement with those we obtained in EXAMPLE 6.2-1.

For negative K our angle criterion becomes

$$\underline{\big|s^{n-d}} = 2k\pi$$

and so we obtain

$$\alpha_{as} = \underline{\big|s} = \frac{2k\pi}{n - d} = \frac{2k(180°)}{n - d}$$

for $k = 0, 1, \ldots |n - d| - 1$. Since it is not often that we are interested in root locus behavior for negative gain we will not incorporate statements concerning behavior of the locus for negative K into our root locus rules but will mention these in our discussion of the rules.

8. *Origin of the Asymptotic Behavior of Loci for Large Gain.* The origin of the asymptotes which gives the centroid for the asymptotic behavior of the closed loop poles is on the real axis at a point given by

$$\text{OA} = \frac{a_1 - b_1}{d - n} = \frac{\displaystyle\sum_{i=1}^{n} n_i - \sum_{i=1}^{d} d_i}{d - n} \qquad (6.2\text{-}17)$$

where a_1, b_1, n_i, d_i, n, and d are defined by Eqs. (6.2-2) and (6.2-6).

The proof of rule 8 is relatively straightforward. We may rewrite Eq. (6.2-2) as

$$G(s) = \frac{K}{s^{d-n} + (b_1 - a_1)s^{d-n-1} + \ldots} \qquad (6.2\text{-}18)$$

by long division of the frequency dependent portion of the numerator of Eq. (6.2-2) into the denominator. The root loci are determined by setting $G(s) + 1 = 0$ and this yields, using Eq. (6.2-18),

$$s^{d-n} + (b_1 - a_1)s^{d-n-1} + \ldots + K = 0 \qquad (6.2\text{-}19)$$

For large s this equation will behave as a polynomial in s^{s-n}.* It is well known in algebra, and it can be easily proven, that the coefficient of the next to the highest order polynomial is the negative sum of the roots. Thus we can compute the negative sum of the $d - n$ roots as $b_1 - a_1$. Far away from the origin, which occurs for large K and large s, it will appear that all roots started from a centroid that is the average location of a root, that is the negative of $(b_1 - a_1)$ divided by the number of poles less the number of zeros or

---

*There will generally be negative powers in s in Eq. (6.2-19). This is the reason why we restrict behavior to large s as the terms involving negative powers of s will go to zero.

$$OA = -\frac{a_1 - b_1}{n - d} = -\frac{\sum_{i=1}^{n} n_i - \sum_{i=1}^{d} d_i}{n - d}$$

and this verifies Eq. (6.2-17). The origin of the asymptotes is thus the negative sum of the open loop zero locations plus the sum of the pole locations divided by the difference between the number of zeros and the number of poles.

EXAMPLE 6.2-5. Simple calculations of the origin of the asymptotes for EXAMPLE 6.2-1 yields

$$OA = \frac{0 + 1 + 3}{-3} = -\frac{4}{3}$$

as we have obtained before.

9. *Breakaway Points and/or Re-entry Points.* Candidate breakaway points may be determined by finding the roots of

$$dG^{-1}(s)/ds = 0 \qquad\qquad (6.2\text{-}20)$$

We may easily calculate the points where a locus will break away from the real axis. In order for there to be a breakaway point there must be at least two poles of the closed loop transfer function at a point on the real axis. The poles start at two separate locations in the real axis and as the gain K is increased they move together. At some value of gain K_m the two poles touch each other and a larger value of gain drives them off the real axis. Thus the gain K at which the breakaway point occurs is a maximum gain point. The root locus must satisfy

$$1 + G(s) = 1 + KG_n(s) = 0$$

where $G_n(s)$ is defined by Eq. (6.2-14). Since K, which is an implicit function of s in the foregoing, is a maximum at a breakaway point we must have

$$\frac{dK}{ds} = \frac{dG_n^{-1}(s)}{ds} = 0$$

There are several ways in which we may write this expression. Since

$$\frac{dG_n^{-1}(s)}{ds} = \frac{1}{G_n^2(s)} \frac{dG_n(s)}{ds}$$

$$\frac{dG_n(s)}{ds} = \frac{dG(s)/K}{ds} = \frac{K \dfrac{dG(s)}{ds} - G(s) \dfrac{dK}{ds}}{K^2}$$

we see that the breakaway points may also be determined from

$$\frac{dG_n(s)}{ds} = 0 \qquad \frac{dG^{-1}(s)}{ds} = 0$$

or

$$\frac{dG(s)}{ds} = 0$$

Either of these equations yield necessary conditions for breakaway and arrival points. We should examine rule 6 to determine real axis root loci conditions to see whether candidate points resulting from Eq. (6.2-20) are possible breakaway points or possible reentry points. Candidate points on the real axis which satisfy Eq. (6.2-20) will be either breakaway or reentry points. Then we should examine the characteristic equation $1 + G(s) = 0$ to insure that any candidate complex conjugate breakaway points also satisfy this condition.

In EXAMPLE 6.2-1 we used an alternate approach to evaluate the breakaway point. This method is normally quite a convenient one to use. It is based on use of the angle criterion and the arctangent approximation. We find a section of the real axis on which valid roots may occur and where two poles must coincide and break away from the real axis. We write the angle criterion for this particular point and use the arctangent approximation to obtain a polynomial relationship involving the breakaway point which can be solved.

EXAMPLE 6.2-6. The open transfer function for EXAMPLE 6.2-1 is

$$G(s) = \frac{K}{s(s + 1)(s + 3)}$$

and setting $dG^{-1}(s)/ds = 0$ easily leads to

$$(s + 1)(s + 3) + s(s + 3) + s(s + 1) = 0$$

which becomes

$$3s^2 + 8s + 3 = 0$$

The solutions to this equation are $s = 2.22$ and $s = -0.45$. We accept the $s = -0.45$ solution as a valid breakaway point since this value satisfies rule 6. It turns out that the value $s = 2.22$ is the value of s which is a breakaway point for negative gain K.

We have actually used the alternate approach for computation of the breakaway point in our discussion of EXAMPLE 6.2-1 and there is no need to repeat that calculation here.

10. *Breakaway Angles and Reentry Angles.* The angles of departure and angles of arrival on the real axis are separated by an angle of $180°/m$ where m represents the number of root locus branches intersecting at the point in question.

Formal use of the angle criterion at a breakaway point would indicate that, from Eq. (6.2-12)

$$\sum_{i=1}^{n} \underline{|s + n_{bi}} - \sum_{i=1}^{d} \underline{|s + d_{bi}} = \pi$$

where s is the breakaway point under consideration and n_{bi} and d_{bi} are the zeros and poles of $H(s) = G(s)/1 + G(s)$ at that K where two or more poles coincide. This relation can be used to establish rule 10. But this is a more general relationship and so we state it as rule 11.

11. *Angles of Departure and Arrival.* The angle of departure of an open loop pole or the arrival angle of a pole at an open loop zero location can be determined by use of the angle criterion

$$\sum_{i=1}^{n} \underline{|s + n_i} - \sum_{i=1}^{d} \underline{|s + d_i} = \pi \qquad (6.2\text{-}21)$$

where s is a given pole or zero location. This rule is a little more difficult to apply for real axis breakaway and reentries and rule 10 is simpler to apply. For complex conjugate poles, however, rule 10 is inapplicable and rule 11 becomes the appropriate rule. This rule is easily stated for negative gains as

$$\sum_{i=1}^{n} \underline{|s + n_i|} - \sum_{i=1}^{d} \underline{|s + d_i|} = 0$$

EXAMPLE 6.2-7. We consider the open loop transfer function

$$G(s) = \frac{K(s + 2)}{s(s + 1 + j4)(s + 1 - j4)}$$

a pole zero plot of which is shown in Fig. (6.2-3). The angle of departure from the complex conjugate pole at $s = -1 - j4$ can easily be calculated from Eq. (6.2-21) as

$$\theta_2 - \theta_1 - \theta_3 - \theta_d = 180°$$

where the angles are identified in Fig. (6.2-3). We have $\theta_2 = 75.96°$, $\theta_1 = 104.04°$, $\theta_3 = 90°$, and so we compute the angle of departure $\theta_d = -298.08°$ or $\theta_d = 61.92°$.

It is straightforward to show that the angle of arrival at the zero located at $s = -2$ is 0.

EXAMPLE 6.2-8. We consider determination of the breakaway points for EXAMPLE 6.2-1. At some value of gain K there are two roots of the closed loop system that coincide. The poles coincide at $s = -0.45$. We can find the gain required and the other pole locations from the closed loop transfer function

$$H(s) = \frac{K}{s^3 + 4s^2 + 3s + K} \tag{1}$$

by writing the denominator of this expression as $(s + 0.45)^2 (s + P)$ which has a double pole at $s = -0.45$. We equate coefficients in

$$s^3 + (0.90 + P)s^2 + (0.2025 + 0.9P)s + 0.2025P = s^3 + 4s^2 + 3s + K$$

and obtain $P = 3.10$, $K = 0.63$. Of course these values could have been obtained directly from the root locus rules except that we have not explicitly discussed the rule for gain determination. Thus with a gain $K = 0.63$ the closed loop system transfer function is

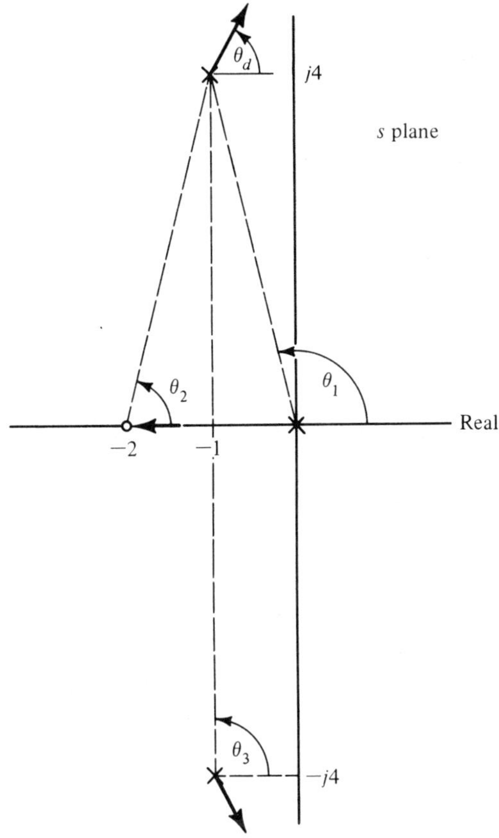

Figure 6.2-3 Illustration of calculation of angles of departure and arrival

$$H(s) = \frac{0.63}{(s + 0.45)^2 (s + 3.10)} \qquad (2)$$

and we can apply rule 11 to easily determine that the angle of departure from one of poles at $s = -0.45$ is 90°.

It should be apparent that we can replace $G(s)$ by the transfer function

$$G'(s) = K'H(s) = \frac{0.63K'}{(s + 0.45)^2 (s + 3.10)} \qquad (3)$$

and plot the root locus for $G'(s)$ with varying K'. We will get the original closed loop pole zero locations for $H(s)$ which occur for $K > 0.63$.

If we examine a more general version of this example we will obtain a very satisfactory proof of rule 10.

12. *Imaginary Axis Intersections.* The root locus diagram crosses the imaginary axis at points where the characteristic equation, $1 + G(s) = 0$, is satisfied. The Routh Hurwitz Criterion or the Bode diagram may be used to obtain the gain and frequency of the sinusoidal oscillation

This rule is nothing but a restatement of the angle criterion and the magnitude criterion with respect to the imaginary axis. The angle criterion supplies all of the information needed to determine the frequency at which the imaginary axis is crossed and the magnitude criterion need be used only if we wish to calibrate the root locus crossing of the imaginary axis in terms of gain K.

EXAMPLE 6.2-9. We can easily determine the imaginary axis intersections for (6.2-1). To use the Routh Hurwitz test we form the characteristic polynomial $s^3 + 4s^2 + 3s + K$ and use the procedures of Sec. 4.4. The gain K must be positive to satisfy the Hurwitz test. The first two Routh rows are

$$
\begin{array}{ccc}
s^3 & 1 & 3 \\
s^2 & 4 & K
\end{array}
$$

We obtain the coefficients for the third Routh row from

$$a = \frac{4(3) - 1(K)}{4} = \frac{12 - K}{4}$$

Since there is but a single entry in the third Routh row the test is concluded and we see that $K = 12$ is the gain which puts poles on the $j\omega$ axis. We could determine the pole locations from the characteristic equation

$$s^3 + 4s^2 + 3s + 12 = (s + j\omega)(s - j\omega)(s + \tau)$$

and equate coefficients of like powers in s to obtain $\tau = 4$ and $\omega = 3$.

Alternately we could use the Bode diagram approach and note that $\beta(\omega_c) = 180°$, if there are poles on the $j\omega$ axis such that the system is oscillatory. Crossover must occur on the -2 slope for a $0°$ phase margin. The crossover frequency is obtained from the open loop transfer function*

$$G(s) = \frac{K/3}{s(1+s)(1+s/3)}$$

as

$$|G(\omega_c)| = 1 = \frac{K/3}{\omega_c^2}$$

where we assume that $1 < \omega_c < 3$. Thus we obtain

$$K = 3\omega_c^2$$

The arctangent approximation for the phase shift at crossover is

$$\beta(\omega_c) = -\frac{\pi}{2} - \left(\frac{\pi}{2} - \frac{1}{\omega_c}\right) + \frac{\omega_c}{3}$$

and to obtain $\beta(\omega_c) = -\pi$ we see from the foregoing that

$$\frac{1}{\omega_c} - \frac{\omega_c}{3} = 0$$

or $\omega_c = 3$ and $K = 9$. This represents a modest error and is caused by the use of the asymptotic approximations. The break frequencies at $\omega = 1$ and 3 are simply too close to the crossover frequency of $\omega_c = 3$. The actual poles corresponding to $K = 9$ are not far from being on the $j\omega$ axis as a simple computation will show.

13. *Sum of System Closed Loop Poles.* If the open loop pole zero excess is at least two the sum of the closed loop poles is a constant. In these cases the center of gravity of the poles is preserved.

---

*Note that the transfer functions are written in slightly different forms for the Bode diagram and the root locus approach.

This rule follows in much the same manner as did rule 8 concerning the origin of the asymptotes. We obtain the closed loop transfer function from Eq. (6.2-2) as

$$H(s) = \frac{G(s)}{1 + G(s)} = \frac{K(s^n + a_1 s^{n-1} + \ldots + a_n)}{s^d + b_1 s^{d-1} + \ldots + b_d + K(s^n + a_1 s^{n-1} + \ldots + a_n)}$$

Now if $d \geqslant n + 2$ then the two highest order terms in s in the foregoing are still $s^d + b_1 s^{d-1}$. b_1 is the negative of the sum of the open loop poles and we see that b_1 will also be the negative sum of the closed loop poles as well. This will not occur if $d < n + 2$. The primary use of this rule is to quickly check a root locus diagram that has already been constructed to verify that it has been constructed correctly. Also it allows us to determine all pole locations but one and then determine this single pole location by application of rule 13. This rule shows that as some loci move to the right of the s plane others must move to the left in order that the sum of the closed loop poles be constant.

14. *Determination of System Gain.* At any complex frequency s that satisfies the angle criterion the gain K can be determined such that the magnitude criterion is satisfied from

$$K = \left| \frac{1}{G_n(s)} \right| \tag{6.2-22}$$

We have demonstrated that Eq. (6.2-22) is a valid relation earlier in this section. Since

$$G_n(s) = \frac{\displaystyle\prod_{i=1}^{n} (s + n_i)}{\displaystyle\prod_{i=1}^{d} (s + d_i)}$$

we can obtain the gain K by obtaining the product of the loci.

Often this is a bit more complex a rule than is desired. Often it is easier to obtain the gain K by assuming a closed loop pole and then finding the value of gain K which will produce that closed loop pole.

EXAMPLE 6.2-10. We again consider EXAMPLE 6.2-1. The product of the three pole locations is simply $-K$. This equation

$$s(s + 1)(s + 3) = -K$$

is simply the numerator of the closed loop characteristic equation. Use of the magnitude relation simply says that

$$|s|\,|s + 1|\,|s + 3| = |K|$$

We may assume values of s permitted by the angle criterion and find the value of K. We may establish the following table, for example

s	K
0	0
−0.2	0.45
−0.4	0.62
−0.45	0.88
−0.5	0.63
−0.75	0.42
−1	0
−3	0
−3.1	0.65
−3.2	1.41
−4	12

Also, we may use permitted complex values of s. This poses no problem if the locus has been accurately plotted. Often we are not especially concerned with great accuracy when plotting the root locus. Thus it is often convenient to factor the roots obtained in the previous table out of the closed loop transfer function and then find these roots using the quadratic rule. For example when a closed loop root is at -3.2 and the gain is 1.41, we have

$$
\begin{array}{r}
s^2 + 0.8s^2 + 0.44 \\
s + 3.2\,\overline{\smash{)}\,s^3 + 4s^2 + 3s + 1.41} \\
\underline{s^3 + 3.2s^2} \\
0.8s^2 + 3s \\
\underline{0.8s^2 + 2.56s} \\
0.44s + 1.41 \\
0.44s + 1.41
\end{array}
$$

and see that the other roots are at $s^2 + 0.8s^2 + 0.44 = 0$ or $s = -0.4 \pm j\,0.53$. An advantage to this procedure is that it is entirely analytical and we use the graphical sketch of the loci only to assist by steering our analytical calculations.

We have just presented 14 rules for root locus construction. All of them are suited to analytical calculations for systems that are reasonably low in order. With the use of hand held calculators analytical calculation of root loci for systems with three or four closed loop poles are not at all difficult. Rapid sketches of root loci may be made of systems of even higher order with ease. For systems of high order accurate root locus diagram plotting is difficult. Use of a graphical tool to replace analytical calculations is suggested.* Alternately those zeros and poles located far away from the main cluster of zeros and poles about the origin can be neglected. The digital computer is very useful for analytical construction of root locus diagrams. It is a relatively straightforward matter to develop a digital computer routine for root locus plotting.**

The root locus diagram presents information that compliments, in many ways, the information obtained by Bode diagram design. The two can be used to nicely compliment one another in the design process as we shall see in our next section.

EXERCISE 6.2-2.　Derive rule 10. This can be accomplished by extension of EXAMPLE 6.2-8.

EXERCISE 6.2-3.　Restate those root locus rules which change for negative K.

EXERCISE 6.2-4.　Show that the root locus plot of $G'(s)$ in EXAMPLE 6.2-8 does lead to the same closed loop poles and zero as does the root locus plot of $G(s)$. What is the relationship between $G'(s)$ and $G(s)$ that causes this to happen?

---

*An inexpensive plastic aid for constructing root locus diagrams known as the Spirule is available from the Spirule Company, 9728E, Venado, Whittier, Calif.

**See (1) Ash, R. H. and G. R. Ash, "Numerical Computation of Root Loci Using the Newton-Raphson Technique," *IEEE Transactions on Automatic Control*, Vol. AC-13, No. 5, October 1968, pp. 576–582.

(2) Klagsbrunn, Z. and Y. Wallach, "On Computer Implementation of Analytical Root Locus Plotting," *IEEE Transactions on Automatic Control*, Vol. AC-13, No. 6, December 1968, pp. 744–745.

(3) Melsa, J. L. *Computer Programs for Computational Assistance in the Study of Linear Control Theory*, McGraw Hill Book Co., New York, 1970, for discussion of computer implementations of root locus plotting.

EXERCISE 6.2-5. Construct root locus diagrams for the following transfer functions

a. $\quad G(s) = \dfrac{K}{(s+1)(s+2+j2)(s+2-j2)}$

b. $\quad G(s) = \dfrac{K(s+2)}{(s+4)(s+1+j1)(s+1-j1)}$

c. $\quad G(s) = \dfrac{K}{s(s+1)(s+5)(s+8)}$

d. $\quad G(s) = \dfrac{K(s+10)}{s(s+5)(s+100)(s+2+j2)(s+2-j2)}$

6.3 DESIGN USING THE BASIC ROOT LOCUS TECHNIQUE

In this section we will examine several design and analysis examples using the root locus technique. We will relate many of these to the Bode diagram frequency domain approach of the previous chapter.

Prior to examining these examples it is helpful to detail an approach to root locus design. We will assume that we are not using a spirule or making detailed calculations but wish to obtain a reasonable picture of the locus by using the rules we obtained in our last section. Figure (6.3-1) presents a DELTA Chart of a suggested approach to analysis using the root locus. Figure (6.3-2) illustrates the use of this analysis approach for series equalizer design. We will discuss the use of the root locus for minor loop equalization in our next section.

The root locus diagrams we obtain by the rules of the last section and the procedures suggested here result in a useful graphical representation of the roots of the characteristic equation which are the poles of the closed loop system. For a second order system the damping ratio, damped frequency, and undamped natural frequency completely describe the response characteristics as we recall from Sec. 4.2. From that section the graph of Fig. (6.3-3), which shows curves of constant response characteristics of a second order system with no finite zeros, results. Circles about the origin in this figure correspond to roots with the same constant undamped natural frequency. If we wished to design a system with a minimum undamped natural frequency, such as to insure a minimum speed of response, we would have to be sure that our closed loop system did not have any poles located inside a circle

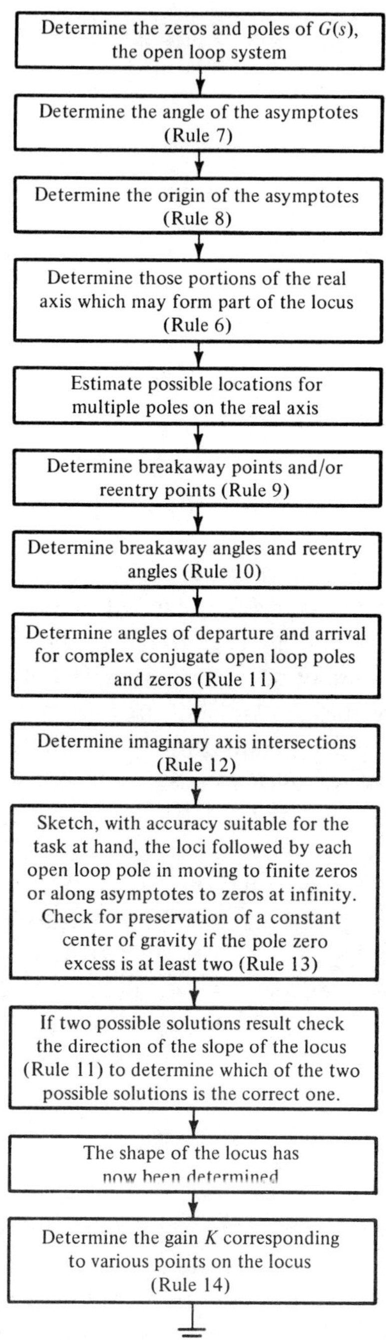

Figure 6.3-1 DELTA Chart for suggested procedure for root locus analysis

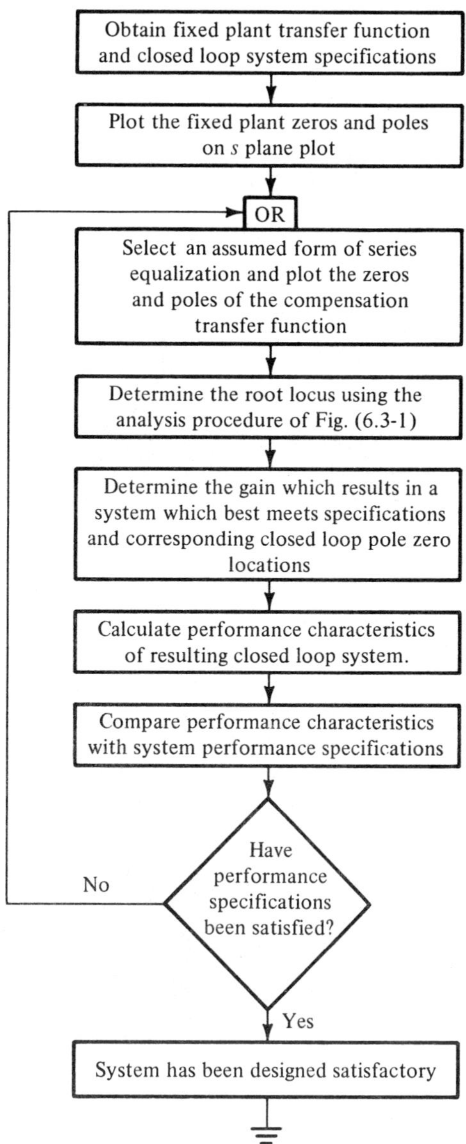

Figure 6.3-2 DELTA Chart of the basic series compensation design process using the root locus

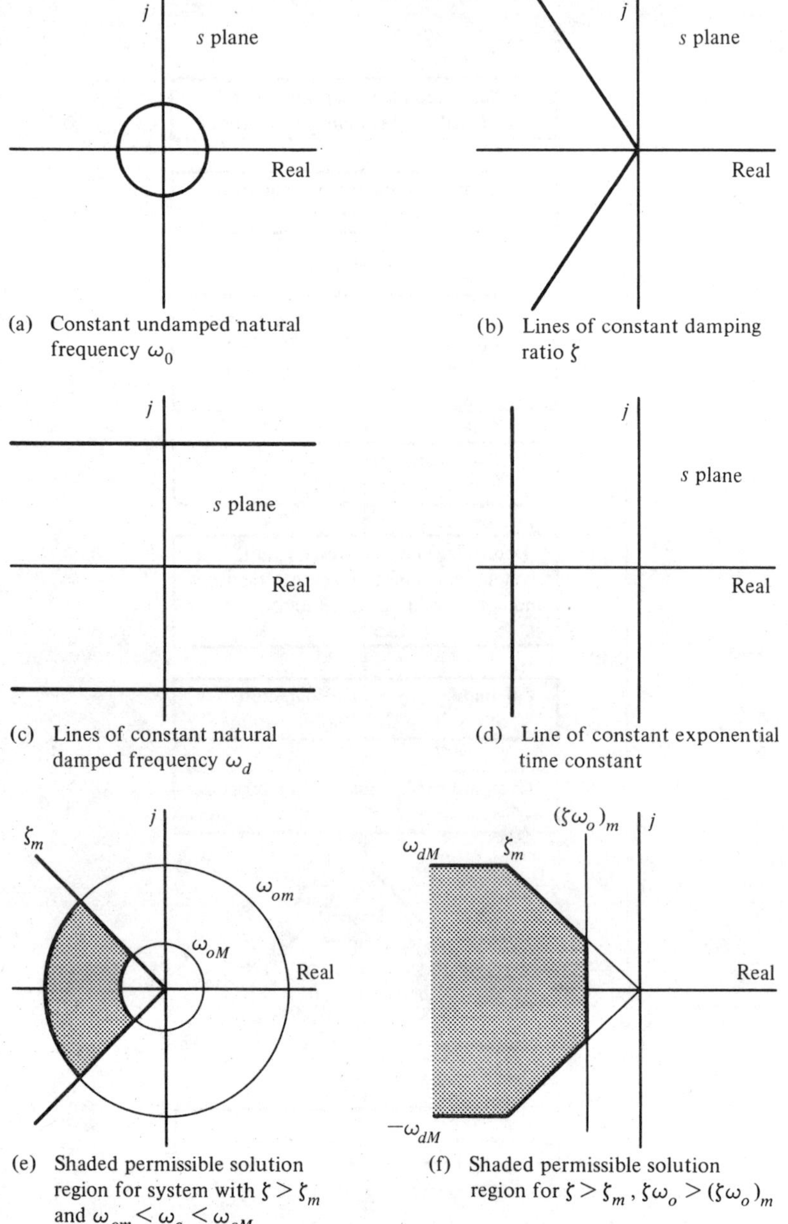

Figure 6.3-3 Constant response characteristic curves—second order system

from the origin are lines of constant damping ratio. If we specify a minimum damping ratio then we must be sure that the dominant response poles of the closed loop system lie to the left of the corresponding radial line for this damping ratio. A damping ratio of zero corresponds to an unstable system and specifying that the closed loop system be stable restricts all closed loop poles to lie in the left half plane.

Straight lines with constant imaginary part correspond to systems with the same natural damped frequency. Straight lines with constant real part correspond to systems with the same exponential time constant. Several of these constraints may be combined together and a permissible region for closed loop roots determined as in Fig. (6.3-3).

In addition to the relationships shown in Fig. (6.3-3) we have a large number of other responses and performance specifications which are presented in Chap. 4. All of these, together with experience, provide a means for interpretation of the root locus diagram and allow us to interrelate this diagram to other design approaches in linear systems control.

EXAMPLE 6.3-1. Let us reconsider the type two system with a simple lead network that we examined in Chap. 5. The fixed plant transfer function is

$$G_f(s) = \frac{K}{s^2} \tag{1}$$

and we assume a single stage lead network

$$G_c(s) = \frac{1 + s/\omega_1}{1 + s/\omega_2} \qquad \omega_1 < \omega_2 \tag{2}$$

such that the open loop transfer function becomes

$$G(s) = G_f(s)G_c(s) = \frac{K\omega_2}{\omega_1} \frac{(s + \omega_1)}{s^2(s + \omega_2)} \tag{3}$$

It is convenient to normalize this open loop transfer function in order to obtain the most general results possible for this example. There are a number of ways we might normalize this transfer function. One particularly useful and physically significant way is to let

$$\omega_2 = a\omega_1 \qquad a > 1 \tag{4}$$

such that a represents the ratio of break frequencies in the lead network and the active gain required. The open loop transfer function then becomes

$$G(s) = \frac{aK(s + \omega_1)}{s^2(s + a\omega_1)} \tag{5}$$

Now we normalize by change of time scale

$$t = \frac{\tau}{\omega_1} \tag{6}$$

or

$$s = p\omega_1 \tag{7}$$

where τ and p are the normalized time and frequency variables. The normalized open loop transfer function becomes

$$G(p) = \frac{aK}{\omega_1^2} \frac{p + 1}{p^2(p + a)} = \frac{K_n(p + 1)}{p^2(p + a)} \tag{8}$$

where the normalized gain is

$$K_n = \frac{aK}{\omega_1^2} \tag{9}$$

Before turning to an analysis of the closed loop transfer function corresponding to the normalized open loop system of Eq. (8) let us re-examine some of the Bode diagram type performance characteristics of the system. First we assume that we wish to find the gain K_n to yield maximum phase margin for a fixed a. The crossover frequency will be assumed to occur on the -1 slope such that we can obtain the crossover frequency from

$$1 = \frac{K_n}{a\omega_c}$$

as

$$\omega_c = \frac{K_n}{a} \qquad (10)$$

The phase shift on the -1 slope becomes, using the arctangent approximation,

$$\beta(\omega) = -\pi + \left(\frac{\pi}{2} - \frac{1}{\omega}\right) - \frac{\omega}{a} \qquad (11)$$

To maximize the phase margin we set

$$\left.\frac{\partial\beta(\omega)}{\partial\omega}\right|_{\omega=\omega_c} = 0 = + \frac{1}{\omega_c^2} - \frac{1}{a}$$

to obtain

$$\omega_c = a^{1/2} \qquad (12)$$

From Eqs. (10) and (12) we obtain the optimum value of gain for maximum phase margin

$$\hat{K}_n = a^{3/2} \qquad (13)$$

and the corresponding maximum value of the phase margin, from Eq. (11)

$$\hat{\text{PM}} = \frac{\pi}{2} - \frac{2}{a^{1/2}} \qquad (14)$$

As we should expect, the phase margin continues to increase for increases in a.

Our root locus analysis for the open loop transfer function of Eq. (8) proceeds as follows. We follow the order indicated in the DELTA Chart of Fig. (6.3-1).

1. The zeros and poles of the open loop system are given by Eq. (8).

2. The angle of the asymptotes is determined from rule 7 and Eq. (6.2-15) as

$$\alpha_{as} = \frac{(1 + 2k)180°}{1 - 3} = -90°, -270° \tag{15}$$

3. The origin of the asymptotes is determined from rule 8 and Eq. (6.2-17) as

$$OA = -\frac{1 - a}{1 - 3} = -\frac{a - 1}{2} \tag{16}$$

4. Those portions of the real axis which may form part of the root locus are determined from Eq. (8) as that part of the real axis from $-a < \sigma < -1$.

5. Possible locations for multiple poles on the real axis are those points in the interval $-a < \sigma < -1$. A reentry and breakaway point are not guaranteed to exist here. We must determine whether or not they do. Our first thought is that there are no real axis breakaways but it is very dangerous to not examine this further, as we will soon see.

6. Breakaway points and/or reentry points are determined using rule 9 and Eq. (6.2-20)

$$\frac{dG^{-1}(s)}{ds} = 0$$

as the roots of

$$2p^3 + p^2(3 + a) + 2ap = 0 \tag{17}$$

One root is $p = 0$ and this is a trivial solution since the open loop transfer function contains the two poles at $p = 0$. They do, of course, break away from the origin as K_n increases from zero. The other roots are

$$p = \frac{-3}{4} - \frac{a}{4} \pm \left[\frac{(3 + a)^2 - 16a}{16}\right]^{1/2} \tag{18}$$

For values of a in the interval $1 < a < 9$ the values of real axis breakaway or reentry points are complex. This makes no sense at all and so we conclude that no breakaway or reentry points exist for $1 < a < 9$. For $a \geqslant 9$ there exists a single breakaway point and a single reentry point and they are given by Eq. (18). We can compute from Eq. (18) the table

a	p breakaway	p reentry
9	−3	−3
10	−4	−2.50
15	−6.79	−2.21
30	−14.42	−2.08
∞	$-\infty$	−2.00

7. Breakaway and reentry angles are easily determined, from rule 10, as separated by 90°.

8. Since there are no open loop poles and zeros off of the real axis there is no need to apply rule 11.

9. The closed loop system is stable for all positive values of K and so there are no imaginary axis intersections.

Figure (6.3-4) illustrates the root locus of this example for several different values of a. The closed loop system is, from Eq. (8),

$$H(p) = \frac{G(p)}{1 + G(p)} = \frac{K_n(p + 1)}{p^3 + ap^2 + K_n p + K_n} \tag{19}$$

In the region where there is a single real root at $s = -\sigma$ and two complex conjugate roots located at $s^2 + 2\zeta\omega_0 s + \omega_0^2$, the closed loop transfer function can be written as

$$H(s) = \frac{\omega_0^2 \sigma(p + 1)}{(p^2 + 2\zeta\omega_0 p + \omega_0^2)(p + \sigma)}$$

$$= \frac{\omega_0^2 \sigma(p + 1)}{p^3 + (2\zeta\omega_0 + \sigma)p^2 + (2\zeta\omega_0\sigma + \omega_0^2)p + \omega_0^2\sigma} \tag{20}$$

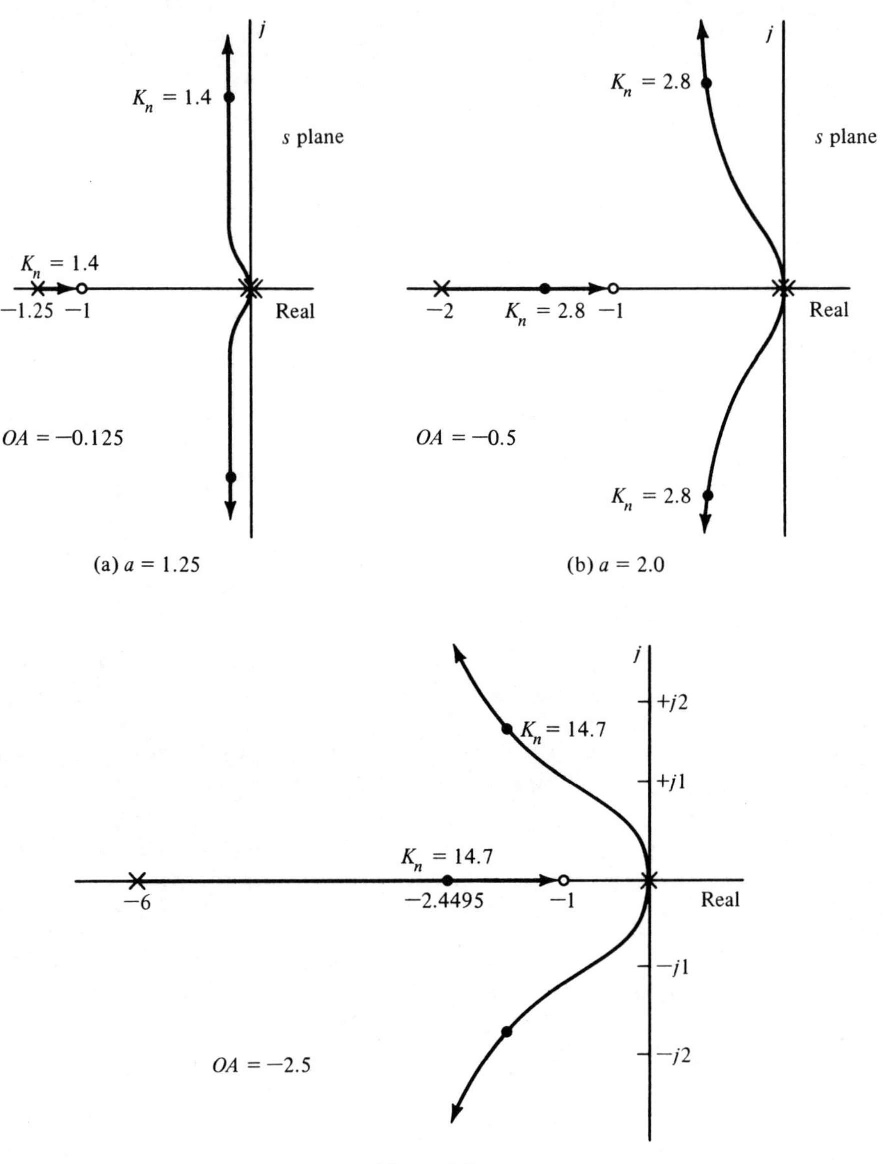

Figure 6.3-4 Root locus diagrams for EXAMPLE 6.3-1 for several values of a

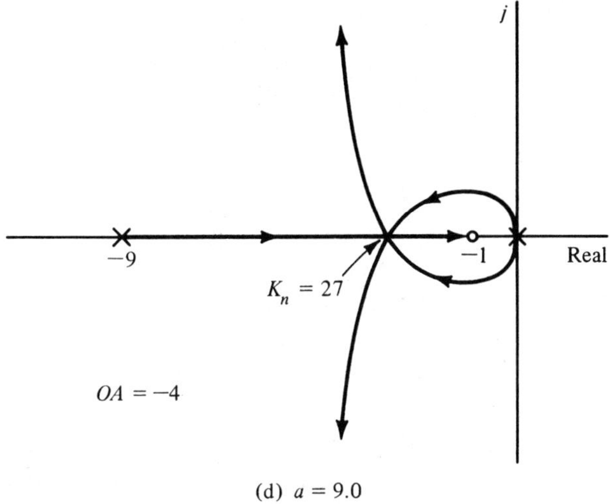

(d) $a = 9.0$

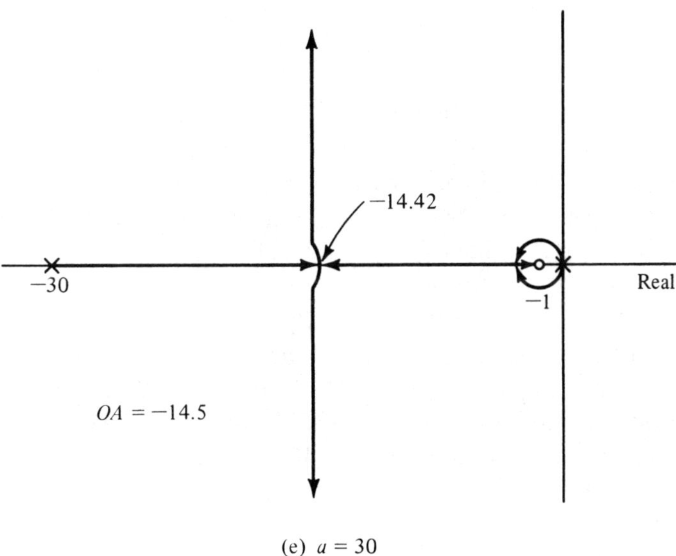

(e) $a = 30$

Figure 6.3-4 (cont'd)

By comparing coefficients in Eqs. (19) and (20) we obtain

$$\zeta = \frac{1}{2}\left[\frac{(a-\sigma)(\sigma-1)}{\sigma}\right] \qquad (21)$$

$$\omega_0 = \frac{2\zeta\sigma}{\sigma-1} = \left[\frac{(a-\sigma)(\sigma)}{\sigma-1}\right]^{1/2} \qquad (22)$$

and we can obtain the damped natural frequency from $\omega_d = \omega_0(1-\zeta^2)^{1/2}$. These equations allow us to obtain the damping ratio and natural frequency in terms of the a value for a given lead network and the real closed loop pole location. These equations are of much assistance in the construction of the root locus of Fig. (6.3-4).

The value of σ which leads to maximum ζ is $\sigma = a^{1/2}$. The damping ratio ζ will always go to zero at $\sigma = a$ and $\sigma = 1$. Thus we see that there is a gain setting which leads to maximum ζ and this is

$$K_n = \frac{\sigma^2(a-\sigma)}{\sigma-1} = a^{3/2} \qquad (23)$$

The relations obtained from Figs. (6.3-5) and (6.3-6) have been used to assist in sketching the root loci of Fig. (6.3-4). They are useful in their own right as design aids as well. Figure (6.3-5) illustrates maximum damping ratio and the corresponding ω_d as a function of a whereas Fig. (6.3-6) illustrates maximum damping ratio and maximum phase margin obtained as a function of the optimum gain $\hat{K}_n = a^{3/2}$ which leads to maximum phase margin. These last two design charts are strongly reinforcing of our previous discussions concerning performance specifications in Chap. 4. It is extremely interesting and of much practical significance that the gain $\hat{K}_n$ which maximizes the phase margin also maximizes the damping ratio. This indicates, very clearly, the strong compatibility of the root locus and the Bode diagram design approaches.

Inspection of Figs. (6.3-4) through (6.3-6) indicates that the system response will not be very good unless a is at least 4 and preferably as large as 5. There is essentially no point in having a value of a greater than $a = 9$ from the point of view of relative stability as the phase margin becomes quite large at $a = 9$, and $K_n = 27$. For the case where $a = 9$ all the poles are on the real axis. It is not at all true that the system response will not overshoot unless there are poles off of the

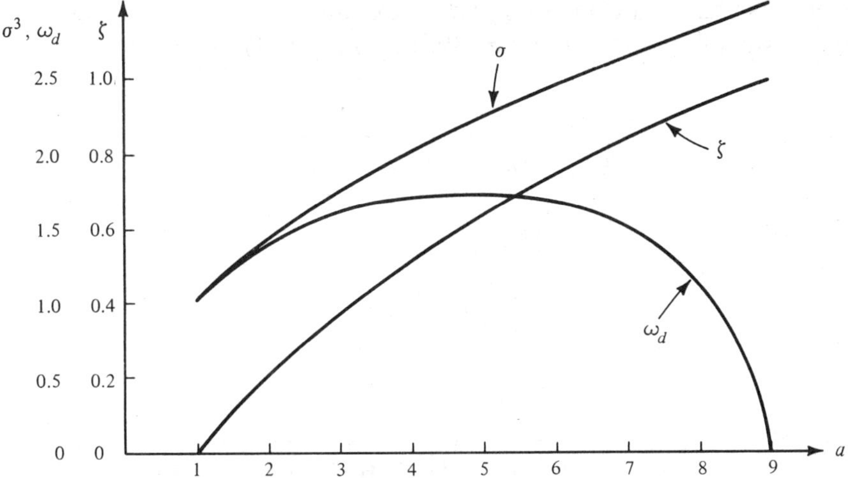

Figure 6.3-5 Real axis pole location, maximum damping ratio, and the corresponding damped natural frequency as a function of a

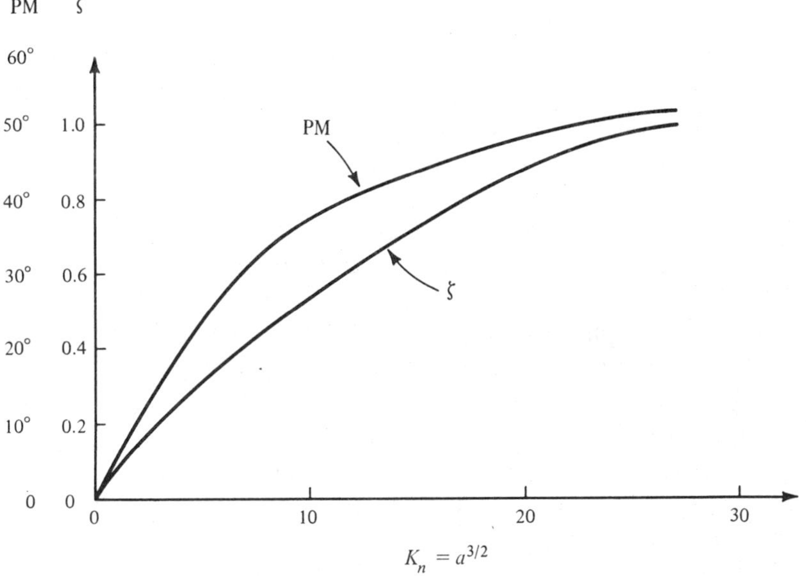

Figure 6.3-6 Damping ratio and phase margin as a function of gain K_n

real axis. As a case in point let us examine the unit step response of this system for $K_n = 27$ and $a = 9$. From Eq. (19) we have

$$\frac{Z(p)}{U(p)} = \frac{27(p+1)}{(p+3)^3} = \frac{(1+p)}{(1+p/3)^3} \tag{24}$$

and for $U(p) = 1/p$ we have

$$Z(p) = \frac{27(p+1)}{p(p+3)^3} =$$

$$= \frac{1}{p} + \frac{18}{(p+3)^3} - \frac{3}{(p+3)^2} - \frac{1}{p+3}$$

The normalized time response of the system is

$$z(t) = 1 + (18t^2 - 3t - 1)e^{-3t}$$

This response is indicated in Fig. (6.3-7) and shows that there is a 74% overshoot of the step response even though there are poles only on the negative real axis. Figure (6.3-8) illustrates the closed loop frequency response and indicates that there is also peaking of the system frequency response as the closed loop gain magnitude is 1.3 at 1.73 radians per second even though there are no poles off of the real axis. Both the asymptotic and actual frequency response curves are shown in this figure.

Values of a between 5 and 9 are quite reasonable for this system. Either a Bode diagram approach or the root locus approach leads to this conclusion. The basic approach to root locus design suggested here in Fig. (6.3-2) implies, for this example, that we would pick a value of a and then plot the root loci. The best value of K to yield maximum damping would be found and the performance of the closed loop system calculated. If that is not acceptable, as would be the case if we had selected $a = 2$ initially, the value of a would be changed and the root locus for varying K_n would again be determined. It would be desirable to be able to vary a and plot a root locus for varying a, and perhaps also K. It is generally possible to modify the root locus approach slightly such that this is possible as we will see in our next section.

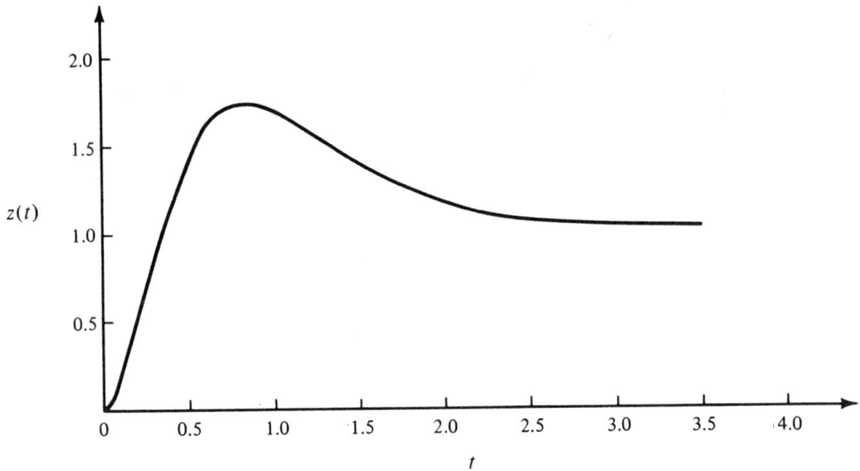

Figure 6.3-7 Step response of the system with $a = 9$

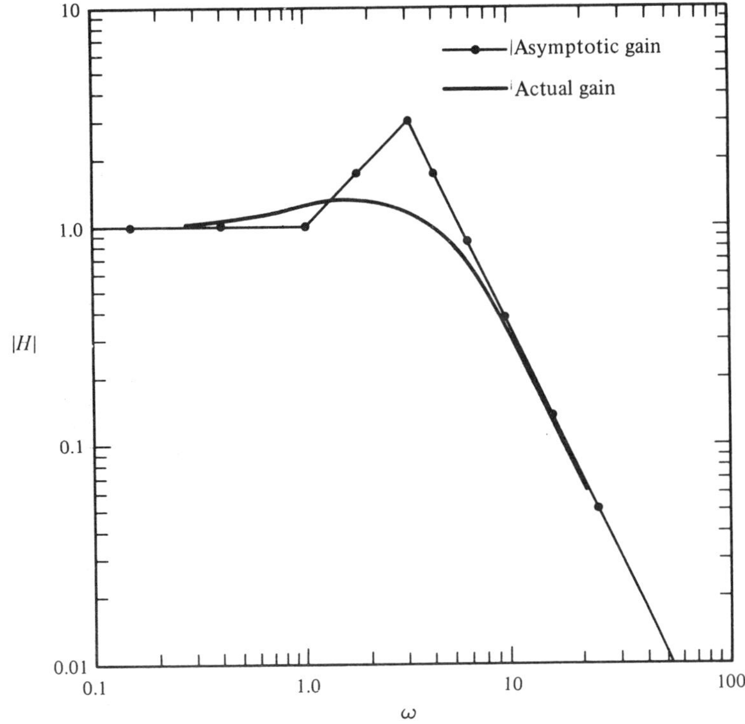

Figure 6.3-8 Closed loop transfer function frequency response magnitude

This very general lead network design example has been rather exhaustively discussed in order to indicate, in as thorough a fashion as possible, the techniques and the advantages of root locus design. We have normalized the time variable, as indicated by Eq. (6), and it is a simple matter to "unnormalize" the response to fit any given time scale.

EXAMPLE 6.3-2. Although the Bode diagram approach to series compensation appears to indicate considerable difference between lead and lag network design, it turns out that there are many similarities as this example will show. We consider the open loop transfer function

$$G_f(s) = \frac{1000}{s\left(1 + \dfrac{s}{10}\right)} \tag{1}$$

Figure (6.3-9) illustrates "optimum"

    a.   Lead network compensation,

    b.   Lag network compensation, and

    c.   Lag-Lead network compensation

to achieve a 45° phase margin. We use the word "optimum" in the same sense that it was used in Chap. 5.

The "optimum" open loop transfer function for the lead network is obtained by analytical solution of the equations

$$10^4 = \omega_1 \omega_c$$

$$\omega_c = [\omega_2(\omega_1 - 10)]^{1/2}$$

$$\beta(\omega_c) = -\frac{\pi}{2} - 2[(\omega_1 - 10)/\omega_2]^{1/2} = -\frac{3\pi}{4}$$

as

$$G_{\text{lead}}(s) = \frac{1000(1 + s/\omega_1)}{s(1 + s/10)(1 + s/\omega_2)} \tag{2}$$

where

$$\omega_1 = 63, \qquad \omega_2 = 477, \qquad \omega_c = 159$$

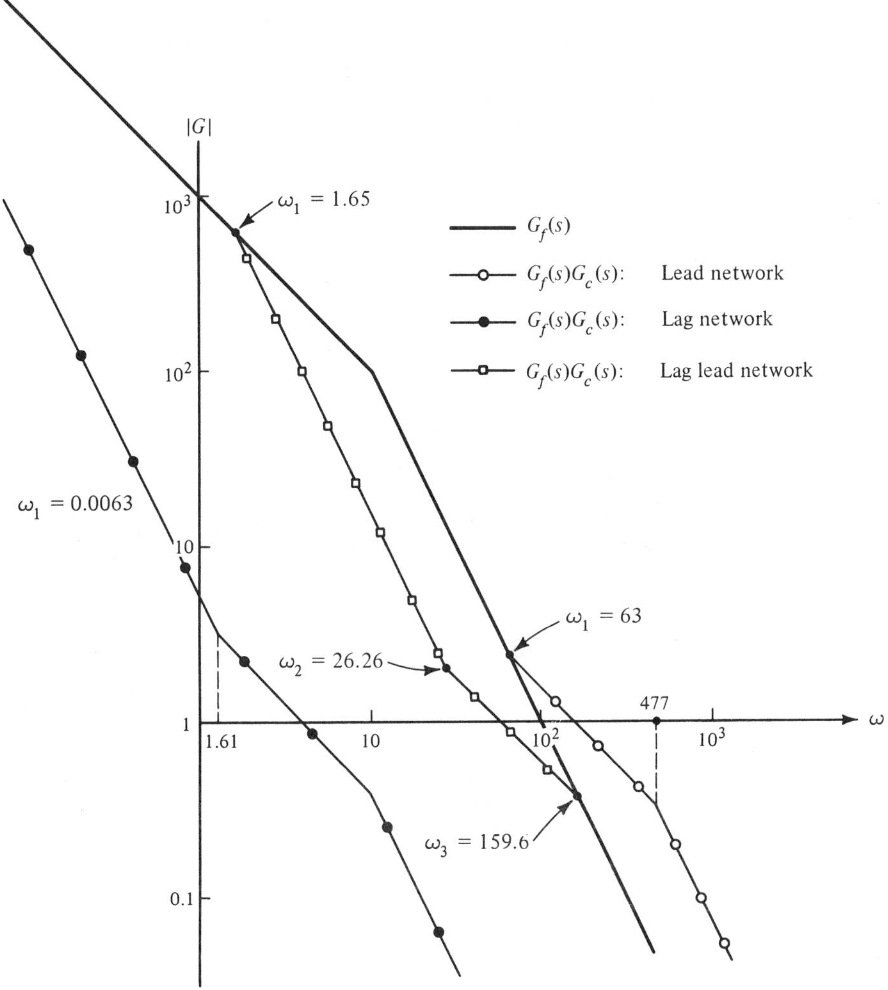

Figure 6.3-9 Uncompensated and compensated open loop transfer function gain magnitude for EXAMPLE 6.3-2

The optimum open loop transfer function for the lag network is obtained by analytical solution of the equations

$$1000\omega_1 = \omega_c\omega_2$$

$$\omega_c^2 = 10(\omega_2 - \omega_1)$$

$$\beta(\omega_c) = \frac{-\pi}{2} - 2[(\omega_2 - \omega_1)/10]^{1/2} = -\frac{\pi}{2} - \frac{2\omega_c}{10} = \frac{-3\pi}{4}$$

as

$$G_{\text{lag}}(s) = \frac{1000(1 + s/\omega_2)}{s(1 + s/\omega_1)(1 + s/10)} \tag{3}$$

where

$$\omega_1 = 0.0063, \qquad \omega_2 = 1.61, \qquad \omega_c = 3.93$$

The "optimum" open loop transfer function for the lag lead network is

$$G_{\text{lag lead}}(s) = \frac{1000(1 + s/\omega_2)}{s(1 + s/\omega_1)(1 + s/\omega_3)} \tag{4}$$

where the unknown equalizer parameters are obtained from simultaneous solution of

$$1000\omega_1 = \omega_c\omega_2$$

$$\omega_c^2 = \omega_3(\omega_2 - \omega_1)$$

$$\beta(\omega_c) = \frac{-\pi}{2} - 2[(\omega_2 - \omega_1)/\omega_3] = \frac{-3\pi}{4}$$

$$\omega_1\omega_3 = 10\omega_2$$

as

$$\omega_1 = 1.65, \qquad \omega_2 = 26.26, \qquad \omega_3 = 159.6, \qquad \omega_c = 62.7$$

These three compensated open loop transfer functions are

$$G_{\text{lead}}(s) = \frac{7.57 \times 10^4 (s + 63)\alpha}{s(s + 10)(s + 477)} \tag{5}$$

$$G_{\text{lag}}(s) = \frac{39.1(s + 1.61)\alpha}{s(s + 0.0063)(s + 10)} \tag{6}$$

$$G_{\text{lead lag}}(s) = \frac{10^4(s + 26.26)\alpha}{s(s + 1.65)(s + 159.6)} \tag{7}$$

where α is a nominal gain which should be one but will be varied in order to sketch the root locus.

If we normalize each transfer function by a time scale change $s = \omega_c p$, where ω_c is the crossover frequency for the particular compensated system, then the normalized crossover frequency for each system becomes unity. We have

$$G_f(p) = \frac{10}{p(p + 0.1)} = \frac{10}{p(1 + 10p)} \tag{8}$$

$$G_{\text{lead}}(p) = \frac{2.99(p + 0.40)\alpha}{p(p + 0.063)(p + 3)} = \frac{6.32(1 + 2.5p)\alpha}{p(1 + 15.87p)(1 + p/3)} \tag{9}$$

$$G_{\text{lag}}(p) = \frac{2.53(p + 0.41)\alpha}{p(p + 0.0016)(p + 2.545)} = \frac{255(1 + 2.44p)\alpha}{p(1 + 623p)(1 + p/2.545)} \tag{10}$$

$$G_{\text{lag lead}}(p) = \frac{2.54(p + 0.42)\alpha}{p(p + 0.026)(p + 2.546)}$$

$$= \frac{16.12(1 + 2.38p)\alpha}{p(1 + 38.46p)(1 + p/2.546)} \tag{11}$$

In the vicinity of the crossover frequency, the transfer function of the compensated systems all are approximately the same

$$G_{\text{compensated}}(p) \cong \frac{K(1 + p/\omega_a)}{p^2(1 + p/\omega_b)} \tag{12}$$

where appropriate values of K, ω_a, and ω_b are given in the table for $0.07 < \omega < \infty$.

Type of Compensating Element	K	ω_a	ω_b
Lead	0.40α	0.40	3
Lag	0.41α	0.41	2.545
Lag Lead	0.42α	0.42	2.546

Thus we will expect the normalized performance of all three systems to be almost the same.

    Figure (6.3-10) represents root locus diagrams for the unnormalized systems of Eqs. (1) through (4).

    The closed loop compensated system transfer functions are all of the form (where $\alpha = 1$)

$$H(s) = \frac{G(s)}{1 + G(s)} = \frac{\dfrac{\sigma_2 \omega_0^2}{\sigma_1}(s + \sigma_1)}{(s + \sigma_2)(s^2 + 2\zeta\omega_0 s + \omega_0^2)} \tag{13}$$

where, for the unnormalized system

Compensating Network Type	σ_1	σ_2	ζ	ω_0
Lead	63.6	225.0	0.89	146.3
Lag	1.61	4.33	0.745	3.804
Lag Lead	26.33	66.84	0.753	62.95

For the normalized system, the corresponding table is

Compensating Network Type	σ_1	σ_2	ζ	ω_0
Lead	0.40	1.415	0.890	0.920
Lag	0.41	1.101	0.745	0.968
Lag Lead	0.42	1.066	0.753	1.004

By use of the transform methods of Chap. 2 we easily obtain for the unit step response of the system of Eq. (13)

$$z(t) = 1 + \frac{\sigma_2}{\sigma_1}\left[\frac{\sigma_1^2 - 2\zeta\sigma_1\omega_0 + \omega_0^2}{(1 - \zeta^2)(\sigma_2^2 - 2\zeta\sigma_2\omega_0 + \omega_0^2)}\right]^{1/2} e^{-\zeta\omega_0 t} \sin[\omega_0(1 - \zeta^2)^{1/2}t + \theta]$$

$$+ \frac{\omega_0^2(\sigma_2 - \sigma_1)}{\sigma_1(\sigma_2^2 - 2\sigma_2\omega_0 + \omega_0^2)}e^{-\sigma_2 t} \tag{14}$$

where

$$\theta = \tan^{-1}\left[\frac{\omega_0(1-\zeta^2)^{1/2}}{\sigma_1 - \zeta\omega_0}\right] - \tan^{-1}\left[\frac{\omega_0(1-\zeta^2)^{1/2}}{\sigma_2 - \zeta\omega_0}\right]$$

$$+ \tan^{-1}\left[\frac{(1-\zeta^2)^{1/2}}{\zeta}\right]$$

It is a straightforward but tedious task to plot the response time function using Eq. (14). There are many differential equation solving routines available on digital computers which will solve the differential equation corresponding to Eq. (13). Alternately the methods to be discussed in Sec. (7.2) may be used. The step response of the three normalized compensated systems is shown in Fig. (6.3-11). We again recall that the actual system time response is different than that shown in the figure because of the normalization, which amounts to a time scale change for this example. The lead network compensated system response is 159 times faster than that shown, the lag network compensated system response is 3.93 times faster than that shown, and the lag lead network system response is 62.7 times faster than that shown.

We have attempted to take a fairly general approach in this example to illustrate the versatility of the root locus approach. Obviously a set of system specifications for this example would certainly include something about closed loop bandwidth or rise time to a step input or some other input time function and this would typically eliminate two of the three compensating network types used here.

EXERCISE 6.3-1. Determine and compare actual and asymptotic Bode diagrams for the compensated closed loop transfer functions of EXAMPLE 6.3-2.

EXERCISE 6.3-2. Discuss compensation of the open loop unstable fixed plant $G_f(s) = \dfrac{10}{s(s-1)}$ by means of a lead network $G_c(s) = \dfrac{1 + s/\omega_1}{1 + s/\omega_2}$. Leave the velocity error coefficient of 10 unchanged and design the system for minimum ω_2/ω_1 in order to not seriously reduce the ability of the system to reject high frequency noise. The minimum allowable damping ratio is $\zeta = 0.5$.

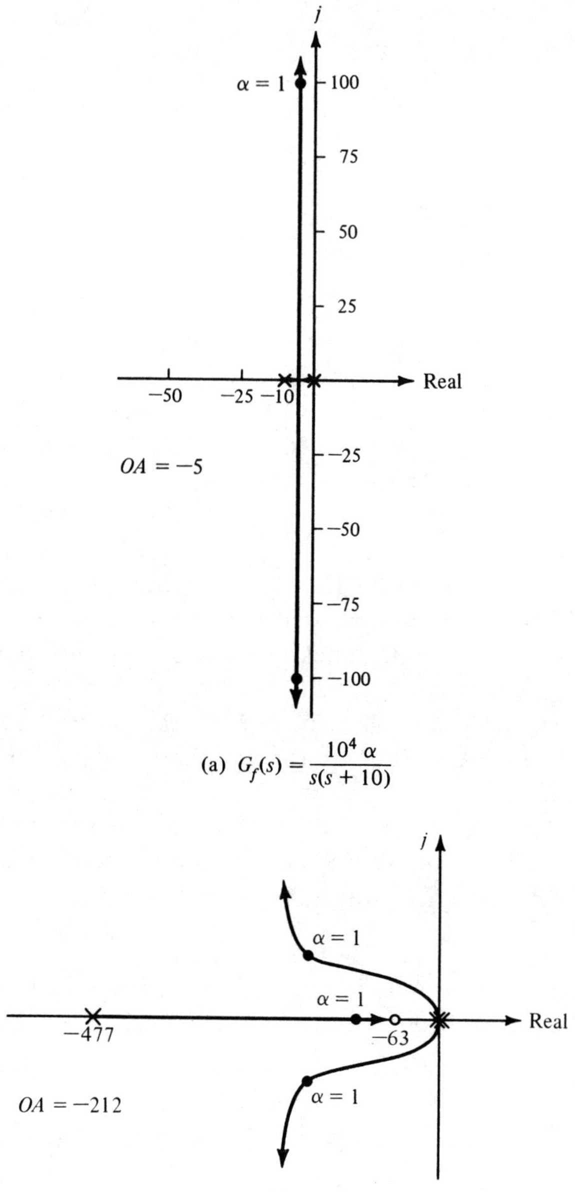

(a) $G_f(s) = \dfrac{10^4\,\alpha}{s(s + 10)}$

(b) Lead network compensated system

Figure 6.3-10   Root locus diagrams of unnormalized system of EXAMPLE 6.3-2

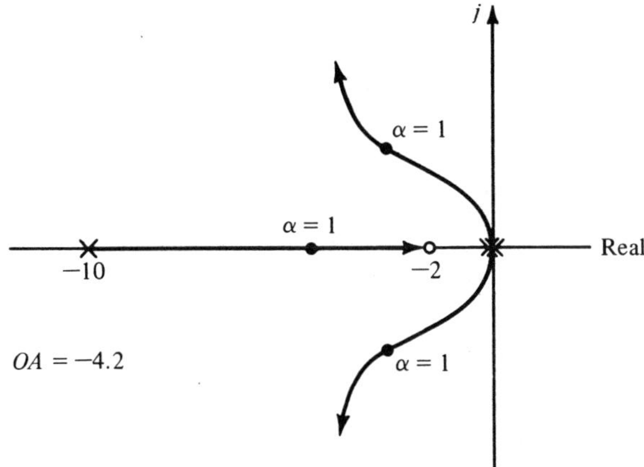

(c) Lag network compensated system

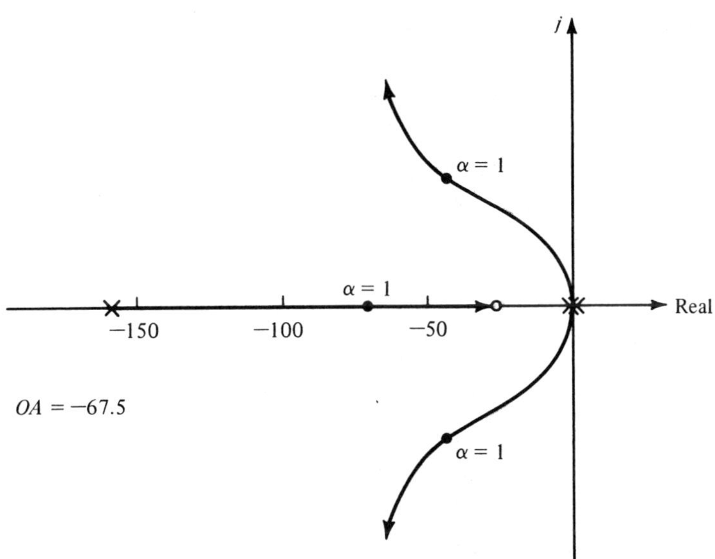

(d) Lag lead network compensated system

Figure 6.3-10 (cont'd)

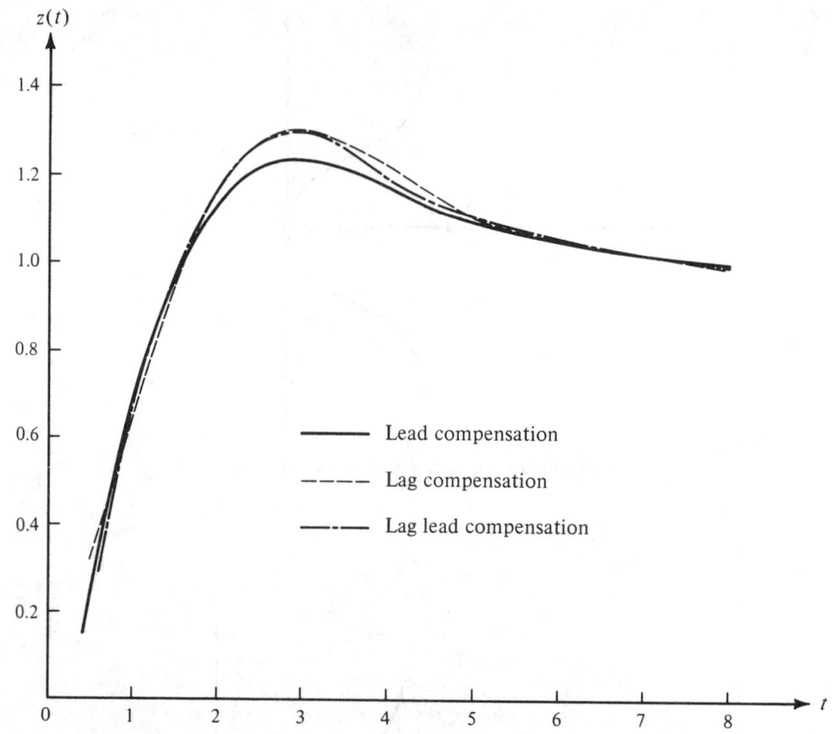

Figure 6.3-11 Step response of the normalized compensated systems of EXAMPLE 6.3-2

6.4 EXTENSIONS OF THE BASIC ROOT LOCUS TECHNIQUE

In this section we will consider several extensions of the basic root locus approach. While our discussions thus far cover the major topics of interest with respect to root locus theory and applications, there are some cases where the basic theory needs to be extended somewhat in order for it to be more fully useable.

Negative Gain Loci. One of these extensions concerns the complimentary root locus or root locus for negative gain. We have developed most of these rules but have not actually applied them to examples of interest. A particular case where the complimentary root locus is needed is that which occurs when the open loop transfer function is nonminimum phase and contains zeros in the right half of the s plane.

EXAMPLE 6.4-1. We consider the open loop fixed plant transfer function

$$G_f(s) = \frac{10(1-s)}{s(1+s)}$$

The velocity constant must be left at 10. The closed loop system is unstable as we see from the observation that at the crossover frequency of $\omega_c = 10$ the phase shift is considerably more negative than $-\pi$ radians. We wish to compensate the system using a lag network for "reasonable" stability and transient performance.

In order to obtain the system root locus we write the open loop transfer function in the form

$$G_f(s) = \frac{-10\alpha(s-1)}{s(s+1)} \tag{1}$$

where α is nominally 1. Since the open loop gain is negative we need to obtain the complimentary root locus. We easily obtain the following:

1. The open loop poles are located at $s = 0, -1$, and the open loop zero at $s = +1$.

2. The angle of the asymptotes is $0°$ and $180°$.

3. The origin of the asymptotes is not a meaningful expression.

4. The portions of the real axis which may form part of the locus are $-1 < s < 0$ and $s > 1$.

5. Multiple poles may exist in these regions.

6. Breakaway points and reentry points are located at $dG^{-1}(s)/ds = 0$ and these points are given by $s = 1 \pm \sqrt{2}$.

7. The imaginary axis intersections are $s = \pm j1$.

The loci may now be sketched as in Fig. (6.4-1). The gain corresponding to various points on the locus may now be determined. As we see the system is indeed unstable.

We could now try a variety of lag networks, plot the root locus and iterate this procedure until we find a satisfactory solution. Our suggested approach is to accomplish a preliminary design using the Bode diagram approach and then check the final design using the root locus

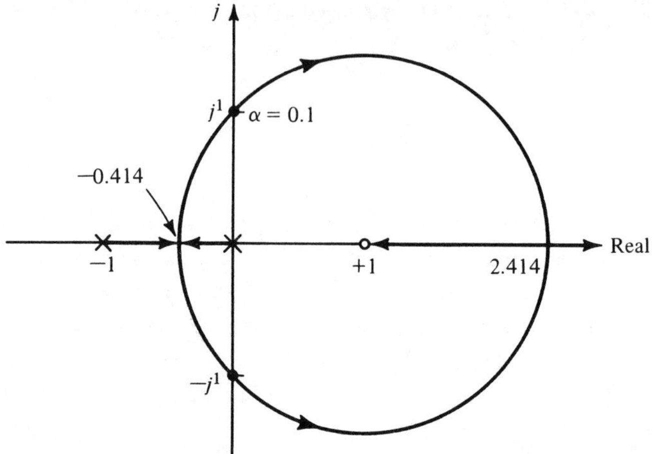

Figure 6.4-1 Root locus of uncompensated closed loop system

to see whether the design obtained using the Bode diagram approach is satisfactory from a closed loop root location viewpoint. The compensated system transfer function becomes

$$G(s) = G_f(s)G_c(s) = \frac{10(1-s)(1+s/\omega_2)}{s(1+s)(1+s/\omega_1)} \qquad (2)$$

The crossover frequency is defined by

$$1 = \frac{10\omega_1}{\omega_2\omega_c}$$

The minimum attenuation lag network is obtained when

$$\omega_c^2 = (\omega_2 - \omega_1)/2$$

and a 45° phase margin is obtained when

$$\frac{-3\pi}{4} = -\frac{\pi}{2} - 2\frac{\omega_c}{1} - \left(\frac{\pi}{2} - \frac{\omega_1}{\omega_c}\right) + \left(\frac{\pi}{2} - \frac{\omega_2}{\omega_c}\right)$$

Thus our compensating network parameters are

$$\omega_1 = 0.00148 \qquad \omega_2 = 0.07 \qquad \omega_c = 0.20$$

Figure (6.4-2) illustrates the root locus for the compensated system. As we see from inspection of the locus the relative system stability is quite reasonable. The closed loop transfer function, with $\alpha = 1$, is

$$\frac{Z(s)}{U(s)} = \frac{-0.0448(s-1)}{(s+0.135)(s+0.328+j0.049)(s+j0.328-j0.049)} \quad (3)$$

Except for the action of the "dipole" around the origin we see that the behavior of the compensated system root locus is similar to that for the uncompensated system. We have two distinct frequency modes of system operation. At low frequencies the root locus behaves essentially as if it were that of a closed loop system with open loop transfer function of the form

$$G'(s) = \frac{10(1+s/0.07)\alpha}{s(1+s/0.00148)} \approx \frac{0.21\alpha}{s}$$

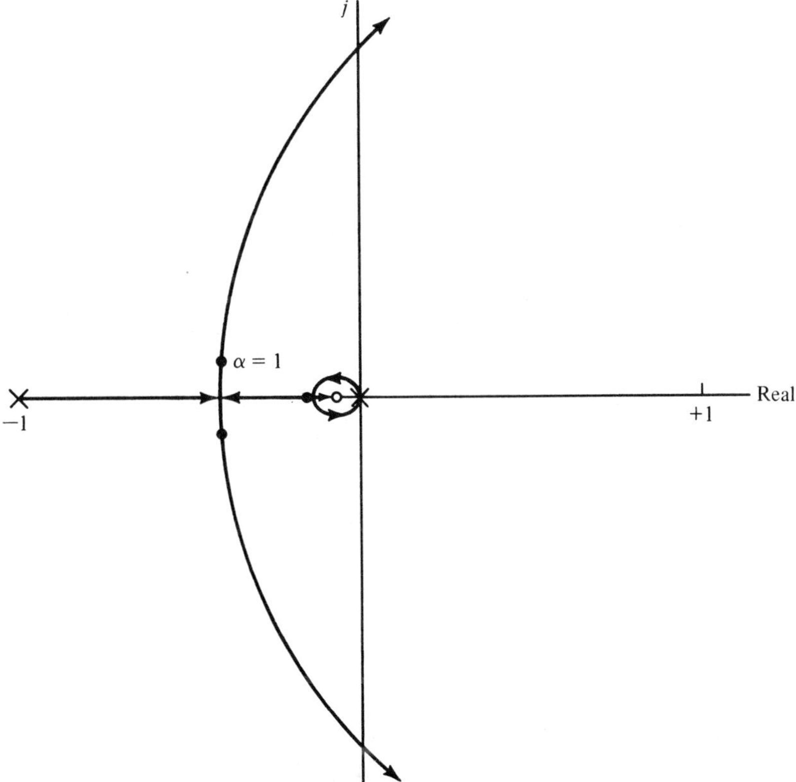

Figure 6.4-2 Root locus for the compensated transfer function of EXAMPLE 6.3-1

The two poles move in a circle about the zero as the gain α is increased. They meet on the negative real axis to the left of the zero. One pole moves towards the zero and the other moves towards infinity. The interested reader may wish to sketch a root locus diagram showing this behavior. At $\alpha = 1$ this closed loop system has the transfer function

$$H'(s) = \frac{G'(s)}{1 + G'(s)} = \frac{0.21(s + 0.07)}{(s + 0.106 + j0.058)(s + 0.106 - j0.058)}$$

As far as the high frequency behavior of the locus is concerned $G'(s)$ behaves as if it were just a simple integrator with $G'(s) \approx 0.21\alpha/s$. Thus we see that at high frequencies the lag network acts as if it were an attenuation of 0.021 which, we realize, from our previous studies in Chap. 5, is a correct interpretation. Thus the high frequency portion of the systems is essentially that of a system with open loop transfer function

$$G''(s) = \frac{-0.21\alpha(s - 1)}{s(s + 1)}$$

The closed loop transfer function corresponding to this high frequency portion of the locus is, for $\alpha = 1$,

$$H''(s) = \frac{G''(s)}{1 + G''(s)} = \frac{-0.21(s - 1)}{(s + 0.395 + j0.232)(s + 0.395 - j0.232)}$$

EXERCISE 6.4-1.   Determine and plot the actual and asymptotic closed loop gain magnitude for the compensated system of EXAMPLE 6.4-1. Determine the unit step response of this system and contrast and compare this with the results you obtain using the approximations of Chap. 4 for pertinent time response parameters.

EXERCISE 6.4-2.   Determine root loci for the following transfer functions for $K > 0$ and $K < 0$.

$$\text{a.} \quad G(s) = \frac{K(1 - s)}{s(1 + s/4)(1 + s/40)}$$

$$\text{b.} \quad G(s) = \frac{K(1 - s)}{s(1 + s/0.1)(1 + s)(1 + s/10)}$$

Nonrational Transfer Functions. The root locus approach is valid for any linear system. Only for linear constant coefficient systems with no pure time delays will the system transfer function be a ratio of rational polynomials in s. We will discuss the particular case of a unity feedback ratio system in which the open loop transfer function is of the form

$$G(s) = Ke^{-sT}F(s) \tag{6.4-1}$$

where $F(s)$ is a ratio of rational polynomials in s. The e^{-sT} transfer function is the Laplace transform of a pure time delay which has an impulse response $\delta(t - T)$ and which reproduces an input to the time delay $a(t)$ with a delayed version $a(t - T)$ of this input. Time delay systems occur in many industrial processes such as steel and chemical processes. We will consider a relatively simple example and will then state some general rules for time delay systems.

EXAMPLE 6.4-2. We consider the time delay system with open loop transfer function

$$G(s) = \frac{Ke^{-sT}}{s} \tag{1}$$

There is a single open loop pole at $s = 0$ and an uncountably large number at $s = -\alpha \pm j\omega$ for any ω. There are an uncountably large number of zeros at $s = +\alpha \pm j\omega$ for any ω.

Let us attempt to determine the root locus by inspection of our rules in Sec. (6.1) and manipulation of the transfer function $G(s)$. The closed loop transfer function is

$$H(s) = \frac{G(s)}{1 + G(s)} = \frac{\dfrac{Ke^{-sT}}{s}}{1 + \dfrac{Ke^{-sT}}{s}} = \frac{Ke^{-sT}}{s + Ke^{-sT}} \tag{2}$$

and we see that we do have difficulty in determining closed loop pole locations. There exist zeros of the characteristic equation where

$$\frac{Ke^{-sT}}{s} = -1 \tag{3}$$

As before there are two parts to this equation since we are dealing with complex numbers. The magnitude condition yields, for $s = \sigma + j\omega$

$$\left| \frac{Ke^{-sT}}{s} \right| = \frac{Ke^{-\sigma t}}{\omega} = 1 \tag{4}$$

For any value of gain K Eq. (4) may be used to plot a curve in the s plane. When this curve, obtained from the magnitude condition, crosses the root locus curve, obtained from the angle condition, we have obtained a calibrated gain point on the root locus curve.

The angle criterion applied to Eq. (3) yields, for $s = \sigma + j\omega$,

$$+ \omega T + \tan^{-1} \frac{\omega}{\sigma} = \pi + 2k\pi \tag{5}$$

A graph of this relation yields the root loci. The real axis to the left of $s = 0$ (i.e., $s < 0$) is a valid location for poles. As immediately seen from Eq. (1) or Eq. (5) with $s = \sigma + j\omega = \sigma$ the angle criterion is satisfied along the negative real axis.

The breakaway points may be determined from $dG^{-1}(s)/ds = 0$ to be given by the real axis roots of

$$e^{sT} + sTe^{sT} = 0$$

which are $s = -1/T$ and $s = -\infty$.

For σ very small ($\sigma \ll 0$) we may use Eq. (5) to determine asymptotes for the root locus as

$$\omega T = 2k\pi$$

For σ very large ($\sigma \gg 0$) we may use Eq. (5) to establish the asymptotes for this region of the s plane as

$$\omega T = \pi + 2k\pi$$

at $\sigma = 0$ we have $\tan^{-1} \omega/\sigma - \pi/2$ for positive ω and $\pi/2$ for negative ω. Thus we have the imaginary axis intersections

$$\omega T = \frac{\pi}{2} + 2k\pi \qquad \omega > 0, \quad k = 0, 1, 2, \ldots$$

$$\omega T = -\frac{\pi}{2} - 2k\pi \qquad \omega < 0, \quad k = 0, 1, 2, \ldots$$

Armed with this information we can plot the root locus as in Fig. (6.4-3). As we see, the root locus is quite complicated due to the infinite number of open loop poles and zeros.

EXERCISE 6.4-3. Determine the root locus diagrams for the unity ratio feedback systems with the following open loop transfer functions

$$\text{a.} \quad G(s) = \frac{Ke^{-sT}}{s^2}$$

$$\text{b.} \quad G(s) = \frac{Ke^{-sT}}{s(s+1)}$$

EXERCISE 6.4-4. Determine the root locus diagram for the unity ratio feedback system with transfer function

$$G(s) = \frac{K(s+1)(s-1)}{s}$$

EXERCISE 6.4-5. Develop a set of root locus construction rules that apply to systems with open loop transfer functions containing a time delay as in Eq. (6.4-1).

We could now proceed to develop a set of root locus construction rules applicable to time delay systems. While these rules are of interest they are limited in applicability by the facts that:

1. Root locus diagrams for time delay systems are quite complex.

2. Serious design interest exists only in cases where the phase shift due to the time delay is fairly small, generally $45°$ or less.

We suggest use of an approximation to a time delay of the form

$$e^{-sT} = [e^{-sT/n}]^n = \left[\frac{e^{-sT/2n}}{e^{sT/2n}}\right]^n = \frac{(1-sT/2n)^n}{(1+sT/2n)^n} \tag{6.4-2}$$

Other approximations are certainly possible although this one will be sufficient for our interests here.

If we use the approximation of Eq. (6.4-2) the basic root locus construction rules of the previous section can be used.

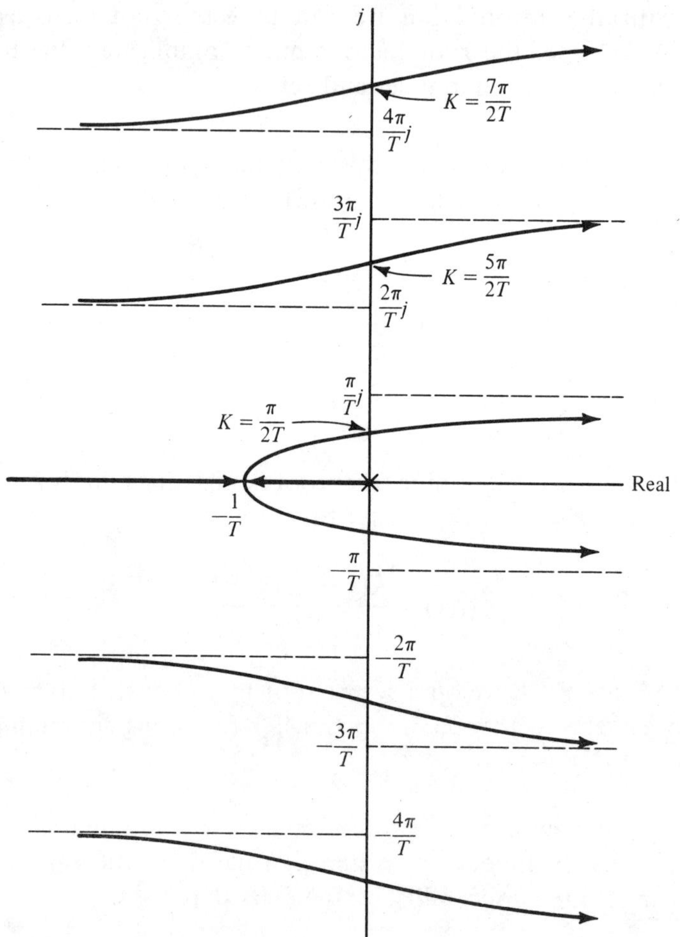

Figure 6.4-3 Root locus for open loop transfer function $\dfrac{Ke^{-st}}{s}$

EXAMPLE 6.4-3. We will reinvestigate EXAMPLE 6.4-2 using the approximation of Eq. (6.4-2) for $n = 1$ and 2. For $n = 1$ we have

$$G(s) = \frac{Ke^{-sT}}{s} \cong \frac{K(1 - sT/2)}{s(1 + sT/2)}$$

which we recognize as the same transfer function we used for EXAM-

unstable at $\omega_c T/2 = 1$ and $K = \omega_c$. There is a modest error here but when we recall that satisfactory design should result in a phase margin of about $45°$ we see that the error is of little significance. For a $45°$ phase margin the true system parameters are $\omega_c T = \pi/4$ and $K = \omega_c$ whereas the approximate system parameters with $n = 1$ are $\omega_c T/2 = 0.41$ and $K = \omega_c$. Thus there is very little error between the approximate system and the actual system for phase margins representative of reasonable system stability considerations.

For $n = 2$ the approximate system becomes an even closer representation of the actual system. We have

$$G(s) = \frac{Ke^{-sT}}{s} \cong \frac{K(1 - sT/4)^2}{s(1 + sT/4)^2}$$

and the root locus is as shown in Fig. (6.4-4).

EXERCISE 6.4-6. What is the error at $0°$ phase margin and $45°$ phase margin between conditions leading to these phase margins for the approximate system of EXAMPLE 6.4-3 with $n = 2$? Calibrate the root locus of Fig. (6.4-4) with these values.

EXERCISE 6.4-7. Examine the root locus behavior of the open loop transfer function

$$G(s) = \frac{Ke^{-0.2s}}{s(1 - s)}$$

using the approximation of Eq. (6.4-2) for $n = 1$ and $n = 2$.

EXERCISE 6.4-8. Examine the time delay approximations

a. $e^{-sT} = [e^{-sT/n}]^n \cong (1 - sT/n)^n$

b. $e^{-sT} = \dfrac{1}{[e^{sT/n}]^n} \cong \dfrac{1}{(1 + sT/n)^n}$

c. $e^{-sT} \cong 1 - sT + \dfrac{s^2 T^2}{2!} - \dfrac{s^3 T^3}{3!} + \ldots$

and indicate why these are inferior to Eq. (6.4-2). You may use $G(s) = Ke^{-sT}/s$ for example calculations.

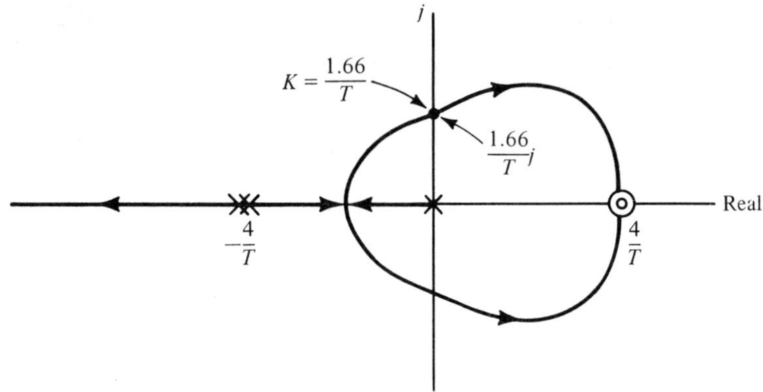

Figure 6.4-4   Root locus diagram for $n = 2$ approximation

NonUnity Ratio Feedback Systems.   Up to this point in our development of the root locus we have assumed a single unity ratio feedback
loop around an open loop transfer function $G(s)$. It is a relatively
straightforward procedure to extend this to nonunity ratio feedback
control systems. Figure (6.4-5) illustrates a linear system with single
nonunity ratio feedback loop. The closed loop transfer function of the
system represented by this block diagram is

$$\frac{Z(s)}{U(s)} = H(s) = \frac{G_1(s)}{1 + G_1(s)G_2(s)} \tag{6.4-3}$$

We shall manipulate this transfer function into a form such that our
basic root locus rules are applicable. The specific procedure to be used
depends upon whether the gain element to be varied is in the forward
path or the feedback path. If we assume that it is in the forward path
we may let

$$G_1(s) = KG_{n1}(s) \tag{6.4-4}$$

where we assume that $G_{n1}(s)$ and $G_2(s)$ are completely known. Using
Eq. (6.4-4) we may rewrite Eq. (6.4-3) as

$$H(s) = \frac{KG_{n1}(s)G_2(s)}{1 + KG_{n1}(s)G_2(s)} \frac{1}{G_2(s)} \tag{6.4-5}$$

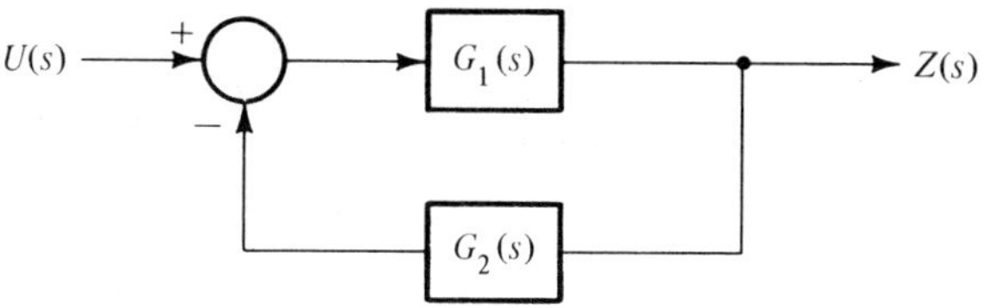

Figure 6.4-5 Nonunity ratio feedback system

A block diagram corresponding to this is shown in Fig. (6.4-6). The feedback portion of this block diagram with unity ratio feedback and open loop transfer function

$$G(s) = K_1 G_{n1}(s) G_2(s) \tag{6.4-6}$$

is precisely in a form such that our basic root locus rules are applicable. When we generate the root locus diagram using $G(s)$ we do not obtain the closed loop transfer function $H(s)$. To obtain the $H(s)$ transfer function we must delete the poles and zeros of $G_2(s)$ in the final transfer function. In other words we must multiply the transfer function we obtain using the root locus procedure by $1/G_2(s)$.

An often expressed statement for the closed loop zeros and poles of a system with the block diagram of Fig. (6.4-5) is as follows. The closed loop zeros of $H(s)$ are the open loop zeros of $G_1(s)$ and the open loop poles of $G_2(s)$. The closed loop poles of $H(s)$ are the zeros of $1 + G_1(s)G_2(s)$ and those open loop poles of $G_1(s)$ cancelled by zeros of $G_2(s)$. The zeros of $1 + G(s)$ are the roots we obtain in plotting the root locus.

This statement is also applicable to the case where the adjustable gain element is in the feedback path for then we assume

$$G_2(s) = K_2 G_{n2}(s) \tag{6.4-7}$$

where $G_1(s)$ and $G_{n2}(s)$ are presumed completely known. The closed loop transfer function may be written, using Eqs. (6.4-3) and (6.4-7) as

$$H(s) = \frac{G(s)}{1 + G(s)} = \frac{KG_1(s)G_{n2}(s)}{1 + KG_1(s)G_{n2}(s)} \frac{1}{KG_{n2}(s)} \tag{6.4-8}$$

Figure (6.4-6) indicates the block diagram manipulations appropriate to yield Eq. (6.4-8). We see that we obtain the closed loop poles and zeros

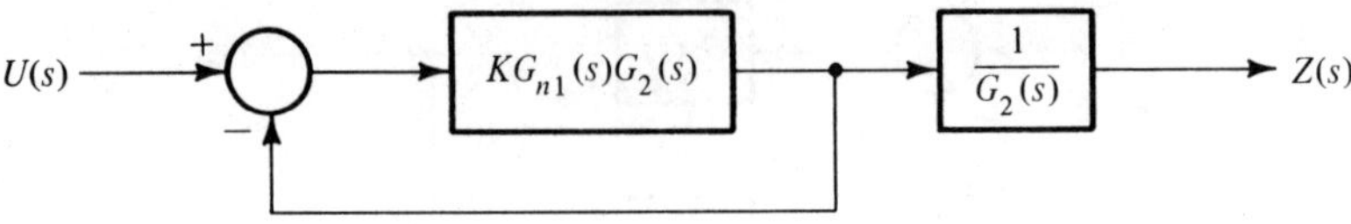

(a) Adjustable gain in forward path

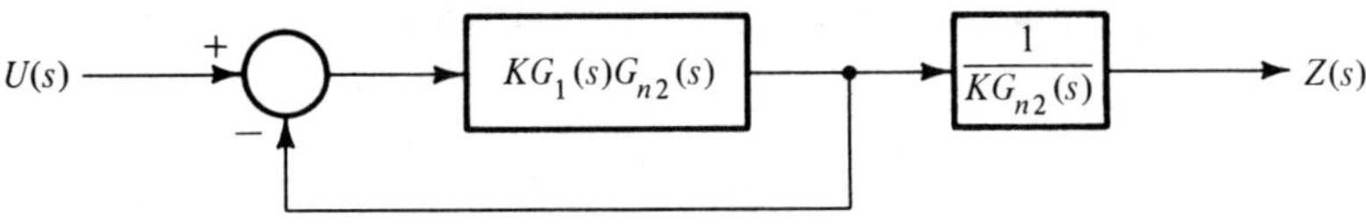

(b) Adjustable gain in feedback path

Figure 6.4-6 Equivalent unity ratio feedback system

by finding the root locus for the open loop $G(s)$ and multiply this transfer function by $1/KG_{n2}(s)$.

EXAMPLE 6.4-4. Suppose that $G_1(s)$ and $G_2(s)$ in Fig. (6.4-5) are given by

$$G_1(s) = \frac{K}{s^2(s+2)} \qquad G_2(s) = s$$

Figure (6.4-7) illustrates the root locus for this system. This is actually the root locus for $H(s)G_2(s)$ and so we see that we must multiply the transfer function obtained from this root locus by $1/s$ to get the true $H(s)$ transfer function. For example at $K = 1$ the transfer function of the system is

$$H(s) = \frac{2}{s^2 + 2s + 2} \frac{1}{s} = \frac{2}{s(s+1+j1)(s+1-j1)}$$

EXERCISE 6.4-9. Find the closed loop transfer function for the system of Fig. (6.4-5) where $G_1(s) = \dfrac{1}{s^2(s+2)}$ and $G_2(s) = Ks$, $K = 1$.

EXERCISE 6.4-10. Find the root locus and the closed loop system transfer function for the block diagram of Fig. (6.4-5) where $G_1(s) =$

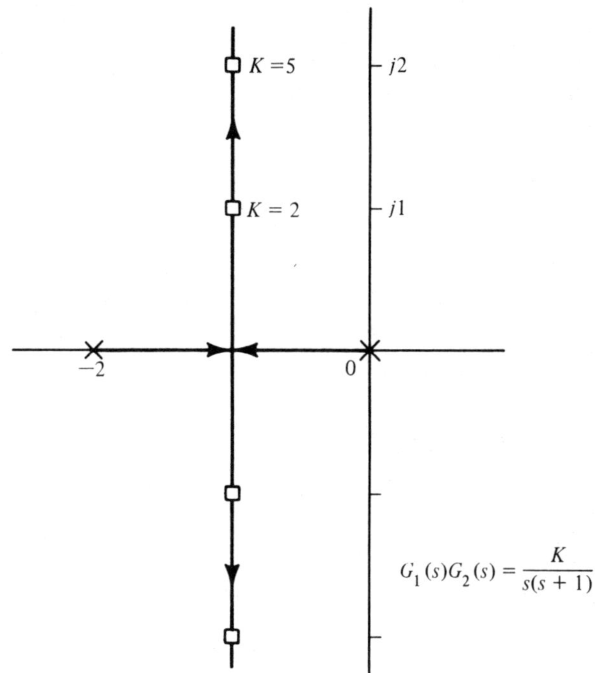

Figure 6.4-7 System root locus for EXAMPLE 6.4-4

K/s and $G_2(s) = e^{-2s} \cong (1 - s)/(1 + s)$. Find K to give a 45° phase margin and K to give a $\zeta = 0.707$ and determine the closed loop system transfer function for these two cases. Contrast and compare the unit step response for these two values of K.

Root Locus with General Parameter Variations. It will often occur that we desire to obtain a root locus diagram by varying parameters other than the loop gain. There are two primary reasons for this:

1. In design we are usually more interested in determining equalizer break frequencies for satisfactory stability margin. In fact we have shown in Chap. 5 that compensation by gain adjustment often does not lead to very satisfactory system design.

2. We are often interested in determining the sensitivity of a system to changes in the parameters assumed constant and known for system design purposes.

The procedure for determining system root loci for general parameter variations is conceptually a very simple one. We have derived our basic root locus rules for a characteristic equation of the form, where $G(s) = KG_n(s)$ is the open loop transfer function of a unity feedback ratio linear system,

$$1 + KG_n(s) = 0 \tag{6.4-9}$$

$G_n(s)$ is assumed known. The zeros of this equation are the poles of the closed loop system transfer function. We resolve problems involving general parameter variations by recasting the system characteristic equation into the form of Eq. (6.4-9). Actually we have already used this procedure in our treatment of systems with nonunity ratio feedback. Once a system has been cast into the form of Eq. (6.4-9), where K is the varying parameter, we use the root locus approach developed in our previous efforts in this chapter.

EXAMPLE 6.4-5. We consider the unity ratio feedback system with

$$G(s) = \frac{4}{s(1 + s/8)} = \frac{32}{s(s + 8)} \tag{1}$$

We suppose that the break frequency of 8 is subject to variation and we wish to examine the variation in the closed loop system to changes in this parameter. Thus we replace 8 by 8α where α will be converted to a variable gain parameter.

The closed loop system zeros are the same as the open loop zeros of $G(s)$. There are, therefore, no finite closed loop zeros. The closed loop poles are the same as the zeros of the characteristic equation or the zeros of

$$1 + G(s) = 1 + \frac{32}{s(s + 8\alpha)} = 0 \tag{2}$$

We rewrite this as

$$s(s + 8\alpha) + 32 = 0 \tag{3}$$

and rearrange Eq. (3) such that the coefficients multiplying α are separated from those that do not multiply α. This yields

$$\alpha(8s) + s^2 + 32 = 0 \tag{4}$$

Next we rearrange this expression into the form of Eq. (6.4-9)

$$\frac{\alpha(8s)}{s^2 + 32} + 1 = 0 = \alpha G_n(s) + 1 \tag{5}$$

We need to determine the root locus of $\alpha G_n(s)$. Figure (6.4-8) illustrates the root locus of

$$\alpha G_n(s) = \frac{\alpha 8s}{s^2 + 32} \tag{6}$$

The zeros of the characteristic equation are the poles shown in the root locus plot of Fig. (6.4-8). These are the poles of the closed loop system transfer function

$$H(s) = \frac{G(s)}{1 + G(s)} = \frac{32}{s^2 + 8\alpha s + 32} \tag{7}$$

There are no finite zeros in $G(s)$ and so no finite zeros in $H(s)$. Thus we have determined the root locus for the closed loop system of Eq. (1) where the break frequency 8 is replaced by 8α and allowed to vary.

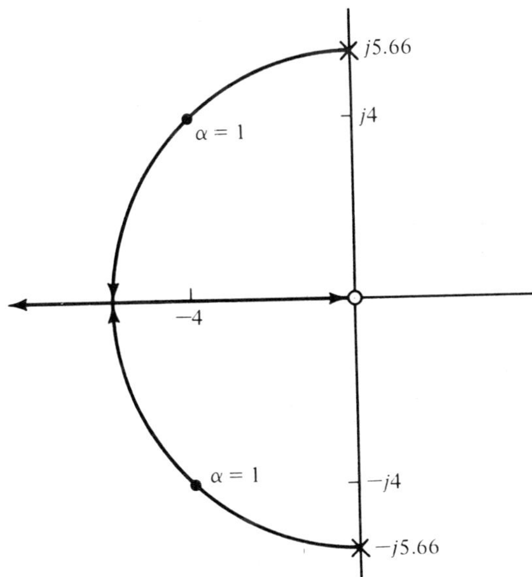

Figure 6.4-8   Root locus for EXAMPLE 6.4-5

We can obtain this same result by block diagram manipulation into a form suitable for application of the basic root locus rules. We may rewrite Eq. (7) as

$$H(s) = \frac{\alpha \dfrac{8s}{s^2 + 32}}{\alpha \dfrac{8s}{s^2 + 32} + 1} \frac{4}{\alpha s}$$

and thus see that we may realize the system in the form of Fig. (6.4-6b) in which

$$KG_1(s)G_{n2}(s) = \frac{\alpha 8s}{s^2 + 32} \qquad KG_{n2}(s) = \frac{\alpha s}{4}$$

With the root locus plot of Fig. (6.4-8) we can easily visualize the variation in the closed loop response with respect to a variation in α. As we see from the figure, the system damping is sensitive to variations in α. As α varies from 0.9 to 1.1, a $\pm 10\%$ variation, the closed loop pole locations vary from $-3.6 \pm j4.36$ to $-4.4 \pm j3.56$. The closed damping ratio and natural resonant frequency raises from 0.64 to 5.65 radians per second to 0.78 and 5.66 radians per second whereas for the nominal value of $\alpha = 1$ these values are 0.707 and 5.66.

EXAMPLE 6.4-6. To illustrate how we might use this approach in linear systems control design we consider compensation of the fixed plant

$$G_f(s) = \frac{100}{s^2} \tag{1}$$

by a 6:1 lead network compensating network

$$G_c(s) = \frac{1 + s/\omega_1}{1 + s/6\omega_1} \tag{2}$$

Our task is to find the value of ω_1 which yields "most reasonable" or "best" system performance.

In Chap. 5 we developed the Bode diagram approach to finding the value of ω_1 which yields maximum phase margin. We accomplished this by finding the crossover frequency, the frequency where the gain magnitude is unity

$$1 = \frac{100}{\omega_c \omega_1}$$

and then obtain the value of ω_1 to maximize the phase shift at crossover given by the arctangent approximation

$$\beta(\omega_c) = -\pi + \left(\frac{\pi}{2} - \frac{\omega_1}{\omega_c}\right) - \frac{\omega_c}{6\omega_1}$$

$$= -\frac{\pi}{2} - \frac{\omega_1^2}{100} - \frac{100}{6\omega_1^2}$$

which is obtained from $\partial\beta(\omega_c)/\partial\omega_1 = 0$ as $\omega_1 = 10/6^{1/4} = 6.39$. This leads to a phase margin of about $43°$ and a damping ratio for the dominant poles of 0.76, and a natural damped resonant frequency of 1.7 (6.39) = 10.86 radians per second and a pole located at $s = a^{1/2}\omega_1 = -15.65$ radians per second as can easily be obtained from the design curves of Figs. (6.3-5) and (6.3-6).

The open loop transfer function is given by the product of Eqs. (1) and (2) as

$$G(s) = G_c(s)G_f(s) = \frac{600(s + \omega_1)}{s^2(s + 6\omega_1)} \tag{3}$$

The closed loop zeros are the same as the open loop zeros. Thus we can see that there is a single finite closed loop zero located at $s = -\omega_1$. The closed loop poles are located at the zeros of the characteristic equation

$$1 + G(s) = 0 = 1 + \frac{600(s + \omega_1)}{s^2(s + 6\omega_1)} \tag{4}$$

We arrange Eq. (3) in standard form. We multiply both sides of Eq. (4) by $s^2(s + 6\omega_1)$ to obtain

$$600(s + \omega_1) + s^2(s + 6\omega_1) = 0$$

This is equivalent to

$$\omega_1(600 + 6s^2) + 600s + s^3 = 0$$

In standard form $KG_n(s) + 1 = 0$ this becomes

$$\omega_1 \frac{6(s^2 + 100)}{s(s^2 + 600)} + 1 = 0$$

Thus we see that we need to obtain the root locus of

$$KG_n(s) = \omega_1 \frac{6(s^2 + 100)}{s(s^2 + 600)}$$

which is shown in Fig. (6.4-9).

The gain is, for this example, the frequency ω_1. We wish to select ω_1 for good performance. As we see from Fig. (6.4-9), $\omega_1 = 6.39$ does give rise to the system with maximum damping.

The example just concluded demonstrates the several advantages of using the root locus directly as a design tool in order to determine a break frequency location for a lead network equalizer. In many ways this is a more useful approach than the basic root locus method in which we vary a loop gain. It suffers from the disadvantage however that the modified open loop pole zero plots are often not those for which the root locus is as easily and intuitively sketched as would be the case with the basic root locus approach. Nevertheless it is a very useful approach as we have seen in this example.

EXERCISE 6.4-11.   Repeat EXAMPLE 6.4-5 for the case where the lead network of Eq. (2) is replaced by

a. $\quad G_c(s) = \dfrac{1 + s/\omega_1}{1 + s/9\omega_1}$

b. $\quad G_c(s) = \dfrac{1 + s/\omega_1}{1 + s/20\omega_1}$

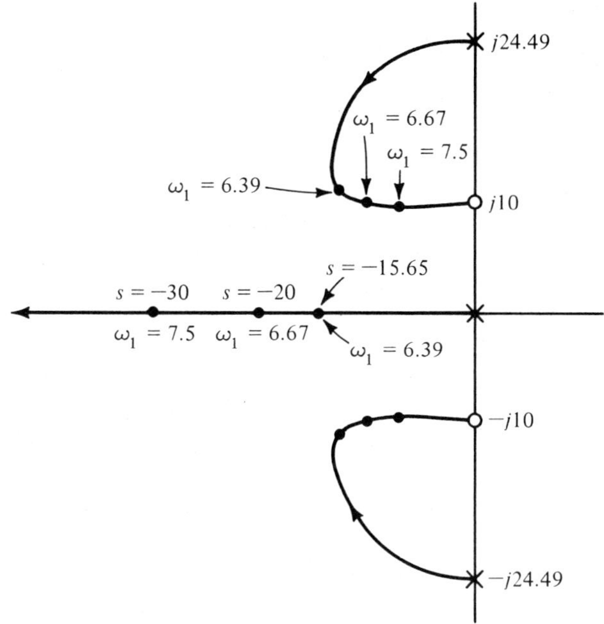

Figure 6.4-9 Root locus for varying break frequency in lead network equalizer

Multiple Feedback Loop Systems and Minor Loop Design. We may approach the problem of multiple loop feedback systems in at least two ways:

1. We may attempt to write the system closed loop transfer function in the form

$$H(s) = \frac{G_1(s)}{1 + G_1(s)G_2(s)} = \frac{F(s)}{1 + G(s)}$$

    and examine the root locus of the characteristic equation $1 + G(s)$.

2. We may use minor loop root locus constructions to determine the transfer function of a given minor loop. At a particular parameter setting the closed loop transfer function of the minor loop is determined. These poles and zeros become the open loop poles and zeros of another feedback loop. The root locus of this feedback loop is determined and the process repeated until the overall system root locus is determined.

Each approach has its advantages. The first is generally simpler in terms of plotting the root loci. The second approach allows us to consider a single varying parameter (or gain) in each loop and allows more physical insight into the effect on the overall loop of parameter changes within minor loops. The second approach requires that more than a single root locus be obtained and is, therefore, more difficult or tedious to apply than the first approach. Since it is generally desirable to design systems with stable minor loops and since the approach allows us to discern minor loop stability, this approach will often be the preferred one.

EXAMPLE 6.4-7. As an example to illustrate the root locus technique for multiple loop feedback systems, we consider the two loop system represented by the block diagram of Fig. (6.4-10). The basic fixed plant is a double integrator (type two) system but a minor loop feedback through a time delay of one second reduces the overall effect to that of a type 1 system. Suppose that the object of our system design is to determine K_1 and K_2 for reasonable system performance.

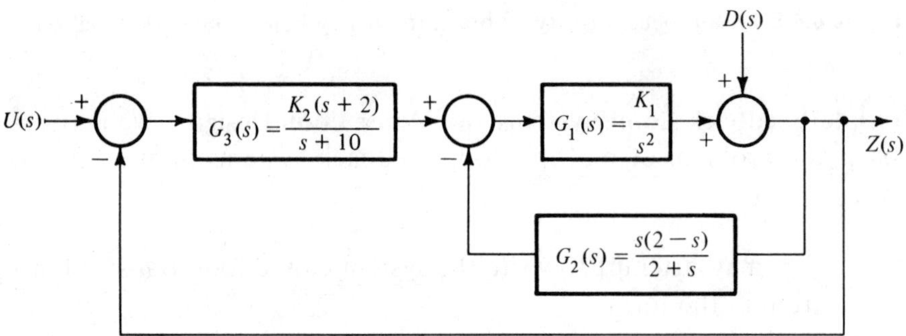

Figure 6.4-10 Multiple loop block diagram for EXAMPLE 6.4-7

Let us first assume that we pick $K_2 = 50$ and adjust K_1 for this "best reasonable" performance using the root locus approach. Our first approach will be the quickest in that we will rearrange the characteristic equation $1 + G(s) = 0$, which becomes

$$1 + G_1(s)G_2(s) + G_1(s)G_3(s) = 0 \qquad (1)$$

for the block diagram of Fig. (6.4-10), into standard form. Equation (1) becomes, using the transfer functions of Fig. (6.4-10),

$$1 + \frac{K_1}{s^2}\left[\frac{s(2-s)}{2+s} + \frac{50(s+2)}{s+10}\right] = 0$$

or

$$1 + K_1\left[\frac{-s^3 + 42s^2 + 220s + 200}{s^2(s+2)(s+10)}\right] = 0$$

Thus we need to obtain the root locus diagram for the open loop transfer function

$$G(s) = \frac{K_1(-s^3 + 42s^2 + 220s + 200)}{s^2(s+2)(s+10)}$$

$$= \frac{-K_1(s+1.185)(s+3.608)(s-46.793)}{s^2(s+2)(s+10)} \tag{2}$$

The standard root locus rules outlined in Sec. (6.2), as discussed in the first part of this section concerning negative loop gain and the complimentary root locus, may be used to determine the root loci illustrated in Fig. (6.4-11). The interested reader will find it very helpful to verify that the root locus shown is that obtained from our root locus rules. The complete root locus is not shown due to scale difficulties and Fig. (6.4-11) shows that portion of the locus of major interest. As the gain K_1 is increased the complex conjugate poles move into the right half plane and intersect the real axis to the right of the zero at 46.793. This portion of the locus, where the system is unstable, is of little design interest and showing it would necessarily crowd, by scale reduction, that portion of the locus that is of interest. It would appear that setting gain K_1 between 1 and 2.2 would result in the most acceptable damping as the damping ratio of the dominant closed loop poles will vary from 0.38 to 0.23. This represents damping ratios that are less than normally desirable but the best obtainable with this particular system and choice of $K_2 = 50$. The gain magnitude for the open loop transfer function $G(s)$ of Eq. (2) is obtainable from

$$G(s) = \frac{10K_1(1 + s/1.185)(1 + s/3.608)(1 + s/46.793)}{s^2(1 + s/2)(1 + s/10)} \tag{3}$$

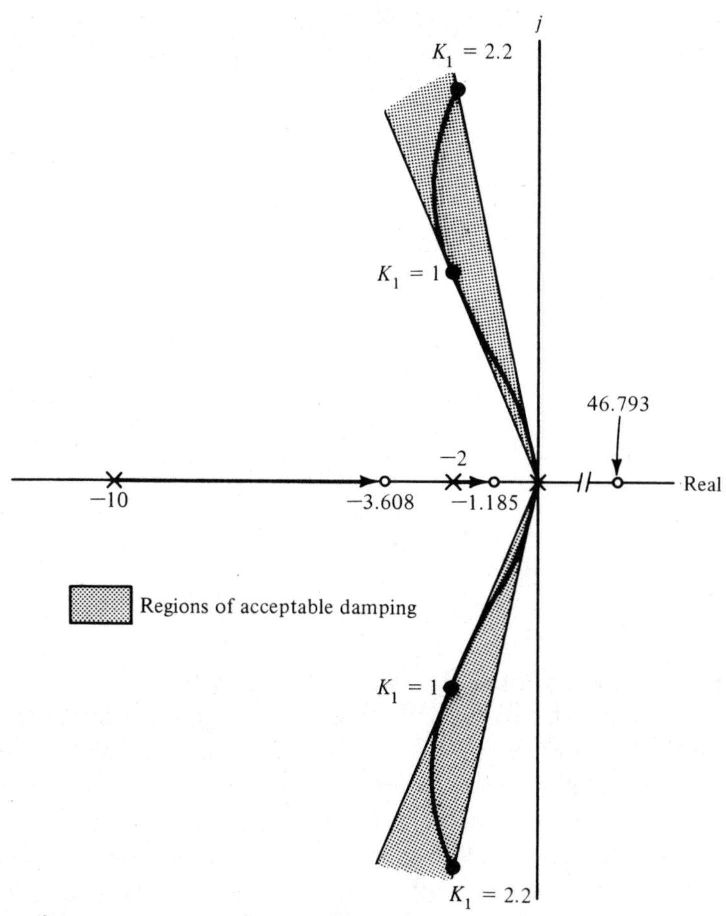

Figure 6.4-11   Root locus of the characteristic equation for varying K_1 with $K_2 = 50$

The phase margin is 42.2° and the crossover frequency varies from 5 to 8.5 radians per second as the gain varies from 1 to 2.2. The closed loop transfer functions for these two gain settings are

$$\frac{Z(s)}{U(s)} = \frac{50(s + 2)^2}{(s + 1.276)(s + 5.829)(s + 1.948 + j4.806)(s + 1.948 - j4.806)} \tag{4}$$

for $K_1 = 1$ and

$$\frac{Z(s)}{U(s)} = \frac{110(s + 2)^2}{(s + 1.225)(s + 4.301)(s + 2.137 + j8.885)(s + 2.137 - j8.885)} \tag{5}$$

for $K_1 = 2.2$. Figure (6.4-12) illustrates the magnitude of the closed loop gain versus frequency curve for these two cases. The M_p is 1.6 for $K_1 = 1$ and 2.2 for $K_1 = 2.2$. The closed loop bandwidth is about 50% greater for $K_1 = 2.2$ however and if the increased speed of response is desirable it can be had at the expense of greater peaking in the frequency response curve. Suppose that we do need the greater bandwidth and more rapid transient response design and so we select $K_1 = 2.2$.

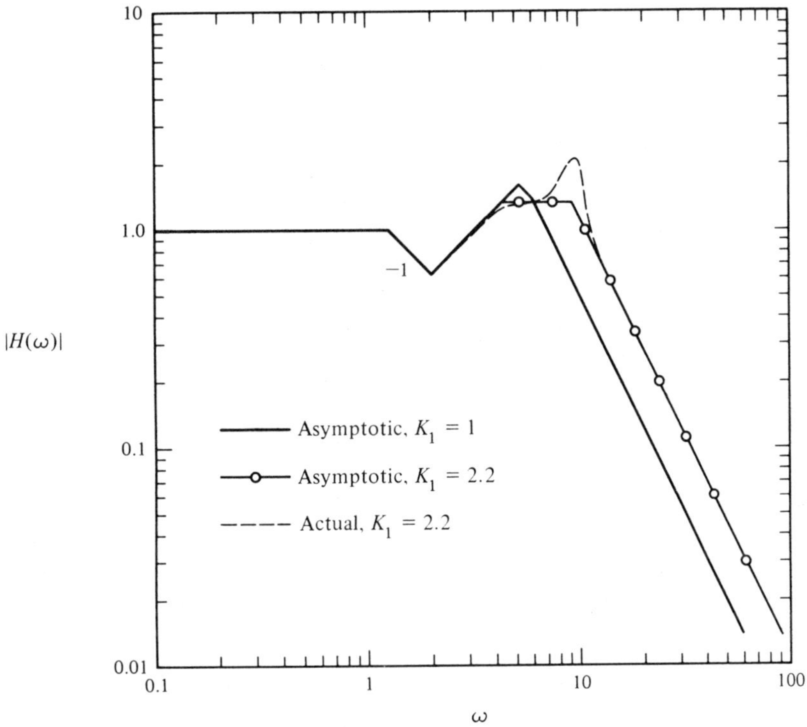

Figure 6.4-12 Closed loop system frequency response, $K_2 = 50, K_1 = 1$, and $K_1 = 2.2$

Is the system we have just designed satisfactory? No—not entirely. However, there is no way that this will become apparent to us from the approach we have just taken.

The minor loop is unstable with $K_1 = 2.2$. We can obtain this if we follow suggested approach number 2 and use minor loop root locus constructions in order to determine the transfer function of the minor loop. Let us do this and demonstrate that we obtain equivalent closed loop transfer functions although we obtain root loci that are quite different.

Figure (6.4-13) shows the root locus for the minor loop characteristic equation

$$G(s) = \frac{-K_1(s-2)}{s(s+2)}$$

and shows that it is indeed unstable at $K_1 = 2.2$ which we use as the minor loop gain. With $K_1 = 2.2$ the minor loop input output transfer function is

$$\frac{G_1(s)}{1 + G_1(s)G_2(s)} = \frac{2.2(s+2)}{s(s^2 - 0.2s + 4.4)} \tag{6}$$

We use this minor loop transfer function and the transfer function of the series equalizer $G_3(s)$ to obtain the characteristic equation

$$G'(s) = \frac{G_1(s)G_3(s)}{1 + G_1(s)G_2(s)} = \frac{2.2K_2(s+2)^2}{s(s+10)(s^2 - 0.2s + 4.4)} \tag{7}$$

transfer function for the center loop. The overall system transfer function is then

$$\frac{Z(s)}{U(s)} = \frac{G'(s)}{1 + G'(s)} \tag{8}$$

Figure (6.4-14) illustrates the root locus for the outer loop and also indicates the zero pole locations for the closed loop system given by Eq. (8). Although the root loci of Figs. (6.4-13) and (6.4-14) do not resemble that of Fig. (6.4-11) we see that we obtain, as of course we should, the same closed loop zeros and poles when $K_2 = 50$ (and $K_1 = 2.2$).

In concluding this example let us attempt a redesign for improved minor loop performance. We see from Fig. (6.4-14) that the outer loop is stable even for an unstable minor loop. So we see that the key to a better design lies in redesigning the minor loop. If we pick a minor loop gain of $K_1 = 0.5$ we obtain a minor loop phase margin of 62° and a closed loop transfer function of

$$\frac{G_1(s)}{1 + G_1(s)G_2(s)} = \frac{15(s+2)}{s(s^2 + 1.5s + 1)} \tag{9}$$

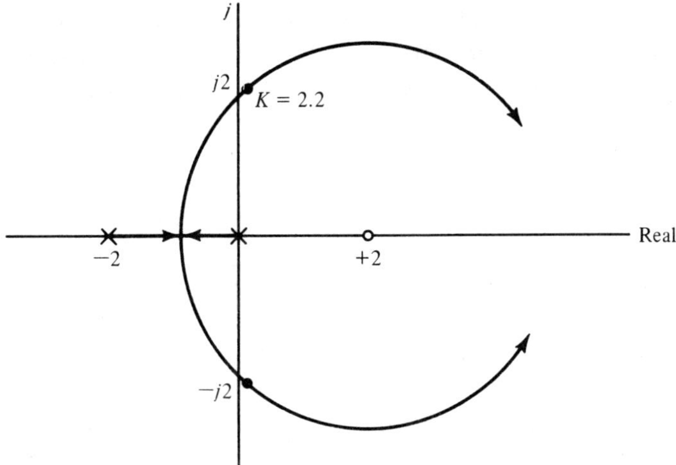

Figure 6.4-13 Root locus of characteristic $G(s)$ for minor loop

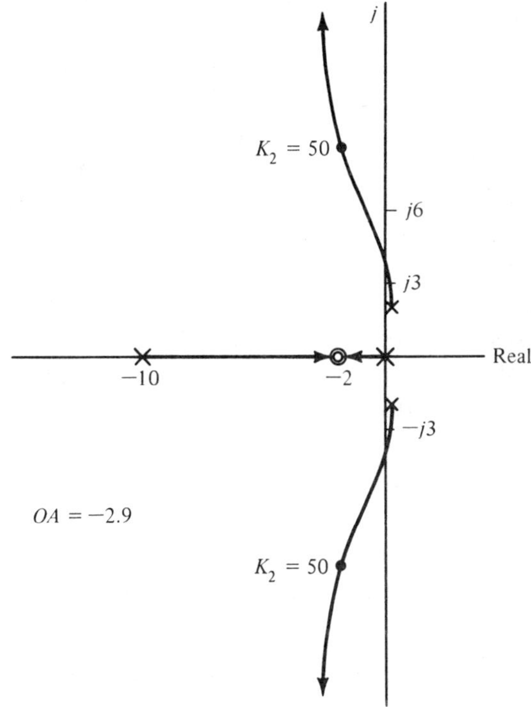

Figure 6.4-14 Outer loop root loci for characteristic equation $G'(s)$ of Eq. (7)

This minor loop is perhaps more stable than need be the case but let us use this value of K_1 and determine the outer loop root locus from the characteristic equation transfer function of Eq. (7) which becomes

$$G'(s) = \frac{0.5K_2(s + 2)^2}{s(s^2 + 1.5s + 1)(s + 10)} \tag{10}$$

Figure (6.4-15) illustrates the root locus for the closed loop system with the characteristic equation transfer function of Eq. (10). We see that approximate maximum damping is obtained for $K_2 = 100$. This yields a phase margin of approximately $37°$ at a crossover frequency of 5 radians per second. The dominant pole damping ratio is 0.24. The closed loop transfer function, with $K_1 = 0.5$ and $K_2 = 100$ is

$$\frac{Z(s)}{U(s)} = \frac{50(s + 2)^2}{(s + 1.497)(s + 5.134)(s^2 + 4.87s + 26.04)} \tag{11}$$

As the reader can verify, the step response of the systems defined by Eqs. (4) and (11) are very nearly the same. The step response of Eq. (5) is a bit faster but there is more overshoot. The minor loop systems of Eqs. (4) and (11) are stable where that of Eq. (5) is unstable. The reader may question whether or not the damping ratios obtained in this example are too low. The answer would, of course, depend upon the specific application under consideration but generally the answer is no. The low frequency closed loop pole is a determinant of response roughly equivalent in weight to that of the complex conjugate poles. Thus we do not get the considerable overshoot that would exist in a simple second order system with the same low damping ratio. For this particular example phase margin and M_p are more indicative of system behavior than damping ratio alone.

EXERCISE 6.4-12. Determine the three transfer functions for the gain values obtained in EXAMPLE 6.4-7 for $Z(s)/D(s)$. Here $D(s)$ would represent an output disturbance. Which of the three compensations is best from the point of view of output disturbance rejection?

EXERCISE 6.4-13. Suppose that we compensate the fixed plant transfer function

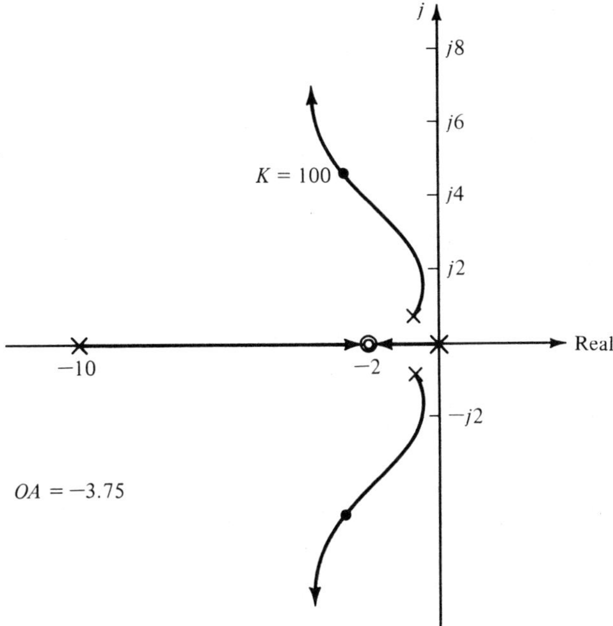

Figure 6.4-15 Major loop root locus with $K_1 = 0.5$

$$G_f(s) = \frac{K}{s(s/10 + 1)} \qquad K = 100$$

by means of

    a.   a series lag lead network

$$G_c(s) = \frac{(1 + s/10)(1 + s/\omega_3)}{(1 + s/\omega_1)(1 + s\omega_1/10\omega_3)}$$

    b.   an equivalent minor loop compensator

$$G_m(s) = \frac{s/\omega_1}{(1 + s/\omega_3)(1 + s/10)}$$

    1.   Determine ω_1 and ω_3 for a minimum attenuation lag lead network and a $45°$ phase margin.

2. Use the root locus method to determine the sensitivity of the closed loop poles to variations in K about the nominal value of 100. Which is best in terms of lower sensitivity: the series compensation method or the minor loop compensation approach?

3. What are the merits of each compensation method in terms of attenuation of an output disturbance?

6.5 SUMMARY

This chapter has concerned the root locus approach. We have examined motivation for the root locus approach as well as detailed rules for construction of the locus. Suggested design approaches have been given and illustrated with a number of examples. We have seen that it is possible to sketch and visualize the behavior characteristics of a closed loop system in a relatively straightforward manner which allows us to obtain a quick estimate of the transient response modes of a system and to determine the sensitivity of closed loop pole locations to gain and other parameter variations. Our examples and suggested reader exercises should do much to develop practical experience using the technique and to enhance appreciation of the digital computer routines available for plotting root loci once preliminary engineering decisions have been made relative to system structure. As we have seen in our many examples of this chapter the combination of the root locus approach and the Bode diagram approach leads to a very potent and useful set of linear systems control design tools.

6.6 REFERENCES

There are a large number of textbook references to the root locus approach. Sound presentations of the root locus technique are available in many elementary control texts including:

Truxal, J. G., *Control System Synthesis*, McGraw-Hill Book Co., New York, N. Y., 1955.

Grabbe, E. M., Ramo, S., and D. E. Woolridge, *Handbook of Automation Computation and Control*, Vol. 1, John Wiley and Sons, Inc., New York, N. Y., 1958.

Cannon, R. H. Jr., *Dynamics of Physical Systems*, McGraw-Hill Book Co., New York, 1967.

Melsa, J. L. and D. G. Schultz, *Linear Control Systems*, McGraw-Hill Book Co., New York 1969.

The interested reader may find it especially motivating to read the original journal papers which discuss the root locus approach as well as a text based, in part, on the papers:

Evans, W. R., "Graphical Analysis of Control Systems," *Transactions American Institute of Electrical Engineers*, Vol. 67, 1948, pp. 547-551.

Evans, W. R., "Control System Synthesis by Root Locus Method," *Transactions American Institute of Electrical Engineers*, Vol. 69, 1950, pp. 67-69.

Evans, W. R., *Control System Dynamics*, McGraw-Hill Book Co., New York, N. Y., 1954.

6.7 PROBLEMS

1. Plot the root locus diagram for each of the following open loop transfer functions of a unity feedback ratio system. Be sure to compute and label critical points on the root locus.

    a. $\quad G(s) = \dfrac{K(s + 2)}{s(s + 1)(s^2 + 4s + 10)}$

    b. $\quad G(s) = \dfrac{K}{(s^2 + 4s + 10)(s^2 + 12s + 45)}$

    c. $\quad G(s) = \dfrac{K(s + 3)(s^2 + 2s + 10)}{s(s + 1)(s + 2)(s + 3)(s + 4)(s^2 + 10s + 29)}$

    d. $\quad G(s) = \dfrac{K(s^2 + 6s + 25)}{s(s + 1)(s + 2)}$

2. Determine design parameters for lag network series equalization to compensate the fixed plant

    $$G_f(s) = \frac{100}{s(1 + s/100)^3}$$

    to yield 45° phase margin and reasonable damping. Use the root locus approach to verify and validate any use of the Bode diagram approach. Be certain to label critical points on your sketch of the locus.

3. Outline a procedure to determine partial fraction coefficients for Laplace transform inversion using the root locus approach.

4. Determine root loci for the following transfer function for both positive and negative K.

$$G_f(s) = \frac{K(1 - s/10)}{s^2(1 + s/10)}$$

   Assume that an attempt is made to stabilize the system with a lead network

$$G_c(s) = \frac{1 + s/\omega_1}{1 + s/\omega_2}$$

   For $K = 4$ find a suitable minimum gain lead network for a 45° phase margin. Determine the system root locus and contrast and compare root locus information with that obtained from the Bode diagram.

5. Design a minimum gain lead network equalizer to compensate the unstable fixed plant

$$G_f(s) = \frac{10}{s(s - 1)}$$

   such that the damping ratio is $\zeta = 0.707$. Leave the velocity constant at $K_v = 10$. What is the phase margin of the resulting system? What is the unit step response of the system?

6. A fixed plant with transfer function

$$G_f(s) = \frac{2(1 - s/2)}{s(1 + s/2)}$$

   is compensated by a lag network

$$G_c(s) = \frac{1 + s/6\omega_1}{1 + s/\omega_1}$$

   Find the value of ω_1 which yields maximum damping. What is this value of damping? What is the unit step response of the compensated closed loop system?

7. Investigate the use of the first two Taylor series terms

$$e^{-sT} = \frac{1 - sT/2 + \dfrac{s^2 T^2}{4}}{1 + sT/2 + \dfrac{s^2 T^2}{4}}$$

as an approximation to a time delay. Determine the approximate root loci of the unit ratio system with $G(s) = \dfrac{Ke^{-sT}}{s}$ where this approximation is used.

8. Find reasonable values of K_2 and K_3 which yield acceptable phase margin, reasonable damping, and maximum bandwidth for a system described by the block diagram of Fig. (6.4-10) where

$$G_1(s) = \frac{0.1}{s(1 + s/100)^2} \qquad G_2(s) = K_2 s \qquad G_3(s) = K_3$$

9. Find the value of ω_1 which yields a damping ratio for the dominant closed loop poles of 0.5 for the unity feedback ratio system with

$$G(s) = \frac{40}{s(1 + s/10)^2(1 + s/\omega_1)}$$

10. We desire a closed loop transfer function

$$H(s) = \frac{G(s)}{1 + G(s)} = \frac{10{,}000}{(s + 50)(s^2 + 20s + 200)}$$

Find $G(s)$ by two methods:

a. direct solution of the foregoing equation for $G(s)$.

b. replacement of the unity feedback gain value of 1 by a value α. When $\alpha = 1$ we get the closed loop transfer function and when $\alpha = 0$ we get the open loop transfer function. Thus use root locus techniques to determine the open loop transfer function.

11. Discuss the use of the root locus method for the purpose of factoring polynomials. In particular show that if we write the polynomial

$$P(s) = s^n + a_{n-1}s^{n-1} + \ldots + a_1 s + a_0 = 0$$

by division by $s^n + a_{n-1}s^{n-1} + a_{n-2}s^{n-2}$

$$0 = 1 + \frac{a_{n-3}s^{n-3} + a_{n-4}s^{n-4} + \ldots + a_1 s + a_0}{s^{n-2}(s^2 + a_{n-1}s + a_{n-2})}$$

Now if n is no greater than 5 then there will be at most a quadratic in the numerator so it can easily be factored. Show that this approach can be used to obtain a root locus which when plotted will easily yield the roots of the 5th order (or less) polynomial. How can this be extended to higher order polynomials?

7

DIGITAL OPTIMAL AND ADAPTIVE LINEAR SYSTEMS CONTROL

The ubiquity of the digital computer both as a systems analysis and synthesis tool as well as a component in control systems makes it highly desirable, even in an introductory text, to present some discussion of digital and sampled data control. We will present a brief discussion of sampled data and digital control systems and will show that essentially all of our design practices for continuous time systems have a discrete time equivalent.

Then we will turn our attention to an introduction to optimum linear systems control. In our discussions in earlier chapters we have often used simple optimization techniques, such as obtaining maximum phase margin from a system by adjusting compensating network parameters. But we have not considered the introduction of a performance index directly involving system error, the difference between system input and system output, which is often what we really would like to have remain small and perhaps even be minimized in a mathematical sense. Here we will introduce the concept of a performance index expressed as the integral of some function of the system error and will determine either system parameters or control inputs, or both, to minimize this error.

A potential use of optimum systems control concepts is to instrument a system such that it tunes itself to an optimum condition by measuring its own environment, input, and disturbances. Such systems are known as adaptive systems, learning systems, or selforganizing systems and we will conclude our presentation with a discussion of the adaptive control problem.

Throughout our discussions we will attempt to illustrate relation-ships between digital optimal and adaptive control concepts and the Bode diagram and root locus techniques we have discussed in our previous two chapters.

7.1 SAMPLED DATA AND DIGITAL SYSTEMS CONTROL

We will call systems which operate on information obtained at discrete time points, denoted sampling points, sampled data systems or digital control systems. Figure (7.1-1) illustrates basic components of a linear sampled data system. There are both continuous time dynamic elements in sampled data systems as well as one or more samplers. The samplers may operate in a periodic or aperiodic fashion or even a random fashion. Here we will assume that the sampler is periodic and that the amplitude of the output of the sampler is linearly proportional to the amplitude of the input to the sampler.

Sampling. We will denote the input to a sampler as a time function $u(t)$ and the output as a sampled time function $u_\tau^*(t)$. This process is illustrated in Fig. (7.1-2). At $t = 0, T, 2T, \ldots$, the sampler closes for a time interval τ and the output of the sampler equals the input during this time interval. For reasons which will soon become apparent we place an amplifier with gain $1/\tau$ after the sampler. We obtain, at the output of the sampler, a periodic series of pulses of width τ and height $u(nT)/\tau$, $n = 0, 1, 2, \ldots$. In most practical samplers the pulse width is very small, such that $\tau \ll T$ and thus it is reasonable to approximate the sampling time function by an impulse. Thus we approximate

$$i(t) \cong \sum_{n=0}^{\infty} \delta(t - nT) \tag{7.1-1}$$

where δ represents the Dirac delta or impulse function. The output of the sampler is then

$$u^*(t) = \lim_{\tau \to 0} u_\tau^*(t) = \sum_{n=0}^{\infty} u(t)\delta(t - nT) \tag{7.1-2}$$

or

$$u^*(t) = u(t)i(t) = \sum_{n=0}^{\infty} u(nT)\delta(t - nT) \tag{7.1-3}$$

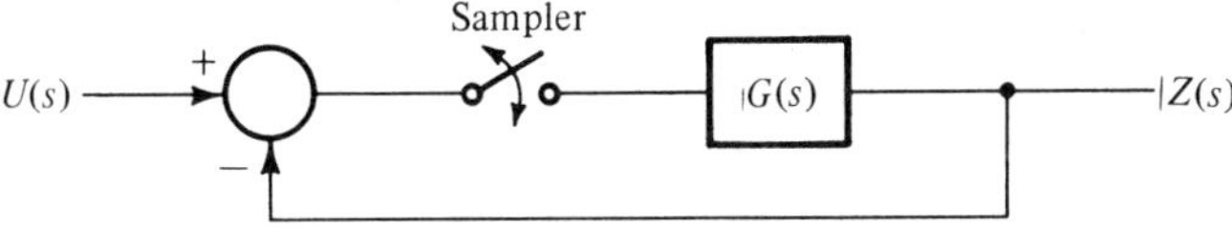

Figure 7.1-1 Basic sampled data block diagram

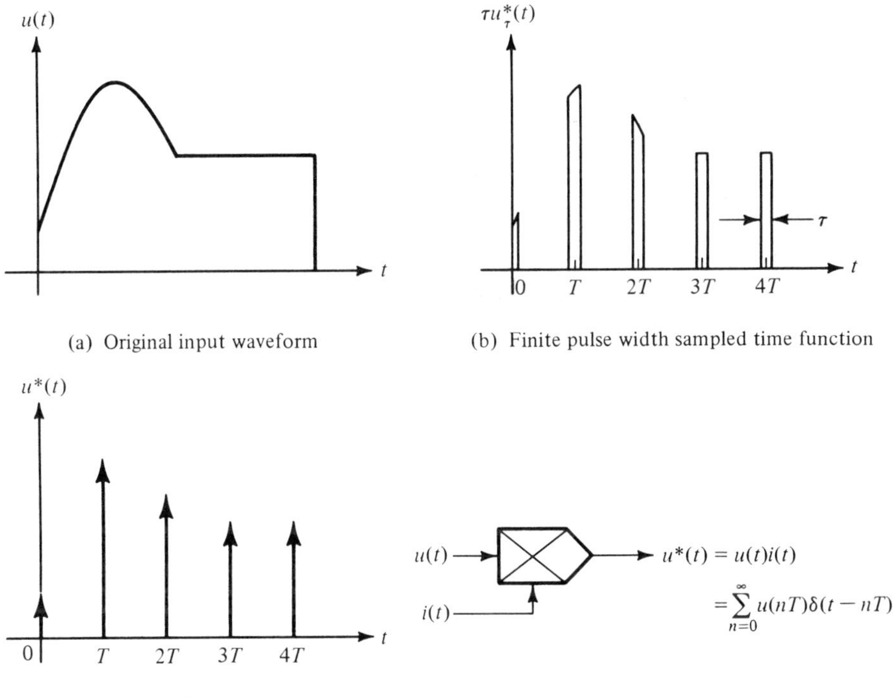

(a) Original input waveform

(b) Finite pulse width sampled time function

(c) Impulse sampled time function

Figure 7.1-2 Sampled waveforms and block diagram of impulse sampler

The Laplace transform of the unit impulse sampler of Eq. (7.1-1) is

$$I(s) = \int_0^\infty i(t)e^{-st}\, dt = \sum_{n=0}^\infty e^{-snT} \tag{7.1-4}$$

We can also write this in an alternate form by using a Taylor series approximation to obtain

$$I(s) = \sum_{n=0}^{\infty} e^{-snT} = 1 + e^{-sT} + e^{-2sT} + e^{-3sT} + \ldots$$

$$= \frac{1}{1 - e^{-sT}} \tag{7.1-5}$$

The poles of $I(s)$ can easily be determined. We set the denominator of Eq. (7.1-5) equal to zero and see that there are poles whenever

$$e^{-sT} = 1 \tag{7.1-6}$$

or whenever

$$s = \pm \frac{j\,2\pi n}{T} \tag{7.1-7}$$

where n is an integer. We will define the sampling frequency, in accordance with Fig. (7.1-2), as

$$\omega_s = \frac{2\pi}{T} \tag{7.1-8}$$

Thus we see that the poles of a sampled pulse train are located from Eqs. (7.1-7) and (7.1-8) as

$$s = \pm jn\omega_s \tag{7.1-9}$$

An alternate form of Eq. (7.1-4) is of value. Suppose that we "guess" that it is possible to rewrite the definition of the ideal impulse sampler as

$$i(t) = \frac{1}{T} \sum_{m=-\infty}^{\infty} e^{-jm\omega_s t} \tag{7.1-10}$$

To see that this definition is equivalent to that of Eq. (7.1-1) we may show that the two expressions have the same Laplace transform. From Eq. (7.1-10) we have

$$I(s) = \int_0^{\infty} \frac{1}{T} \sum_{m=-\infty}^{\infty} e^{-jm\omega_s t}\, e^{-st}\, dt = \frac{1}{T} \sum_{m=-\infty}^{\infty} \frac{1}{s + jm\omega_s} \tag{7.1-11}$$

The poles of Eq. (7.1-11) are at $s = \pm jm\omega_s$ and these are just the pole locations we have found for the Laplace transform of Eq. (7.1-4). Neither Laplace transform has any zeros. This proves our assertion that Eqs. (7.1-1) and (7.1-10) are equivalent time functions. A much more sophisticated proof is possible using Fourier series but ours, which is perhaps really much more of a plausibility argument, is quite adequate for our needs.

The Laplace transform of the sampled time function is, from Eq. (7.1-3),

$$U^*(s) = \int_0^\infty u^*(t)e^{-st}\,dt = \int_0^\infty \sum_{n=0}^\infty u(nT)\delta(t-nT)e^{-st}\,dt$$

$$= \sum_{n=0}^\infty u(nT)\int_0^\infty \delta(t-nT)e^{-st}\,dt = \sum_{n=0}^\infty u(nT)e^{-snT} \qquad (7.1\text{-}12)$$

This is a very important relationship as we can use it to establish a table of sampled transforms. An expression equivalent to Eq. (7.1-12) can be obtained using the alternative definition of the impulse train given by Eq. (7.1-10). We have

$$U^*(s) = \int_0^\infty u(t)i(t)e^{st}\,dt = \frac{1}{T}\int_0^\infty u(t)\sum_{m=-\infty}^\infty e^{jm\omega_s t}\,e^{st}\,dt$$

$$= \frac{1}{T}\sum_{m=-\infty}^\infty U(s + jm\omega_s) \qquad (7.1\text{-}13)$$

This expression is very important because it can be used to establish the sampling theorem. Figure (7.1-3) illustrates two transforms of frequency limited or band limited functions.* In Fig. (7.1-3a) the sampling frequency is greater than $\omega_s/2$ and so the magnitude of the Laplace transform of the sampled time function does not overlap from one band to the other. In Fig. (7.1-3b) separation of the primary signal from $U^*(s)$ is not possible. If the frequency spectra or frequency components of the time function $u(t)$ exist for $\omega > (\omega_s/2)$ then the Laplace transform of the sampled time function shows overlap of the frequency components of the primary signal $u(t)$ as well that of the complimentary signal, i.e. $U(s)$ and $U(s + j\omega_s)$ overlap. Thus it will not be possible to filter $U^*(s)$ such as to recover $U(s)$. This verifies the sampling theorem

---

*Strictly speaking a band limited function does not exist.

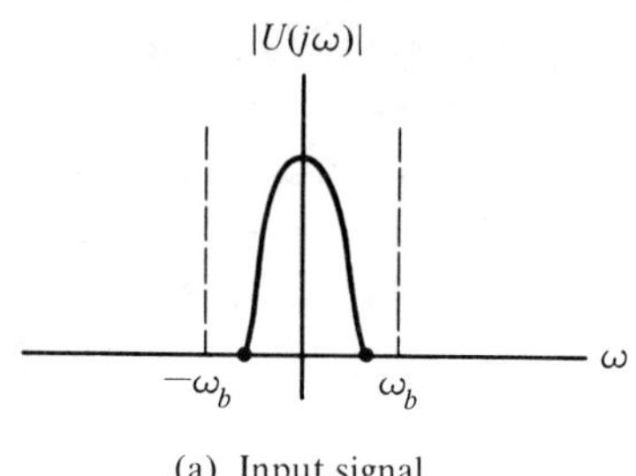

(a) Input signal

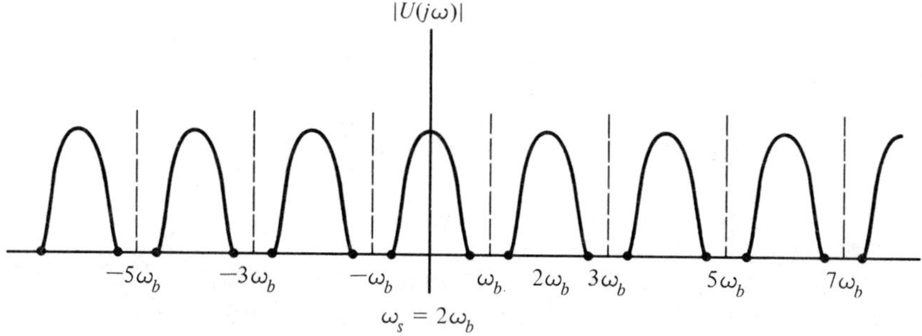

(b) Sampled signal from which $u(t)$ can be recovered

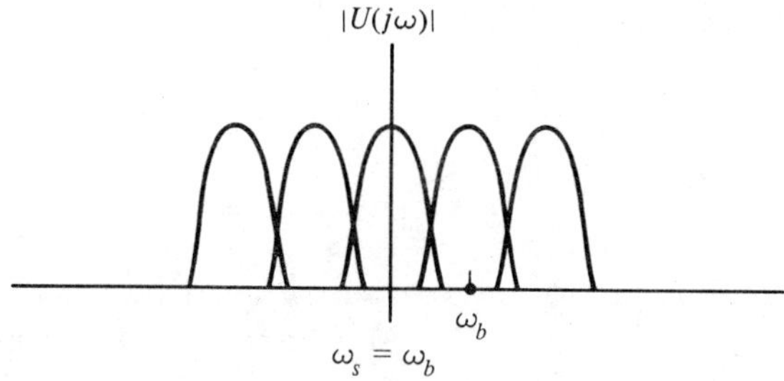

(c) Sampled signal from which $u(t)$ cannot be recovered

Figure 7.1-3 Spectra of sampled signals

which states that we should sample at a sampling frequency that is at least twice as large as the largest frequency components of interest in the primary signal. Since some very ideal assumptions are used in deriving the results of the sampling theorem, such as band limited input functions and a perfect recovery mechanism, we will generally have to sample at a considerably higher frequency than just twice the highest input frequency of significance.

Two results of special significance have been obtained in our discussions thus far. One of these is the sampling theorem indicating conditions under which we may recover the original signal from the sampled signal and the second is the result that we may use Laplace transform techniques to work with sampled signals. There are a number of reasons why we may wish to use a sampled data or digital control signal. It may be possible to use simpler more economical digitally actuated controllers. Of perhaps greater importance is the fact that we can use a digital computer in the control loop. Thus this sampling theorem and the fact that we may use Laplace transforms for analysis purposes are very important as we seek to develop design technique.

The z transform. It is somewhat more convenient to use the symbol z defined by

$$z = e^{sT} \qquad (7.1\text{-}14)$$

to replace the variable s in our definitions used thus far. We see that the z transform of a sampled time function $u^*(t)$ is given, from Eq. (7.1-12), as

$$U(z) = U^*(s)\big|_{e^{sT}=z} = \sum_{n=0}^{\infty} u(nT)z^{-n} \qquad (7.1\text{-}15)$$

Unlike the Laplace transform there is not a unique relationship between a time function $u(t)$ and the z transform of the sampled time function $u^*(t)$. In particular any time function which is zero at the sampling instants, $u(nT) = 0$, $n = 0, 1, 2, \ldots$ can be added to $u(nT)$ in Eq. (7.1-15) without altering the resulting z transform. This is usually not a disadvantage to use of the z transform if we sample at a rapid enough rate. This only says that we are in trouble if we sample a sine wave $\sin \omega t$ at a frequency $\omega_s = \pi/T = \omega/2$ which is only half of the bandwidth of the sine wave. As we have indicated, we should sample at a much higher frequency than this.

We may use Eq. (7.1-15) to establish the z transform of a variety of time functions. The combined use of series summations and partial fraction expansions will lead to the z transform of almost any exponential type time function.

EXAMPLE 7.1-1. The z transform of a unit step is easily obtained from the series summation method. We have

$$u(t) = 1 \qquad t \geqslant 0$$

$$u(nT) = 1$$

and from Eq. (7.1-15) we have

$$U(z) = \sum_{n=0}^{\infty} u(nT)z^{-n} = \sum_{n=0}^{\infty} z^{-n} = 1 + \frac{1}{z} + \frac{1}{z^2} + \frac{1}{z^3} + \ldots$$

We recognize the form of the infinite series as that of $(1 + a)^{-1} = 1 - a + a^2 - a^3 + \ldots$ which converges only for $|a| \leqslant 1$. So we have

$$U(z) = \frac{1}{1 - z^{-1}} = \frac{z}{z - 1}$$

and this converges within the unit circle where $|z| < 1$.

EXAMPLE 7.1-2. The transform of the simple exponential is obtained in a similar manner. We have

$$u(t) = e^{-at} \qquad t > 0$$

so

$$u(nT) = e^{-anT}$$

and

$$U(z) = \sum_{n=0}^{\infty} u(nT)z^{-n}$$

$$= \sum_{n=0}^{\infty} e^{-anT}z^{-n}$$

$$= 1 + e^{-aT}z^{-1} + e^{-2aT}z^{-2} + e^{-3aT}z^{-3} + \ldots$$

$$= \frac{1}{1 - e^{-aT}z^{-1}}$$

EXAMPLE 7.1-3. The z transform of $f(t) = 1 - e^{-at}$ is easily found by combining the results of the previous two examples. We have

$$F(z) = \frac{1}{1 - z^{-1}} - \frac{1}{1 - e^{-aT}z^{-1}} = \frac{z^{-1}(1 - e^{-aT})}{(1 - z^{-1})(1 - e^{-aT}z^{-1})}$$

Often we wish to find the z transform corresponding to a given Laplace transform. For this purpose, partial fractions and a table of z transforms are especially useful. Thus to obtain the z transform corresponding to

$$F(s) = \frac{a}{s(s + a)}$$

we expand $F(s)$ in partial fractions as

$$F(s) = \frac{1}{s} - \frac{1}{s + a}$$

and so

$$f(t) = 1 - e^{-at} \qquad t \geqslant 0$$

and then use the z transform tables to find the z transform corresponding to the terms in $F(s)$ or $f(t)$ and add these. We get the same result as we have just obtained, of course.

TABLE 7.1-1 presents a short list of z transforms obtained by use of Eq. (7.1-15). We should be very careful to remember that the shortened expression "find the z transform of $f(t)$" is not quite fully descriptive of the operations involved. By this expression we mean "find the sampled sequence $f(nT)$ corresponding to $f(t)$ and take the z transform of this sampled sequence." More elaborate z transform tables than these are available but this brief one will be adequate for our purposes in this text.

EXERCISE 7.1-1. Derive the z transform relations in TABLE 7.1-1.

TABLE 7.1-1. Brief Table of z Transforms

Time Function	Description	Laplace Transform	z Transform
Impulse	$\delta(t) = u_0(t)$	1	does not exist
Step	$u_{-1}(t) = 1$	$\dfrac{1}{s}$	$\dfrac{1}{1 - z^{-1}}$
Ramp	$u_{-2}(t) = t$	$\dfrac{1}{s^2}$	$\dfrac{Tz^{-1}}{(1 - z^{-1})^2}$
Acceleration	$u_{-3}(t) = t^2/2$	$\dfrac{1}{s^3}$	$\dfrac{1}{2} T^2 \dfrac{z^{-1}(1 + z^{-1})}{(1 - z^{-1})^3}$
Exponential	e^{-at}	$\dfrac{1}{s + a}$	$\dfrac{1}{1 - e^{-aT} z^{-1}}$
Step and exponential	$1 - e^{-at}$	$\dfrac{a}{s(s + a)}$	$\dfrac{z^{-1}(1 - e^{-aT})}{(1 - z^{-1})(1 - e^{-aT} z^{-1})}$
Sine wave	$\sin \omega t$	$\dfrac{\omega}{s^2 + \omega^2}$	$\dfrac{z^{-1} \sin \omega T}{1 - 2z^{-1} \cos \omega T + z^{-2}}$
Cosine wave	$\cos \omega t$	$\dfrac{s}{s^2 + \omega^2}$	$\dfrac{1 - z^{-1} \cos \omega T}{1 - 2z^{-1} \cos \omega T + z^{-2}}$
Damped sine wave	$e^{-at} \sin \omega t$	$\dfrac{\omega}{(s + a)^2 + \omega^2}$	$\dfrac{z^{-1} e^{aT} \sin \omega T}{e^{2aT} - 2z^{-1} e^{aT} \cos \omega T + z^{-2}}$
Exponential translation theorem	$e^{-at} f(t)$	$F(s + a)$	$F(e^{aT} z)$
Real translation theorem	$f(t - mT)$	$e^{-mTs} F(s)$	$z^{-m} F(z)$

EXERCISE 7.1-2. What is the z transform which corresponds to the Laplace transform

$$F(s) = \frac{8(s + 1)}{s^2(s + 2)(s^2 + 2s + 4)}$$

We may establish a number of very useful theorems for z transforms. These play essentially the same role as the corresponding Laplace transform theorems. Let us examine a few of these.

EXAMPLE 7.1-4. In this example we will establish the final value theorem for z transforms. If the final value theorem is to have meaning the function in question must become a constant as time becomes infinite. Thus we can write the time function as a constant plus a sum of exponentials*

$$f(t) = K + \sum_j \beta_j e^{-\alpha_j t} \qquad \alpha_j > 0$$

The z transform corresponding to this is

$$F(z) = \frac{K}{1 - z^{-1}} + \sum_j \frac{\beta_j}{1 - \exp(-\alpha_j T)z^{-1}}$$

Now if we multiply the foregoing by $(1 - z^{-1})$ and let z approach 1 we have

$$K = \lim_{z \to 1} (1 - z^{-1})F(z)$$

since we can show that

$$\lim_{z \to 1} (1 - z^{-1}) \sum_j \frac{\beta_j}{1 - \exp(-\alpha_j T)z^{-1}} = 0$$

since α_j is always greater than zero. But K is just the limiting behavior of $f(t)$ as t becomes infinite or $f(nT)$ as n becomes infinite. So we have the final value theorem

$$\lim_{n \to \infty} f(nT) = \lim_{z \to 1} (1 - z^{-1})F(z)$$

EXERCISE 7.1-3. Derive the initial value theorem

$$\lim_{n \to 0} f(nT) = \lim_{z \to \infty} F(z)$$

by direct expansion of Eq. (7.1-15).

---

*We could consider more complex exponentials such as

$$f(t) = K + \sum_j \beta_j e^{-a_j t} + \sum_j \sum_k \Delta_j t^k e^{-\gamma_j t}$$

but this would only complicate our proof.

EXERCISE 7.1-4. What is the z transform relation corresponding to the difference equation

$$x(n) + \alpha x(n-1) + \beta x(n-2) = \gamma u(n-1) + \delta u(n-2)$$

Inverse z transforms. There are at least four methods we may use to obtain an inverse z transform. The first method we consider, the residue method, is of theoretical value but not of much practical use in actually finding the time sequence corresponding to a given z transform. If we multiply both sides of the defining relation for the z transform, Eq. (7.1-15) by z^{m-1} we obtain

$$z^{m-1} U(z) = \sum_{n=0}^{\infty} u(nT)z^{m-n-1} \tag{7.1-16}$$

It is possible to show, using the theory of complex variables, that

$$\frac{1}{2\pi j} \oint \frac{1}{z} \, dz = 1$$

$$\frac{1}{2\pi j} \oint \frac{1}{z^r} \, dz = 0 \qquad r > 1$$

where the contour is a closed contour enclosing the origin. Thus if we integrate Eq. (7.1-16) over a closed contour encircling all poles in $U(z)$ we obtain

$$\frac{1}{2\pi j} \oint \sum_{n=0}^{\infty} u(nT)z^{m-n-1} \, dz = \frac{1}{2\pi j} \oint z^{m-1} U(z) \, dz$$

But we have just demonstrated that only one of these integrals, that for $m = n$, will yield a non zero result. So we have for the foregoing, with $m = n$, where the closed contour is the unit circle, $|z| = 1$,

$$u(nT) = \frac{1}{2\pi j} \oint z^{n-1} U(z) \, dz \tag{7.1-17}$$

Equations (7.1-15) and (7.1-17) constitute a z transform pair. Equation (7.1-17) can be used to obtain the inverse time sequence from a given z transform but extensive use of this integral requires a familiarity with residue theory and complex variables which we do not assume in our presentation.

It is generally advantageous, just as was the case with respect to the Laplace transform, to use partial fraction expansion and a table of transforms to replace formal use of the contour integral Eq. (7.1-17). It is usually more convenient to expand a partial fraction expansion in terms of a polynomial in z rather than z^{-1}. Because of the form in which z transforms occur we suggest partial fraction expansion of $F'(z)/z$ such that we can obtain desired expressions in terms of z^{-1}. A simple example will illustrate the partial fraction expansion method.

EXAMPLE 7.1-5. We wish to find the inverse z transform of

$$F(z) = \frac{0.09z^{-1}(1 + 0.716z^{-1})}{(1 - z^{-1})(1 - 0.607z^{-1})^2} = \frac{0.09z(z + 0.716)}{(z - 1)(z - 0.607)^2} \tag{1}$$

We expand the function $F'(z) = F(z)/z$ in partial fractions as

$$F'(z) = \frac{F(z)}{z} = \frac{a}{z - 1} + \frac{b}{(z - 0.607)^2} + \frac{c}{(z - 0.607)}$$

where

$$a = \frac{0.09(z + 0.716)}{(z - 0.607)^2}\bigg|_{z=1} = 1.0$$

$$b = \frac{0.09(z + 0.716)}{z - 1}\bigg|_{z=0.607} = -0.30$$

$$c = \frac{d}{dz}\left[\frac{0.09(z + 0.716)}{z - 1}\right]\bigg|_{z=0.607} = -1.0$$

Thus we have

$$F(z) = \frac{z}{z - 1} - \frac{0.3z}{(z - 0.607)^2} - \frac{z}{(z - 0.607)} \tag{2}$$

We recognize 0.607 as e^{-aT} and 0.3 as Te^{-aT}. Thus $a = 1$ and $T = 0.5$ is an acceptable set of parameters for use of our z transform tables. A time sequence corresponding to Eqs. (1) or (2) is obtained from TABLE 7.1-1 as

$$f(nT) = 1 - (1 + nT)e^{-anT} \tag{3}$$

This could correspond to the time function, if we accept $a = 1$ and $T = 0.5$,

$$f(t) = 1 - (1 + t)e^{-t} \tag{4}$$

and will generally correspond to this time function unless there is a periodic component present which is zero at the sample times such as $f(t) + \sin \dfrac{\pi t}{T}$. It is unreasonable that we sample at such a low frequency and so $f(t)$ is usually accepted as *the* time response of the system.

Regardless of the values of a and T we can write a correct sample sequence for the response as

$$f(n) = 1 - (1 + 0.5n)(0.607)^n$$

Figure (7.1-4) presents a sketch of the time sequence $f(n)$.

A third method for inversion of the z transform is a *difference equation method* in which we convert the z transform to a difference equation and solve this. In so doing we will obtain a difference equation of the form

$$a_0 f(nT) + a_1 f(\overline{n-1}\ T) + \ldots + a_{m-1} f(\overline{n-m+1}\ T)$$

$$+ a_m f(\overline{n-m}\ T) = b_0 u(nT) + b_1 u(\overline{n-1}\ T) \tag{7.1-18}$$

$$+ \ldots + b_{m-1} u(\overline{n-m+1}\ T) + b_m u(\overline{n-m}\ T)$$

where the denominator of $F(z)$ is a polynomial in z of order m. $u(kT)$ is an input equal 1 if $k = 0$ and zero otherwise. Many control systems are assumed at initial rest for analysis purposes and so it is usually possible to assume that all initial conditions in Eq. (7.1-18) are zero. Generally the b_k coefficients are zero for large k.

Often it will be more convenient to recast Eq. (7.1-18) into state space form. This is especially convenient if a digital computer is used to

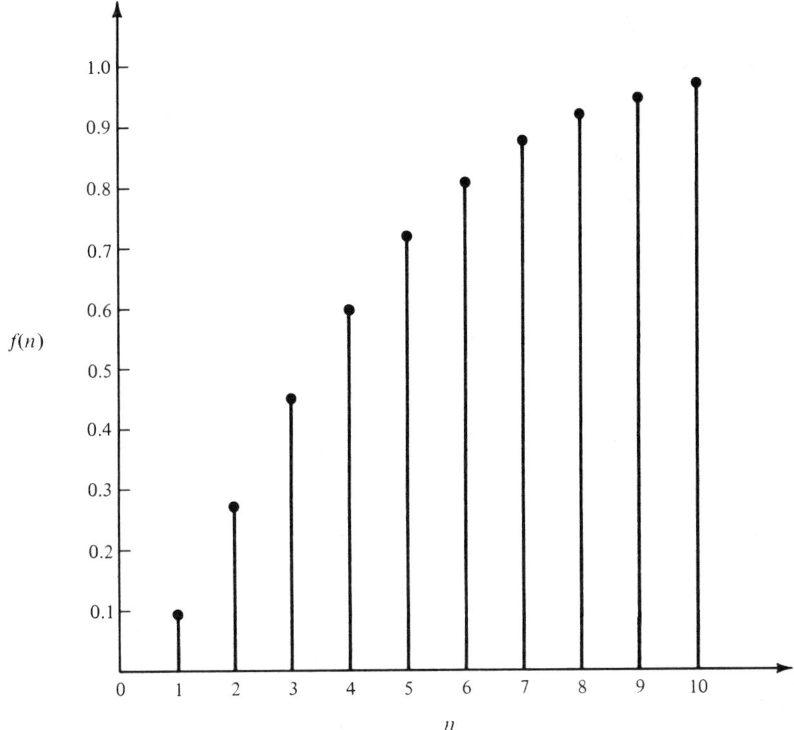

Figure 7.1-4 Time sequence $f(n)$

solve the differential equation. We will then replace the z transform by a vector set of equations of the form

$$\mathbf{x}(nT) = \mathbf{A}\mathbf{x}(\overline{n-1}\ T) + \mathbf{b}u(\overline{n-1}\ T)$$

$$\mathbf{z}(nT) = \mathbf{c}\mathbf{x}(nT) \tag{7.1-19}$$

EXAMPLE 7.1-6. We again consider the z transform

$$F(z) = \frac{0.09z^{-1}(1 + 0.716z^{-1})}{(1 - z^{-1})(1 - 0.607z^{-1})^2} \tag{1}$$

which is equivalent to

$$[1 - 2.214z^{-1} + 1.5824z^{-2} - 0.3684z^{-3}]F(z)$$

$$= 0.09z^{-1} + 0.0644z^{-2} \tag{2}$$

The difference equation corresponding to this is

$$f(nT) - 2.214f(\overline{n-1}\ T) + 1.5824f(\overline{n-2}\ T)$$

$$- 0.3684f(\overline{n-3}\ T) = 0.09u(\overline{n-1}\ T) + 0.064u(\overline{n-2}\ T) \quad (3)$$

We can easily solve this equation and obtain the time function sketched in Fig. (7.1-4).

A state variable formulation of this difference equation can be obtained in any of several ways as we have indicated for continuous time systems in Chap. 2. A convenient formulation is obtained by letting

$$x_1(nT) = x_2(\overline{n-1}\ T) \qquad\qquad\qquad (4)$$

$$x_2(nT) = x_3(\overline{n-1}\ T) + \lambda u(\overline{n-1}\ T)$$

$$x_3(nT) = x_3(\overline{n-1}\ T) + \beta x_2(\overline{n-1}\ T) + \gamma x_1(\overline{n-1}\ T) + \Delta u(\overline{n-1}\ T)$$

and then setting

$$x_3(nT) = f(nT)$$

$$\alpha = 2.214$$

$$\beta = 1.5824$$

$$\gamma = 0.3684$$

$$\Delta = 0.09$$

$$\lambda = -0.0573$$

such that Eq. (4) is equivalent to Eq. (3). In vector matrix form we have

$$\mathbf{A} = \begin{bmatrix} 0 & 1 & 0 \\ 0 & 0 & 1 \\ \alpha & \beta & \gamma \end{bmatrix}$$

$$\mathbf{b} = \begin{bmatrix} 0 \\ \lambda \\ \Delta \end{bmatrix}$$

$$\mathbf{c} = \begin{bmatrix} 0 & 0 & 1 \end{bmatrix}$$

where the vector matrix difference equation is given by Eq. (7.1-19).

For hand held calculator computation as well as for digital computer solution the *direct division method* will often be the most convenient method to use to obtain numerical inversion of a z transform. The method is extraordinarily simple and indicates the considerable ease with which the numerical inversion of z transform relations can be obtained. We will denote our next section to exploiting this ease as we develop very useful methods to obtain numerical solution of control system differential equations.

The direct division methods proceeds simply by long division of the polynomials in $F(z)$ to obtain

$$F(z) = c_0 + c_1 z^{-1} + c_2 z^{-2} + c_3 z^{-3} + \dots$$

We can replace z by e^{st} and then obtain the inverse Laplace transform to obtain

$$f^*(t) = c_0 \delta(t) + c_1 \delta(t - T) + c_2 \delta(t - 2T) + \dots$$

In other words the coefficients of z obtained by this long division process will be the sample sequence of the time function we desire to obtain.

EXAMPLE 7.1-7. We again consider the z transform of the previous example which we write as

$$F(z) = \frac{0.09 z^{-1} + 0.0644 z^{-2}}{1 - 2.214 z^{-1} + 1.5824 z^{-2} - 0.3684 z^{-3}}$$

Long division yields

$$F(z) = 0.0895z^{-1} + 0.26310z^{-2} + 0.44088z^{-3} + 0.59274z^{-4}$$

$$+ 0.71159z^{-5} + 0.79993z^{-6} + 0.866338z^{-7} + 0.90786z^{-8}$$

$$+ 0.93847z^{-9} + 0.95926z^{-10} + \ldots$$

The values we have just obtained are the output sequence for the function $f(n)$ as indicated in Fig. (7.1-4).

EXERCISE 7.1-5. What is the sample sequence for the following z transforms? Use the partial fraction method.

$$\text{a.} \quad F(z) = \frac{2}{(1 - z^{-1})(1 - 0.4z^{-1})}$$

$$\text{b.} \quad F(z) = \frac{z^{-1}(1 + 2z^{-1} + z^{-2})}{(1 - z^{-1} + z^{-2})(1 + z^{-1} + 2z^{-2})}$$

EXERCISE 7.1-6. Repeat EXERCISE 7.1-5 using the difference equation method.

EXERCISE 7.1-7. Repeat EXERCISE 7.1-5 using the direct division method.

EXERCISE 7.1-8. Solve EXAMPLE 7.1-5 using the residue method. This is an especially meaningful exercise if you have had previous study of complex variable theory.

Transfer Functions for Closed Loop Digital Control Systems. We now consider the determination of transfer functions for linear digital control systems. It turns out that sampler location makes quite a bit of difference with respect to the transfer function of a system. Consider the four block diagrams shown in Fig. (7.1-5). Each of the systems depicted in block diagram form has the same input $u(t)$. The continuous time transfer functions $G_1(s)$ and $G_2(s)$ are the same for each system. Yet the transfer functions are different. Let us see why this is the case. All samplers are assumed to operate periodically and to be perfectly synchronized.

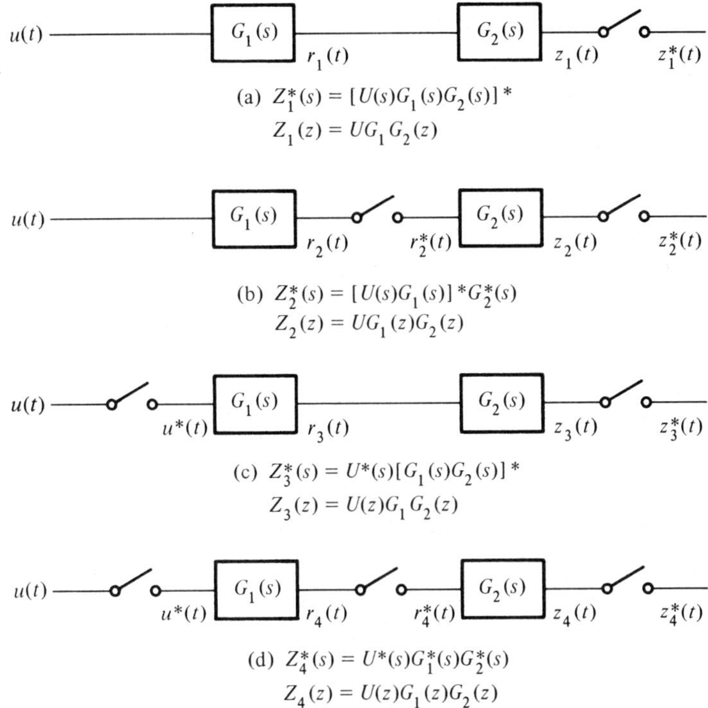

Figure 7.1-5 Open loop systems showing different sampler locations

The output of the system of Fig. (7.1-5a) is easily shown to be

$$Z_1(s) = U(s)G_1(s)G_2(s) \qquad (7.1\text{-}20)$$

Upon taking the sampled transform we have

$$Z_1^*(s) = [U(s)G_1(s)G_2(s)]^* \qquad (7.1\text{-}21)$$

We write this in z transform notation as

$$Z_1(z) = UG_1G_2(z) \qquad (7.1\text{-}22)$$

In this particular case it is not possible to find the pulsed or discrete transfer function $Z_1^*(s)/U^*(s)$ unless we specify U, G_1, and G_2. The reason for this is that we are really dealing with a time varying system here because of the sampler.

For Fig. (7.1-5b) we easily obtain

$$R_2(s) = U(s)G_1(s) \tag{7.1-23}$$

and

$$R_2^*(s) = [U(s)G_1(s)]^* \tag{7.1-24}$$

as the output of the $G_1(s)$ block. For the second block we have

$$Z_2(s) = R_2^*(s)G_2(s) \tag{7.1-25}$$

and we now wish to find $Z_2^*(s)$. This is most easily accomplished by using the definition of the pulsed transform given by Eq. (7.1-13)

$$C^*(s) = \sum_{m=-\infty}^{\infty} C(s + jm\omega_s) \tag{7.1-26}$$

Using the relations in Eq. (7.1-24) and Eq. (7.1-25) we have

$$Z_2^*(s) = \sum_{m=-\infty}^{\infty} R_2^*(s + jm\omega_s)G_2(s + jm\omega_s)$$

But we already have for $R_2^*(s)$, using Eq. (7.1-26),

$$R_2^*(s) = \sum_{n=-\infty}^{\infty} R_2(s + jn\omega_s)$$

and so we see that

$$R_2^*(s + jm\omega_s) = R_2^*(s)$$

Equation (7.1-26) becomes, using this identity

$$Z_2^*(s) = R_2^*(s)G_2^*(s) \tag{7.1-27}$$

and so we have finally

$$Z_2^*(s) = [U(s)G_1(s)]^* G_2^*(s) \tag{7.1-28}$$

or in z transform notation

$$Z_2(z) = UG_1(z)G_2(z) \tag{7.1-29}$$

It is now a simple matter, using these relationships, for the reader to verify the given transform relationships for Fig. (7.1-5c) and (7.1-5d).

EXERCISE 7.1-9.   Verify the transform relations given in Fig. (7.1-5c, d).

EXAMPLE 7.1-8.   It is of interest to find the output time sequence for each of the block diagrams of Fig. (7.1-5) for a particular case. We assume

$$U(s) = G_1(s) = G_2(s) = \frac{1}{s} \tag{1}$$

For the block diagram of Fig. (7.1-5a) we have

$$Z_1^*(s) = [1/s^3]^* \tag{2}$$

and from TABLE 7.1-1 we obtain

$$Z_1(z) = \frac{1}{2} T^2 \frac{z^{-1}(1 + z^{-1})}{(1 - z^{-1})^3} \tag{3}$$

For the block diagram of Fig. (7.1-5b) we have

$$Z_2^*(s) = [1/s^2]^* [1/s]^* \tag{4}$$

and from TABLE 7.1-1 we obtain

$$Z_2(z) = \frac{Tz^{-1}}{(1 - z^{-1})^3} \tag{5}$$

It is easy for us to show also that

$$Z_3(z) = \frac{Tz^{-1}}{(1 - z^{-1})^3} \tag{6}$$

and

$$Z_4(z) = \frac{1}{(1 - z^{-1})^3} \tag{7}$$

We may relate each of the z transforms in terms of $Z_4(z)$ as

$$Z_2(z) = Z_3(z) = Tz^{-1}Z_4(z) \tag{8}$$

and

$$Z_1(z) = \frac{1}{2}T^2 z^{-1}(1 + z^{-1})Z_4(z) \tag{9}$$

From the results of the real translation theorem of TABLE 7.1-1 we see that

$$z_2(nT) = z_3(nT) = Tz_4(\overline{n-1}\,T) \tag{10}$$

and

$$z_1(nT) = \frac{1}{2}T^2[z_4(\overline{n-1}\,T) + z_4(\overline{n-2}\,T)] \tag{11}$$

Thus we see that $z_2(nT)$ and $z_3(nT)$ are just one sample delayed versions of $z_4(nT)$ whereas $z_1(nT)$ is an average of $z_4(nT)$ delayed one sample and $z_4(nT)$ delayed two samples.

We can obtain the inverse transform of $Z_4(z)$ in several ways as we have previously mentioned. Here we will illustrate two of these. This is not a standard form in TABLE 7.1-1 and so let us manipulate it into a standard form. We have

$$Z_4(z) = \frac{1}{(1 - z^{-1})^3} \tag{12}$$

We see from the form of $Z_4(z)$ that the sample response will be of the form n^2 and possible polynomials of lower order than n^2. So we might guess

$$z_4(nT) = a + bnT + c(nT)^2 \tag{13}$$

and attempt to determine a, b, and c to satisfy Eq. (12). The z transform of Eq. (13) is

$$Z_4(z) = \frac{a}{(1-z^{-1})} + \frac{Tbz^{-1}}{(1-z^{-1})^2} + \frac{cT^2(z^{-1}+z^{-2})}{(1-z^{-1})^3}$$

$$= \frac{a + (-2a + bT + cT^2)z^{-1} + (a - bT + cT^2)z^{-2}}{(1-z^{-1})^3} \qquad (14)$$

Upon equating Eqs. (12) and (14) we obtain

$$a = 1 \qquad bT = \frac{3}{2} \qquad cT^2 = \frac{1}{2}$$

such that we have

$$z_4(nT) = 1 + \frac{3}{2}n + \frac{1}{2}n^2 \qquad (15)$$

and from Eq. (10)

$$z_2(nT) = z_3(nT) = Tz_4(\overline{n-1}\,T) = \frac{T}{2}(n^2 + n) \qquad (16)$$

Finally we have from Eq. (11)

$$z_1(nT) = \frac{n^2 T^2}{2} \qquad (17)$$

The inverse z transform of Eq. (3) yields the sample sequence of Eq. (17) and so we see that we have indeed obtained the correct answer. It is interesting to note that $z_4(nT)$ is not a function of T whereas $z_2(nT)$ and $z_3(nT)$ are linear functions of T and $z_1(nT)$ is a function of T^2. For a continuous time system one would expect the response of $z_1(nT)$, that is we integrate a step twice and obtain $\frac{1}{2}t^2$. So we see that the presence of samplers can change things a great deal. We introduce hold circuits into sample data systems to ameliorate this change as we will see in our next subsection.

EXERCISE 7.1-10. Repeat EXAMPLE 7.1-8 for the case where

$$G_1(s) = \frac{1}{s+1} \qquad G_2(s) = \frac{2}{s+2} \qquad U(s) = \frac{1}{s}$$

It is now a relatively straightforward task to extend these concepts regarding open loop system transfer functions to closed loop system transfer functions. Figure (7.1-6) represents some common configurations for closed loop systems and associated transfer functions. Let us examine the relations for one of the closed loop systems in Fig. (7.1-6). For the system of Fig. (7.1-6a) we have the following transfer relations

$$Z(s) = G_1(s)E^*(s) \tag{7.1-30}$$

$$E(s) = U(s) - G_2(s)Z^*(s) \tag{7.1-31}$$

We sample these relations and obtain

$$Z^*(s) = G_1^*(s)E^*(s) \tag{7.1-32}$$

and

$$E^*(s) = U^*(s) - G_2^*(s)Z^*(s) \tag{7.1-33}$$

Substituting Eq. (7.1-33) into Eq. (7.1-32) we obtain after some algebra

$$Z^*(s) = \frac{G_1^*(s)U^*(s)}{1 + G_1^*(s)G_2^*(s)} \tag{7.1-34}$$

and

$$E^*(s) = \frac{U^*(s)}{1 + G_1^*(s)G_2^*(s)} \tag{7.1-35}$$

We substitute Eq. (7.1-35) into (7.1-30) to obtain one of the desired transfer relations

$$Z(s) = \frac{G_1(s)U^*(s)}{1 + G_1^*(s)G_2^*(s)} \tag{7.1-36}$$

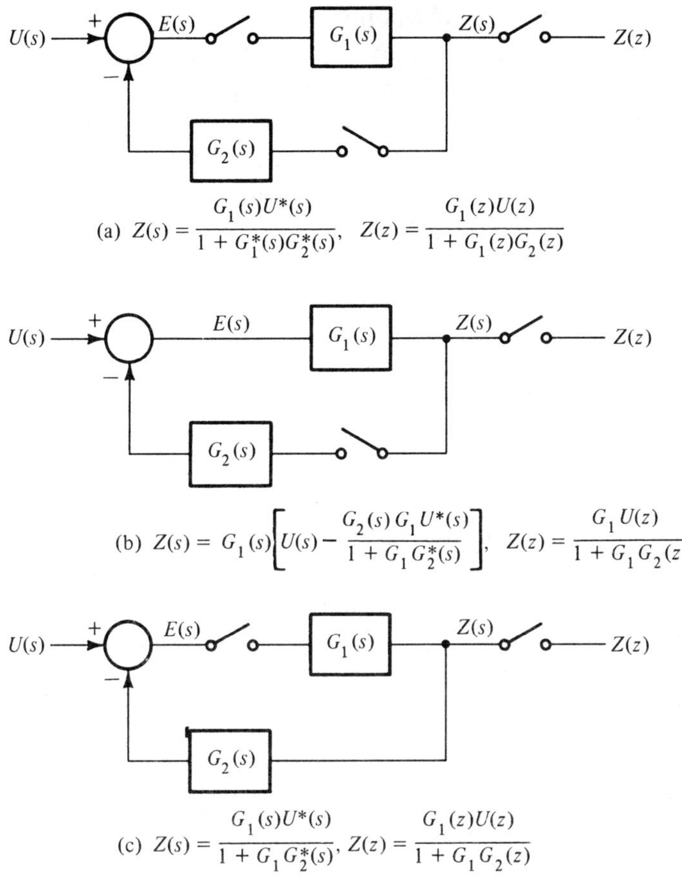

$$\text{(a)} \quad Z(s) = \frac{G_1(s)U^*(s)}{1 + G_1^*(s)G_2^*(s)}, \quad Z(z) = \frac{G_1(z)U(z)}{1 + G_1(z)G_2(z)}$$

$$\text{(b)} \quad Z(s) = G_1(s)\left[U(s) - \frac{G_2(s)G_1U^*(s)}{1 + G_1G_2^*(s)}\right], \quad Z(z) = \frac{G_1U(z)}{1 + G_1G_2(z)}$$

$$\text{(c)} \quad Z(s) = \frac{G_1(s)U^*(s)}{1 + G_1G_2^*(s)}, \quad Z(z) = \frac{G_1(z)U(z)}{1 + G_1G_2(z)}$$

Figure 7.1-6 Common single loop digital control block diagrams and their transfer reflections

We rewrite Eq. (7.1-34) in z transform notation as the other transfer relation

$$Z(z) = \frac{G_1(z)U(z)}{1 + G_1(z)G_2(z)} \tag{7.1-37}$$

We will base our sampled data design approach upon these relations for closed loop system transforms and transfer functions.

EXERCISE 7.1-11. Derive the transfer relations for Fig. (7.1-6b).

EXERCISE 7.1-12. Derive the transfer relations for Fig. (7.1-6c).

Data Hold or Data Reconstruction Systems. Since ideal samplers are difficult to realize and the sharply peaked high frequency output from an ideal sampler is often an undesirable input to a continuous system, sample hold circuits and systems are often used. A hold or data reconstruction system is one in which a time domain extrapolation technique is used in which the output between samples is assumed to be a constant in time, a linearly increasing or decreasing function of time, or any other convenient and useful function of time.

The simplest type of hold or data reconstruction system is the zero order hold in which the output of the hold or data reconstructor over a given sample period is assumed to be the input to the hold or data reconstruction system at the last sample. In other words if the input to the hold devise is $u(nT)$ then the output is equal to this value for time between nT and $(n + 1)T$. In other words

$$z(t + nT) = u(nT) \qquad 0 \leqslant t \leqslant T \qquad (7.1\text{-}38)$$

is the mathematical relation describing a zero order hold.

Of more importance is the transfer function of the continuous system which, when combined with an ideal sampler, yields a zero order hold. Figure (7.1-7) illustrates the general block diagram representation for a sample hold system which includes the zero order hold as a special case. When the input to the sampler $u(t)$ is a unit step function the output function should also be a unit step function. From Fig. (7.1-7) we see that if

$$U(s) = \frac{1}{s} = Z(s) \qquad (7.1\text{-}39)$$

then, from TABLE 7.1-1 we have for the pulsed transforms,

$$Z^*(s) = U^*(s) = \frac{1}{1 - e^{-sT}} \qquad (7.1\text{-}40)$$

and so we see that

$$G_{\text{HO}}(s) = \frac{Z(s)}{U^*(s)} = \frac{1 - e^{-sT}}{s} \qquad (7.1\text{-}41)$$

is the transfer function of the zero order hold. This is a rather interesting transfer function. Figure (7.1-8) illustrates the impulse response and

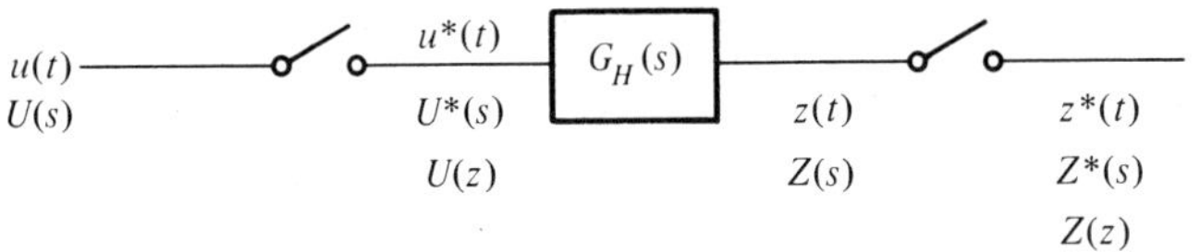

Figure 7.1-7 Hold system block diagram

frequency response of the zero order hold and compares it with a "cardinal hold" which is a "perfect" hold circuit with transfer function $G_{HO}(s)$ which would have a gain magnitude of 1 over the frequency range $\omega_c = \pm 2\pi/T$ and zero elsewhere and a phase shift of zero within the frequency band $\pm\omega_c$. As we see, the gain curve of the zero hold nicely approximates that of the ideal cardinal hold filter which would act as a perfect data reconstruction in that it would perfectly filter everything outside of a frequency range equal to twice the sampling frequency. This filter is unrealizable as its impulse response exists before the time the impulse arrives. Thus it is anticipative and non-causal. It cannot be built. This is one reason why we must sample at a frequency greater than twice the frequency of twice the highest signal frequency in the input signal. We must use approximate data reconstructors which are physically reliable such as zero order data holds.

The first order data reconstructor or first order data hold is a system which will extrapolate a sampled ramp input up to one sample in the future with zero error. In order to do this we will need to know not only the value of the sampled input at time nT but also the derivative at time nT. Thus we will extrapolate using the relation

$$z(t + nT) = u(nT) + u'(nT)(t) \qquad 0 \leqslant t \leqslant T$$

where $u'(nT)$ is the input derivative at time nT. Since we wish perfect extrapolation for a ramp only we are fortunate in that the first backward difference yields a perfect measurement of the derivative. So we use

$$u'(nT) = \frac{u(nT) - u(\overline{n-1}\,T)}{T}$$

and obtain for the first order data extrapolator

$$z(t + nT) = u(nT) + \frac{t}{T}\,[u(nT) - u(\overline{n-1}\,T)] \qquad 0 < t < T \quad (7.1\text{-}42)$$

The transfer function of the first order hold circuit can be obtained by specifying the input and output desired for the block diagram of Fig. (7.1-7). We might attempt to specify that an input $u(t) = t$ result in an output $z(t) = t$. This will lead to

$$U^*(s) = \frac{Te^{-sT}}{(1 - e^{-sT})^2} \qquad (7.1\text{-}43)$$

$$Z(s) = \frac{1}{s^2}$$

and

$$G_{\text{H1}}(s) = \frac{Z(s)}{U^*(s)} = \frac{(1 - e^{-sT})^2 e^{sT}}{Ts^2}$$

This is a noncausal, anticipative, and unrealizable transfer function. The impulse response corresponding to this $G_{\text{H1}}(s)$ is

$$g_{\text{H1}}(s) = \begin{cases} 0 & t < -T \\ \dfrac{t + T}{T} & -T < t < 0 \\ 1 - \dfrac{t}{T} & 0 < t < T \\ 0 & T < t \end{cases}$$

The physical problem here is that when the input starts at $t = 0$ there is no way that the data reconstruction circuit can measure the derivative to determine the slope of the input at a single sample point. We should modify the output to make it more reasonable. We thus specify that the output should only be equal to the input after the sample at $t = T$. Thus we specify

$$z(t) = tu_{-1}(t - T) = (t - T)u_{-1}(t - T) + Tu_{-1}(t - T) \qquad (7.1\text{-}44)$$

where $u_{-m}(t)$ is the singularity function with $U_{-m}(s) = 1/s^m$ and so we have

$$Z(s) = \frac{1}{s^2}e^{-sT} + \frac{T}{s}e^{-sT} = \frac{1 + Ts}{s^2}e^{-sT} \qquad (7.1\text{-}45)$$

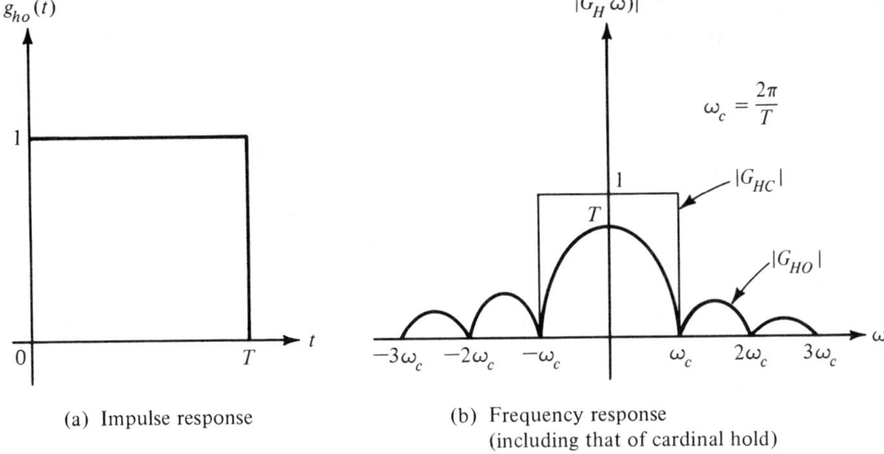

(a) Impulse response

(b) Frequency response
(including that of cardinal hold)

Figure 7.1-8 Response characteristics of zero order hold

and from Eqs. (7.1-43) and (7.1-45) and Fig. (7.1-7),

$$G_{H1}(s) = \frac{Z(s)}{U^*(s)} = \frac{(1 - e^{-sT})^2 (1 + Ts)}{Ts^2} \tag{7.1-46}$$

The impulse response corresponding to this is

$$g_{H1}(t) = \begin{cases} 0 & t < 0 \\ t + T & 0 < t < T \\ T - t & T < t < 2T \\ 0 & 2T < t \end{cases}$$

and this is physically realizable although not by any simple analog computer realization due to the two time delays inherent in Eq. (7.1-46).
A further difficulty with the first order hold is that the step response
shows considerable overshoot. From Eq. (7.1-42) we may easily determine the step response of the system by letting $u(nT) = 1$ for $n \geq 0$.
We see that the response is

$$z(t) = 1 + t/T \qquad 0 < t < T$$

$$z(t + nT) = 1 \qquad n \geq 1, 0 < t < T$$

and thus we see that there is 100% overshoot of the step response and this occurs at the first sample instant. Figure (7.1-9) illustrates the response of a zero order hold and a first order hold to a step function input and a ramp function input. It seems clear that the first order hold may give rise to potential difficulties if the sample period is relatively large and the system subjected to sudden changes, such as step function inputs.

It is possible to implement various fractional order holds as illustrated by EXERCISE 7.1-13 and higher order holds as illustrated by Problem 7.1. Control systems generally do not use other than zero order sample hold circuits because of the difficulty of implementation and the excessive phase lag introduced by the higher order holds.

EXERCISE 7.1-13. A fractional order hold is one in which the impulse response of the continuous element in the hold circuit is

$$g_{Hf}(t) = \begin{cases} 0 & t < 0 \\ T + kt & 0 < t < T \\ k(T - t) & T < t < 2T \\ 0 & 2T < t \end{cases}$$

Find the response of the fractional order hold to a step function input and a ramp function input.

EXERCISE 7.1-14. What is the step response of a unity feedback ratio sampled data system with the block diagram of Fig. (7.1-6c) in which $G_2(s) = 1$ and $G_1(s)$ consists of a zero order hold in cascade with an integrator with gain K such that $G_1(s) = (1 - e^{-sT})K/s^2$?

Relations between the s plane and the z plane. In our efforts in earlier chapters we have developed a considerable facility in dealing with continuous time linear systems. In this subsection we wish to examine some of the relationships between continuous time and sampled data systems. We can learn quite a bit about discrete system response by determining regions in the z plane which correspond to regions in the s plane which possess known response types.

A linear system is stable if its poles lie in the left half of the s plane. The relationship between the s plane and the z plane is

$$z = e^{sT}$$

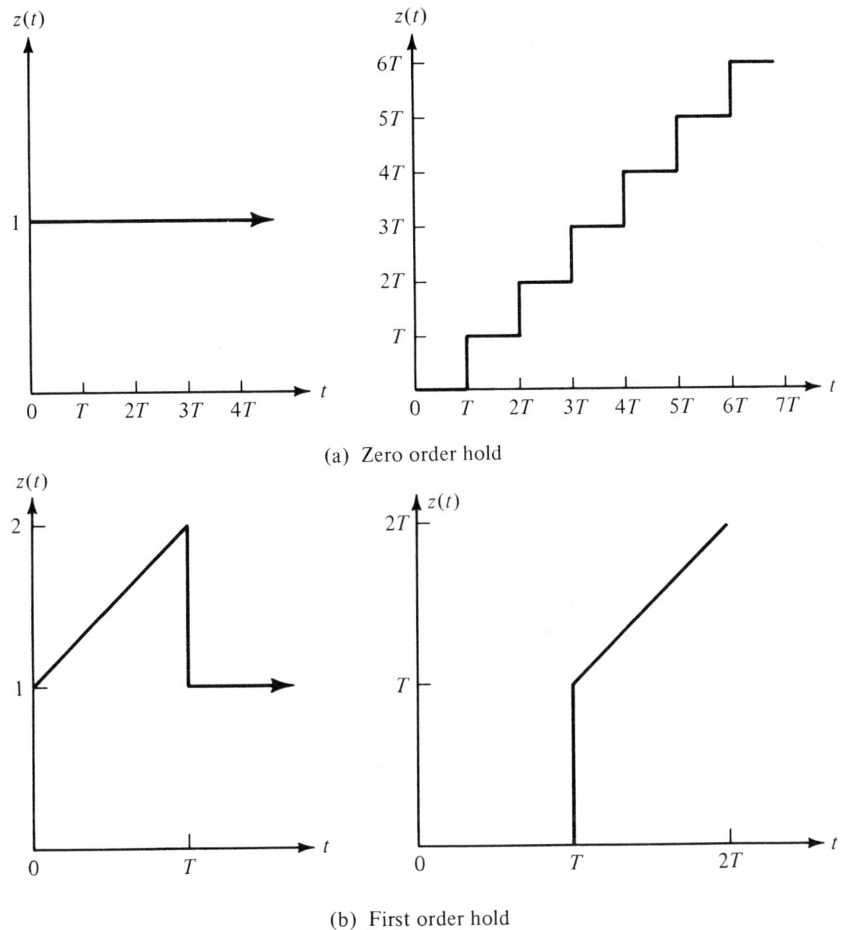

(a) Zero order hold

(b) First order hold

Figure 7.1-9 Step and ramp responses of hold circuits

The boundary between the left and right half of the s plane is described by $s = j\omega$. We see that $s = j\omega$ corresponds to $z = e^{j\omega T}$ and $|z| = 1$. Since $s = \infty$ corresponds to $z = 0$ we see that the left half of the s plane is described by the interior of the unit circle as indicated in Fig. (7.1-10). There are a number of tests of a polynomial in z to determine whether its roots lie inside the unit circle. We shall not develop these here since our Bode diagram and root locus design procedures to follow will guarantee system stability.

We know that poles in the s plane located at $s = -\sigma$ will give rise to response of the form $e^{-\sigma t}$. As σ increases the response quickens. In the z plane we have $z = e^{sT} = e^{-\sigma T}$. For $\sigma = 0$ the pole is at $z = 1$ and

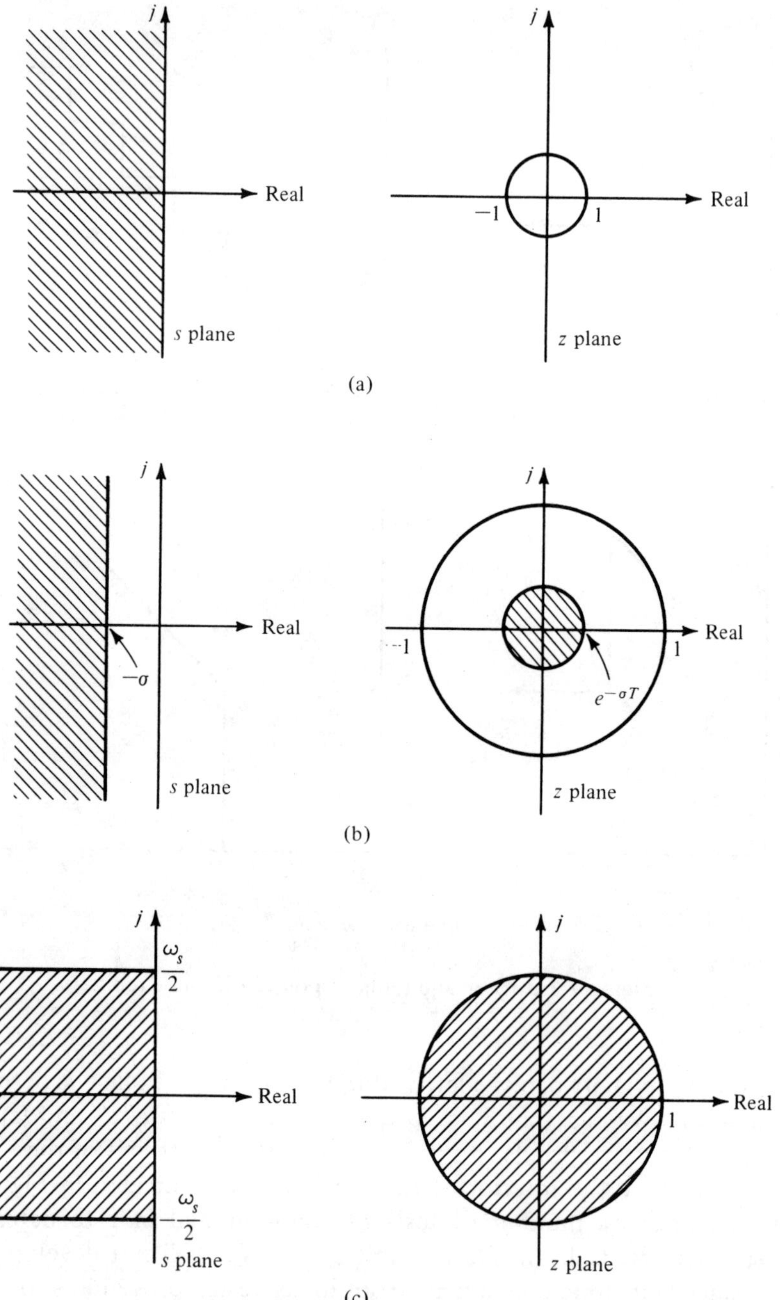

Figure 7.1-10   Relations between *s* plane and *z* plane

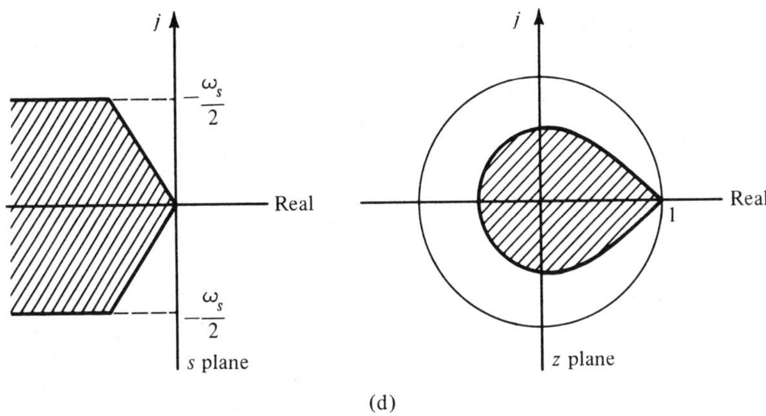

(d)

Figure 7.1-10 (continued)

for $\sigma = \infty$ the pole is at the origin or $z = 0$. Thus we see that rapid response systems are those with poles near the origin in the z plane.

Figure (7.1-10b) indicates how the region $s < -\sigma$ maps into the z plane and clearly illustrates this behavior.

For a pole located at $s = -\sigma + j\omega$* we have a transient response of the form $e^{-\sigma t} \cos \omega t$. In the z plane this corresponds to $z = e^{sT} = e^{-\sigma T} e^{j\omega T}$. Thus a line of constant σ in the s plane corresponds to a circle of radius less than 1 in the z plane as indicated in Fig. (7.1-10) which shows the relationships between a constant decrement factor in the s plane and the corresponding z plane locus. When $\omega T = \pi$ the pole will be located on the negative real axis of the z plane and this will correspond to a response of the form $e^{-\sigma t} \cos \dfrac{\pi t}{T} = e^{-\sigma n T} \cos n\pi$ and thus the response will alternate in sign at successive sample points. This is generally an undesirable behavior pattern for it indicates that the sampling frequency is too low compared with the bandwidth of the continuous time system elements.

Lines of constant damping ratio are important in the s plane and thus it is important to indicate a locus of constant damping in the z plane. This is unfortunately complicated because there are several strips in the s plane which map into the same region of the z plane. The $s = j\omega$ locus in the z plane corresponds to $z = e^{j\omega T}$ and thus a complete unit circle is traversed as ω varies from $-\omega_s/2$ to $+\omega_s/2$ where ω_s is

---

*And, of course, for a conjugate complex pole at $s = -\sigma - j\omega$ also.

the sampling frequency or $\omega_s = 2\pi/T$. Figure (7.1-10) illustrates this primary strip relationship. For constant damping we know that poles in the s plane are located at $s = \zeta\omega + j\sqrt{1 - \zeta^2}\,\omega$ and the complex conjugate of this location. The corresponding z plane root locations is

$$z = e^{sT} = e^{-\zeta\omega T}\exp\left(j\sqrt{1 - \zeta^2}\,\omega T\right)$$

For $\omega = 0$ we have $z = 1$. As ω increases the $e^{-\zeta\omega T}$ term decreases and the $\exp\left(j\sqrt{1 - \zeta^2}\,\omega T\right)$ term causes a phase shift. At $\omega T\sqrt{1 - \zeta^2} = \pi$ the phase shift in the z plane is $180°$ and the corresponding magnitude is $e^{-\zeta\omega T} = \exp\left(-\pi\zeta/\sqrt{1 - \zeta^2}\right)$. Figure (7.1-10) also indicates the logarithmic spiral nature of the constant damping location in the z plane.

We can make polar plots of open transfer functions $G^*(s)$ or $G(z)$ for sampled data systems. Since the characteristic equation of a sampled data system, $1 + G^*(s)$ or $1 + G(z) = 0$, is the same in form as for a continuous time system, we may draw Nyquist diagrams for sampled data systems much as we do for continuous time systems. The Bode diagram approach is not directly applicable since our entire theory of Bode diagram requires that the frequency variable moves up and down the imaginary axis of the s plane and not around a unit circle. We will introduce a simple transformation which will allow us to use our very powerful Bode diagram approach.

The root locus approach of Chap. 6 is directly applicable except that we must draw the root locus in the z plane and give the roots and the locus the proper interpretation. Our interpretation will be aided considerably by our efforts in this subsection and especially by Fig. (7.1-10). Let us now turn our attention to the design of linear sampled data systems.

EXERCISE 7.1-15. Draw a Nyquist diagram for the system of EXERCISE 7.1-14. What requirements on the gain K must be imposed for stability of the system?

Linear Systems Design. We continue our presentation with a discussion of a linear discrete systems control design approach. Since we have already denoted much effort to a presentation of Bode and root locus design in Chaps. 5 and 6 our discussion here can be intentionally brief.

All of our root locus rules and procedures in Chap. 6 carry over directly to sampled data systems. In one respect the root locus design of sampled data systems is easier than that for continuous time systems.

Time delays pose considerable problems in using the continuous time root locus and we have introduced rational transfer function approximations to ameliorate these. Time delays that are an integer number of sampling periods do not pose problems at all in sampled data systems, as the time delay transfer function e^{-sT_d} simply becomes z^{-n} where $n = T_d/T$.

Bode diagram design does pose a problem using z transforms however since a central feature of the Bode diagram is that it is a gain magnitude plot for $s = j\omega$. In order to use the Bode diagram for sampled data systems we must somehow transform the unit circle into the $j\omega$ axis. Certainly the transformation $z = e^{sT}$ or $sT = \ln z$ will do this but this transformation is multivalued as we have seen. Either of the bilinear transformations

$$r = \frac{z + 1}{z - 1} \qquad z = \frac{r + 1}{r - 1} \tag{7.1-47}$$

or

$$w = \frac{z - 1}{z + 1} = \frac{1}{r} \qquad z = \frac{1 + w}{1 - w} \tag{7.1-48}$$

will map the unit circle in the z plane into the $j\omega$ axis. The r plane transformation, while acceptable, does not produce transfer functions which behave as we would expect corresponding continuous time transfer functions to behave. The w plane transformation does lead to acceptable results and w is a frequency variable with real and imaginary parts.

EXAMPLE 7.1-9. The z transform corresponding to a sampler followed by an integrator is

$$G^*(s) = [1/s]^* = \frac{1}{1 - e^{-sT}} \tag{1}$$

$$G(z) = \frac{1}{1 - z^{-1}} = \frac{z}{z - 1} \tag{2}$$

and using Eq. (7.1-47) leads to

$$G(r) = \frac{r + 1}{2} \tag{3}$$

While sketching the Bode diagram for this is a straightforward task it simply doesn't look like the Bode diagram for an integrator should look.

The *w* transformation is much more reasonable however in that we obtain, using Eq. (7.1-48)

$$G(w) = \frac{1 + w}{2w} \tag{4}$$

The phase shift is $-90°$ for low frequencies and increases to $0°$ phase shift at high frequencies according to

$$\beta(w = jv) = -\frac{\pi}{2} + \tan^{-1} v \tag{5}$$

The gain magnitude of Eq. (4) is

$$|G(w = jv)| = \frac{\sqrt{1 + v^2}}{2v} \tag{6}$$

The starred transfer function $G^*(s)$ may be written as, for $s = j\omega$

$$G^*(s = j\omega) = \frac{1}{1 - \cos \omega T + j \sin \omega T} \tag{7}$$

The magnitude of this is

$$|G^*(\omega)| = \frac{1}{2(1 - \cos \omega T)} \tag{8}$$

and the phase shift curve is

$$\beta(\omega) = -\tan^{-1}\left[\frac{\sin \omega T}{1 - \cos \omega T}\right] \tag{9}$$

The gain magnitude of both Eq. (6) and Eq. (8) is infinite for small v or small ω. For $\omega = \pi/2T = \omega_s/4$ the gain magnitude for

$|G^*(\omega)|$ reaches its maximum value of 0.707 and then decreases. Again we see reasons why we should not attempt input frequencies near the sampling frequency. The phase shift characteristics of Eqs. (6) and (9) both begin at $-90°$ phase shift. The phase shift of Eq. (9) becomes $-\pi/4$ at $\omega = \pi/2T = \omega_s/4$ whereas the phase shift of Eq. (5) becomes $-\pi/4$ at $v = 1$. The phase shift of Eq. (5) becomes zero for $\omega = \omega_s/2$ and Eq. (5) yields zero degrees phase shift at $v = \infty$. We shall adopt the w plane transformation since it is more appealing than the r plane transformation and use it as the basis for Bode diagram design.

EXERCISE 7.1-16. Verify that the transformations of Eqs. (7.1-47) and (7.1-48) will map the interior of the unit circle into the left half plane for the w plane transformation. What is the mapping for the r plane transformation?

Let us investigate further the relations between the s plane, the z plane, and the w plane. For $s = j\omega$ the w plane transformation can be written as

$$w = \frac{z-1}{z+1} = \frac{e^{sT}-1}{e^{sT}+1} = \frac{e^{j\omega T}-1}{e^{j\omega T}+1} = \frac{e^{j\omega T/2}-e^{-j\omega T/2}}{e^{j\omega T/2}+e^{-j\omega T/2}}$$

$$= \frac{j\sin(\omega T/2)}{\cos(\omega T/2)} = j\tan(\omega T/2)$$

As we would hope the value obtained for w in the foregoing are purely imaginary. We thus let $w = jv$ and then have the relation between frequency ω and frequency v

$$v = \tan(\omega T/2) \tag{7.1-49}$$

The frequency v is a periodic function of actual frequency ω. The period corresponds to the system sampling frequency $\omega_s = 2\pi/T$. For $\omega = \omega_s/2$ the frequency v becomes infinite. Again we see that we must keep the input frequencies lower than $1/2$ the sampling frequency or considerable difficulties will result.

We suggest the approach of the DELTA Chart of Fig. (7.1-11) for design of linear discrete time systems using the w plane Bode diagram and the root locus technique. We will illustrate this design procedure

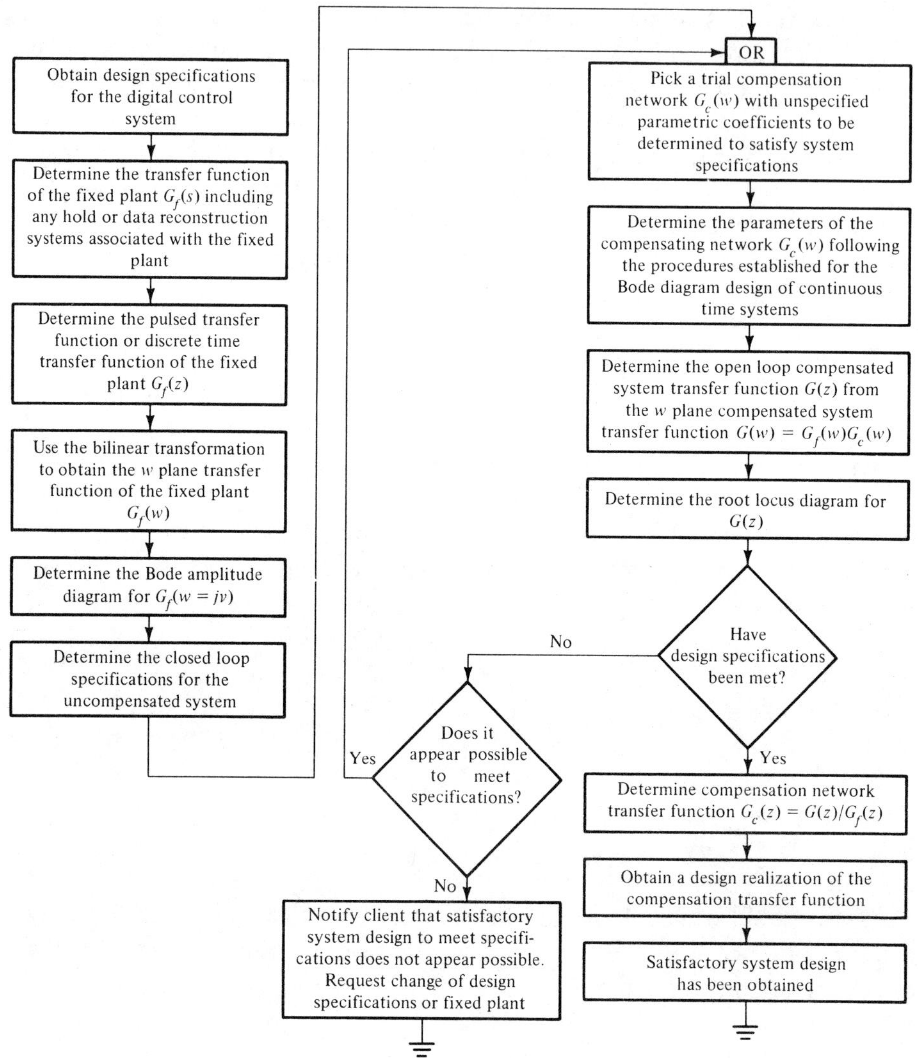

Figure 7.1-11   DELTA Chart of linear sampled data design procedure

with several examples. Although we emphasize the unity feedback ratio single loop case, the extension to multiple loop systems is relatively straightforward.

EXAMPLE 7.1-10. We consider the discrete data control system of Fig. (7.1-12). The continuous time transfer function $G(s)$ is composed of an integrator and gain K. The open loop transfer function is

$$G(s) = \frac{K}{s} \qquad G^*(s) = \frac{K}{1 - e^{-sT}} \qquad G(z) = \frac{K}{1 - z^{-1}} \tag{1}$$

Using the w plane transformation we have

$$G(w) = \frac{K(1 + w)}{2} \tag{2}$$

The single adjustable parameter is the gain K. Figure (7.1-13) illustrates the Bode gain magnitude $|G(jv)|$ for this open loop system. Crossover occurs where

$$K\sqrt{1 + v_c^2} = 2v_c \tag{3}$$

and the phase shift at crossover is, for the open loop system of Eq. (1),

$$\beta(v_c) = -\frac{\pi}{2} + \tan^{-1} v_c \tag{4}$$

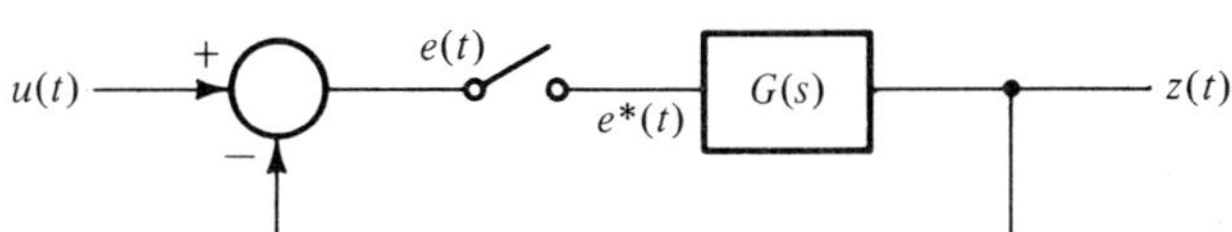

Figure 7.1-12   Block diagram of simple digital control system

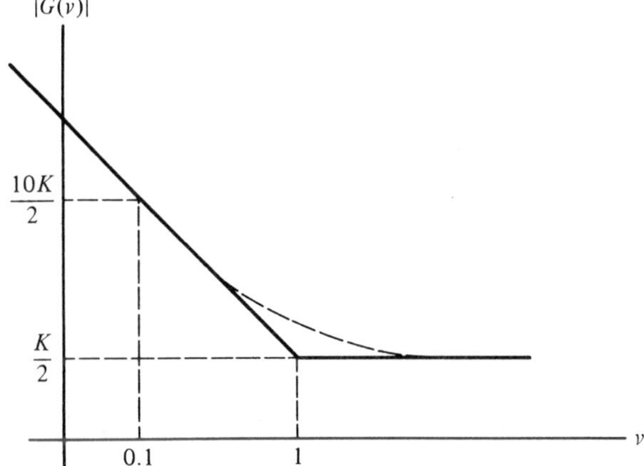

Figure 7.1-13   Gain magnitude $|G(jv)|$

Figure (7.1-14) illustrates the variation of the phase margin and the crossover frequency versus the gain K. From Fig. (7.1-13) we see that the crossover frequency is infinite unless $\dfrac{K}{2} < 1$. Now this is not the actual crossover frequency ω_c but the transformed crossover frequency v_c. From Eq. (7.1-49) we see that when $\omega = \omega_c = \pi/T$ the transformed frequency v_c is indeed infinite.

Let us find the output response $z(t)$ for a unit step input. With the transfer functions of Eq. (1) and the block diagram of (7.1-12) we have

$$\frac{Z(z)}{U(z)} = \frac{G(z)}{1 + G(z)} = \frac{K}{1 + K - z^{-1}} \tag{5}$$

and for a unit step input

$$Z(z) = \frac{1}{1 - z^{-1}} \cdot \frac{K}{1 + K - z^{-1}} = \frac{\dfrac{Kz^2}{1 + K}}{(z - 1)\left(z - \dfrac{1}{1 + K}\right)}$$

We use partial fractions to obtain

$$\frac{Z(z)}{z} = \frac{1}{z - 1} - \frac{1/(1 + K)}{z - 1/(1 + K)}$$

We look up the inverse transforms of

$$Z(z) = \frac{1}{1 - z^{-1}} - \frac{1/(1 + K)}{1 - z^{-1}/(1 + K)}$$

and obtain

$$z(nT) = 1 - \frac{1}{1 + K}\left[\frac{1}{1 + K}\right]^n \tag{6}$$

It might appear that the response improves as we increase K and that is indeed the correct interpretation we obtain from Eq. (6). But Eq. (6) only represents the response at the sample times and we will sometimes be concerned with the response between samples.

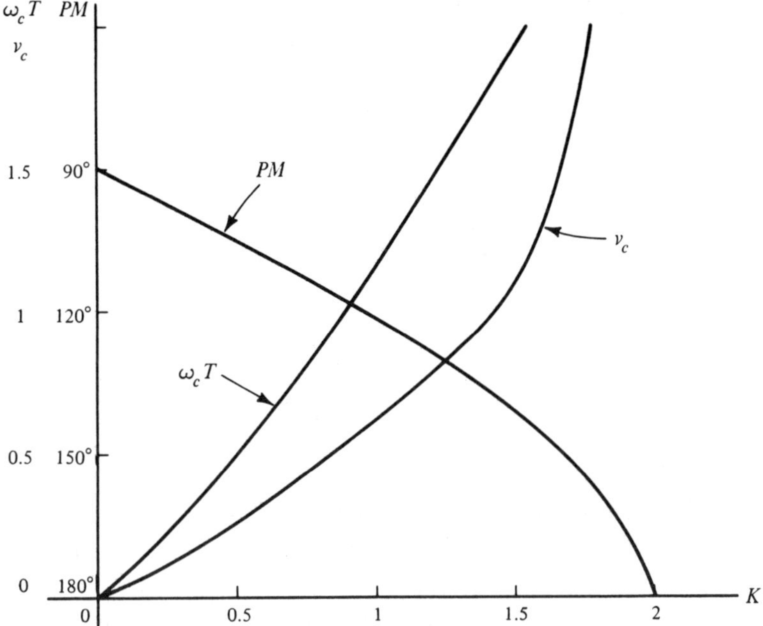

Figure 7.1-14   Crossover frequency and phase margin as a function of loop gain K

The Laplace transform of the output of Fig. (7.1-12) is

$$Z(s) = \frac{G(s)U^*(s)}{1 + G^*(s)}$$

$$= \frac{K}{s(1 + K - e^{-sT})} \tag{7}$$

We may rewrite a portion of Eq. (7) using long division as

$$\frac{K}{1 + K - e^{-sT}} = \sum_{i=0}^{\infty} \frac{K}{(1 + K)^{i+1}} e^{-isT}$$

Thus we have

$$Z(s) = \frac{1}{s} \sum_{i=0}^{\infty} \frac{K}{(1 + K)^{i+1}} e^{-isT}$$

and upon taking the inverse transform we obtain

$$z(t) = \sum_{i=0}^{\infty} \frac{K}{(1 + K)^{i+1}} \, u_{-1}(t - iT) \tag{8}$$

As we see Eqs. (6) and (8) lead to equivalent results at the sampling instants.

We may easily obtain the root locus of this system by using the open loop transfer function of Eq. (1)

$$G(z) = \frac{K}{1 - z^{-1}} = \frac{Kz}{z - 1}$$

and as we see from the root locus diagram of Fig. (7.1-15) the system is always stable and relative stability improves as K increases.

In many ways this is an unrealistic system and this shows up, in part, in our observation that there is no crossover frequency if $K > 2$. The problem here is that we have a system with no pole zero excess and so we have finite system gain at infinite frequency. Even though the integrator does increase the attenuation as the frequency increases there is a sampler in front of the integrator and the resulting $G^*(s)$ does not have a pole zero excess. It would be much more realistic if we used a zero order hold in front of the integrator. If the integrator were a high power output device the jerks in the output response would certainly make it a very expensive device to build.

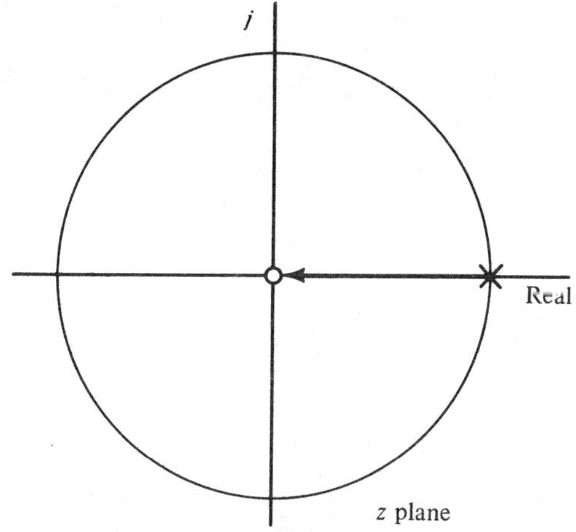

Figure 7.1-15   Root locus for EXAMPLE 7.1-10

EXAMPLE 7.1-11. In this example we consider a zero order sample hold followed by an integrator with gain K such that the open loop transfer function of Fig. (7.1-12) is

$$G(s) = \frac{1 - e^{-sT}}{s} \frac{K}{s}$$

$$G^*(s) = \frac{KTe^{-sT}}{1 - e^{-sT}}$$

$$G(z) = \frac{KT}{z - 1} \tag{1}$$

Using the w plane transformation of Eq. (7.1-48) we obtain

$$G(w) = \frac{KT(1 - w)}{2w} \tag{2}$$

The crossover frequency for the open loop system of Eq. (2) is given by

$$2v_c = KT\sqrt{1 + v_c^2} \tag{3}$$

and the phase shift at crossover is given by

$$\beta(v_c) = -\frac{\pi}{2} - \tan^{-1} v_c$$

Figure (7.1-16) presents curves of crossover frequency and phase margin as a function of K. As we note from Eq. (3) the crossover frequency v_c becomes infinite for $KT = 2$. Again this does not imply that the actual crossover frequency ω_c is infinite but only that the transformed crossover frequency is infinite.

Figure (7.1-17) presents the root locus diagram for this system. The closed loop transfer function is

$$\frac{Z(z)}{U(z)} = \frac{KT}{z + KT - 1} \tag{4}$$

and we see from this transfer function and from the root locus that the system becomes unstable for $-1 > (1 - KT) > 1$ or $0 > KT > 2$. We

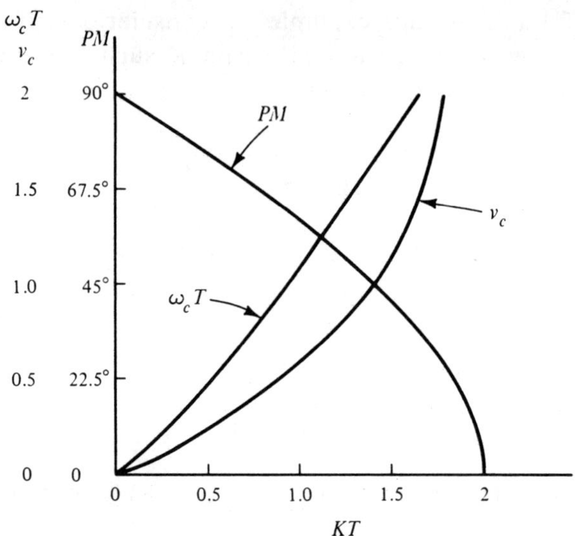

Figure 7.1-16 Crossover frequency and phase margin versus K

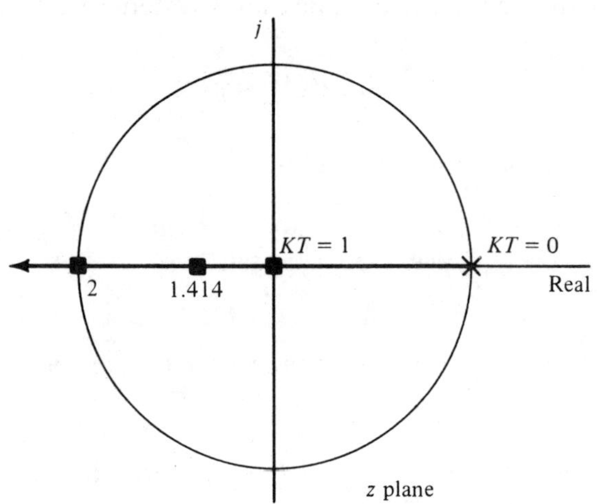

Figure 7.1-17 Root locus of $G(z)$

note that $KT = 2$ yields $v_c = \infty$ and this corresponds to $0°$ phase margin or borderline instability. We cannot consider the case where $KT > 2$ using the w plane transformation results of Eqs. (3) and (4). But this is of little practical consequence as we surely desire a stable system and thus surely wish $KT < 2$.

A "nice" response appears to result if we set $KT = 1$ for then the closed loop transfer function of Eq. (1) is just

$$\frac{Z(z)}{U(z)} = z^{-1} \tag{5}$$

and is just that of a one period delay. This yields a crossover frequency v_c of 0.58 radians per second and an actual crossover frequency given by $\omega_c T = 1.047$. This is a reasonable crossover frequency as $\omega_c = 0.1667\omega_s$ is well below 1/2 the sampling frequency. The phase margin corresponding to $KT = 1$ is $60°$ and thus there is a good degree of relative stability.

The continuous time step response of the system is given by

$$Z(s) = \frac{G(s)}{1 + G^*(s)} U^*(s) = \frac{K(1 - e^{-sT})}{s^2 [1 + (KT - 1)e^{-sT}]}$$

For $KT = 1$ we have

$$Z(s) = \frac{1 - e^{-sT}}{Ts^2}$$

and

$$z(t) = \begin{cases} 0 & t < 0 \\ t/T & 0 < t < T \\ 1 & T < t \end{cases}$$

which is indeed quite a smooth response as indicated in Fig. (7.1-18). This is sometimes called a "dead beat" response, a term used to denote the fact that the transient is over after a minimum number of sample periods and there is zero error thereafter.

If we wish to design the system for $45°$ phase margin we choose $v_c = 1$ and $KT = 1.414$. This yields a pole at $z = 0.414$ and closed loop transfer function

$$\frac{Z(z)}{U(z)} = \frac{1.414}{z + 0.414}$$

which gives the step response transform

$$Z(z) = \frac{1}{1 - z^{-1}} - \frac{1}{1 + 0.414z^{-1}}$$

and the sequence

$$z(nT) = 1 - (-0.414)^n$$

The continuous time response corresponding to this is also shown in Fig. (7.1-18). The response is a bit faster which corresponds to the higher crossover frequency and has more overshoot which corresponds to the lower phase margin.

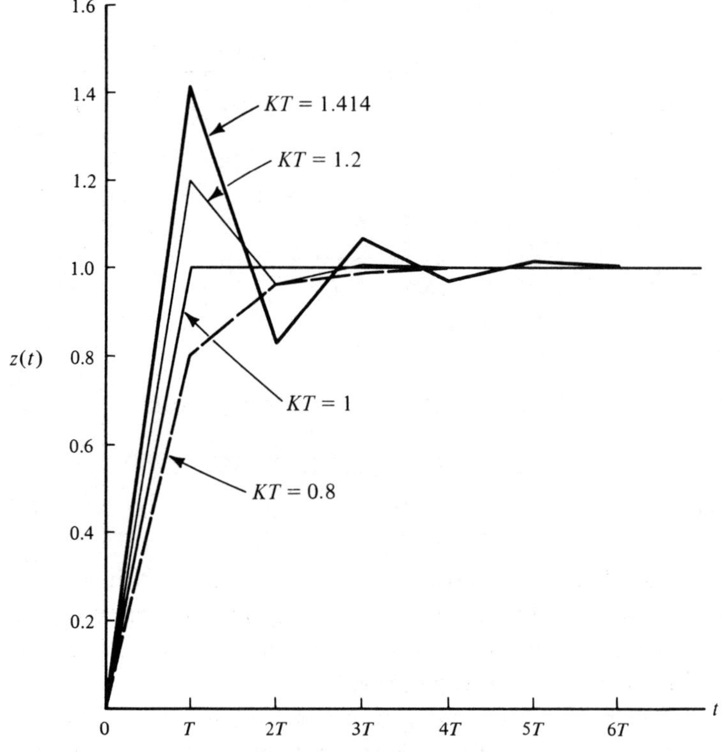

Figure 7.1-18 Step response of the system of EXAMPLE 7.1-11

Since the general response of this system to a step input is

$$z(nT) = 1 - (1 - KT)^n$$

we see that the maximum response occurs at $n = 1$ when $KT > 1$. This maximum response is

$$z(nT) = KT$$

and so we see that the percentage overshoot is $(KT - 1)/100$. To obtain a 20% overshoot we set $KT = 1.2$. This corresponds to a phase margin of $53°$. For this particular system and for the majority of sampled data systems it turns out that a phase margin of $45°$ will usually lead to a step response with somewhat more overshoot than is generally the case for a continuous time system.

EXAMPLE 7.1-12. As an example of system design which requires an equalizer we consider the system shown in block diagram form in Fig. (7.1-19). The fixed plant transfer function, not including the hold circuit, is

$$G_f'(s) = \frac{K}{s^2} \tag{1}$$

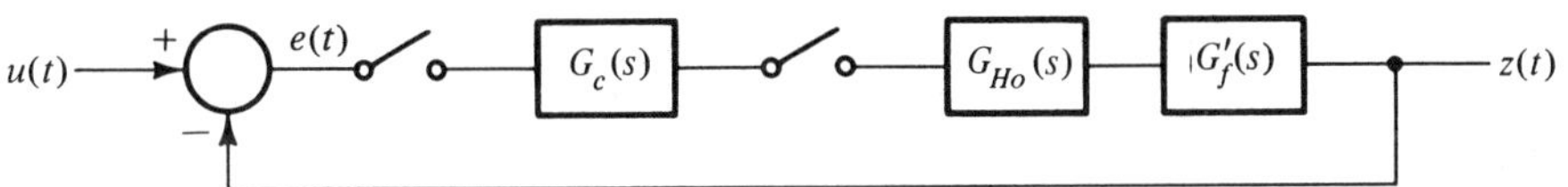

Figure 7.1-19 Sample data system for EXAMPLE 7.1-12

and this is preceded by a zero order sample hold, the hold circuit having the transfer function

$$G_{HO}(s) = \frac{1 - e^{-sT}}{s} \tag{2}$$

For a basic sampled data system such as that shown in Fig. (7.1-12) the error transfer function is

$$\frac{E(z)}{U(z)} = \frac{1}{1 + G(z)} \tag{3}$$

The final value theorem of EXAMPLE 7.1-4 enables us to write

$$\lim_{n \to \infty} e(nT) = \lim_{z \to 1} \frac{(1 - z^{-1})U(z)}{1 + G(z)} \tag{4}$$

and we may use this relation to establish the discrete system error coefficients. In particular when the input is a unit acceleration, we obtain

$$U(t) = \frac{1}{2} t^2$$

$$U(s) = \frac{1}{s^3}$$

$$U^*(s) = \frac{T^2(1 + z^{-1})}{2(1 - z^{-1})^3} \tag{5}$$

In order to obtain finite steady state error we see that we must have $G(z)$ of the form

$$G(z) = \frac{F(z)}{(1 - z^{-1})^2} \tag{6}$$

that is $G(z)$ must contain a double integrator. Combining Eqs. (4) and (6) leads to

$$\lim_{n \to \infty} e(nT) = \lim_{z \to 1} \left[\frac{\dfrac{T^2(1 + z^{-1})}{2(1 - z^{-1})^2}}{1 + G(z)} \right] = \lim_{z \to 1} \left[\frac{T^2}{(1 - z^{-1})^2 G(z)} \right] = \frac{1}{K_a} \tag{7}$$

which verifies our intention stated in Eq. (6). As with the continuous time systems we define the acceleration error coefficient and have

$$K_a = \frac{1}{T^2} \lim_{z \to 1} [(1 - z^{-1})^2 G(z)] \tag{8}$$

as the acceleration error coefficient. For the particular fixed plant and hold circuit used here

$$G_f(s) = \frac{K(1 - e^{-sT})}{s^3}$$

$$G_f(z) = \frac{KT^2 z^{-1}(1 + z^{-1})}{2(1 - z^{-1})^2} \tag{9}$$

and so we see that $F(z) = KT(1 + z^{-1})/2$ and we have the steady state error, from Eq. (7),

$$\frac{1}{K_a} = \lim_{n \to \infty} e(nT) = \lim_{z \to 1} \frac{2}{K(1 + z^{-1})} = \frac{1}{K}$$

Thus we see that the acceleration error coefficient is

$$K_a = K$$

just as it would be in a continuous time system.

Suppose that we specify $K_a = 4.0$, $T = 0.1$, phase margin $= 50°$, and reasonable damping and overshoot. We will use these design specifications and obtain a corresponding network to achieve the desired performance. We will follow the suggested design procedure given by the DELTA Chart of Fig. (7.1-11).

The fixed plant transfer function is given by Eq. (9). In the w plane this becomes, using $K = 4.0$ and $T = 0.1$

$$G_f(w) = \frac{0.01(1 - w)}{w^2} \tag{10}$$

Figure (7.1-20) illustrates the w plane gain magnitude for the uncompensated system. The uncompensated closed loop crossover frequency is $v_c = 0.1$ from the asymptotic gain plot. The phase shift at crossover is $-185.71°$ and so the closed loop uncompensated system is unstable.

We will use a lead network in the w plane to compensate the system. The compensated system Bode diagram is illustrated in Fig. (7.1-20). We must be careful to not insert compensation network break frequencies at too high a value of w. From the relationship $\omega = \frac{2}{T} \tan^{-1} v$ which expresses the relationship between a frequency in the

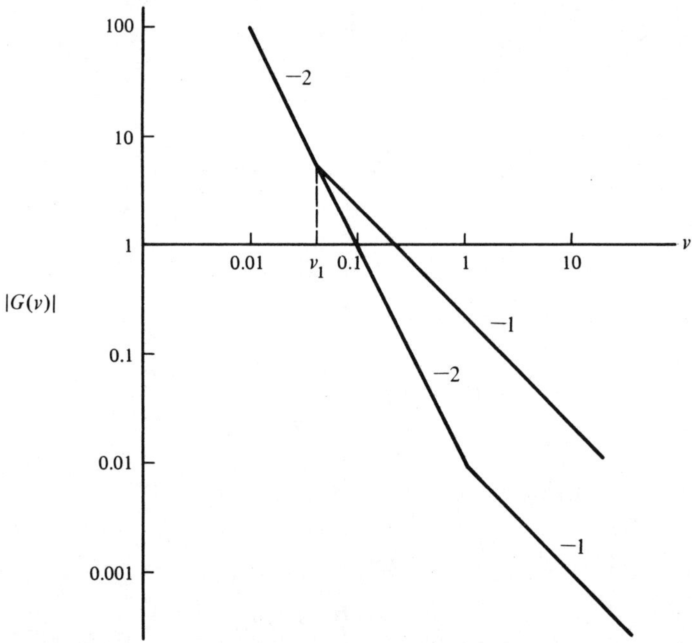

Figure 7.1-20 w plane Bode amplitude diagram

w plane and one in the s plane we see that $v = 1$ corresponds to $0.25\omega_{\mathrm{s}}$ and we should keep break frequencies in the compensating network considerably less than the sampling frequency.

We assume a compensating network of the form

$$G_{\mathrm{c}}(w) = \frac{1 + w/v_1}{1 + w} \tag{11}$$

such that the compensated system transfer function becomes

$$G(w) = \frac{0.1(1 - w)(1 + w/v_1)}{w^2(1 + w)} \tag{12}$$

The crossover frequency occurs at

$$v_{\mathrm{c}} = \frac{0.01}{v_1} \tag{13}$$

and the phase shift at crossover is given by

$$\beta(v_c) = -\pi + \tan^{-1}\frac{v_c}{v_1} - 2\tan^{-1} v_c$$

$$\cong -\frac{\pi}{2} - \frac{v_1}{v_c} - 2v_c \tag{14}$$

Substituting Eq. (13) into Eq. (14) yields for the arctangent phase shift approximation

$$\beta\left(v_c = \frac{0.01}{v_1}\right) = -\frac{\pi}{2} - 100v_1^2 - \frac{0.02}{v_1} \tag{15}$$

Figure (7.1-21) illustrates the variation of the phase margin $\pi + \beta(v_c)$ as a function of v_1. The maximum phase margin is given by

$$\frac{dPM}{dv_1} = 0 = 200v_1 - \frac{0.02}{v_1^2} = 0 \tag{16}$$

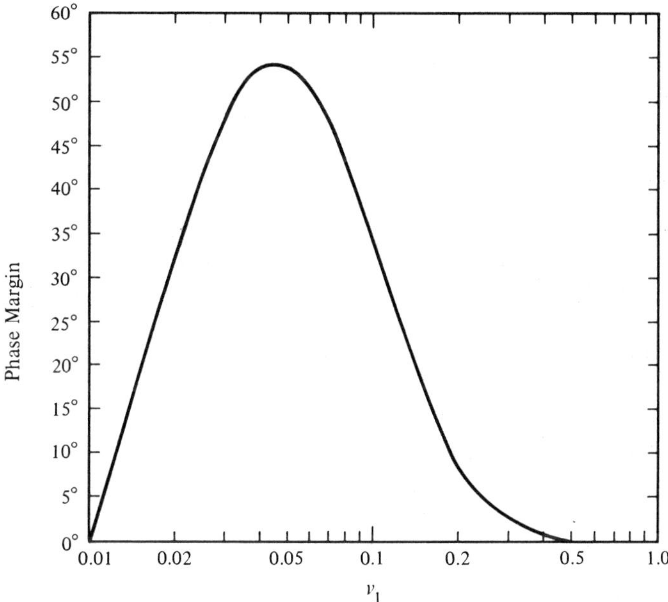

Figure 7.1-21 Phase margin as a function of break frequency v_1

or $v_1 = 0.04642$ and this yields a phase margin of $53.53°$ and a cross-over frequency of $v_c = 0.2154$. This phase margin is so close to our specified value of $50°$ that we accept this design as having met our phase margin specification.

We might have attempted a more complicated design procedure in which we assumed a compensating network of the form

$$G_c(w) = \frac{1 + w/v_1}{1 + w/v_2} \tag{17}$$

We would find v_1 and v_2 not only to yield the desired phase margin but also to result in a minimum gain lead network. The procedures would be just as we have outlined in Chap. 5. There is little to be gained in this specific example since setting v_2 at its largest desirable value of 1 gives a phase margin that just barely meets specifications. We might obtain a slightly larger v_1 and a v_2 a littler smaller than one by this procedure but the values we will obtain will be so close to those obtained for the design procedure followed here that it is not worth our efforts to make this small refinement.

The compensated system has the transfer function, from Eq. (12) and using the determined v_1,

$$G(z) = \frac{0.2254(z - 0.9113)(z + 1)}{(z - 1)^2(z)} \tag{18}$$

Thus the compensating network transfer function is

$$G_c(z) = \frac{G(z)}{G_f(z)} = \frac{11.27(z - 0.913)}{z} = 11.27(1 - 0.913z^{-1}) \tag{19}$$

Also we obtain this directly from $G_c(w)$ by using the transformation relation between w and z. The root locus for the open loop transfer function of Eq. (18) is illustrated in Fig. (7.1-22) which also shows the root locus for the uncompensated system transfer function of Eq. (9). As we see, the location of the closed loop poles on the root locus appears entirely satisfactory. The closed loop transfer function is

$$\frac{Z(z)}{U(z)} = \frac{0.2254(z - 0.9113)(z + 1)}{(z - 0.49)(z - 0.494)(z - 0.791)} \tag{20}$$

The step response of the system is illustrated in Fig. (7.1-23) and as we would expect, shows little overshoot. This step response can be obtained from any of the four procedures we have discussed earlier.

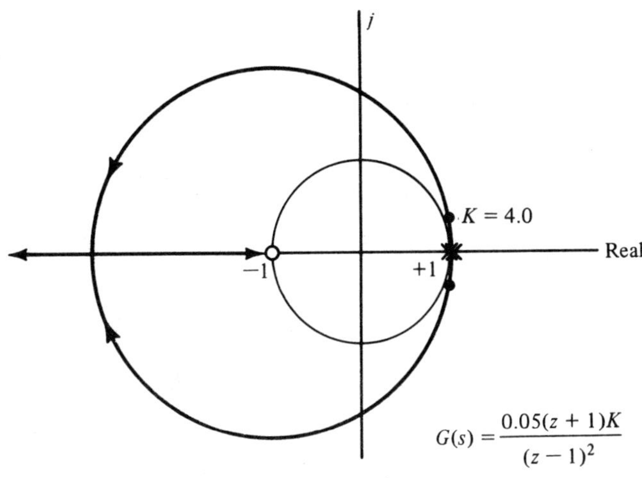

$$G(s) = \frac{0.05(z + 1)K}{(z - 1)^2}$$

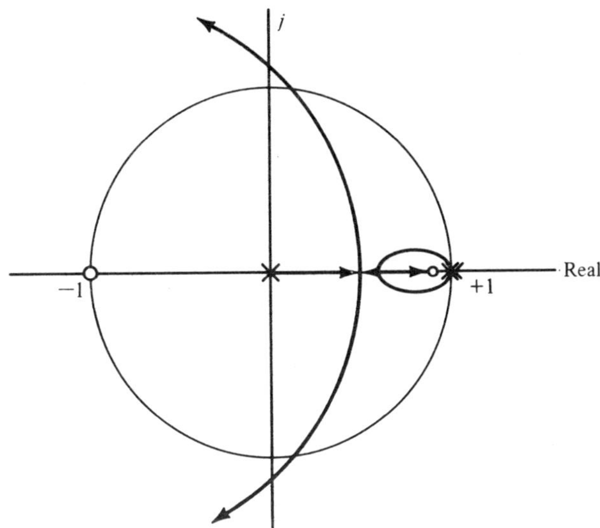

Figure 7.1-22   Root locus diagram for uncompensated and compensated systems

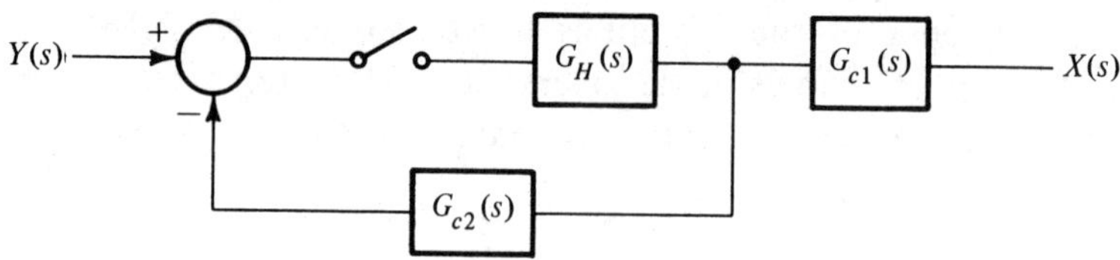

Figure 7.1-23　Realization of discrete time transfer function by means of continuous elements

Using the long division approach we obtain for

$$Z(z) = \frac{0.2254(z)(z - 0.9113)(z + 1)}{(z - 1)(z^3 - 1.7746z^2 + 1.02z - 0.2054)} \tag{21}$$

the sequence

$$Z(z) = 0.2254z^{-1} + 0.6454z^{-2} + 0.9554z^{-3}$$

$$+ 1.1194z^{-4} + 0.9154z^{-5} + \ldots$$

and so we see that the peak overshoot is 12% and this occurs on the 5th sample.

It is of interest to determine methods of realization of the compensating network transfer function

$$G_c(z) = 11.27(1 - 0.913z^{-1}) \tag{22}$$

There are at least three possible methods to accomplish this:

1. If a digital computer is part of the system we can program a simple difference equation

$$x(nT) = 11.27[y(nT) - 0.913y(\overline{n - 1}\,T)]$$

where y is the input sequence to the compensating network and x is the output of the compensating network.

2. If we have delay lines available we can realize the transfer function

$$G_c(s) = \frac{Y(s)}{X(s)} = 11.27(1 - 0.913e^{-sT}) \tag{24}$$

3. We can realize the compensating network by use of a sampler and hold followed by a continuous time dynamic system. For the particular $G_c(z)$ desired here we could attempt an equalizer consisting of series and parallel elements as in Fig. (7.1-23). The transfer function of the equalizer, assuming a zero order hold, is

$$G_c(z) = \frac{X(z)}{Y(z)} = \frac{(1 - z^{-1}) \; Z\left[\dfrac{G_{c1}(s)}{s}\right]}{1 + (1 - z^{-1}) \; Z\left[\dfrac{G_{c2}(s)}{s}\right]}$$

$$= G_{cs}(z)G_{cp}(z) \tag{25}$$

The approach we use is to rewrite the compensatory network transfer function in the form of Eq. (25). $G_{cs}(z)$ is the "series" compensating network

$$G_{cs}(z) = (1 - z^{-1}) \; Z\left[\frac{G_{c1}(s)}{s}\right] \tag{26}$$

and $G_{cp}(z)$ is the "parallel" transfer function

$$G_{cp}(z) = \frac{1}{1 + (1 - z^{-1}) \; Z\left[\dfrac{G_{c2}(s)}{s}\right]} \tag{27}$$

In order to preserve realizability only the poles of $G_c(z)$ which are positive, real, simple, and within the unit circle should be assigned to the series compensator $G_{cs}(z)$ or we will not be able to realize it with stable resistance capacitance elements. Only zeros of $G_c(z)$ which are positive, real, simple, and within the unit circle should be assigned to $G_{cp}(z)$ or it cannot be realized with stable resistance capacitance elements.

For our particular $G_c(z)$ of Eq. (22) we can let $G_{cs}(z) = 11.27$ and $G_{cp}(z) = 1 - 0.913z^{-1}$. We have, from Eq. (27)

$$Z\left[\frac{G_{c2}(s)}{s}\right] = \frac{0.913z^{-1}}{(1 - z^{-1})(1 - 0.913z^{-1})}$$

which can be realized with

$$G_{c2}(s) = 10.492\left(1 - \frac{s}{s + 0.910}\right)$$

This completes this reasonably detailed example of discrete linear systems control design.

This concludes our presentation of basic linear sampled data or digital system design and analysis principles. Our discussion has been brief since we have been able to rely upon our knowledge of linear continuous time systems control which applies with minor modification to the discrete time case.

EXERCISE 7.1-17. Repeat the details of EXAMPLE 7.1-11 for $G_f'(s) = \dfrac{K}{s(1 + s/0.1)}$, $T = 0.1$, $K = 5$. Leave the gain constant unchanged and design a minimum attenuation lag network to achieve a 45° phase margin.

EXERCISE 7.1-18. Investigate the possibility of inserting a sample hold circuit in cascade with a transfer function

$$G_f(s) = \frac{316.2(1 + s/10)}{s^2(1 + s/100)}$$

of a unity ratio feedback system. Determine the phase margin and system root locus as a function of the sample period T. Over what sample period range is the performance reasonable?

Digital Simulation of Continuous Systems. We conclude our presentation in this section with a discussion of the digital simulation of continuous time systems. While this is not a fundamental topic in sampled data control systems, many methods for the digital simulation of continuous time systems are based on sampled data theory. Also it is sometimes desired to approximate continuous time compensating networks on a digital computer. Since this topic is a very important one for practical analysis and since it follows logically from our presentation of sampled data theory we present it here.

In digital computer simulation, simulation is carried out by some form of numerical integration or of replacing a continuous time differential or difference differential equation by a difference equation. In this subsection we will be concerned with development of numerical integration and digital simulation techniques and a discussion of the accuracy of these methods when compared with ideal continuous time integration.

In the classical approach to developing integration rules, a function is approximated by a polynomial over a short time interval t and then the polynomial is integrated rather than the original time function. Newton's formula for representation of a function over the interval from $t_0 \leqslant t \leqslant t_0 + nT$ is the polynomial*

$$P(t) = x_0 + u\Delta x_0 + \frac{u(u-1)}{2!}\Delta^2 x_0 + \frac{u(u-1)(u-2)}{3!}\Delta^2 x_0$$

$$+ \frac{u(u-1)(u-2)(u-3)}{4!}\Delta^2 x_0 + \ldots \tag{7.1-50}$$

Here $x(t)$ is the function being approximated, $P(t)$ the approximating polynomial, T is the sampling rate, and

$$u = \frac{t - t_0}{T}$$

is the normalized time variable.

$$\Delta x_0 = x_1 - x_0$$

is the first backward difference.

$$\Delta^2 x_0 = \Delta x_1 - \Delta x_0 = x_2 - 2x_1 + x_0$$

is the second backward difference and where $x_0, x_1, x_2, \ldots, x_n$ are the values of $x(t)$ at $t_0, t_0 + T, t_0 + 2T, \ldots, t_0 + nT$. The value of the integral

$$y(t) = \int_{t_0}^{t_0 + nT} x(t)\, dt \ldots$$

*See Problem 7.3 for suggestions concerning how to derive Newton's formula.

is then approximately

$$y(t) = \int_{t_0}^{t_0+nT} x(T)\,dt \approx \int_{t_0}^{t_0+nT} P(t)\,dt \ldots \tag{7.1-51}$$

If Eq. (7.1-50) is substituted into Eq. (7.1-51) and integrated term by term the result is

$$\int_{t_0}^{t_0+nT} x(t)\,dt \approx T\left[nx_0 + \frac{n^2\,\Delta x_0}{2} + \left(\frac{n^3}{3} - \frac{n^2}{2}\right)\frac{\Delta^2 x_0}{2!}\right.$$
$$\left. + \left(\frac{n^4}{4} - n^3 + n^2\right)\frac{\Delta^3 x_0}{3!}\right] \tag{7.1-52}$$

Various classical integration rules are obtained from Eq. (7.1-52). The trapezoidal rule results when $n = 1$ and Simpson's rule is obtained when $n = 2$.

As n is increased in Eq. (7.1-52), the approximation to the true value of the integral usually becomes better. The error involved in this approximation is known as a truncation error and will be the only type of error considered in detail here. Another source of error in digital integrators is known as round-off or quantization error which has an approximate root mean square value of $q/\sqrt{12}$ where q is the quantization level.

The differential equations describing system dynamics may be solved by any of the standard numerical integration methods obtainable from Eq. (7.1-52). Among discrete methods for approximating continuous system response, such integration schemes do provide a convenient means of obtaining results for the continuous system, especially in the nonlinear case. One of the best known approaches to integrating differential equations is the Runge-Kutta method, for which there are many modifications. A commonly employed formulation is shown here. We assume a differential equation

$$\frac{d\mathbf{x}}{dt} = \mathbf{f}(t, \mathbf{x}), \qquad \mathbf{x}(t_n) = \mathbf{x}(nT) \tag{7.1-53}$$

where T is the increment in t, the sampling interval, and $\mathbf{x}$ is an n-vector. We compute a set of coefficients α_1, where

$$\alpha_1 = T\mathbf{f}(nT, \mathbf{x}(nT))$$

$$\alpha_2 = T\mathbf{f}\left(nT + \frac{T}{2}, \mathbf{x}(nT) + \frac{\alpha_1}{2}\right)$$

$$\alpha_3 = T\mathbf{f}\left(nT + \frac{T}{2}, \mathbf{x}(nT) + \frac{\alpha_2}{2}\right)$$

$$\alpha_4 = T\mathbf{f}(\overline{n + 1}\ T, \mathbf{x}(nT) + \alpha_3)$$

The new value of **x** is then computed from

$$\mathbf{x}(\overline{n + 1}\ T) = \mathbf{x}(nT) + \frac{1}{6}(\alpha_1 + 2\alpha_2 + 2\alpha_3 + \alpha_4) \qquad (7.1\text{-}54)$$

This formulation is that of a fourth-order Runge-Kutta method, having a truncation error proportional to T^5. Selection of a sufficiently small increment in the independent variable produces a solution of the desired accuracy; however, the very small increment size sometimes required for a suitable result and the calculation of a complete new set of coefficients at every iteration combine to frustrate any simple attempt to utilize such a technique for real-time digital simulation for most applications. This is a very popular method for general purpose digital computer simulation however.

Knowledge of z transform analysis of linear sampled-data system is fundamental to all the approaches to digital simulation. Only features immediately applicable are mentioned in this subsection. The transform variable z is taken as $z = e^{sT}$ and represents a unit time advance operator as used in the analysis methods to be discussed. z transforms of continuous transfer functions are often readily obtained from available transform tables. Application of z transform concepts to the discrete approximation of continuous systems can be made in two different ways. Substitution may be made for integration operators s^{-n} of a transfer function, or the pulse transfer function may be obtained for the complete continuous time transfer function.

Also there are a number of ways to generate approximate transform representations of integrations. We will discuss some of these.

1. *z transforms*   The pulse transfer function resulting from the z transform operation may be implemented on a digital computer as a difference equation relating the input and output variables of the system under study. Consider a system represented by Fig. (7.1-24) where the input $u(t)$ and output $z(t)$ are sampled; hence the pulse transfer function may be represented as $H(z) = Z(z)/U(z)$, or

$$H(z) = \frac{a_0 + a_1 z^{-1} + a_2 z^{-2} + \ldots + a_n z^{-n}}{1 + b_1 z^{-1} + b_2 z^{-2} + \ldots + b_q z^{-q}} = \frac{Z(z)}{U(z)} \qquad (7.1\text{-}55)$$

Since $z^{-n} F(z) = [f(\overline{t-n}\ T)]^*$, where $[\ \ ]^*$ represents the z transform of the function within the brackets and $F(z) = [f(t)]^*$, we are led to a difference equation relating $Z(kT)$ and $U(kT)$ when $H(z)$ is replaced by $Z(z)/U(z)$ and the inverse transform obtained for the resulting expression.

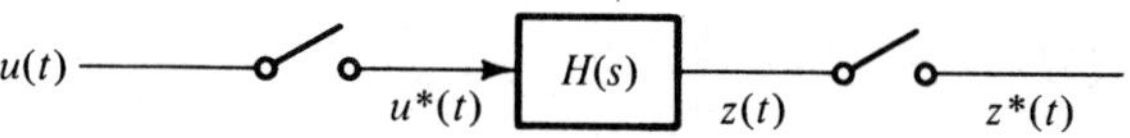

Figure 7.1-24   Basic continuous time system with input and output sampling

We may rewrite this difference equation as a recursion formula for $z(kT)$

$$z(kT) = a_0 u(kT) + a_1 u(\overline{k-1}\ T) + \ldots + a_n u(\overline{k-n}\ T)$$

$$- b_1 z(\overline{k-1}\ T) - \ldots - b_q z(\overline{k-q}\ T) \qquad (7.1\text{-}56)$$

which is our desired relationship for digital computer implementation of the discrete approximation resulting from the z transform method. The exact form of Eq. (7.1-56) employed for computer solution depends upon the available computer memory storage. We may need to recast Eq. (7.1-56) in state space notation.

The z transform expressions for integration operators s^{-n} may be employed in an alternate approach to discretization of a continuous system transfer function. Having expressed a transfer function $H(s)$ as a ratio of polynomials in s^{-1}, substitution is made for the s^{-n} by the corresponding z transforms. Typical z transforms for integration operators are given in TABLE 7.1-2. This approach to discretization of a continuous system transfer function is often termed integrator substitution, or internal substitution, and is an approach commonly taken. Multiplication of the z transform of s^{-n} by T is required to obtain integration operators as can easily be shown.

2. *Tustin Method*   Originally presented as a general approach to linear system analysis through the representation of time functions

in terms of sequences of numbers, the practical application of the Tustin method may be reduced to use of Tustin's definition of the differentiating and integrating operators. A linear transfer function $H(s)$ expressed as a ratio of polynomials in s is readily digitized by substituting for s^n the Tustin operator expressed as

$$s^n = \left[\frac{2}{T} \frac{1 - z^{-1}}{1 + z^{-1}} \right]^n \tag{7.1-57}$$

where T is the sampling interval. We note that this corresponds to repeated usage of the trapezoidal integration rule. The Tustin relationship is easily obtained also from our previously used time delay approximation

$$z^{-1} = e^{-sT} \cong \frac{1 - sT/2}{1 + sT/2}$$

3. *Madwed-Truxal Method* Madwed extended the Tustin time series approach to system analysis and developed higher order integrating operators of increased accuracy in his comprehensive treatment of this technique. The complex notation developed by Madwed for the polygonal approximation of time functions was clarified by Truxal in

TABLE 7.1-2 Examples of Discrete Forms for Integrating Operators

	Operators		
Methods of Approximation	$1/s$	$1/s^2$	$1/s^3$
z transform	$\dfrac{Tz^{-1}}{1 - z^{-1}}$	$\dfrac{T^2 z^{-1}}{(1 - z^{-1})^2}$	$\dfrac{T^3 z^{-1}(1 + z^{-1})}{(1 - z^{-1})^3}$
Tustin	$\dfrac{T(1 + z^{-1})}{2(1 - z^{-1})}$	$\left[\dfrac{T(1 + z^{-1})}{2(1 - z^{-1})} \right]^2$	$\left[\dfrac{T(1 + z^{-1})}{2(1 - z^{-1})} \right]^3$
Madwed-Truxal	$\dfrac{T(1 + z^{-1})}{2(1 - z^{-1})}$	$\dfrac{T^2(1 + 4z^{-1} + z^{-2})}{6(1 - z^{-1})^2}$	$\dfrac{T^3(1 + 11z^{-1} + 11z^{-2} + z^{-3})}{24(1 - z^{-1})^3}$
Boxer-Thaler	$\dfrac{T(1 + z^{-1})}{2(1 - z^{-1})}$	$\dfrac{T^2(1 + 10z^{-1} + z^{-2})}{12(1 - z^{-1})^2}$	$\dfrac{T^3(z^{-1}(1 + z^{-1}))}{2(1 - z^{-1})^3}$

his formulation of the numerical convolution operation for system analysis using z transform notation. The first order integration operator of Madwed is the trapezoidal rule encountered in the Tustin method; however, higher-order Madwed integration operators assume different forms as shown by the examples of TABLE 7.1-2. To digitize a transfer function using this approach, the integrator substitution technique is employed, and the appropriate Madwed integrating operator substituted for s^{-n} to obtain $G(z)$.

4. *Boxer-Thaler Method* The Boxer-Thaler method was presented as a technique for numerical inversion of Laplace transforms. The procedure for application of this approach follows that of the previous methods in that substitutions are made for the complex variable s, but the integrating operators employed are the "z forms" developed by Boxer and Thaler. These special forms were developed in the frequency domain in contrast to the time domain development of Tustin and Madwed. It was noted that polynomial approximation for $s^{-1} = T/\ln z$ could be obtained by expanding $\ln z$ in a rapidly convergent series and then expressing the operator s^{-1} as

$$s^{-1} = \frac{T}{\ln z} = \frac{T}{2(u + u^3/3 + u^5/5 + \ldots)}$$

where

$$u = \frac{1 - z^{-1}}{1 + z^{-1}}$$

From this expression there results by synthetic division

$$s^{-1} = \frac{T}{2}(u^{-1} - u/3 - 4u^3/45 - \ldots)$$

which leads to z forms for s^{-n} when both sides of the above expression are raised to the nth power and the constant term and principal part of the resulting series retained. TABLE 7.1-2 contains z forms for several orders of integrating operators. A linear system transfer function may be discretized by integrator substitution employing the appropriate z forms for the s^{-n}.

EXAMPLE 7.1-13. We consider digital simulation of the single unity ratio feedback system with

$$G_1(s) = \frac{6}{s+6}$$

$$G_2(s) = \frac{25}{6s}$$

$$G(s) = G_1(s)G_2(s) = \frac{25}{s(s+6)}$$

In order to arrange this transfer function into an appropriate form we write it as

$$G(s) = \frac{\dfrac{25}{s^2}}{\dfrac{6}{s}+1} \tag{1}$$

and then use one of the approximations for $1/s$ and $1/s^2$ from TABLE 7.1-2.

The system response to a given input may be found by obtaining the z transform of the input, computing the closed loop transfer function by using the z form equivalent of Eq. (1) and standard inverse z transform computer methods.

5. *Fowler Method* The approach to digital simulation developed by Fowler employs root locus techniques in conjunction with the z transforms for the continuous system under consideration. It applies to both linear and nonlinear systems but basically it is an extension of linear methods to cope with nonlinear situations. For a nonlinear system such as in Fig. (7.1-25), the z transforms of the individual linear transfer functions are first determined. The nonlinearity is replaced by a representative gain, frequently unity, and the poles of the resulting closed loop pulse transfer function

$$\frac{Z(z)}{U(z)} = \frac{G_1(z)G_2(z)}{1 + G_1(z)G_2(z)G_3(z)} \tag{7.1-58}$$

made equal to the poles of the z transform of the closed loop continuous linearized system

$$H(z) = \left[\frac{G_1(s)G_2(s)}{1 + G_1(s)G_2(s)G_3(s)} \right]^* \qquad\qquad (7.1\text{-}59)$$

by adjusting gain parameters in the forward and feedback paths. This may be accomplished by equating the denominator terms of Eqs. (7.1-58) and (7.1-59) and solving for the unknown gain parameters. The use of the notation $F_1(z)$ and $F_2(z)$ in Fig. (7.1-26) reflects the changed character of $G_1(z)$ and $G_2(z)$ when the gain parameters are inserted.

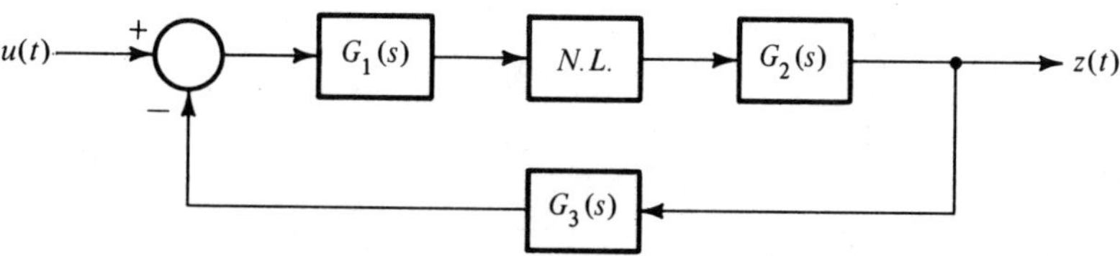

Figure 7.1-25   Nonlinear system with a separable nonlinearity

Additional requirements of steady state gain and system input approximation are met by determining an input pulse transfer function $F_i(z)$ of Fig. (7.1-26) so that the product of this transfer function and the closed loop expression of Eq. (7.1-58) equals the z transform of the product of the desired input approximation or data hold and the linearized closed loop transfer function. Thus we have

$$\frac{F_i(z)F_1(z)F_2(z)}{1 + F_1(z)F_2(z)} = \left[\frac{G_{HO}(s)G_1(s)G_2(s)}{1 + G_1(s)G_2(s)G_3(s)} \right]^* \qquad (7.1\text{-}60)$$

where $G_{HO}(s)$ is the data hold transfer function. With the discrete model completed, it can be implemented by a digital computer program as we indicated earlier for the case of pulse transfer functions.

EXAMPLE 7.1-14.   We again consider the system of EXAMPLE 7.1-13 which is also in the form of Fig. (7.1-25) with

$$G_1(s) = \frac{6}{s + 6}, \qquad G_2(s) = \frac{25}{6s}$$

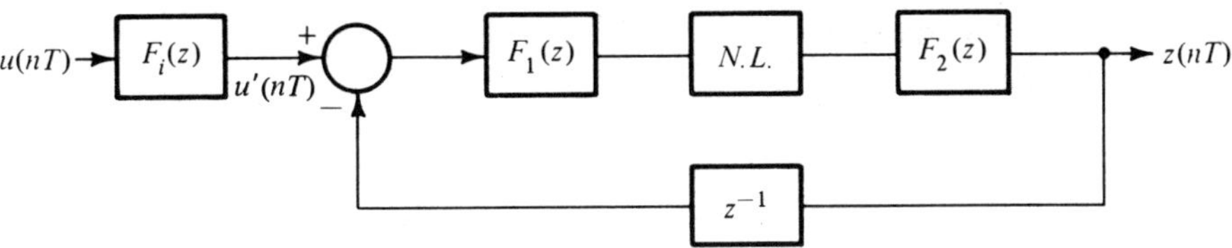

Figure 7.1-26 Discrete time model approximation to the continuous time system

These have z transforms

$$G_1(z) = \frac{6}{1 - e^{-6T}z^{-1}}, \qquad G_2(z) = \frac{25}{6(1 - z^{-1})}$$

The expression for $G_2(z)$ is multiplied by the sampling interval T to obtain an integration operator. The numerator of $G_1(z)$ is replaced by a gain parameter K to be adjusted for the correct closed loop transfer function condition stated above. The expressions for the discrete system operators of Fig. (7.1-26) now appear as

$$F_1(z) = \frac{K}{1 - e^{-6T}z^{-1}}, \qquad F_2(z) = \frac{25T}{6(1 - z^{-1})}$$

where the change in notation reflects the changed character of the transfer functions. By setting $G_3(z) = z^{-1}$ in Eq. (7.1-58), the above $F_1(z)$ and $F_2(z)$ yield the closed loop expression for Fig. (7.1-26),

$$\frac{Z(z)}{U'(z)} = \frac{KT(25/6)}{1 + z^{-1}(25/6KT - 1 - e^{-6T}) + z^{-2}e^{-6T}} \tag{1}$$

Equation (7.1-59), the true z transform of the closed loop system, becomes

$$H(z) = \frac{z^{-1}25e^{-3T}\sin 4T}{1 - z^{-1}2e^{-3T}\cos 4T + z^{-2}e^{-6T}} \tag{2}$$

The gain parameter K of $F_1(z)$ is now sought to make the roots of the denominators of Eq. (1) and Eq. (2) equal. Thus we pick K such that

$$25/6KT - 1 - e^{-6T} = -2e^{-3T}\cos 4T$$

Solution of this equation for the gain K yields

$$K = (6/25T)(1 - e^{-6T} - 2e^{-3T} \cos 4T) \tag{3}$$

Additional requirements of steady state gain and system input approximation are met by determining an input pulse transfer function so that the product of this transfer function, and the closed loop expression and Eq. (7.1-58) equals the z transform of the product of the desired input approximation or data hold and the linearized closed loop transfer function. We expressed this requirement as Eq. (7.1-60). In the present example a zero-order data hold is employed and an approximation is desired to give a half sample period lead. For more compact presentation the result is shown for $T = 0.1$ second, so that

$$\left[e^{-0.5sT} G_{\text{Ho}}(s)H(s) \right]^* = \frac{0.0277 + 0.1375z^{-1} + 0.0189z^{-2}}{1 + 1.364z^{-1} + 0.5488z^{-2}} \tag{4}$$

The desired input transfer function $F_1(z)$ may be obtained by dividing the numerator of Eq. (3) by the numerator of Eq. (1) using K as determined above, with the result

$$F_i(z) = A + Bz^{-1} + Cz^{-2}$$

$$= 0.1504 + 0.7469z^{-1} + 0.1027z^{-2} \tag{5}$$

With the results of Eqs. (3) and (5), information is complete for the discrete approximation which we may express as a set of difference equations for digital computer implementation. The input transfer function may sometimes be simplified or even neglected. Use of the z transform approximation requires introduction of a delay in the feedback loop to obtain a realizable formulation for any single closed loop. It appears that such arbitrary inclusions of delay may in some cases degrade the simulation performance, but they are necessary for computational realizability. Alternately one of the previous integrator substitution forms from TABLE 7.1-2 may be used.

A comparison of the methods for digital simulation which we have presented here offers significant insight. To obtain data for a critical comparison of discrete modeling methods, and to better define approximation technique, a program of experiments was developed for the digital computer. The experiments consist of the digital simulation of an example system by each of the several discrete modeling methods we discussed earlier, and of determining the simulation sensitivity to changing sampling intervals and different inputs.

EXAMPLE 7.1-15. Since all the discrete modeling techniques have been developed primarily for linear transfer function approximation, a fundamental basis for comparison of the different methods should be their ability to model a linear system. The example chosen for this purpose is the second-order linear system shown in Fig. (7.1-27). The differential equation describing the system dynamics is

$$\frac{d^2z}{dt} + 6\frac{dz}{dt} + 25z = 25u(t) \tag{1}$$

which describes a position servomechanism with a damping factor of 0.6 and an undamped natural frequency of 5 radians per second. The difference equations for simulation of the example system on the digital computer will be developed for each method. After we obtain the several representations for the approximating model, results for the complete group of experiments will be indicated.

The system closed loop transfer function $H(s)$ given by

$$H(s) = \frac{25}{s^2 + 6s + 25} \tag{2}$$

is placed in the form

$$H(s) = \frac{25s^{-2}}{1 + 6s^{-1} + 25s^{-2}} \tag{3}$$

in preparation for obtaining the pulse transfer function by means of integrator substitution.

The Tustin approximation for the example transfer function is obtained through integrator substitution employing the trapezoidal rule integrator form

$$s^{-n} = \left[\frac{T}{2} \frac{1 + z^{-1}}{1 - z^{-1}} \right]^n$$

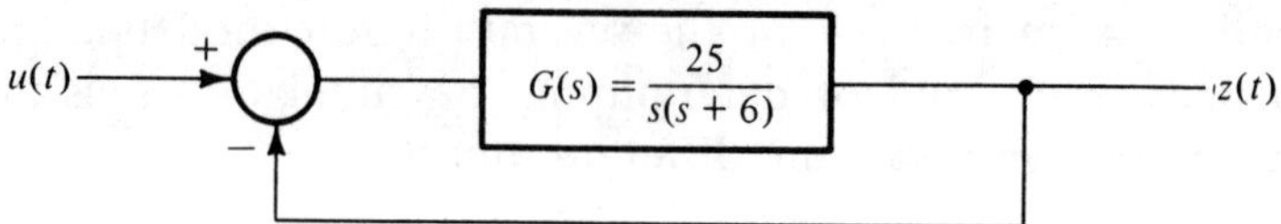

Figure 7.1-27 Continuous time system

in Eq. (3) which yields the pulse transfer function

$$H(z) = \frac{\beta_1 + \beta_2 z^{-1} + \beta_3 z^{-2}}{1 - \alpha_1 z^{-1} - \alpha_2 z^{-2}} \tag{4}$$

where

$$\alpha_1 = (8 - 50T^2)\Delta$$

$$\alpha_2 = (12T - 25T^2 - 4)\Delta$$

$$\beta_1 = 25T^2\Delta$$

$$\beta_2 = 2\beta_1$$

$$\beta_3 = \beta_1$$

$$\Delta = (4 + 12T + 25T^2)^{-1}$$

The desired difference equation can now be determined from Eq. (4) as

$$z(\overline{n+1}\ T) = \alpha_1 z(nT) + \alpha_2 z(\overline{n-1}\ T) + \beta_1 u(\overline{n+1}\ T) + \beta_2 u(nT)$$

$$+ \beta_3 u(\overline{n-1}\ T) \tag{5}$$

Alternately we may simply choose to use long division in Eq. (4) with the z transform of an assumed input.

The Madwed-Truxal form for the pulse transfer function approximating $H(s)$ of Eq. (2) is obtained by integrator substitution in Eq. (3) where we make use of the integrator forms

$$s^{-1} = \frac{T}{2}\frac{1 + z^{-1}}{1 - z^{-1}}$$

and

$$s^{-2} = \frac{T^2}{6} \frac{1 + 4z^{-1} + z^{-2}}{6(1 - z^{-1})^2}$$

The resulting pulse transfer function is given by

$$H(z) = \frac{\beta_1 + \beta_2 z^{-1} + \beta_3 z^{-1}}{1 - \alpha_1 z^{-1} - \alpha_2 z^{-2}} \tag{6}$$

where

$$\alpha_1 = (12 - 100T^2)\Delta$$

$$\alpha_2 = (18T - 25T^2 - 6)\Delta$$

$$\beta_1 = 25T^2\Delta$$

$$\beta_2 = 4\beta_1$$

$$\beta_3 = \beta_1$$

$$\Delta = (6 + 18T + 25T^2)^{-1}$$

The related difference equation then assumes the form of Eq. (5) with the α_i and β_i as defined here.

Integrator substitution utilizing the z forms of Boxer and Thaler in Eq. (3) requires our substitution of

$$s^{-1} = \frac{T}{2} \frac{1 + z^{-1}}{1 - z^{-1}}$$

and

$$s^{-2} = \frac{T^2}{12} \frac{1 + 10z^{-1} + z^{-2}}{(1 - z^{-1})^2}$$

which yields the pulse transfer function

$$H(z) = \frac{\beta_1 + \beta_2 z^{-1} + \beta_3 z^{-2}}{1 - \alpha_1 z^{-1} - \alpha_2 z^{-2}}$$

where we have

$$\alpha_1 = (24 - 250T^2)\Delta$$

$$\alpha_2 = (36T - 25T^2 - 12)\Delta$$

$$\beta_1 = 25T^2\Delta$$

$$\beta_2 = 10\beta_1$$

$$\beta_3 = \beta_1$$

$$\Delta = (12 + 36T + 25T^2)^{-1} \tag{7}$$

The Boxer-Thaler approximation may then be completed by implementing the difference equation of Eq. (5) with our α_i and β_i of Eq. (7).

For a completely linear system, the application of Fowler's method reduces to the derivation of the z transform for the linear system transfer function with some desired input data hold or input approximation. If the input data hold for the present example is taken as a zero-order data hold $G_{\mathrm{Ho}}(s)$ with an adjusting lead of one half sample period to compensate for the lag of the basic hold operation the pulse transfer function sought to model $H(s)$ of Eq. (2) is $[G_{\mathrm{Ho}}(s)e^{0.5sT}H(s)]^*$. By carrying out the indicated z transformation, we obtain the pulse transfer function

$$H(z) = \frac{\beta_1 + \beta_2 z^{-1} + \beta_3 z^{-2}}{1 - \alpha_1 z^{-1} - \alpha_2 z^{-2}}$$

for which the related difference equation is

$$z(\overline{n+1}\ T) = \alpha_1 z(nT) + \alpha_2 z(\overline{n-1}\ T) + \beta_1 u(\overline{n+1}\ T) + \beta_2 u(nT)$$

$$+ \beta_3 u(\overline{n-1}\ T) \tag{8}$$

where

$$\alpha_1 = 2e^{-3T}\cos 4T$$

$$\alpha_2 = -e^{-6T}$$

$$\beta_1 = 1 - e^{-1.5T}(\cos 2T + 0.75 \sin 2T)$$

$$\beta_2 = e^{-1.5T}(\cos 2T + 0.75 \sin 2T) + e^{-4.5T}(\cos 2T - 0.74 \sin 2T) - \alpha_1$$

$$\beta_3 = -\alpha_2 - e^{-4.5T}(\cos 2T - 0.75 \sin 2T)$$

Solution of the difference equations derived above for the several approximation methods was carried out by programming them for a general purpose computer. A comparison of the discrete models developed for the example system was made by determining the response of the model to a unit step function input at a number of sampling intervals. The approximate model response was compared to the solution of the system differential equation obtained by a fourth-order Runge-Kutta integration method for a small sample interval, 0.001 second, which yields an essentially error free result. The standard of comparison for the different methods was chosen as a normalized sum of error squared NSES between the approximate model response and that obtained via the Runge-Kutta integration scheme. The NSES criterion is given by*

$$\text{NSES} = \frac{1}{N-1} \sum_{k=0}^{N-1} ||z_i(kT) - z_d(kT)||^2_{R(kT)} \tag{9}$$

where $z_i(kT)$, an n-vector, is the continuous system state vector at $t = kT$, the n-vector $z_d(kT)$ is the discrete model state vector at the same sample instant, $R(kT)$ is an $n \times n$ positive semi-definite weighting matrix, and N is the number of sample points over the observation interval. The observation interval for the experiments was taken as five seconds. An analytical determination of the sum of error squared in Eq. (9) is possible for linear systems for the case where the weighting matrix R is constant and where $N = \infty$. A bilinear transformation for a mapping from the z plane into the w plane

$$z = \frac{1+w}{1-w}$$

maps the contour of integration Γ, the unit circle, into an integration about the left half of the w plane as we have seen in our previous efforts here. The integration in the w plane can be accomplished by

---

*The symbol $||\ \ ||$ represents the norm such that $||\alpha||^2_\beta = \alpha^T \beta \alpha$

employing the integral tables found in our next section. The NSES criterion is stated in general form here for convenient application to systems other than that of the immediate example. For the present examples the weighting matrix is the $n \times n$ identity matrix and we use the computer to calculate Eq. (9).

Results of the computer experiments for sampling intervals ranging from 0.01 to 0.3 seconds are shown in Fig. (7.1-28).* The classical approximate methods and the Fowler method yield quite the same error

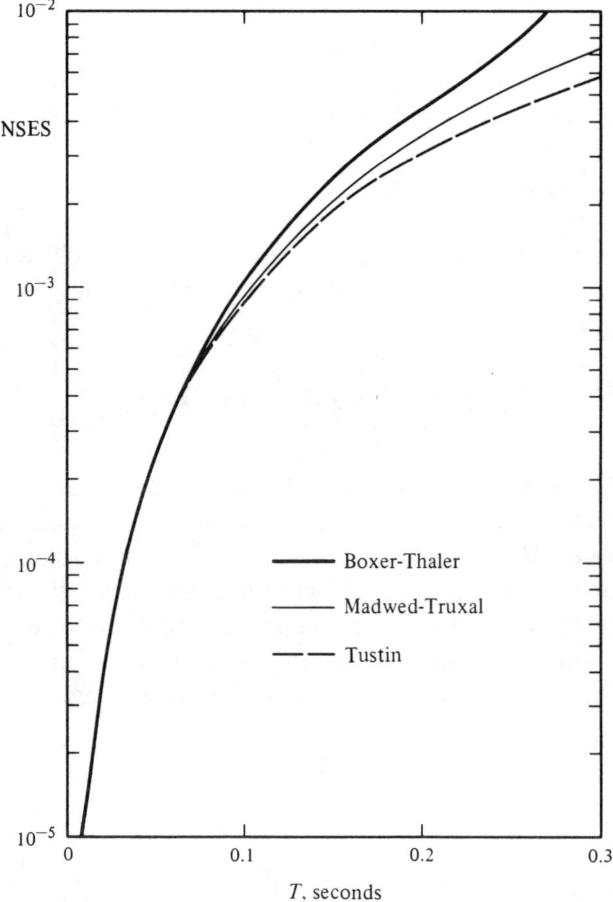

Figure 7.1-28 Normalized sum of error squared as a function of sample size

*For a more complete comparison including a comparison with optimum methods see: Sage, A. P. and Smith, S. L., "Real Time Digital Simulation for Systems Control," *Proceedings of the IEEE*, Vol. 54, No. 12, December 1966, pp. 1802–1812.

in the simulation result for sampling intervals within the range that would normally be considered practical. In order to adequately display the system dynamic behavior for inputs with high frequency content, we would choose the sampling interval not greater than 0.2 of the system rise time, and perhaps as small as 0.1 of the rise time. In this range of sampling interval size, the majority of the discrete models yield essentially the same result for the step input. For sampling periods greater than 0.1 second, the methods become increasingly distinctive in response, though still not greatly different, with exception of the Boxer-Thaler approximation, which demonstrates increasing sensitivity to change in the sampling interval and was found to be unstable for a sampling period of 0.5 second and larger.

Figure (7.1-29) illustrates the step response of the continuous system and the discrete approximations. For the sampling interval of 0.1 second employed for the case shown, the response of the models determined by the Tustin, Madwed-Truxal, Boxer-Thaler, and Fowler methods are nearly indistinguishable on the scale chosen. Each method is started with all initial conditions zero.

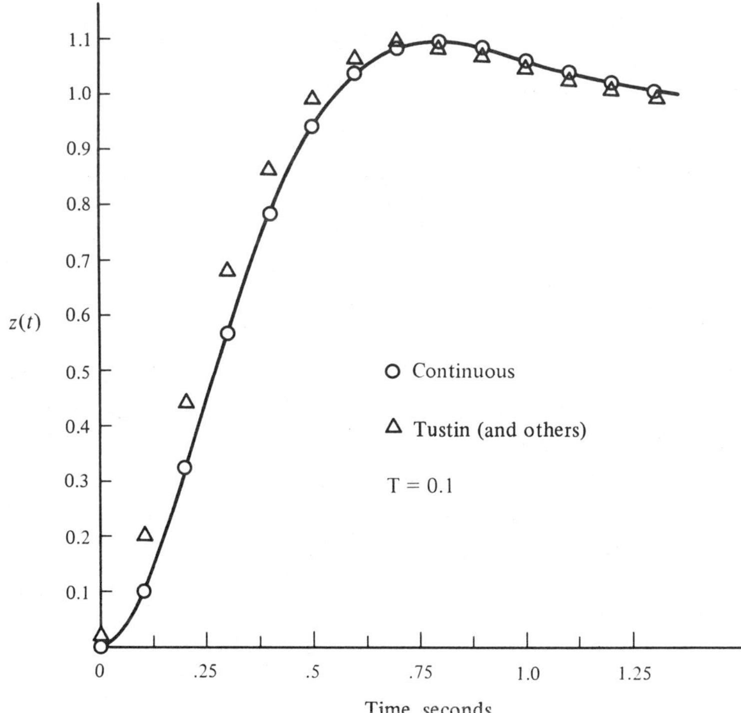

If a conclusion can be drawn from these examples it would be that the Tustin method is essentially as accurate as any simple method available and, in part because of its especially simple form, it has much to recommend it.

EXERCISE 7.1-19. Probably the most demanding function required would be the generation of sine waves. Far less demanding, for example, would be the generation of a decaying exponential. In this exercise you are asked to investigate the use of the z forms of TABLE 7.1-2 to generate sine waves. A suggested approach is to consider the continuous time system of Fig. (7.1-27) and let $G(s) = K/s^2$ where $K = \omega_0^2$. The output from this system is $\sin \omega_0 t$ for appropriate initial conditions. Use the root locus approach to determine the usefulness of the sine waves generator using the z forms.

7.2 OPTIMUM LINEAR SYSTEMS CONTROL

In this section we wish to provide an introduction to the optimum control of linear systems. This has been a very active area for theoretical research involving both control engineers and applied mathematicians for the past fifteen to twenty years. We shall not be able to do much more than dent this very large area and shall present some of the theory of optimum control for linear systems that appears most useful for practical applications while paying little or no attention to the many interesting developments that relate more to pure and applied mathematics than to engineering results of utility for applications. Also we shall not discuss the many uses of optimum systems control for coping with stochastic or random phenomena. Sadly we shall not be able to discuss optimization of nonlinear systems in any depth.

Our efforts will first concern optimum linear systems control in fixed configuration systems. We will then consider constraints posed by fixed system elements such as controllers with limited linear range. Finally we will very briefly mention free configuration optimization. The idea behind all optimization approaches is basically a simple one. There are four basic parts to an optimum systems control problem.

1. A goal or objective function is defined.

2. We determine our current position with respect to the goal.

3. We determine all of the environmental and other factors affecting system transition from the past to the present and the future.

4.   From the goal definition in (1) and a knowledge of the system dynamics and system disturbances in (3), and our present position with respect to the goal in (2) we determine an optimum system input or control policy.

In our work to follow we assume that the current position with respect to the goal is known. Also we assume that there are no stochastic or environmental parameters disturbing the system and assume that the system dynamics are known. Then the problem becomes one of finding control input $\mathbf{u}(t)$ and/or parameters $\mathbf{p}$ to minimize a cost function

$$J = \int_{t_0}^{t_f} \phi[\mathbf{x}(t), \mathbf{u}(t), \mathbf{p}, t] \; dt \qquad (7.2\text{-}1)$$

where ϕ is a scalar function of the control state $\mathbf{x}(t)$, the control vector $\mathbf{u}(t)$, and the system fixed but adjustable parameters $\mathbf{p}$. It is possible, of course, to maximize a cost function as well as minimize one. But maximization of a function is just minimization of the negative of the function. Also we could consider minimization of a vector function but this typically results in a very difficult problem and one much outside the scope of our efforts here. The time interval of optimization t_0 to t_f depends upon the physical problem being resolved as does the precise nature of the function ϕ.

The control system is assumed to be described in state vector form

$$\dot{\mathbf{x}} = \mathbf{f}[\mathbf{x}(t), \mathbf{u}(t), \mathbf{p}, t], \; \mathbf{x}(t_0) = \mathbf{x}_0 \qquad (7.2\text{-}2)$$

in which the structural form of the system is known such that $\mathbf{f}$ is a known function. $\mathbf{u}(t)$ is the control input and $\mathbf{p}$ is a vector of adjustable parameters. Since $\mathbf{p}$ is assumed to be a constant we will often need the vector differential equation

$$\dot{\mathbf{p}} = 0 \qquad (7.2\text{-}3)$$

to describe this fact. $\mathbf{x}(t_0)$ is the system initial condition which is assumed known.

Here we will first consider the case where the time interval of optimization is infinite such that we may set $t_0 = 0$ and $t_f = \infty$. The system input $\mathbf{u}(t)$ will be assumed known. The error function ϕ will be assumed a quadratic function of the state and the system dynamics are linear. Our task is to make an off line determination of the best parameter vector. This is known as the fixed configuration, known input, infinite time interval, parameter optimization problem. We shall resolve it in this section using frequency domain techniques. Then we shall briefly consider the case where the parameter vector is known but the control input is unknown. This is known as the optimum control problem. In our next section we will consider the case where we determine the parameter vector $\mathbf{p}$ in an online or real time fashion as the system operates with normal operating inputs. This is the adaptive or learning system problem.

Optimum Parameter Determination in Fixed Configuration Systems. Perhaps the best way to motivate parameter optimization in linear systems is to consider a simple specific system and then make some observations for this system which are applicable to more general cases and then discuss these general cases.

EXAMPLE 7.2-1. In this example we wish to consider finding the best damping ratio for the standard second order system with closed loop transfer function

$$\frac{Z(s)}{U(s)} = H(s) = \frac{1}{s^2 + 2\zeta s + 1} \tag{1}$$

such that the integral of error squared

$$J = \int_0^\infty e^2(t)\, dt \tag{2}$$

for a step input is a minimum. The error is defined to be the difference between input and output or

$$e(t) = u(t) - z(t) \tag{3}$$

The solution to this example is relatively straightforward. For $U(s) = 1/s$ we have, from Eq. (1)

$$Z(s) = \frac{1}{s(s^2 + 2\zeta s + 1)} \tag{4}$$

The error is, from the Laplace transform of Eq. (3) and Eq. (4),

$$E(s) = U(s) - Z(s) = \frac{s + 2\zeta}{s^2 + 2\zeta s + 1} \tag{5}$$

Use of our table of inverse Laplace transforms in Chap. 2 leads to the time function of error

$$e(t) = \frac{e^{-\zeta t}}{\sqrt{1 - \zeta^2}} \sin\left(\sqrt{1 - \zeta^2}\, t + \theta\right)$$

$$\theta = \tan^{-1} \sqrt{(1 - \zeta^2)/\zeta} \tag{6}$$

Use of a table of definite integrals leads directly to the evaluation of Eq. (2), using the error function of Eq. (6), as

$$J = \frac{1 + 4\zeta^2}{4\zeta} \tag{7}$$

For $\zeta = 0$ and $\zeta = \infty$ the cost function becomes infinitely large. We may plot J as in Fig. (7.2-1). From this figure or by use of $\partial J / \partial \zeta = 0$ we see that the best value of J to give minimum integral square error is $\zeta = 0.5$.

As we know from our efforts in Chaps. 5 and 6 where we examined systems using the Bode diagram and root locus, we know that this is a reasonable value of ζ. As we see from Fig. (7.2-1) the cost function is not especially sensitive to changes of ζ between 0.32 and 0.78. The cost function only increases by 10% for ζ varying over this range. The open loop transfer function of a unity ratio feedback system which yields the closed loop transfer function of Eq. (1) is

$$G(s) = \frac{1}{s(s + 2\zeta)} = \frac{1/2\zeta}{s\left(\dfrac{s}{2\zeta} + 1\right)}$$

The crossover frequency of the system is given by

$$\frac{1}{2\zeta} = \omega_c \sqrt{(\omega_c^2/4\zeta^2) + 1}$$

and the phase margin is

$$PM = \frac{\pi}{2} - \tan^{-1} \frac{\omega_c}{2\zeta}$$

Figure (7.2-1) also illustrates the variation of phase margin versus damping ratio ζ. Values of ζ between 0.32 and 0.78 yield phase margins between $38°$ and $77°$ and phase margins in this range are generally considered acceptable.

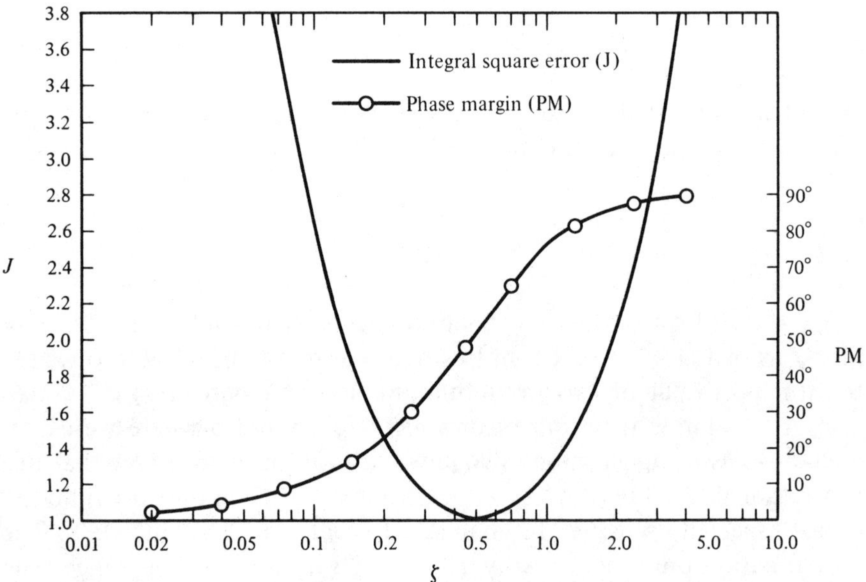

Figure 7.2-1 Integral square error and phase margin versus damping ratio for a second order system

Figure (7.2-1) shows that while phase margins is a monotone increasing function of damping the integral square error reaches a minimum at a reasonable value of ζ. Thus we see that minimization of integral square error appears to be a meaningful criterion whereas extremization of phase margin certainly does not. This example serves to counter any belief we might have that increasing the phase margin of a system will yield better performance. It does not! Increasing the phase margin will certainly increase the damping and relative stability of a system but it may well reduce system bandwidth to the point that performance is impaired as it does in this example. Thus maximization of phase margin cannot be used as a meaningful error criterion whereas minimization of integral square error appears to be a reasonable single figure of merit scalar objective function, at least for this example.

Integral Square Error Design. Let us now attempt to restate the results of this example in more general form. Figure (7.2-2) represents a block diagram appropriate for our discussion. We wish to minimize the square of the error by choosing unspecified parameters in the system transfer function. We write the cost function as

$$J = \int_0^\infty e^2(t)\, dt \tag{7.2-4}$$

as we have seen from EXAMPLE 7.2-1 this integral will often be tedious to evaluate. It turns out that the frequency domain equivalent of this is rather easily tabulated for rational $E(s)$. Parceval's theorem states* that

$$\frac{1}{2\pi j}\int_{-\infty}^{\infty} E(s)E(-s)\, ds = \int_0^\infty e^2(t)\, dt \tag{7.2-5}$$

The advantage to this frequency domain integral is that it is relatively easy to evaluate it using the residue theorem for $E(s)$ of the form

$$E(s) = \frac{c_0 + c_1 s + c_2 s^2 + \ldots}{d_0 + d_1 s + d_2 s^2 + \ldots}$$

$$= \frac{\displaystyle\sum_{i=0}^{n-1} c_i s^i}{\displaystyle\sum_{i=0}^{n} d_i s^i}$$

---

*This theorem is easily proved. If we use the inverse transform definition of $e(t)$ we have

$$J = \int_0^\infty e^2(t)\, dt = \int_0^t e(t)\left[\frac{1}{2\pi j}\int_{-j\infty}^{j\infty} E(s)e^{st}\, ds\right] dt$$

We interchange the order of integration and have

$$J = \frac{1}{2\pi j}\int_{-j\infty}^{j\infty} E(s)\left[\int_0^t e(t)e^{st}\, dt\right] ds$$

Now we note that the expressions within brackets is just the definition of $E(-s)$ so we have

$$J = \int_0^\infty e^2(t)\, dt = \frac{1}{2\pi j}\int_{-j\infty}^{j\infty} E(s)E(-s)\, ds$$

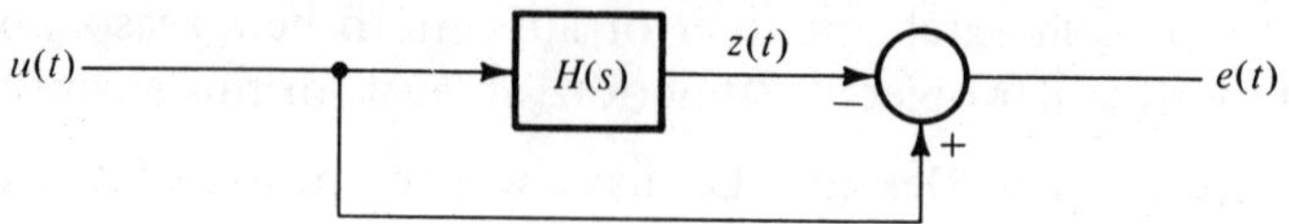

Figure 7.2-2   Appropriate block diagram for generation of error signal

We have, assuming that the poles of $E(s)$ are in the left half plane, the results of TABLE 7.2-1. This table may be extended to values of n greater than 4 but the expressions for J_n become extraordinarily complex.*

Using the result of Eq. (7.2-5) we can write for the integral square error ISE

$$J = \int_0^\infty e^2(t)\, dt = \frac{1}{2\pi j} \int_{-j\infty}^{j\infty} E(s)E(-s)\, ds \tag{7.2-6}$$

where

$$E(s) = [1 - H(s)]\, U(s) \tag{7.2-7}$$

$U(s)$ is the known input. $H(s)$ is the structurally known transfer function which contains unspecified parameters which we determine by minimizing the cost function of Eq. (7.2-6).

EXAMPLE 7.2-2.   We consider minimization of the ISE for the nth order system

$$H(s) = \frac{1}{s^n + a_{n-1}s^{n-1} + \ldots + a_1 s + 1} \tag{1}$$

with a unit step input. Since we may frequency scale by the transformation $s = p/\omega_0$ we see that this can represent a relatively general nth order system. The error transform is

$$E_n(s) = \frac{s^{n-1} + a_{n-1}s^{n-2} + \ldots + a_2 s + a_1}{s^n + a_{n-1}s^{n-1} + \ldots + a_1 s + 1} \tag{2}$$

*For a very complete table for $n \leqslant 10$ see pp. 372–381 of Newton, G. C., Jr., Gould, L. A. and Kaiser, J. F., *Analytical Design of Linear Feedback Controls*, John Wiley & Sons, New York 1957.

TABLE 7.2-1　**Integral Table for** $J_n = \int_0^t e_n^2(t)\, dt,\ E_n(s) = \dfrac{\sum_{i=0}^{n-1} c_i s^i}{\sum_{i=0}^{n} d_i s^i}$

n	J_n
1	$\dfrac{c_0^2}{2 d_0 d_1}$
2	$\dfrac{c_0^2 d_2 + c_1^2 d_0}{2 d_0 d_1 d_2}$
3	$\dfrac{c_0^2 d_2 d_3 + (c_1^2 - 2 c_0 c_2) d_0 d_3 + c_2^2 d_0 d_1}{2 d_0 d_3 (d_1 d_2 - d_0 d_3)}$
4	$\dfrac{c_0^2 (d_2 d_3 d_4 - d_1 d_4^2) + (c_1^2 - 2 c_0 c_2) d_0 d_3 d_4 + (c_2^2 - 2 c_1 c_3) d_0 d_1 d_4 + c_3^2 (d_0 d_1 d_2 - d_0^2 d_3)}{2 d_0 d_4 (d_1 d_2 d_3 - d_0 d_3^2 - d_1^2 d_4)}$

for various values of n this becomes

$$E_1(s) = \frac{1}{s + 1} \tag{3}$$

$$E_2(s) = \frac{s + a_1}{s^2 + a_1 s + 1} \tag{4}$$

$$E_3(s) = \frac{s^2 + a_2 s + a_1}{s^3 + a_2 s^2 + a_1 s + 1} \tag{5}$$

$$E_4(s) = \frac{s^3 + a_3 s^2 + a_2 s + a_1}{s^4 + a_3 s^3 + a_2 s^2 + a_1 s + 1} \tag{6}$$

We can now use the results of TABLE 7.2-1 to evaluate the ISE. For $n = 1$ we have

$$J_1 = \frac{c_0^2}{2 d_0 d_1}$$

where

$$c_0 = d_0 = d_1 = 1$$

and we obtain $J = 1$. For a first order system the frequency normalization is such that there is no free parameter to optimize.

For $n = 2$ we have

$$J_2 = \frac{c_0^2 d_2 + c_1^2 d_0}{2 d_0 d_1 d_2} = \frac{a_1^2 + 1}{2a_1}$$

and by setting $dJ_1/da_1 = 0$ we obtain $a_1 = 1$. This result corresponds precisely to that obtained in the previous example.

For $n = 3$ we have

$$J_3 = \frac{a_1^2 a_2 + a_2^2 - a_1}{2(a_1 a_2 - 1)}$$

We have two parameters to determine. We set $\partial J_3/\partial a_1 = 0$ and obtain

$$a_1^2 a_2^2 - 2a_1 a_2 + 1 - a_2^3 = 0 \tag{7}$$

We also set $\partial J_3/\partial a_2 = 0$ and obtain

$$a_1 a_2 = 2 \tag{8}$$

We now proceed to solve Eqs. (7) and (8) for the parameters a_1 and a_2. We see one disadvantage of this analytical optimization approach. Even though the system is linear and the cost function quadratic we must generally solve nonlinear algebraic equations to determine the optimum parameter settings. There are, of course, a number of numerical analysis algorithms, such as the gradient method, which allow solution of equations such as these on a digital computer.

We may illustrate a solution here as follows. Eq. (8) may be inserted into Eq. (7) and we obtain $a_2 = 1$ as the optimum value of a_2. Equation (8) then yields the optimum value of a_1 as 2.

The optimum closed loop system transfer function is

$$H(s) = \frac{1}{s^3 + s^2 + 2s + 1}$$

The closed loop roots may be obtained by factoring this expression to yield

$$H(s) = \frac{1}{(s + \sigma)(s^2 + 2\zeta\omega_0 s + \omega_0^2)}$$

where

$$\sigma = 0.57$$

$$\omega_0 = 1.32$$

$$\zeta = 0.16$$

The effects of the low damping ratio are not as pronounced as might be expected because of the real pole located at $-\sigma$. The primary effect is not a large overshoot but rather a reasonably long settling time as indicated in Fig. (7.2-3) which illustrates the step response for the system of this example.

The open loop system which yields this closed loop transfer function is

$$G(s) = \frac{1}{s(s^2 + s + 2.0)}$$

$$= \frac{0.50}{s\left(1 + \dfrac{s}{0.50 + j1.32}\right)\left(1 + \dfrac{s}{0.50 - j1.32}\right)}$$

and we see that the open loop system has complex conjugate poles. The crossover frequency is $\omega_c = 0.56$ and the phase margin is $40°$.

A sometimes indicated design procedure consists of determining a closed loop transfer function which yields good or perhaps optimum response and then modifying the given fixed plant such as to obtain the equivalent open loop transfer function. Thus we would generally need an equalizer with complex conjugate poles if this is to be accomplished

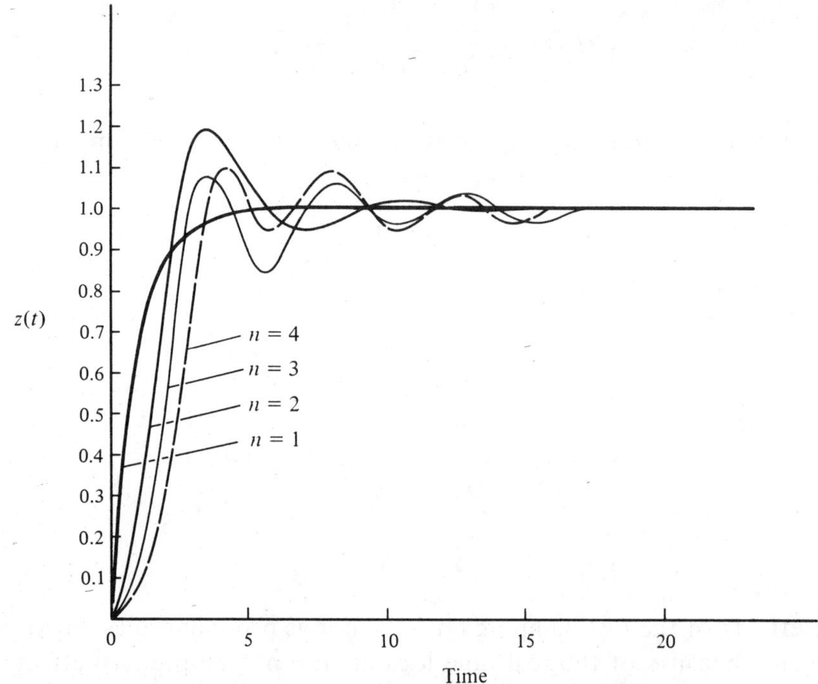

Figure 7.2-3 Step response of normalized system

using cascade compensation. Such equalizers are difficult and expensive to build and this limits the ultimate usefulness of a design approach which does not consider fixed plant and compensator complexity constraints at the beginning of the design procedure. Of course if a digital computer is to be used to realize the compensation then the situation is somewhat different and there is no special problem in realization of the complex conjugate poles required of the compensating network. Also minor loop design techniques may be used. For example, if the fixed plant is

$$G_f(s) = \frac{1}{s(2 + s)^2}$$

then we might use a minor loop feedback compensator as indicated in Fig. (7.2-4).

EXERCISE 7.2-1. Determine the optimum parameters in EXAMPLE 7.2-2 for a fourth order system. What is the phase margin of this system? How might this system be realized if the fixed plant is

$$G_f(s) = \frac{1}{s(1+s)^3}$$

EXAMPLE 7.2-3. We can develop an approach to system design using the integral square error criterion by postulating an equalizer for a specified fixed plant and then determining the equalizer parameters to minimize the integral square error for an assumed input, as this example shows. We have a fixed plant

$$G_f(s) = \frac{300}{s^2}$$

and we postulate a 9 to 1 break frequency ratio lead network

$$G_c(s) = \frac{1 + s/\omega_1}{1 + s/9\omega_1}$$

From our past experience using the Bode diagram we know that the maximum phase margin of $53°$ occurs when $\omega_1 = 10$ and this yields a crossover frequency of $\omega_c = 30$ radians per second. The closed loop transfer function is

$$H(s) = \frac{(1 + s/10)}{(1 + s/30)^3}$$

and the step function response overshoots once as we have previously determined in EXAMPLE 6.3-1.

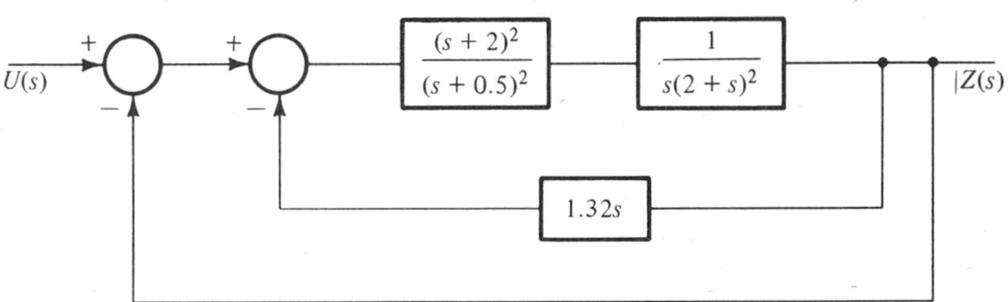

Figure 7.2-4 Minor loop realization for $(n = 3)$

We will now obtain the value of ω_1 to minimize the integral square error for a unit step input. The Laplace transform of the error is easily determined to be

$$E(s) = \frac{s^2 + 9\omega_1 s}{s^3 + 9\omega_1 s^2 + 2700s + 2700\omega_1}$$

From TABLE 7.2-1 we obtain the integral square error for $n = 3$ as

$$J = \frac{2700 + 81\omega_1^2}{43,200\omega_1}$$

The optimum value of ω_1 is obtained from $dJ/d\omega_1 = 0$ as $\omega_1 = 10/\sqrt{3}$ $= 5.77$. This changes the crossover frequency to 41 radians per second and the phase margin to 53.7°. This is less than the maximum value obtainable but the crossover frequency has increased considerably and so the system response is faster than it will be under maximum phase margin conditions. However the overshoot is greater. The integral square error criterion asks for a system response that is fast and so it is not surprising that our system becomes faster but with less damping.

We may also obtain the best ω_1 to minimize the ISE for a ramp input. We obtain, for $U(s) = 1/s^2$,

$$E(s) = \frac{s + 9\omega_1}{s^3 + 9\omega_1 s^2 + 2700s + 2700\omega_1}$$

and the integral square error is

$$J = \mathrm{ISE} = \frac{100 + 27\omega_1^2}{4,320,000\omega_1}$$

The best value of ω_1 is now $\omega_1 = 1.92$. This yields a crossover frequency of $\omega_c = 50$ and a very small phase margin of 17°. However a ramp is a very "gentle" input and the response of this system to a ramp input is quite good as the reader can easily show. Again the ISE criterion is one that asks for fast response to reduce the error to a small quantity quickly and the result we have obtained does achieve this.

EXERCISE 7.2-2. Find the break frequency ω_1 in a 20 to 1 lag network to minimize the integral square error for a step function input for the system with open loop transfer function

$$G(s) = \frac{(1 + s/20\omega_1)}{(1 + s/\omega_1)} \cdot \frac{10}{s(1 + s)}$$

In most systems we cannot exceed a limited linear range constraint at particular physical points within the system or the system becomes nonlinear. Often this will be associated with a deterioration in system performance. We may attempt to control the peak input into a physical point $x_s(t)$ by limiting the integral of the derivative of the signal at this point, i.e.,

$$J = \int_0^\infty \left[\frac{dx_s(t)}{dt}\right]^2 dt$$

Often, for example, we wish to limit the output velocity of a servo-motor. If the output shaft position is the system output $z(t)$ then we will tend to limit the peaks velocity if we minimize the integral square acceleration

$$J_s = \int_0^\infty \left[\frac{d^2 z(t)}{dt^2}\right]^2 dt = \int_0^\infty y^2(t)\, dt$$

We may use Parceval's theorem to write the foregoing expression as

$$J_s = \frac{1}{2\pi j} \int_{-j\infty}^{j\infty} Y(s)Y(-s)\, ds$$

where

$$Y(s) = s^2 Z(s)$$

If we had a way to incorporate the constraint of a limited integral square response at some point in the system we could proceed with our system optimization. Fortunately, the Lagrange multiplier method provides a vehicle to accomplish this. We wish to minimize

$$J = \int_0^\infty e^2(t)\, dt = \frac{1}{2\pi j} \int_{-j\infty}^{j\infty} E(s)E(-s)\, ds \tag{7.2-8}$$

subject to a constraint that

$$J_s = \int_0^\infty y^2(t)\,dt = \frac{1}{2\pi j}\int_{-j\infty}^{j\infty} Y(s)Y(-s)\,ds \qquad (7.2\text{-}9)$$

be less than or equal to the specified value J_s. Generally the equality will result for physical systems. In this case we have that a special form of zero is

$$0 = \frac{1}{2\pi j}\int_{-j\infty}^{j\infty} Y(s)Y(-s)\,ds - J_s \qquad (7.2\text{-}10)$$

We add this special form of zero multiplied by a Lagrange multiplier λ to the cost function of Eq. (7.3-8) and obtain

$$J = \frac{1}{2\pi j}\int_{-j\infty}^{j\infty} [E(s)E(-s) + \lambda Y(s)Y(-s)]\,ds - \lambda J_s \qquad (7.2\text{-}11)$$

Now when the equality constraint of Eq. (7.2-10) is satisfied Eq. (7.2-11) will be precisely the same as Eq. (7.2-8). We can minimize the cost function J of Eq. (7.2-11) and will adjust the λ such that Eq. (7.2-10) is satisfied. Then we will have the best solution to the problem posed which satisfies the given constraint. Figure (7.2-5) illustrates a suggested design approach to analytical parameter optimization. An example will now be given to illustrate the procedure.

EXAMPLE 7.2-4. We first consider optimization of the gain and time constant in the fixed plant

$$G_f(s) = \frac{K}{s(Ts + 1)} \qquad (1)$$

to minimize the ISE for a unit step input. The system is a closed loop unity feedback ratio system with the open loop transfer function of Eq. (1). The Laplace transform of the error is easily shown to be

$$E(s) = \frac{Ts + 1}{Ts^2 + s + K} \qquad (2)$$

and the integral error is

$$J = \frac{c_0^2 d_2 + c_1^2 d_0}{2 d_0 d_1 d_2} = \frac{1 + KT}{2K} = \frac{1}{2K} + \frac{T}{2} \tag{3}$$

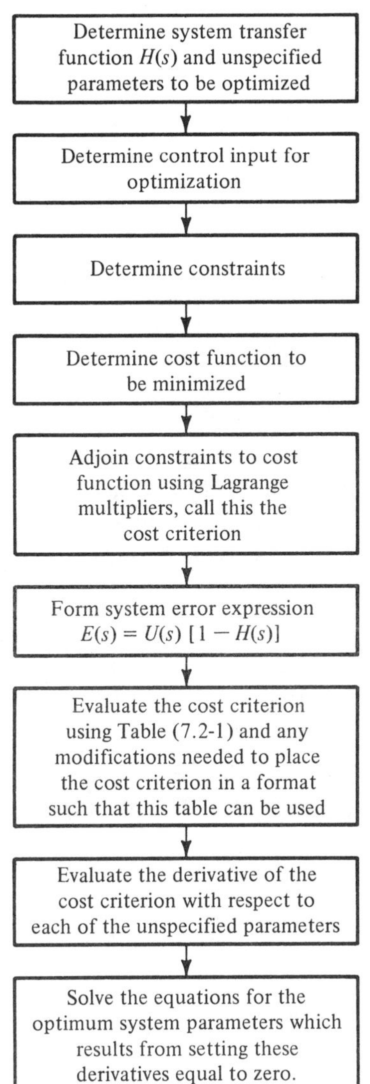

Figure 7.2-5 DELTA Chart of design approach for analytical parameter optimization

As we might expect the best value of T is $T = 0$. Also the cost function decreases with increasing K and the best response is obtained with $K = \infty$. If we adjust both K and T we obtain very meaningful results here.

We might suspect that things would work out differently if we adjusted the system break frequency for optimum behavior rather than the system time constant. That is indeed the case, for if we use the fixed plant transfer function

$$G_f(s) = \frac{K'}{s(s + \omega_1)} \tag{4}$$

then we have for the ISE

$$J = \frac{K' + \omega_1^2}{2K'\omega_1} \tag{5}$$

Now the best ω_1 is given from $\partial J/\partial\omega_1 = 0$ as $\omega_1 = (K')^{1/2}$. This value of ω_1 results in a second order system with a damping ratio of 0.5. This result turns out to be different than that obtained in the first part of this example because the locus of closed loop poles for changing K and T is different from that obtained by changing K' and ω_1. The result from adjusting K' and ω_1 is a very meaningful one. We set the damping ratio to $\zeta = 0.5$ and then make the natural resonant frequency K' just as large as possible.

Suppose that we are free *only* to adjust the gain K or the gain K'. Unfortunately neither result is meaningful as we still get lowest ISE for infinite K or K'. The damping ratio under these conditions approaches zero and the phase margin approaches zero as these gains become infinite. The closed loop transfer function for the system of Eq. (1) becomes

$$H(s) = \frac{K}{Ts^2 + s + K} \tag{6}$$

and for large K the step response is

$$z(t) = 1 - \alpha e^{-t/2T} \sin(\omega_d t + \phi)$$

where

$$\alpha = (1 + K^2)^{1/2}/K$$

Even though $\omega_d = (1 - 4K)^2/2T$ becomes very large as K becomes large the settling time of the system does not change as a function of K. Thus the envelope of the response is invariant with respect to a change in K and a large K does keep the sine function near zero more often than a small K does. Thus our results are intuitively correct and, even though they are physically unreasonable, they follow directly from the assumed error criterion. A similar result is obtained from the fixed plant of Eq. (4) in that

$$H(s) = \frac{K'}{s^2 + s\omega_1 + K'} \tag{7}$$

and the unit step response is

$$z(t) = 1 - \alpha e^{-\omega_1 t/2} \sin(\omega_d + \phi)$$

where $\alpha = (\omega_1^2 + K'^2)^{1/2}/K'$.

A real difficulty here is that we obtain very large shaft velocity and acceleration for large K. If we calculate the integral square output acceleration

$$J_s = \int_0^\infty \left(\frac{d^2 z(t)}{dt^2}\right)^2 dt = \frac{1}{2\pi j} \int_{-j\infty}^{j\infty} Y(s)Y(-s)\, ds \tag{8}$$

where

$$Y(s) = s^2 H(s)U(s) = \frac{Ks}{Ts^2 + s + K} \tag{9}$$

we obtain

$$J_s = \frac{K^2}{2T}$$

and we see that this becomes infinite for infinite K which is a very unreasonable result. For the system of Eq. (4) we obtain the same result in that $J_s = (K')^2/2\omega_1$.

Introducing the constraint of a maximum integral square acceleration converts this into a meaningful problem. For the system of Eq. (1) we have, using the Lagrange multiplier method

$$J = \frac{1 + KT}{2K} + \lambda\left(\frac{K^2}{2T} - J_s\right) \tag{10}$$

To find the best K we set $\partial J/\partial K = 0$ and obtain

$$\frac{2KT - 2(1 + KT)}{4K^2} + \frac{\lambda K}{T} = 0$$

such that K is given by

$$K = (T/2\lambda)^{1/3} \tag{11}$$

The integral square acceleration is given by Eq. (9) as

$$J_s = \frac{K^2}{2T} = \frac{1}{2^{1/3}\lambda^{2/3}T^{1/3}} \tag{12}$$

we obtain the Lagrange multiplier in terms of the constraint J_s as

$$\lambda = \frac{1}{2^{1/2}T^{1/2}J_s^{3/2}} \tag{13}$$

The best gain K is now obtained from Eqs. (11) and (13) as

$$K = \frac{T^{1/2}J_s^{1/2}}{2^{1/6}} \tag{14}$$

and the resulting ISE is given from Eq. (3) or Eq. (10) as

$$J = \frac{1}{2^{5/6}T^{1/2}J_s^{1/2}} + \frac{T}{2} \tag{15}$$

The minimum integral square error for $J_s = \infty$ is $T/2$ and this is increased as indicated by Eq. (15) for noninfinite J_s. We can use these relations to make tradeoffs among system parameters. For example, if we decide to allow an increase by a factor k in the minimum ISE then we have from Eq. (15)

$$\frac{1}{2^{5/6}\, T^{1/2}\, J_s^{1/2}} = k(T/2)$$

From Eq. (14) the foregoing leads to the optimum gain

$$K = \frac{1}{Tk}$$

The damping ratio of the system of Eq. (1) or Eq. (6) is

$$\zeta = \frac{1}{2\sqrt{KT}}$$

and this now becomes

$$\zeta = \frac{k^{1/2}}{2}$$

We can now determine tradeoffs to see how the error increases for various constraint values. If we allow a doubling of the minimum integral square error then $k = 1$ and $\zeta = 0.5$. This results in $KT = 1$ and $J_s = 1/2T^3$. The integral square acceleration would be reduced from an infinite value to a value $1/2T^3$. Clearly all of these parameters are related to motor sizing but we have not developed the details of this example in anywhere enough detail to allow any specific interpretation of these results in terms of physical control system component capacities. This is generally not an easy task but it is of course a necessary one if we are to successfully select appropriate control system components for a given problem.

EXERCISE 7.2-3. Repeat EXAMPLE 7.2-4 for the fixed plant $G_f(s) = K(s + \omega_1)/s^2$ where the input is a) a step, b) a ramp. Both K and ω_1 may be considered as adjustable.

We have observed in some of the examples we have considered thus far in this section that low damping often results. Primarily this occurs because the error criterion we are using is such that the system should act fast to reduce the initially large error and this will cause the system to willingly trade a decrease in damping for an increase in bandwidth. There are several ways in which we could modify the error criterion such as to make it not penalize the large initial error so much. We could consider as an error criterion the integral of the absolute value of the error

$$IAE = \int_0^\infty |e(t)|\, dt \qquad (7.2\text{-}12)$$

and this would not penalize large errors so severely as an error squared criterion. Also we could use the integral of the time multiplied absolute value of the error

$$ITAE = \int_0^\infty t|e(t)|\, dt \qquad (7.2\text{-}13)$$

Unfortunately these error criteria are essentially impossible to cope with in an analytical fashion. They are suitable for computer use such as cost functions in our adaptive controllers to be considered in the next section. Also these nonquadratic error criteria offer essentially no advantage to the quadratic error criteria which we will now consider.

An attractive error criterion is the integral of time multiplied squared error defined by

$$ITSE = \int_0^\infty te^2(t)\, dt \qquad (7.2\text{-}14)$$

Unfortunately the integral tables which we may prepare for this error criterion are very complex. If we write the definition of the Laplace transform of $e^2(t)$

$$F(p) = \int_0^\infty e^2(t) \exp(-pt)\, dt \qquad (7.2\text{-}15)$$

then we see that

$$-\frac{\partial F(p)}{\partial p} = \int_0^\infty te^2(t) \exp(-pt)\, dt$$

and have

$$\int_0^\infty te^2(t)\, dt = -\lim_{p \to 0} \frac{\partial F(p)}{\partial p} \qquad (7.2\text{-}16)$$

We can also write for the definition of $F(p)$

$$F(p) = \int_0^\infty y^2(t)\, dt \qquad (7.2\text{-}17)$$

where

$$y(t) = e(t) \exp(-pt/2) \qquad (7.2\text{-}18)$$

Parveval's theorem may now be used such that we have for Eq. (7.2-17)

$$F(p) = \frac{1}{2\pi j} \int_{-j\infty}^{j\infty} Y(s)Y(-s)\, ds \qquad (7.2\text{-}19)$$

where from Eq. (7.2-18), using the exponential weighting Laplace transform theorem,

$$Y(s) = E(s + p/2) \qquad (7.2\text{-}20)$$

Collecting together Eqs. (7.2-16), (7.2-19), and (7.2-20) and letting $p/2 = \eta$, results in

$$\int_0^\infty t e^2(t)\, dt = -\lim_{\eta \to 0} \frac{\partial F(\eta)}{\partial \eta} \qquad (7.2\text{-}21)$$

where

$$F(\eta) = \frac{1}{4\pi j} \int_{-j\infty}^{j\infty} E(\eta + s)E(\eta - s)\, ds \qquad (7.2\text{-}22)$$

Equations (7.2-21) and (7.2-22) are the end result equations for derivation of tables of ITSE. The TABLE 7.2-1 for ISE can be used to evaluate Eq. (7.2-22) and then Eq. (7.2-21) applied to establish Eq. (7.2-21). Unfortunately, the expressions for ITSE can become quite complex for even low order systems.

EXAMPLE 7.2-5. We consider minimization of the ITSE for the second order system

$$H(s) = \frac{1}{s^2 + 2\zeta s + 1}$$

with a unit step input. The error expression is

$$E(s) = U(s)[1 - H(s)]$$

$$= \frac{s + 2\zeta}{s^2 + 2\zeta s + 1}$$

We easily find $E(\eta + s)$ as

$$E(\eta + s) = \frac{s + (2\zeta + \eta)}{s^2 + 2(\zeta + \eta)s + \eta^2 + 1 + 2\zeta\eta}$$

This is in a form such that the integral table, TABLE 7.2-1 with $n = 2$ is directly applicable. We have

$$F(\eta) = \frac{1}{2} \frac{c_0^2 d_2 + c_1^2 d_0}{2 d_0 d_1 d_2}$$

$$= \frac{(2\zeta + \eta)^2 + (\eta^2 + 1 + 2\zeta\eta)}{4(\zeta + \eta)(\eta^2 + 1 + 2\zeta\eta)}$$

Use of Eq. (7.2-21) leads to

$$\text{ITSE} = -\lim_{\eta \to 0} \frac{\partial F(\eta)}{\partial \eta} = \frac{1 + 8\zeta^4}{4\zeta^2}$$

To minimize the ITSE we set $\partial\text{ITSE}/\partial\zeta = 0$ and obtain $\zeta = 0.59$. Thus, as we would expect the system damping is greater than for the ISE criterion although the response is a bit slower. This is reasonable since we do not penalize errors at small time as much as errors for large time.

EXERCISE 7.2-4. For the unity feedback ratio system with open loop transfer function $G(s) = 100(s + \omega_1)/s^2$ find the optimum value of ω_1 to minimize the ITSE for a unit step input and a unit ramp input. How do the results of this example compare with the results obtained by minimizing the ISE?

EXERCISE 7.2-5. Show that for

$$E(s) = \frac{c_0 + c_1 s}{d_0 + d_1 s + d_2 s^2}$$

that the value of $\int_0^\infty t e^2(t)\, dt$ is given by

$$\text{ITSE} = \frac{c_0^2 d_1^2 + 2c_1^2 d_0^2 + 2c_0^2 d_0 d_2 - 2c_0 c_1 d_0 d_1}{4 d_0^2 d_1^2}$$

EXERCISE 7.2-6. Find the best value of K for the open loop transfer function $G(s) = K/s(s + 1)$ to minimize ITSE where the input is a unit step.

It is sometimes convenient to define error expressions such as the integral of time to the $2N$th power multiplied by error squared

$$\text{ISTNSE} = \int_0^\infty t^{2N} e^2(t) \, dt \qquad (7.2\text{-}23)$$

We can use Parcevals theorem to write for the foregoing

$$\text{ISTNSE} = \int_0^\infty [t^N e(t)]^2 \, dt = \frac{1}{2\pi j} \int_{-j\infty}^{j\infty} Y_N(s) Y_N(-s) \, ds \qquad (7.2\text{-}24)$$

where

$$Y_N(s) = [t^N e(t)] = (-1)^N \frac{\partial^N E(s)}{\partial s^N} \qquad (7.2\text{-}25)$$

as we know from our efforts with the Laplace transform in Chap. 2. Again we may use the integral table, TABLE 7.2-1, to evaluate Eqs. (7.2-21) and (7.2-25).

EXAMPLE 7.2-6. Let us find the optimum damping for the second order system

$$H(s) = \frac{1}{s^2 + 2\zeta s + 1}$$

such as to minimize ISTNSE where the input is a unit step. We have the error transform

$$E(s) = \frac{s + 2\zeta}{s^2 + 2\zeta s + 1}$$

For $N = 1$ we have

$$Y_1(s) = \frac{s^2 + 4\zeta s + 4\zeta^2 - 1}{(s^2 + 2\zeta s + 1)^2}$$

Unfortunately we now must evaluate an integral with $n = 4$ in TABLE 7.2-1. We obtain

$$\text{ISTSE} = \int_0^\infty t^2 e^2(t)\, dt = \frac{(2\zeta)^6 - (2\zeta)^4 + (2\zeta)^2 + 1}{4(2\zeta)^3}$$

and we obtain from $\partial\text{ISTSE}/\partial\zeta = 0$ the fact that the optimum damping ratio is $\zeta = 2/3$.

We can obtain a sometimes more useful form of Eqs. (7.2-24) and (7.2-25) by noting that we can replace Eqs. (7.2-21) and (7.2-22) by

$$\text{ITMSE} = \int_0^\infty t^M e^2(t)\, dt = (-1)^M \lim_{\eta \to 0} \frac{\partial^M F(\eta)}{\partial \eta^M} \qquad (7.2\text{-}26)$$

where

$$F(\eta) = \frac{1}{4\pi j} \int_{-j\infty}^{j\infty} E(\eta + s)E(\eta - s)\, ds \qquad (7.2\text{-}27)$$

It is generally preferred to use the relations of Eqs. (7.2-26) and (7.2-27) rather than Eqs. (7.2-24) and (7.2-25).

EXAMPLE 7.2-7. To evaluate the ITMSE for EXAMPLE 7.2-6 we obtain $E(\eta + s)$ from the $E(s)$ expression of that example or EXAMPLE 7.2-5. Then we evaluate $F(\eta)$ as we have already done in EXAMPLE 7.2-5. Successive derivatives as required by Eq. (7.2-26) yield for the values of $\int_0^\infty t e^2(t)\, dt$ and $\int_0^\infty t^2 e^2(t)\, dt$ the same results obtained in our earlier examples by slightly different approaches.

We can extend the class of analytical performance criteria to general exponential weighting of the form

$$\text{IEMSE} = \int_0^\infty e^2(t) \exp(-2\alpha t)\, dt \qquad (7.2\text{-}28)$$

suggested here. If we write

$$\text{IEMSE} = \int_0^\infty [e(t) \exp(-\alpha t)]^2 \, dt = \int_0^\infty y^2(t) \, dt \qquad (7.2\text{-}29)$$

and use Parcevals theorem to obtain

$$\text{IEMSE} = \frac{1}{2\pi j} \int_{-j\infty}^{j\infty} Y(s) Y(-s) \, ds \qquad (7.2\text{-}30)$$

where

$$Y(s) = E(\alpha + s) \qquad (7.2\text{-}31)$$

we may evaluate IEMSE directly from our integral tables.

EXAMPLE 7.2-8. We can compute the IEMSE of the previous second order example from the relation of Eq. (7.2-31) which becomes

$$Y(s) = \frac{s + (2\zeta + \alpha)}{s^2 + 2(\zeta + \alpha)s + \alpha^2 + 1 + 2\zeta\alpha} \qquad (1)$$

as

$$\text{IEMSE} = \frac{2\alpha^2 + 6\zeta\alpha + 4\zeta^2 + 1}{2(6\zeta\alpha^2 + 4\zeta^2\alpha + 2\zeta + 2\alpha^3 + 2\alpha)}$$

Solving for the optimum value of ζ we obtain

$$\zeta = \frac{1 - \alpha}{2} \qquad (2)$$

There is nothing at all in our work which prohibits α from being negative. In fact when α is positive we are not weighting the error at large time heavily at all. So we would generally desire to use negative values. In the particular performance index we see that we can control the damping by picking appropriate values of α. We must exercise care in doing this however because negative values of α may cause $Y(s)$ in Eq. (1) to have poles in the right half plane. The integral table, TABLE 7.2-1, is derived assuming $Y(s)$ has left half plane poles only. Suppose for example that $\alpha = -1$ such that $\zeta = 1$ is obtained from Eq. (2). Eq. (1) becomes

$$Y(s) = \frac{s + 1}{s^2}$$

and we see that $y(t) = t + 1$ and so $\int_0^\infty y^2(t)\, dt$ becomes infinite. This does not indicate that the basic system $[H(s)]$ is unstable only that the weighting function $\exp(-\alpha t)$ is increasing so rapidly in time that the product of the error and $\exp(-\alpha t)$ does not decrease to zero as time increases. For this particular example we can show that no difficulties result as long as $-1 < \alpha$.

We see that the integral square error expression (ISE) can be evaluated with reasonable ease for systems of order 4 or less. More complicated functions of time multiplied error squared can also be considered but the manipulations involved unfortunately become very difficult to cope with without having recourse to a computer. After a brief discussion of optimal control determination we will present a discussion of adaptive control which includes a discussion of computational techniques to accomplish parameter optimization for either offline use in system studies or online or real time use for implementation in actual operating systems.

EXERCISE 7.2-7.   Evaluate

$$\int_0^\infty e^2(t) \exp(-2\alpha t)\, dt$$

for the unity feedback ratio system with open loop transfer function

$$G(s) = \frac{100(s + \omega_1)}{s^2}$$

Consider the case where the input is a unit step as well as the case where the input is a unit ramp.

Optimum Control Determination.   In our efforts thus far in this section we have assumed that the system input is known and we wished to determine certain constant unspecified system parameters such that the output looks as much as possible like the input.

Alternately it is possible to specify a desired system output and, with a knowledge of the system dynamics, determine the best control input such as to minimize a scalar cost function. Although it may not appear that this is equivalent to the servomechanism problem that we have been discussing it turns out that the two are equivalent. The desired output for the optimum systems control problem is just the input for the classical servomechanism problem. In the classical servo problem we form an error signal by subtracting the input from the output and operate upon this error signal with a compensation network to form an input signal to the fixed plant. In the optimal control problem this input signal to the fixed plant is the control signal. The classical servomechanism problem is inherently a feedback control problem with all of the advantages of feedback. Fundamental results obtained by solving the optimal control problem are often open loop inputs to the fixed plant although we will see that, in some special instances, determination of a closed loop structure is quite possible.

A general development of optimum systems control would take us much too far afield of our intentions in a beginning linear systems control text. As indicated in our reference section, Sec. 7.5, a number of advanced works are available which provide indepth discussion of optimum systems control theory and applications. Here we shall state some of the basic results and apply them to some simple problems of interest.

A fairly general statement of the optimal control problem is that it consists of finding the state $\mathbf{x}(t)$ and the control $\mathbf{u}(t)$ which minimize the cost function

$$J = \theta[\mathbf{x}(t_f)] + \int_{t_0}^{t_f} \phi[\mathbf{x}(t), \mathbf{u}(t), t] \, dt \tag{7.2-32}$$

subject to the equality constraint that a differential equation describing the dynamics of the fixed plant

$$\dot{\mathbf{x}} = \mathbf{f}[\mathbf{x}(t), \mathbf{u}(t), t], \qquad \mathbf{x}(t_0) = \mathbf{x}_0 \tag{7.2-33}$$

be satisfied. The initial condition vector $\mathbf{x}_0$ is assumed known as are the initial and final times t_f and t_0. There are no constraints on the state at the final time other than those implied by Eq. (7.2-32). Further there are no inequality constraints restricting the admissable or allowable regions for the state $\mathbf{x}(t)$ or the control $\mathbf{u}(t)$. Any or all of these restric-

We will now present a very simplified derivation of the celebrated Pontryagin maximum principle which results in the necessary conditions which must be solved to obtain a solution to Eqs. (7.2-32) and (7.2-33). These relations may be stated in scalar form but the vector space format is much more convenient to use and so we will use it. A very special form of zero can be obtained by rewriting Eq. (7.2-33) as

$$0 = \mathbf{f}[\mathbf{x}(t), \mathbf{u}(t), t] - \dot{\mathbf{x}}$$

We wish to adjoin the foregoing relation to the cost function by means of a Lagrange multiplier. Since the foregoing relation is a vector relation and the cost function is a scalar we introduce a vector Lagrange multiplied or adjoint variable $\lambda(t)$ and write the scalar or inner product

$$0 = \lambda^T(t)\{\mathbf{f}[\mathbf{x}(t), \mathbf{u}(t), t] - \dot{\mathbf{x}}\}$$

which we add to Eq. (7.2-32) to obtain

$$J = \theta[\mathbf{x}(t_f)] + \int_{t_0}^{t_f} (\phi + \lambda^T \mathbf{f} - \lambda^T \dot{\mathbf{x}})\, dt \qquad (7.2\text{-}34)$$

where we drop some of the formal symbolism in Eq. (7.2-34) for notational convenience. *If* the equality constant of Eq. (7.2-33) is satisfied then Eq. (7.2-34) is entirely equivalent to Eq. (7.2-32).

For convenience let us define a scalar quantity called the Hamiltonian

$$H[\mathbf{x}(t), \lambda(t), \mathbf{u}(t), t] = \phi[\mathbf{x}(t), \mathbf{u}(t), t]$$

$$+ \lambda^T(t)\mathbf{f}[\mathbf{x}(t), \mathbf{u}(t), t] \qquad (7.2\text{-}35)$$

such that Eq. (7.2-34) becomes

$$J = \theta[\mathbf{x}(t_f)] + \int_{t_0}^{t_f} (H - \lambda^T \dot{\mathbf{x}})\, dt \qquad (7.2\text{-}36)$$

Let us assume that we know the optimal control $\hat{\mathbf{u}}(t)$ and the optimal state $\hat{\mathbf{x}}(t)$. For changes about the optimal we have

$$\mathbf{x}(t) = \hat{\mathbf{x}}(t) + \epsilon \eta_{\mathbf{x}}(t) \qquad (7.2\text{-}37)$$

$$\mathbf{u}(t) = \hat{\mathbf{u}}(t) + \epsilon \eta_{\mathbf{u}}(t) \qquad (7.2\text{-}38)$$

Here the change in $\mathbf{x}(t)$ from the optimal $\hat{\mathbf{x}}(t)$ is $\Delta\mathbf{x}(t) = \epsilon\eta_{\mathbf{x}}(t)$ and the change in $\mathbf{u}(t)$ from the optimal $\hat{\mathbf{u}}(t)$ is $\Delta\mathbf{u}(t) = \epsilon\eta_{\mathbf{u}}(t)$. We could also perturb $\lambda(t)$ but we will invoke Eq. (7.2-33) and not do this. The particularly interesting fact about Eqs. (7.2-37) and (7.2-38) is that, for any $\eta_{\mathbf{x}}(t)$ and $\eta_{\mathbf{u}}(t)$, $\mathbf{x}(t)$ and $\mathbf{u}(t)$ become the optimal values when $\epsilon = 0$. Suppose we pick a specific $\eta_{\mathbf{x}}^{1}(t)$ and $\eta_{\mathbf{u}}^{1}(t)$. We might plot J versus the scalar parameter ϵ and get a result like that shown in Fig. (7.2-6). We might do this for a number of $\eta_{\mathbf{x}}^{i}(t)$ and $\eta_{\mathbf{u}}^{i}(t)$ as also indicated in Fig. (7.2-6). This figure shows that the minimum of J occurs for $\epsilon = 0$ and this same minimum exists for arbitrary $\eta_{\mathbf{x}}$ and $\eta_{\mathbf{u}}$. Thus we see that an appropriate criterion to use for our optimization is that

$$\left.\frac{\partial J}{\partial \epsilon}\right|_{\epsilon=0} = 0 \text{ for all } \eta_{\mathbf{x}}(t),\, \eta_{\mathbf{u}}(t) \tag{7.2-39}$$

We wish to minimize the cost J in Eq. (7.2-32). All Eq. (7.2-39) provides is necessary conditions for a minimum. For sufficient conditions we need to examine the second derivative of J. For most physical problems it is clear that when a solution exists it is the optimum solution which minimizes J. This is also true with respect to our error criteria used earlier in this chapter and we are very fortunate indeed for obtaining the sufficient conditions turns out to be a very difficult task.

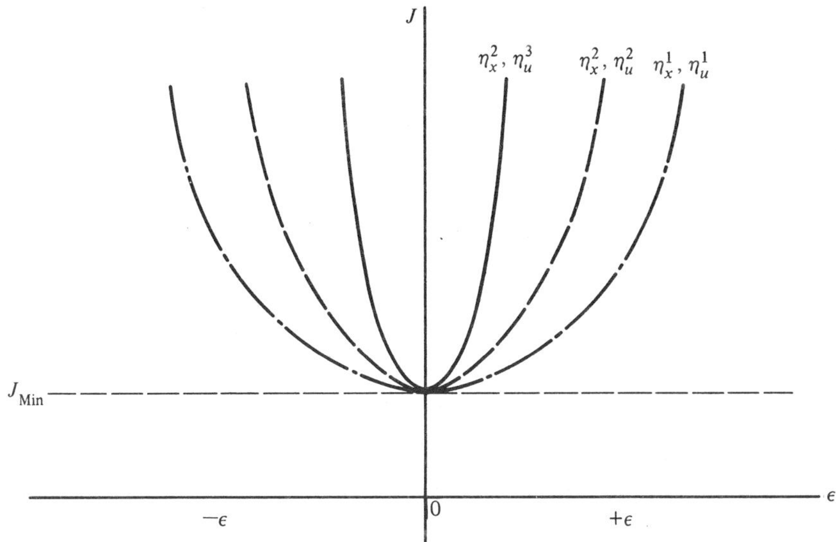

Figure 7.2-6 Cost function for variations in ϵ, η_x, and η_u

It is convenient but not necessary to integrate a portion of Eq. (7.2-36) by parts using

$$\int_{t_0}^{t_f} \lambda^T \dot{\mathbf{x}}\, dt = \lambda^T(t)\mathbf{x}(t)\Big|_{t_0}^{t_f} - \int_{t_0}^{t_f} \dot{\lambda}^T \mathbf{x}\, dt$$

such that we obtain

$$J = \theta[\mathbf{x}(t_f)] - \lambda^T(t_f)\mathbf{x}(t_f) + \lambda^T(t_0)\mathbf{x}(t_0) + \int_{t_0}^{t_f}(H + \dot{\lambda}^T\mathbf{x})\, dt \qquad (7.2\text{-}40)$$

We now insert the variations of Eqs. (7.2-37) and (7.2-38) into Eq. (7.2-40) and perform the derivative operation of Eq. (7.2-39). We obtain, in a rather straightforward way, the requirement that

$$0 = \eta_{\mathbf{x}}^T(t_f)\left\{\left[\frac{\partial\phi[\hat{\mathbf{x}}(t_f)]}{\partial\hat{\mathbf{x}}(t_f)}\right] + \lambda^T(t_f)\right\}$$

$$- \eta_{\mathbf{x}}^T(t_0)\lambda(t_0)$$

$$+ \int_{t_0}^{t_f}\left\{\eta_{\mathbf{x}}^T(t)\left[\dot{\lambda} + \frac{\partial H}{\partial\hat{\mathbf{x}}(t)}\right] + \eta_{\mathbf{u}}^T(t)\frac{\partial H}{\partial\hat{\mathbf{u}}(t)}\right\} dt \qquad (7.2\text{-}41)$$

as the interested reader can easily verify. Here we use the gradient notation $\partial H/\partial\mathbf{u}$ to indicate a column vector

$$\frac{\partial H}{\partial\mathbf{u}} = \begin{bmatrix} \dfrac{\partial H}{\partial\mathbf{u}_1} \\[2ex] \dfrac{\partial H}{\partial\mathbf{u}_2} \\[1ex] \vdots \\[1ex] \dfrac{\partial H}{\partial\mathbf{u}_m} \end{bmatrix} \qquad (7.2\text{-}42)$$

Now Eq. (7.2-41) must be true for *arbitrary* $\eta_{\mathbf{x}}(t)$, $\eta_{\mathbf{u}}(t)$ and $\eta_{\mathbf{x}}(t_f)$. But $\eta_{\mathbf{x}}(t_0)$ must be zero and not at all arbitrary since the initial state is fixed, i.e., $\mathbf{x}(t_0) = \mathbf{x}_0 = \hat{\mathbf{x}}(t_0)$. The only way these will be true is that the terms multiplying the variations be zero or

$$\dot{\lambda} = -\frac{\partial H}{\partial \hat{\mathbf{x}}} \tag{7.2-43}$$

$$\frac{\partial H}{\partial \mathbf{u}} = 0 \tag{7.2-44}$$

$$\lambda^T(t_f) = \frac{\partial \theta[\hat{\mathbf{x}}(t_f)]}{\partial \hat{\mathbf{x}}(t_f)} \tag{7.2-45}$$

Also we must require that Eq. (7.2-33) be valid for the optimal state and control. Thus we must also require the alternate expressions

$$\dot{\hat{\mathbf{x}}} = \frac{\partial H}{\partial \lambda} \tag{7.2-46}$$

$$\hat{\mathbf{x}}(t_0) = \mathbf{x}_0 \tag{7.2-47}$$

These last five equations are known as the primary equations of the Pontryagin maximum principle. They are of much use in optimal control theory and apply to nonlinear as well as linear problems. Unfortunately these equations are in the form of a split boundary value problem since the conditions on the state are specified at the initial time whereas those on the adjoint are specified at the final time. Solution to these equations is normally not a simple task, especially for nonlinear systems, and generally requires the use of digital computers. Figure (7.2-7) illustrates a DELTA Chart of the optimal control approach for the class of problems posed here.

There are many extensions to the basic maximum principle which we have derived here. Some of the many useful extensions include problems where the terminal time is not fixed and problems where there are inequality constraints on the control and/or state. In the case, for example, where there are inequality constraints on the control, Eq. (7.2-44) is replaced by the requirement that

$$H[\hat{\mathbf{x}}(t), \hat{\lambda}(t), \hat{\mathbf{u}}(t), t] \leqslant H[\hat{\mathbf{x}}(t), \hat{\lambda}(t), \mathbf{u}(t), t] \tag{7.2-48}$$

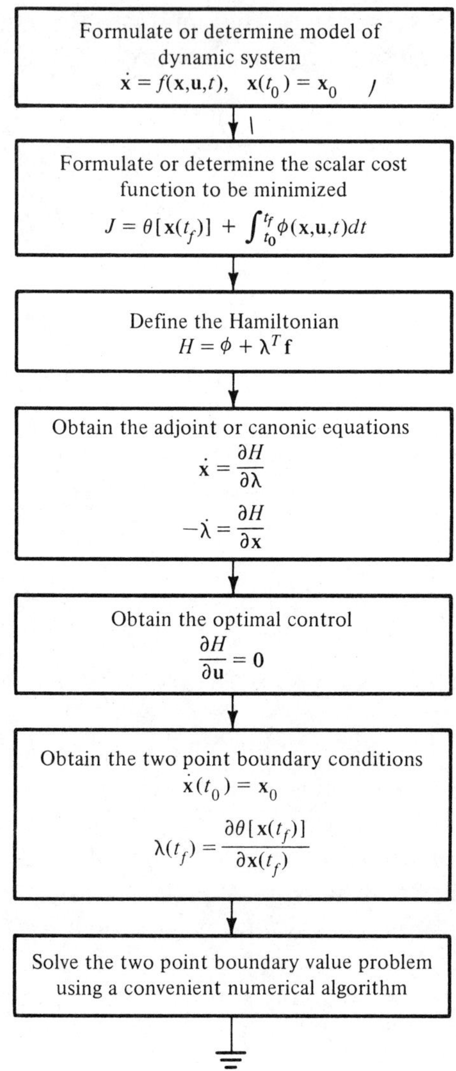

Figure 7.2-7   DELTA Chart of the optimal control approach

for $\mathbf{u}(t)$ and $\hat{\mathbf{u}}(t)$ within the constrained region. Also there are solution difficulties which occur for example when the Hamiltonian is linear in the control variable. These are called singular problems. Extensions such as these are discussed in most modern control theory courses offered at the graduate level. We will now present one example to illustrate the solution technique using the maximum principle and will then discuss the linear feedback regulator and servomechanism problems.

EXAMPLE 7.2-9. We consider optimal control determination for the simplified equation of motion of a mass

$$M \frac{d^2 x}{dt^2} = f \tag{1}$$

where x represents distance, M represents mass, and f represents force. The object is to take the object from some initial position $x(0)$ and velocity $\left. \dfrac{dx(t)}{dt} \right|_{t=0}$ to a point very near $x(t_f) = 0$ and $\dfrac{dx(t_f)}{dt} = 0$ at time t_f while minimizing the integral square force. We may formulate the problem as one of minimizing

$$J = \frac{1}{2} S_{11} x^2(t_f) + \frac{1}{2} S_{22} \left[\frac{dx(t_f)}{dt_f} \right]^2$$

$$+ \frac{1}{2} \int_{t_0}^{t_f} \left[\frac{f(t)}{M} \right]^2 dt \tag{2}$$

subject to the constraint of Eq. (1) and the given initial conditions. While it may appear somewhat arbitrary to assume a quadratic cost function it often turns out that this is a very meaningful cost. The larger we make S_{11} and S_{22} the closer we will force x and $\dot{x}$ to zero at the terminal time.

It is convenient to recast the problem in state space form by letting

$$x_1(t) = x(t)$$

$$\frac{dx_1(t)}{dt} = x_2(t)$$

$$\frac{dx_2(t)}{dt} = \frac{d^2 x_1(t)}{dt^2} = \frac{d^2 x(t)}{dt^2} = \frac{f(t)}{M} = u(t)$$

such that the system dynamics are

$$\dot{\mathbf{x}} = \mathbf{A}\mathbf{x} + \mathbf{b}u \tag{3}$$

where

$$\mathbf{A} = \begin{bmatrix} 0 & 1 \\ 0 & 0 \end{bmatrix} \qquad \mathbf{b} = \begin{bmatrix} 0 \\ 1 \end{bmatrix} \qquad \mathbf{x} = \begin{bmatrix} x_1 \\ x_2 \end{bmatrix}$$

The cost function of Eq. (2) may be written as

$$J = \frac{1}{2} \mathbf{x}^T(t_f) \mathbf{S} \mathbf{x}(t_f) + \frac{1}{2} \int_0^{t_f} u^2(t)\, dt \tag{4}$$

where

$$\mathbf{S} = \begin{bmatrix} S_{11} & 0 \\ 0 & S_{22} \end{bmatrix}$$

and we can use the vector matrix formulation to obtain the maximum principle equations. Alternately we may work directly with the individual equations and minimize

$$J = \frac{1}{2} S_{11} x_1^2(t_f) + \frac{1}{2} S_{22} x_2^2(t_f) + \frac{1}{2} \int_0^{t_f} u^2(t)\, dt \tag{5}$$

subject to

$$\dot{x}_1 = x_2, \qquad x_1(t_0) = x_{10}$$

$$\dot{x}_2 = u, \qquad x_2(t_0) = x_{20} \tag{6}$$

We will use this latter approach here.

The Hamiltonian is defined as

$$H = \phi + \lambda^T \mathbf{f} = \frac{1}{2} u^2 + \lambda_1 x_2 + \lambda_2 u \tag{7}$$

The canonic equations are obtained from

$$\dot{\mathbf{x}} = \frac{\partial H}{\partial \lambda} \tag{8}$$

$$\dot{\lambda} = -\frac{\partial H}{\partial \mathbf{x}} \tag{9}$$

which yield

$$\dot{x}_1 = \frac{\partial H}{\partial \lambda_1} = x_2$$

$$\dot{x}_2 = \frac{\partial H}{\partial \lambda_2} = u$$

$$\dot{\lambda}_1 = -\frac{\partial H}{\partial x_1} = 0$$

$$\dot{\lambda}_2 = -\frac{\partial H}{\partial x_2} = -\lambda_1 \tag{10}$$

The optimal control is obtained from

$$\frac{\partial H}{\partial u} = 0 = u + \lambda_2 \tag{11}$$

and the two point boundary conditions are obtained from the given initial conditions

$$x_1(0) = x_{10}$$

$$x_2(0) = x_{20} \tag{12}$$

and the transversality conditions $\lambda(t_f) = \partial\theta[\mathbf{x}(t_f)]/\partial\mathbf{x}(t_f)$ which become

$$\lambda_1(t_f) = \frac{\partial\theta[\mathbf{x}(t_f)]}{\partial x_1(t_f)} = S_{11}x_1(t_f)$$

$$\lambda_2(t_f) = \frac{\partial\theta[\mathbf{x}(t_f)]}{\partial x_2(t_f)} = S_{22}x_2(t_f) \tag{13}$$

We may combine Eqs. (10) and (11) to eliminate $u(t)$. The resulting equations together with Eqs. (12) and (13) constitute the two point boundary value problem (TPBVP) which we must solve. This is:

$$\dot{x}_1 = x_2 \qquad x_1(t_0) = x_{10} \tag{14}$$

$$\dot{x}_2 = -\lambda_2 \qquad x_2(t_0) = x_{20} \tag{15}$$

$$\dot{\lambda}_1 = 0 \qquad \lambda_1(t_f) = S_{11}x_1(t_f) \tag{16}$$

$$\dot{\lambda}_2 = -\lambda_1 \qquad \lambda_2(t_f) = S_{22}x_2(t_f) \tag{17}$$

We now proceed to solve these equations. The differential equations may be integrated in the order (16), (17), (15), and (14) to yield

$$\lambda_1(t) = a \tag{18}$$

$$\lambda_2(t) = -at + b \tag{19}$$

$$x_2(t) = \frac{at^2}{2} - bt + c \tag{20}$$

$$x_1(t) = \frac{at^3}{6} - \frac{bt^2}{2} + ct + d \tag{21}$$

We must also satisfy the given two point boundary conditions. Since $x_1(0) = x_{10}$, we have $d = x_{10}$. Since $x_2(t_0) = x_{20}$ we have $c = x_{20}$. To obtain $x_1(t_f) = S_{11}x_1(t_f)$ we must have $a = S_{11}x_1(t_f)$. To obtain $\lambda_2(t_f) = S_{22}x_2(t_f)$ we must have $-at_f + b = S_{22}x_2(t_f)$. Unfortunately these values of a and b are not especially useful in that we do not know $x_1(t_f)$ and $x_2(t_f)$. However, these can be evaluated from Eqs. (20) and (21) and we obtain

$$a = S_{11}x_1(t_f) = \left[\frac{at_f^3}{6} - \frac{bt_f^2}{2} + ct_f + d \right] S_{11}$$

and

$$-at_f + b = \left[\frac{at_f^2}{2} - bt_f + c \right] S_{22}$$

We have just found a linear set of equations in the unspecified constants of integration. These are, from the foregoing two equations and the relations $d = x_{10}$ and $c = x_{20}$,

$$a\left[\frac{1}{S_{11}} - \frac{t_f^3}{6}\right] + b\left[\frac{t_f^2}{2}\right] = x_{20} t_f + x_{10}$$

$$a\left[-\frac{t_f}{S_{22}} - \frac{t_f^2}{2}\right] + b\left[\frac{1}{S_{22}} + t_f\right] = x_{20} \tag{22}$$

We may solve these equations for a as

$$a = \frac{\left[\dfrac{1}{S_{22}} + t_f\right] x_{10} + \left[\dfrac{t_f}{S_{22}} + \dfrac{t_f^2}{2}\right] x_{20}}{\dfrac{1}{S_{11} S_{22}} - \dfrac{t_f^3}{6 S_{22}} + \dfrac{t_f}{S_{11}} + \dfrac{t_f^3}{2 S_{22}} + \dfrac{t_f^4}{12}} \tag{23}$$

The items of interest in this example are the optimal control

$$\hat{u}(t) = -at + b \tag{24}$$

and the optimal states

$$\hat{x}_2(t) = \frac{at^2}{2} - bt + x_{20} \tag{25}$$

$$\hat{x}_1(t) = \frac{at^3}{6} - \frac{bt^2}{2} + x_{20} t + x_{10} \tag{26}$$

where a and b are defined by Eqs. (22) and (23).

Special cases of this result are of interest. If we let S_{11} and S_{22} become infinite we are requiring that $x_1(t_f)$ and $x_2(t_f)$ become zero. Otherwise there will be an infinite cost which surely should not be the minimum cost. With $S_{11} = S_{22} = \infty$ we obtain

$$a = \frac{12}{t_f^3} x_{10} + \frac{6}{t_f^2} x_{20} \tag{27}$$

$$b = \frac{6}{t_f^2} x_{10} + \frac{4}{t_f} x_{20} \tag{28}$$

and we can check, using Eqs. (25) and (26), to see that

$$\hat{x}_1(t_f) = \hat{x}_2(t_f) = 0$$

A disappointing result of this example is its complexity. The optimal control itself is a simple time function as indicated by Eq. (24). But the parameters a and b are complex functions of x_{10}, x_{20}, S_{11}, S_{22}, and t_f even though a and b are constants. What is probably worse is the fact that the optimal control $u(t)$ is an open loop control. It is a function of time and not of the state variables. This is quite disappointing because there are many advantages to a feedback solution. For example if we do not know x_{10} and x_{20} rather well, the open loop control will be perhaps quite seriously in error whereas a closed loop control would be much less sensitive, or perhaps not even sensitive at all, to changes in x_{10} and x_{20}.

It turns out for this example that a closed loop control

$$\hat{u}(t) = K_1(t)\hat{x}_1(t) + K_2(t)\hat{x}_2(t) \tag{29}$$

may be found which is indeed the optimal control. It is rather tedious to obtain the gains $K_1(t)$ and $K_2(t)$ for the general case here but for the special case where Eqs. (26) and (27) hold we can easily obtain from Eqs. (24), (27), and (28)

$$u(t) = \frac{6}{t_f^2}\left(1 - \frac{2t}{t_f}\right)x_{10} + \frac{2}{t_f}\left(2 - \frac{3t}{t_f}\right)x_{20} \tag{30}$$

We use the values of a and b in Eqs. (27) and (28) in the optimal state expressions of Eqs. (25) and (26) to obtain

$$\hat{x}_2(t) = \frac{6}{t_f^2}(t^2 - t)x_{10} + \left(1 - \frac{4t}{t_f} + \frac{3t^2}{t_f^2}\right)x_{20}$$

$$\hat{x}_1(t) = \left(\frac{6t^3}{t_f^3} - \frac{3t^2}{t_f^2} + 1\right)x_{10} + \left(\frac{t^3}{t_f^2} - \frac{2t^2}{t_f} + t\right)x_{20} \tag{31}$$

We can solve for x_{10} and x_{20} in terms of $\hat{x}_1$ and $\hat{x}_2$ in Eq. (31) and insert these into Eq. (30) and find that

$$u(t) = \frac{-6x_1(t)}{(t-t_f)^2} - \frac{12x_2(t)}{(t-t_f)^3}$$

is the optimum feedback control law. In our immediate discussions to follow we will see that a much more systematic approach will considerably ease the computational complexities we have encountered here.

EXERCISE 7.2-8. Find the optimal control to minimize

$$J = \frac{1}{2}\int_0^1 (x^2 + qu^2)\,dt$$

for the system

$$\dot{x} = u \qquad x(0) = 1$$

Can you find a closed loop control by taking your optimal control and state $u(t)$ and $x(t)$ and postulating that $u(t) = K(t)x(t)$? What is this closed loop control?

EXERCISE 7.2-9. Use the maximum principle to find the TPBVP which must be solved to find the optimal control to minimize

$$J = \frac{1}{2}Sx^2(1) + \frac{1}{2}\int_0^1 u^2(t)\,dt$$

for the system

$$\dot{x} = -x^3 + u \qquad x(0) = 1$$

Our previous example has demonstrated the possibility of obtaining a closed loop control solution to an optimal control problem in which the cost function is a time integral of a quadratic function of the control and the state. We will now generalize this result to the n vector state variable differential equation describing the system dynamics

$$\dot{\mathbf{x}} = \mathbf{A}(t)\mathbf{x}(t) + \mathbf{B}(t)\mathbf{u}(t) \tag{7.2-49}$$

$$\mathbf{x}(t_0) = \mathbf{x}_0 \tag{7.2-50}$$

with a quadratic, always nonnegative for positive semidefinite **S, Q, R,**

$$J = \frac{1}{2} \mathbf{x}^T(t_f)\mathbf{S}\mathbf{x}(t_f) + \frac{1}{2}\int_{t_0}^{t_f} [\mathbf{x}^T(t)\mathbf{Q}(t)\mathbf{x}(t) + \mathbf{u}^T(t)\mathbf{R}(t)\mathbf{u}(t)]\ dt \quad (7.2\text{-}51)$$

where we may assume without loss of generality that **S, Q,** and **R** are symmetric matrices such that $S_{ij} = S_{ji}$.

Using the procedure advocated in the DELTA Chart of Fig. (7.2-7) we write the Hamiltonian, where we will now drop the time subscripts for notational convenience, as

$$H = \frac{1}{2}\mathbf{x}^T\mathbf{Q}\mathbf{x} + \frac{1}{2}\mathbf{u}^T\mathbf{R}\mathbf{u} + \lambda^T\mathbf{A}\mathbf{x} + \lambda^T\mathbf{B}\mathbf{u} \quad (7.2\text{-}52)$$

The canonic equations are obtained as

$$\dot{\mathbf{x}} = \frac{\partial H}{\partial \lambda} = \mathbf{A}\mathbf{x} + \mathbf{B}\mathbf{u} \quad (7.2\text{-}53)$$

$$\dot{\lambda} = \frac{-\partial H}{\partial \mathbf{x}} = -\mathbf{Q}\mathbf{x} - \mathbf{A}^T\lambda \quad (7.2\text{-}54)$$

The optimal control is obtained as

$$0 = \frac{\partial H}{\partial \mathbf{u}} = \mathbf{R}\mathbf{u} + \mathbf{B}^T\lambda \quad (7.2\text{-}55)$$

The two point boundary conditions are

$$\mathbf{x}(t_0) = \mathbf{x}_0 \quad (7.2\text{-}56)$$

$$\lambda(t_f) = \frac{\partial\theta[\mathbf{x}(t_f)]}{\partial\mathbf{x}(t_f)} = \mathbf{S}\mathbf{x}(t_f) \quad (7.2\text{-}57)$$

Solution of Eqs. (7.2-53) through (7.2-57) leads directly to the optimal control and state. These are linear differential equations but their solution is tedious for any but very low order problems. A worse feature of the solution is that we obtain an open loop control which is a function of time only and not of the state. We desired a closed loop control. Because of the fact that $\lambda(t_f)$ is a linear function of $\mathbf{x}(t_f)$ in

Eq. (7.2-57) we are motivated to "guess" that $\lambda(t)$ might be a linear function of $x(t)$ for all time in the interval t_0 to t_f. This is a universally accepted method of attempting to solve a differential equation. So we try

$$\lambda(t) = P(t)x(t) \tag{7.2-58}$$

and insert this relation into Eq. (7.2-55) to obtain

$$u = -R^{-1}B^T Px \tag{7.2-59}$$

and we see that we do have a linear feedback controller *if* our guess is correct; also we must require R^{-1} to exist and thus we require R to be a positive definite symmetric matrix.

Use of Eq. (7.2-58) in Eq. (7.2-54) leads to

$$\dot{P}x + P\dot{x} = \dot{\lambda} = -Qx - A^T Px \tag{7.2-60}$$

Use of Eq. (7.2-59) in Eq. (7.2-53) leads to

$$\dot{x} = Ax - BR^{-1}B^T Px \tag{7.2-61}$$

$$[\dot{P} + PA - PBR^{-1}B^T P + Q + A^T P]\,x(t) = 0 \tag{7.2-62}$$

We wish our linear feedback control to be valid for any and all states $x(t)$. Thus we see that the only way the foregoing will be satisfied is for us to require the term premultiplying $x(t)$ to be zero or

$$\dot{P} = -PA - A^T P + PBR^{-1}B^T P - Q \tag{7.2-63}$$

This equation is known as a Matrix Riccatti differential equation and is symmetric and positive semi definite. Although it is a nonlinear differential equation it turns out to have a known closed form solution although this is very difficult to obtain. Since this is a differential equation we need an initial condition in order to solve it. From Eqs. (7.2-57) and (7.2-58) we see that we must require

$$P(t_f) = S \tag{7.2-64}$$

Further examination of the sufficiency requirements for a solution will show not only that we require $R(t)$ to be a positive definite but also that we require $\mathbf{S}$ and $\mathbf{Q}(t)$ to be at least positive semi definite. Thus we have indeed found a linear feedback control law for the system and cost function of Eqs. (7.2-50) and (7.2-51). It is given by Eqs. (7.2-59), (7.2-63), and (7.2-64) and illustrated by Fig. (7.2-8). This linear quadratic regulator problem solution is a very important one for optimum systems control theory. One particular use of it is in the "fly by wire" control of nonlinear systems. For example if we know a nominal (optimal) state of trajectory $\mathbf{x}^n(t)$ and the nominal (optimal) control $\mathbf{u}^n(t)$ which produced this nominal (optimal) state then we can expand a nonlinear set of differential equations

$$\dot{\mathbf{x}} = \mathbf{f}(\mathbf{x}, \mathbf{u}, t) \qquad (7.2\text{-}65)$$

describing the system about

$$\mathbf{x}(t) = \mathbf{x}^n(t) + \Delta\mathbf{x}(t) \qquad (7.2\text{-}66)$$

$$\mathbf{u}(t) = \mathbf{u}^n(t) + \Delta\mathbf{u}(t) \qquad (7.2\text{-}67)$$

for small perturbations $\Delta\mathbf{x}(t)$ and $\Delta\mathbf{u}(t)$. Dropping all but zero and first order terms in this expansion results in

$$\dot{\mathbf{x}}^n + \Delta\dot{\mathbf{x}} = \mathbf{f}(\mathbf{x}^n, \mathbf{u}^n, t) + \left[\frac{\partial\mathbf{f}(\mathbf{x}^n, \mathbf{u}^n, t)}{\partial\mathbf{x}^n}\right]\Delta\mathbf{x}(t)$$

$$+ \left[\frac{\partial\mathbf{f}(\mathbf{x}^n, \mathbf{u}^n, t)}{\partial\mathbf{u}^n}\right]\Delta\mathbf{u}(t) \qquad (7.2\text{-}68)$$

Presumably we know $\mathbf{x}^n$ and $\mathbf{u}^n$ and so we know

$$\dot{\mathbf{x}}^n = \mathbf{f}(\mathbf{x}^n, \mathbf{u}^n, t) \qquad (7.2\text{-}69)$$

and can subtract this from the differential equation approximation for $\dot{\mathbf{x}}$ to give

$$\Delta\dot{\mathbf{x}} = \left[\frac{\partial\mathbf{f}(\mathbf{x}^n, \mathbf{u}^n, t)}{\partial\mathbf{x}^n}\right]\Delta\mathbf{x}(t) + \left[\frac{\partial\mathbf{f}(\mathbf{x}^n, \mathbf{u}^n, t)}{\partial\mathbf{u}^n}\right]\Delta\mathbf{u}(t) \qquad (7.2\text{-}70)$$

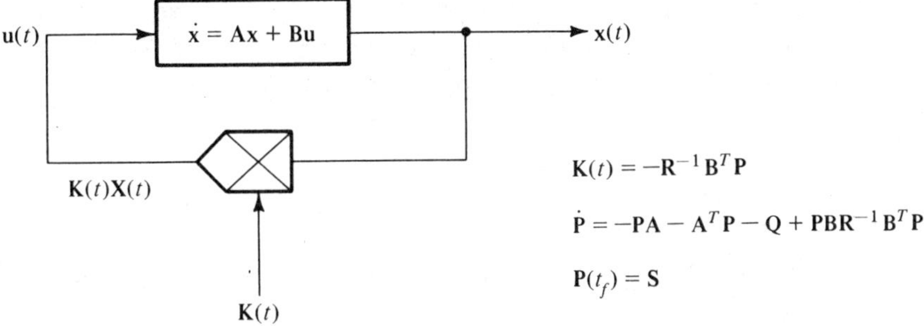

$$K(t) = -R^{-1}B^T P$$

$$\dot{P} = -PA - A^T P - Q + PBR^{-1}B^T P$$

$$P(t_f) = S$$

Figure 7.2-8 Structure of the linear quadratic regulator feedback solution

Now we can postulate a cost function to keep the deviations, $\Delta\mathbf{x}$ and $\Delta\mathbf{u}$, from the nominal (optimal) values, $\mathbf{x}^n$ and $\mathbf{u}^n$, small. A very reasonable one is the quadratic cost function

$$J = \frac{1}{2}\Delta\mathbf{x}^T(t_f)\mathbf{S}\Delta\mathbf{x}(t_f) + \frac{1}{2}\int_{t_0}^{t_f}[\Delta\mathbf{x}^T\mathbf{Q}\Delta\mathbf{x} + \Delta\mathbf{u}^T\mathbf{R}\Delta\mathbf{u}]\ dt \qquad (7.2\text{-}71)$$

and so we see that we can obtain a "fly by wire" closed loop controller for a nonlinear system as indicated in Fig. (7.2-9).

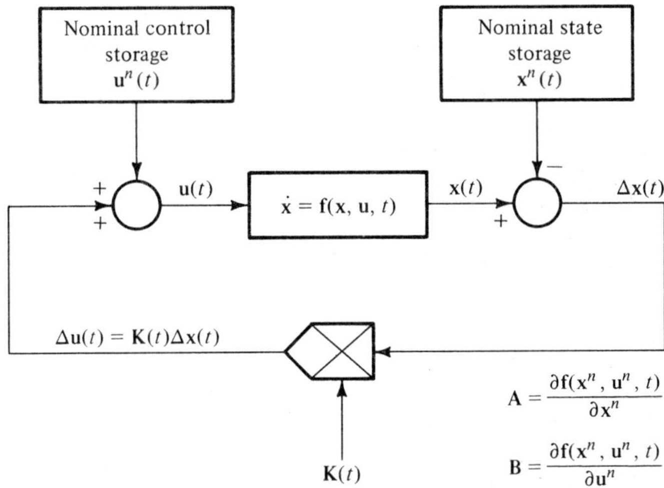

$$A = \frac{\partial f(x^n, u^n, t)}{\partial x^n}$$

$$B = \frac{\partial f(x^n, u^n, t)}{\partial u^n}$$

Figure 7.2-9 Linearized "fly by wire" control of nonlinear system—the gain $K(t)$ is determined from the regulator solution of Fig. (7.2-8)

In order to implement either the linear regulator solution or the linearized fly by wire solution we need to be aware of several items which may not be immediately apparent although they are certainly contained in our mathematical development:

1. We must solve the matrix Riccatti equation backwards in time from t_f to t_0. This poses no real problem in that the equation can be solved off line and the gain matrix $K(t)$ recorded and stored to be reversed in time as the controller is actually instrumented. It is occasionally mentioned that this can be avoided by solving the Riccatti equation backwards in time to find the initial condition $P(t_0)$ which will cause the final condition $P(t_f) = S$ to occur. Then the Riccatti equation is presumably solved forward in time with the initial condition $P(t_0)$ and the storage problem eliminated. Unfortunately this won't work! While the Riccatti equation of Fig. (7.2-8) is quite stable in the backward time direction, it is quite unstable in the forward time direction, and numerical errors in solving the equation combine such as to make this approach unworkable. We need a storage device to contain the precomputed $K(t)$.

2. The nominal state and control $x^n(t)$ and $u^n(t)$ must be known and stored. The $A(t)$ and $B(t)$ matrices for the fly by wire solution are generally time varying since they involve the known time functions $x^n(t)$ and $u^n(t)$. Thus changing nominal conditions while the system is operating, such as to fly a new wire, can involve considerable precomputation of gain matrices.

3. Often all state variables are not available for measurement, only the output variable

$$z = Cx + Du \qquad (7.2\text{-}72)$$

or

$$z = g(x, u)$$

and this can sometimes pose serious problems for practical implementation of modern control results. Fortunately it is possible to build optimal estimators or observers to estimate or recover $x(t)$ from $z(t)$. This estimator or observer is inserted after the physical location of the output variable $z(t)$ and the estimated or reconstructed $x(t)$ is used as the true $x(t)$ for implementation of the regulator solution.

4. Often the **S** matrix is selected as a very large matrix in order to keep the final errors small. It is thus sometimes expedient to solve for a Ξ matrix defined by $\Xi P = I$ where I is the identity matrix. We can differentiate this expression and obtain

$$\dot{\Xi}P + \Xi\dot{P} = 0$$

so that, since $P = \Xi^{-1}$

$$\dot{\Xi} = -\Xi\dot{P}\Xi$$

From Eq. (7.2-63) we obtain the alternate Ricatti equation

$$-\dot{\Xi} = -A\Xi - \Xi A^T + BR^{-1}B^T - \Xi Q\Xi \qquad (7.2\text{-}73)$$

and from Eq. (7.2-64) we have

$$\Xi(t_f) = S^{-1} \qquad (7.2\text{-}74)$$

The controller is then implemented with

$$u(t) = K(t)x(t) \qquad (7.2\text{-}75)$$

where

$$K(t) = -R^{-1}B^T\Xi^{-1}(t) \qquad (7.2\text{-}76)$$

and this will be sometimes easier to implement than the basic regulator solution.

Several digital computer routines have been written to assist in the design of linear regulators by allowing quick display of results for various trial and error combinations of weighting matrices **R**, **Q**, and **S**. A designer experienced in this method will find it to be a very useful asset and complimentary to a full appreciation of the Bode diagram and root locus approaches. This is especially the case for the infinite time interval problem which we will discuss after we present a simple example.

EXAMPLE 7.2-10. Let us reconsider the linear quadratic regulator problem considered in EXAMPLE 7.2-9. For that example we have

$$A = \begin{bmatrix} 0 & 1 \\ 0 & 0 \end{bmatrix}, \quad B = \begin{bmatrix} 0 \\ 1 \end{bmatrix}, \quad S = \begin{bmatrix} S_{11} & 0 \\ 0 & S_{22} \end{bmatrix}, \quad Q = 0, R = 1$$

Thus our regulator solution of Fig. (7.2-8) becomes

$$u(t) = \mathbf{K}(t)\mathbf{x}(t) = K_1(t)x_1(t) + K_2(t)x_2(t) \tag{1}$$

where

$$\mathbf{K}(t) = -\mathbf{R}^{-1}\mathbf{B}^T\mathbf{P}(t) = [-P_{12}(t) - P_{22}(t)] \tag{2}$$

The matrix Riccatti equation becomes, in component form,

$$\left. \begin{aligned} \dot{P}_{11} &= P_{12}, & P_{11}(t_f) &= S_{11} \\[2mm] \dot{P}_{12} &= -P_{11} + P_{22}P_{12}, & P_{12}(t_f) &= 0 \\[2mm] \dot{P}_{22} &= -2P_{12} + P_{22}^2, & P_{22}(t_f) &= S_{22} \end{aligned} \right\} \tag{3}$$

and as we suspected these will not be at all easy to solve analytically.

The alternate solution in terms of the Ξ matrix is actually simpler, at least analytically, in this case. From Eqs. (7.2-73) through (7.2-76) we have

$$u(t) = K_1(t)x_1(t) + K_2(t)x_2(t) \tag{4}$$

where

$$\left. \begin{aligned} \mathbf{K}(t) &= \left\{ \frac{\Xi_{12}(t)}{\Delta(t)} \quad \frac{-\Xi_{11}(t)}{\Delta(t)} \right\} \\[3mm] \Delta(t) &= \Xi_{11}(t)\Xi_{22}(t) - \Xi_{12}^2(t) \end{aligned} \right\} \tag{5}$$

The matrix Riccatti equation of Eqs. (7.2.73) becomes

$$\left. \begin{aligned} \dot{\Xi}_{11} &= -2\Xi_{12} & -\Xi_{11}(t_f) &= S_{11}^{-1} \\[1em] \dot{\Xi}_{12} &= -\Xi_{22} & -\Xi_{12}(t_f) &= 0 \\[1em] \dot{\Xi}_{22} &= 1 & -\Xi_{22}(t_f) &= S_{22}^{-1} \end{aligned} \right\} \tag{6}$$

The solution to this matrix Ricatti equation is easily obtained as

$$\Xi_{22}(t) = -(t - t_f) + S_{22}^{-1}$$

$$\Xi_{12}(t) = -\frac{(t - t_f)^2}{2} - S_{22}^{-1}(t - t_f) \tag{7}$$

$$\Xi_{11}(t) = -\frac{(t - t_f)^3}{3} + S_{22}^{-1}(t - t_f)^2 + S_{11}^{-1}$$

and now we obtain the optimal gains directly from use of Eq. (5). For the particular case in which $S_{11} = S_{22} = \infty$ we obtain

$$K_1(t) = -\frac{6}{(t - t_f)^2} \tag{8}$$

$$K_2(t) = \frac{-4}{(t - t_f)^2} \tag{9}$$

and we observe the interesting fact that the gains become infinite as the time to go $t - t_f$ becomes small. This occurs of course because $S_{11} = S_{22} = \infty$. This will necessitate approximation of the gains for small $t - t_f$ since we cannot physically store infinite valued gain terms.

EXERCISE 7.2-10. What do the results of EXAMPLE 7.2-10 become if $S_{11} = 0$? Show that this leads to a first order system optimization problem.

EXERCISE 7.2-11. Suppose that we wish to drive the system not to the origin but to some set point $\mathbf{x}(t) = \mathbf{x}^d$. The control which will cause $c(t)$ to remain in steady state at the set point $\mathbf{u}(t) = \mathbf{u}^d$ can generally be determined. The linear quadratic regulator problem can be stated, for the nonzero set point case, as that of minimizing

$$J = \frac{1}{2}\,[\mathbf{x}(t_f) - \mathbf{x}^d]^T \mathbf{S}[\mathbf{x}(t_f) - \mathbf{x}^d]$$

$$+ \frac{1}{2}\int_{t_0}^{t_f} \{[\mathbf{x}(t) - \mathbf{x}^d]^T \mathbf{Q}(t)[\mathbf{x}(t) - \mathbf{x}^d]$$

$$+ [\mathbf{u}(t) - \mathbf{u}^d]^T \mathbf{R}(t)[\mathbf{u}(t) - \mathbf{u}^d]\}\, dt$$

For the system

$$\dot{\mathbf{x}} = \mathbf{A}\mathbf{x} + \mathbf{B}\mathbf{u}$$

show that the change of variable

$$\mathbf{y}(t) = \mathbf{x}(t) - \mathbf{x}^d$$

$$\mathbf{v}(t) = \mathbf{u}(t) - \mathbf{u}^d$$

will reduce this problem to the basic linear quadratic regulator problem. How is the feedback controller implemented?

In our earlier work in this section we considered infinite time interval of operation constant coefficient problems. It is of interest to re-examine these problems using the linear quadratic regulator algorithms we have just obtained. From Fig. (7.2-8) we see that if we restrict $\mathbf{A}$, $\mathbf{B}$, $\mathbf{Q}$, and $\mathbf{R}$ to be constants then we should expect that $\mathbf{P}(t)$ should become a constant matrix given by

$$0 = -\mathbf{P}\mathbf{A} - \mathbf{A}^T\mathbf{P} - \mathbf{Q} + \mathbf{P}\mathbf{B}\mathbf{R}^{-1}\mathbf{B}^T\mathbf{P} \qquad (7.2\text{-}77)$$

In fact we can insure this for all time if we pick the value of $\mathbf{S} = \mathbf{P}$ obtained by solution of Eq. (7.2-77). The optimum feedback gain will be a constant matrix given by

$$\mathbf{K} = -\mathbf{R}^{-1}\mathbf{B}^T\mathbf{P} \qquad (7.2\text{-}78)$$

Equation (7.2-77) is still a nonlinear equation. There are many numerical algorithms to factor nonlinear polynomial equations as in Eq. (7.2-77). Alternately the matrix Riccati differential equation may be solved backwards in time and the steady state solution obtained. There are some subtleties involved in the infinite time regulator solution and those of most interest can be illustrated by means of a simple example.

EXAMPLE 7.2-11. We consider the second order system and scalar cost function

$$\dot{x}_1 = x_2$$

$$\dot{x}_2 = u$$

$$J = \frac{1}{2} \int_0^\infty [Q_{11} x_1^2(t) + Q_{22} x_2^2(t) + u^2(t)] \, dt$$

We obtain

$$A = \begin{bmatrix} 0 & 1 \\ 0 & 0 \end{bmatrix}, \quad B = \begin{bmatrix} 0 \\ 1 \end{bmatrix}, \quad Q = \begin{bmatrix} Q_{11} & 0 \\ 0 & Q_{22} \end{bmatrix}, \quad R = 1$$

and so Eqs. (7.2-77) and (7.2-78) become

$$0 = P_{12}^2 - Q_{11}$$

$$0 = -P_{11} + P_{22} P_{12}$$

$$0 = -2P_{12} + P_{22}^2 - Q_{22}$$

We solve these and obtain

$$P_{12} = Q_{11}^{1/2}$$

$$P_{22} = (2Q_{11}^{1/2} + Q_{22})^{1/2}$$

$$P_{11} = Q_{11}^{1/2} (2Q_{11}^{1/2} + Q_{22})^{1/2}$$

and see that the optimal controller is

$$u(t) = P_{12}x_1(t) - P_{22}x_2(t)$$

We may implement this controller as in Fig. (7.2-10a). Shown in this figure are the necessary inputs to drive the state to some prescribed set point $\mathbf{x}_d$ rather than zero. We see that we need have access to all state variables in order to implement this solution. By block diagram manipulation we see that we can realize this as in Fig. (7.2-10b). In our earlier discussions in this chapter we were not concerned with controlling other than a single output variable $x_1(t)$. We can accomplish this here by setting $Q_{22} = 0$. Our result simplifies somewhat in that we obtain $P_{22} = 2^{1/2}P_{12}^{1/2}$. The equivalent single loop, open loop transfer function $G(s)$ for this system is

$$G(s) = \frac{P_{12}}{s(s + P_{22})}$$

and the closed loop transfer function is

$$\frac{X(s)}{X^d(s)} = \frac{P_{12}}{s^2 + P_{22}s + P_{12}} = H(s)$$

By adjusting Q_{11} and Q_{22} we see that we vary the natural resonant frequency

$$\omega_n = P_{12}^{1/2} = Q_{11}^{1/4}$$

and damping ratio

$$\zeta = \frac{P_{22}}{2\omega_n} = \frac{(2Q_{11}^{1/2} + Q_{22})^{1/2}}{2Q_{11}^{1/4}}$$

For the particular case where $Q_{22} = 0$ we have a damping ratio $\zeta = 0.707$ and recognize this as a desirable value. Increasing Q_{22} will increase the damping above this value.

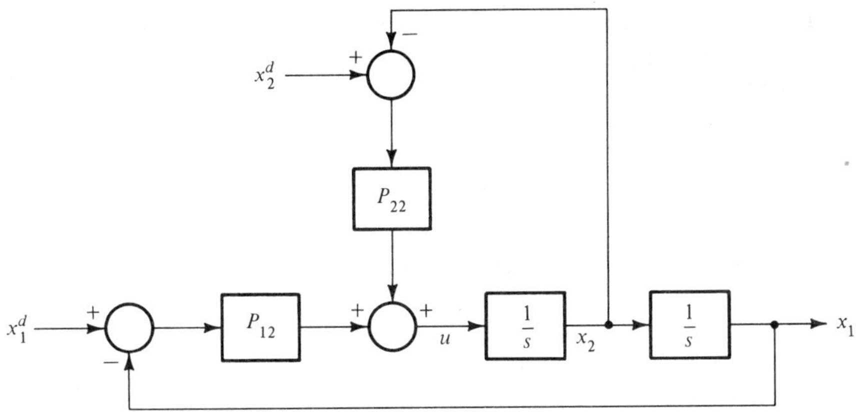

(a) Basic realization requiring that all state variables
be available for measurement

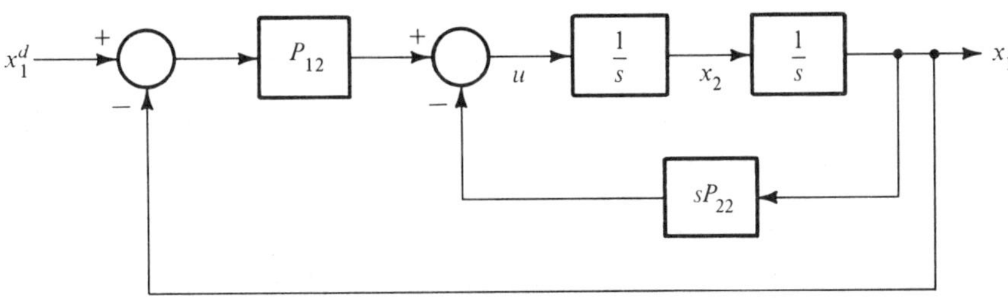

(b) Modified realization using minor loop feedback: $x_2^d = 0$

Figure 7.2-10 Realization of linear feedback control

It is of interest to contrast and compare this result which we have
obtained to that which we would obtain using our earlier frequency
domain approach. Here we minimize, for $Q_{22} = 0$

$$J = \frac{1}{2}\int_0^\infty [Q_{11}(x_1^d - x_1)^2 + u^2(t)]\, dt$$

which is equivalent to the frequency domain relation obtained from
Parceval's theorem

$$J = \frac{1}{4\pi j}\int_{-j\infty}^{j\infty} \{[X_1^d(s) - X_1(s)]Q_{11}[X_1^d(-s) - X_1(-s)] + U(s)U(-s)\}\, ds$$

We must recall here that the only input we are allowed to use for x_d is a set point or constant such that $X_1^d(s) = c/s$. We see that we obtain a very reasonable frequency domain interpretation for this problem. We are minimizing the integral square error where the error is defined by

$$E(s) = X_1^d(s)[1 - H(s)]$$

with a constraint on the integral square input $u(t)$ defined by

$$U(s) = s^2 H(s) X_1^d(s)$$

and we see that this is just a constraint on the integral square output acceleration. There is a considerable difference in philosophy between the two approaches in that we do not assume any specific compensation form in the optimal control approach.

EXERCISE 7.2-12. For the closed loop system with transfer function

$$\frac{Z(s)}{Y(s)} = H(s) = \frac{A}{s^2 + Bs + A}$$

and a unit step input $y(t)$, find the optimum value of A and B to minimize the integral square error where we constrain the output acceleration d^2z/dt^2 integral square value. Compare your results with the result of EXAMPLE 7.2-11.

EXERCISE 7.2-13. Repeat EXERCISE 7.2-12 for the transfer function

$$H(s) = \frac{k}{s^2 T + s + k}$$

Compare with the results of EXERCISE 7.2-13 and EXAMPLES 7.2-4 and 7.2-11.

EXERCISE 7.2-14. It is known that a (nominal) input

$$u^n(t) = 1 - 4t + t^2 \qquad 0 < t < 1$$

to the system

$$\dot{x} = -x^3 + u \qquad x(0) = 1$$

produces the (nominal) output

$$x^n(t) = 1 - t^2$$

Find a fly by wire controller to minimize

$$J = \frac{1}{2} \int_0^1 [Q\Delta x^2(t) + \Delta u^2(t)]\, dt$$

In our linear regulator problem we can assume as a desired state only a constant set point. Often we wish to have the system output $z(t)$ track a desired trajectory $z^d(t)$. The regulator problem where the set point changes in time is commonly known as the servomechanism problem. We consider the system described by

$$J = \frac{1}{2} [z^d(t_f) - z(t_f)]^T S[z^d(t_f) - z(t_f)]$$

$$+ \frac{1}{2} \int_{t_o}^{t_f} [(z^d - z)^T Q(z^d - z) + (u^d - u)^T R(u^d - u)]\, dt \quad (7.2\text{-}79)$$

where $u^d(t)$ is a desired control. The system is described by

$$\dot{x} = Ax + Bu, \qquad x(t_0) = x_0$$

$$z(t) = Cx \tag{7.2-80}$$

This problem reduces to the output regulator problem when z^d and u^d become constant. If $C = I$, the identity matrix, then the problem reduces to the state regulator problem which we have already solved.

To resolve this servomechanism problem we define the Hamiltonian

$$H = \frac{1}{2}(z^d - Cx)^T Q(z^d - Cx) + \frac{1}{2}(u^d - u)^T R(u^d - u)$$

$$+ \lambda^T Ax + \lambda^T Bu \tag{7.2-81}$$

The canonic equations are

$$\dot{x} = \frac{\partial H}{\partial \lambda} = Ax + Bu \tag{7.2-82}$$

$$\dot{\lambda} = \frac{-\partial H}{\partial \mathbf{x}} = \mathbf{C}^T \mathbf{Q}(\mathbf{z}^d - \mathbf{C}\mathbf{x}) - \mathbf{A}^T \lambda \qquad (7.2\text{-}83)$$

and the optimal control is obtained from

$$0 = \frac{\partial H}{\partial \mathbf{u}} = \mathbf{R}(\mathbf{u} - \mathbf{u}^d) + \mathbf{B}^T \lambda \qquad (7.2\text{-}84)$$

The two point boundary conditions are

$$\mathbf{x}(t_0) = \mathbf{x}_0 \qquad (7.2\text{-}85)$$

$$\lambda(t_f) = \mathbf{C}^T \mathbf{S}[\mathbf{C}\mathbf{x}(t_f) - \mathbf{z}^d(t_f)] \qquad (7.2\text{-}86)$$

To attempt a solution we assume, based on the terminal condition of Eq. (7.2-86),

$$\lambda(t) = \mathbf{P}(t)\mathbf{x}(t) - \zeta(t) \qquad (7.2\text{-}87)$$

Substituting Eq. (7.2-87) into Eq. (7.2-83) we obtain

$$\dot{\mathbf{P}}\mathbf{x} + \mathbf{P}\dot{\mathbf{x}} - \dot{\zeta} = \mathbf{C}^T \mathbf{Q}(\mathbf{z}^d - \mathbf{C}\mathbf{x}) - \mathbf{A}^T \mathbf{P}\mathbf{x} + \mathbf{A}^T \zeta \qquad (7.2\text{-}88)$$

We substitute $\mathbf{u}$ from Eq. (7.2-84) into Eq. (7.2-80) and use Eq. (7.2-87) to obtain

$$\dot{\mathbf{x}} = \mathbf{A}\mathbf{x} - \mathbf{B}\mathbf{R}^{-1}\mathbf{B}^T\mathbf{P}\mathbf{x} + \mathbf{B}\mathbf{R}^{-1}\mathbf{B}^T\zeta + \mathbf{B}\mathbf{u}^d \qquad (7.2\text{-}89)$$

Equation (7.2-89) is substituted into Eq. (7.2-88) and we have

$$(\dot{\mathbf{P}} + \mathbf{P}\mathbf{A} - \mathbf{P}\mathbf{B}\mathbf{R}^{-1}\mathbf{B}^T\mathbf{P} + \mathbf{C}^T\mathbf{Q}\mathbf{C} + \mathbf{A}^T\mathbf{P})\mathbf{x} + \mathbf{P}\mathbf{B}\mathbf{R}^{-1}\mathbf{B}^T\zeta$$

$$+ \mathbf{P}\mathbf{B}\mathbf{u}^d - \dot{\zeta} - \mathbf{C}^T\mathbf{Q}\mathbf{z}^d - \mathbf{A}^T\zeta = 0$$

We wish to obtain a solution valid for arbitrary $\mathbf{x}(t)$. The only way that this can happen is that the terms premultiplying $\mathbf{x}(t)$ must sum to zero. Also the other terms must be zero also so we replace the foregoing equation by the requirement that

$$\dot{\mathbf{P}} = -\mathbf{PA} - \mathbf{A}^T\mathbf{P} + \mathbf{PBR}^{-1}\mathbf{B}^T\mathbf{P} - \mathbf{C}^T\mathbf{QC} \qquad (7.2\text{-}90)$$

$$\dot{\boldsymbol{\zeta}} = -(\mathbf{A}^T - \mathbf{PBR}^{-1}\mathbf{B}^T)\boldsymbol{\zeta} + \mathbf{PB}u^d - \mathbf{C}^T\mathbf{Q}z^d \qquad (7.2\text{-}91)$$

The terminal conditions for these two equations are obtained from Eqs. (7.2-86) and (7.2-87) as

$$\mathbf{P}(t_f) = \mathbf{C}^T(t_f)\mathbf{SC}(t_f) \qquad (7.2\text{-}92)$$

$$\boldsymbol{\zeta}(t_f) = \mathbf{C}^T(t_f)\mathbf{S}z^d(t_f) \qquad (7.2\text{-}93)$$

Thus we see that the optimum control is given by

$$\mathbf{u} = \mathbf{R}^{-1}\mathbf{B}^T\mathbf{Px} + \mathbf{R}^{-1}\mathbf{B}^T\boldsymbol{\zeta} + \mathbf{u}^d \qquad (7.2\text{-}94)$$

where $\mathbf{P}(t)$ is obtained by solution of Eq. (7.2-90) with the terminal conditions of Eq. (7.2-92) and $\boldsymbol{\zeta}(t)$ is obtained by solution of Eq. (7.2-91) with the terminal conditions of Eq. (7.2-93). We notice that this solution consists of the linear quadratic regulator feedback solution plus a time varying open loop portion. A complicating feature of this solution is that the equation for $\boldsymbol{\zeta}(t)$ must also be solved backwards in time.

EXAMPLE 7.2-12.   We consider again the second order system

$$\dot{x}_1 = x_2 \qquad x_1(t_0) = x_{10}$$

$$\dot{x}_2 = u \qquad x_2(t_0) = x_{20}$$

The cost function is quadratic

$$J = \frac{1}{2}\int_0^{t_f} [Q_{11}(x_1^d - x_1)^2 + u^2]\, dt$$

The matrix Riccati equation (7.2-90) is the same as in our previous example

$$\dot{P}_{11} = P_{12}^2 - Q_{11} \qquad P_{11}(t_f) = 0$$

$$\dot{P}_{12} = -P_{11} + P_{12}P_{22} \qquad P_{12}(t_f) = 0 \qquad (1)$$

$$\dot{P}_{22} = -2P_{12} + P_{22}^2 \qquad P_{22}(t_f) = 0$$

The differential equation for $\zeta(t)$, Eq. (7.2-91) becomes

$$\dot{\zeta} = P_{12}\zeta_2 - Q_{11}x_1^d \qquad \zeta_1(t_f) = 0$$

$$\zeta_2 = -\zeta_1 + P_{22}\zeta_2 \qquad \zeta_2(t_f) = 0 \qquad (2)$$

and the control is

$$u(t) = P_{12}(t)x_1(t) - P_{22}(t)x_2(t) + \zeta_2(t) \qquad (3)$$

Offline solution of (1) and (2), for a given $x_1^d(t)$, from time t_f to time 0 will produce all gains needed for control determination of Eq. (3). Again the solutions to Eqs. (1) and (2) are obtained backwards in time. They must be stored and then reversed in time as the control system operates in real time.

For the infinite time interval case we can set $\dot{\mathbf{P}} = \mathbf{0}$ and obtain, as we have before,

$$P_{12} = Q_{11}^{1/2}, \qquad P_{22} = 2^{1/2}Q_{11}^{1/4}, \qquad P_{11} = 2Q_{11}^{3/4}$$

and Eq. (2) becomes

$$\dot{\zeta}_1 = Q_{11}^{1/2}\zeta_2 - Q_{11}x_1^d \qquad \zeta_1(t_f) = 0$$

$$\dot{\zeta}_2 = -\zeta_1 + 2^{1/2}Q_{11}^{1/4}\zeta_2 \qquad \zeta_2(t_f) = 0 \qquad (4)$$

Now as x_1^d is a constant and $t_f = \infty$ we easily obtain the equilibrium solution by setting $\dot{\zeta}_1 = \dot{\zeta}_2 = 0$ as $\zeta_2 = Q_{11}^{1/2}x_1^d$, $\zeta_1 = 2^{1/2}Q_{11}^{1/4}\zeta_2 = 2^{1/2}Q_{11}^{3/4}x_1^d$. This is precisely the solution we have obtained in our last example.

If on the other hand

$$x_1^d = 1 - e^{-t} \tag{5}$$

then we can show that as $t_f \to \infty$ we obtain

$$\zeta_2(t) = Q_{11}^{-1/2} - \frac{Q_{11}}{1 + Q_{11}^{1/2} + 2^{1/2} Q_{11}^{1/4}} \tag{6}$$

This is obtained in a straightforward way by writing Eq. (4) as a single second order differential equation

$$\ddot{\zeta}_2 = -Q_{11}^{1/2} \zeta_2 - A_{11} x^d + 2^{1/2} Q_{11}^{1/4} \dot{\zeta}_2$$

and assuming a solution $\zeta_2 = a + b e^{-t}$, substituting in $x_1^d = 1 - e^{-t}$ and equating coefficients and solving for a and b.

We could realize $\zeta_2(t)$ as a time function. Alternately we might wish to generate it from $x^d(t)$. We have

$$\frac{\zeta_2(s)}{X^d(s)} = Q_{11}^{1/2} + \left[\frac{Q_{11}^{-1/2} + 1 + 2^{1/2} Q_{11}^{1/4} - Q_{11}}{1 + Q_{11}^{1/2} + 2^{1/2} Q_{11}^{1/4}} \right] s \tag{7}$$

and we can implement a transfer function to filter $x^d(t)$ to give $\zeta_2(t)$.

Needless to say the complexity of the linear servomechanism, particularly since the $\zeta(t)$ term is very dependent upon the desired output, limits its practical usefulness.

EXERCISE 7.2-15. Investigate the linear servomechanism solution to the problem of minimizing

$$J = \frac{1}{2} \int_0^\infty [Q(x - x^d)^2 + u^2(t)] \, dt$$

for the system

$$\dot{x} = ax + u \qquad x(0) = 0$$

The desired output x^d is $1 - e^{-t}$.

This concludes our presentation of optimum linear systems control. We have attempted to present those introductory portions of optimum control most useful in the design of deterministic linear systems. We have indicated some relations between our optimal solutions and classical solutions to control problems. All of the topics we have presented in this section have a discrete time counterpart but space prevents our development of digital optimal control here.

If the transfer function of a system is unknown it might be desirable to determine it as the system operates and to determine the control based upon the identified parameters. This is the problem of adaptive control, or learning control, or self organizing control, to which we will now turn.

7.3 ADAPTIVE, LEARNING AND SELF ORGANIZING CONTROL SYSTEMS

Often control systems must be designed to operate in an environment such that the dynamics of the system or the inputs to the system are either incompletely known and/or change characteristics in an unpredictable way. Thus the design of a control system which will perform well in the fact of many uncertainties is an attractive possibility. Often when these changes are sufficiently small a conventional controller will be entirely effective. There will exist situations in which such controllers may not be sufficiently effective however.

As a trivial example suppose that the open loop gain of a single unity feedback ratio system is

$$G(s) = \frac{K(s+1)}{s(s-1)(s+4)(s+50)}$$

The root locus behavior of the system is indicated in Fig. (7.3-1). As we see there is a range of gain K over which the system response will be satisfactory. Both at low and high gains the system is unstable however. K might be composed of two parts, an electronic amplifier over which we have precise control and a physical process, such as the change in dynamic atmospheric pressure with altitude, over which we have no control. If we can measure the dynamic pressure and gain then we can adjust the gain of the electronic amplifier such that the product of the two is a constant. This sort of gain scheduling approach is a very simple form of adaptive control and is an often used approach. In other situa-

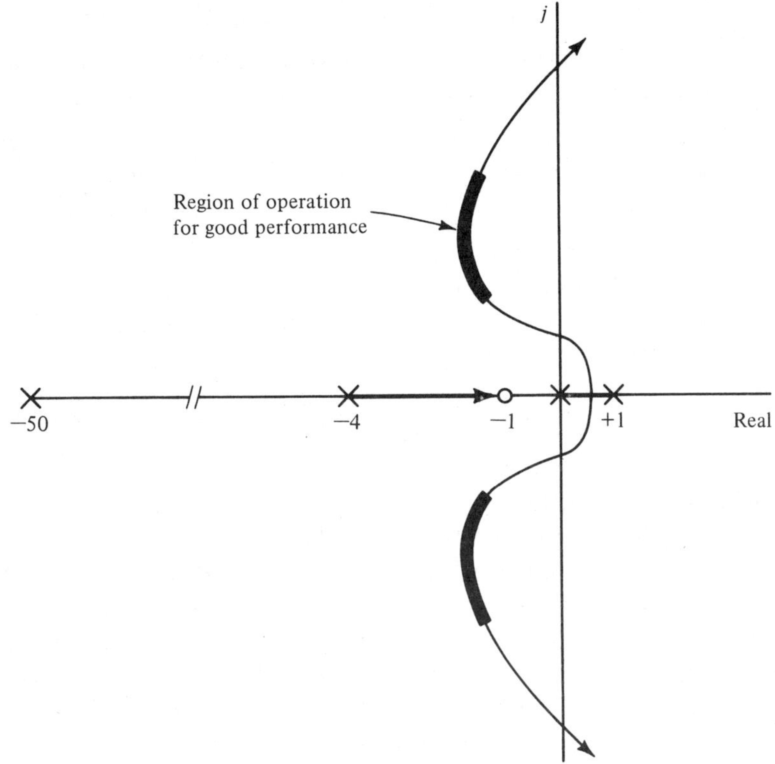

Figure 7.3-1   Root locus of system with $G(s) = K(s + 1)/s(s - 1)(s + 4)(s + 50)$

tions we do not know process relations sufficiently well to employ this type of open loop gain scheduling.

The primary objective of adaptive, learning or self organizing control is to reduce uncertainties concerning knowledge of the environment and system dynamics in an on line or real time fashion and to alter controller performance in an on line or real time fashion such as to cause system operation to continually seek better performance. A number of definitions have been suggested for the terms adaptive, learning, and self organizing and since different disciplinary specialists use these words there is confusion and controversy surrounding many of them. We have described four basic parts to an optimum systems control problem in Sec. 7.2. If we modify these requirements slightly such that parts 2, 3, and 4 occur in real time as the system is operating we then have the four basic parts or functions associated with an adaptive control system:

1. A goal or objective function is defined.

2. We determine our current position with respect to the objective function in real time during normal system operation.

3. We determine, in real time during normal system operation, all of the environmental and other factors affecting system transition from the past to the present and the future.

4. We determine and adjust system parameters and controls in such a fashion as to cause actual system performance to approach the ultimate performance specified by the goal or objective function.

There have been a large number of adaptive systems proposed in recent years. Most of these can be classified as either *performance adaptive*, in which observations of the input and output of the controller are made and the parameters of the controller adjusted by composing the input output performance of the system with a reference standard, or *parameter adaptive*, in which control system parameters are identified by observing control system input output relations and control system compensation networks are modified in an on line fashion in accordance with these changes. Certainly these can be a blending of performance adaptive and parameter adaptive systems which are illustrated in Fig. (7.3-2). We consider two parameter adaptive and one performance adaptive system in our efforts here.

We will first consider a simple parameter adaptive control system. This adaptation of the system of Fig. (7.3-2) is known as an input measurement parameter adaptive control system and was devised originally by R. F. Drenick and R. A. Shahbender. The purpose of the adaptive portion of the system is to measure input characteristics and adjust system parameters within a forward or feedback loop equalizer. This is a very simple adaptive system which only considers input changes and not system dynamics changes and many might consider it as a nonlinear gain system and not even an adaptive system. But then the adaptive concept is more of a mental concept or state of mind than anything else. In fact J. G. Truxal had defined an adaptive system as one designed from an adaptive viewpoint.

EXAMPLE 7.3-1. As a simple example of how we might approach some design aspects for this simple adaptive scheme let us consider the system shown in Fig. (7.3-3). The open loop transfer function is

$$G(s) = \frac{K(1 + s)(9 - s)}{s^2(9 + s)} \tag{1}$$

The input may be either a step or a ramp or a combination of these, i.e.,

$$u(t) = a + bt \qquad t \geqslant 0 \tag{2}$$

and we wish to find the value of the parameter K to minimize the integral square error.

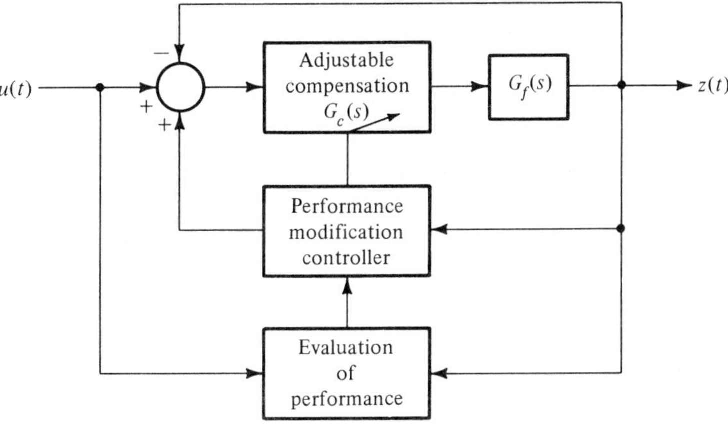

(a) Performance adaptive control system

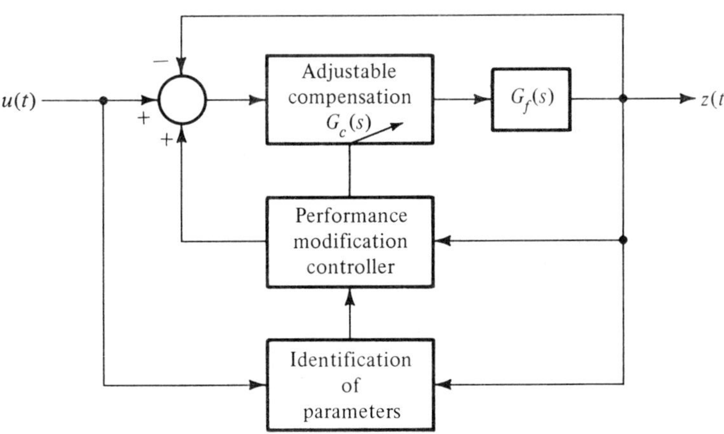

(b) Parameter adaptive control systems

Figure 7.3-2 Two basic types of adaptive systems

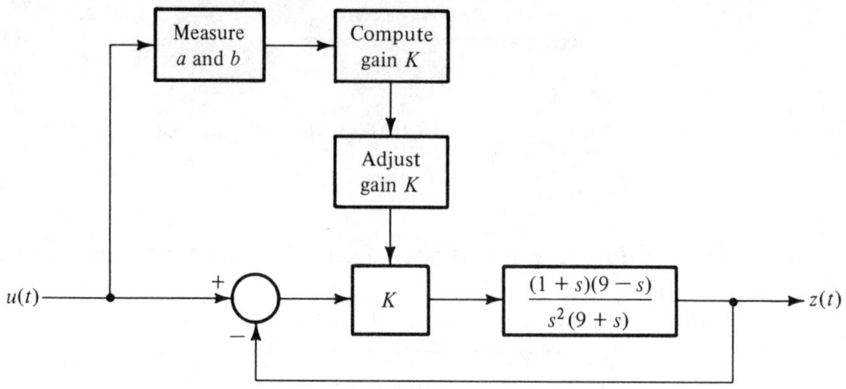

Figure 7.3-3 Block diagram of input adaptive control system

The response error is

$$E(s) = U(s)[1 - H(s)] = \frac{U(s)}{1 + G(s)} \tag{3}$$

and so we obtain

$$E(s) = \frac{as^2 + (9a + b)s + 9b}{s^3 + (9 - K)s^2 + 8Ks + 9K} \tag{4}$$

From TABLE 7.2-1 we obtain the integral square error for this step plus ramp input

$$\text{ISE} = \frac{72a^2 K^2 + (729a^2 - 72b^2)K + 729b}{18K^2(63 - 8K)} \tag{5}$$

The root locus for this system is shown in Fig. (7.3-4). The system becomes unstable at a gain $K = 63/8$. The optimum parameter $\hat{K}$ is determined as the positive value of K between 0 and 63/8 which minimizes Eq. (5). For a ramp input only where $a = 0$ we obtain

$$\hat{K}_{\text{ramp}} = 6.14$$

and for a step input only where $a = 0$ we obtain

$$\hat{K}_{\text{step}} = 3.38$$

and as the input values of a and b change the optimum gain changes between the ranges 3.38 and 6.14 according to the cubic equation obtained from setting $\partial \text{ISE}/\partial K = 0$.

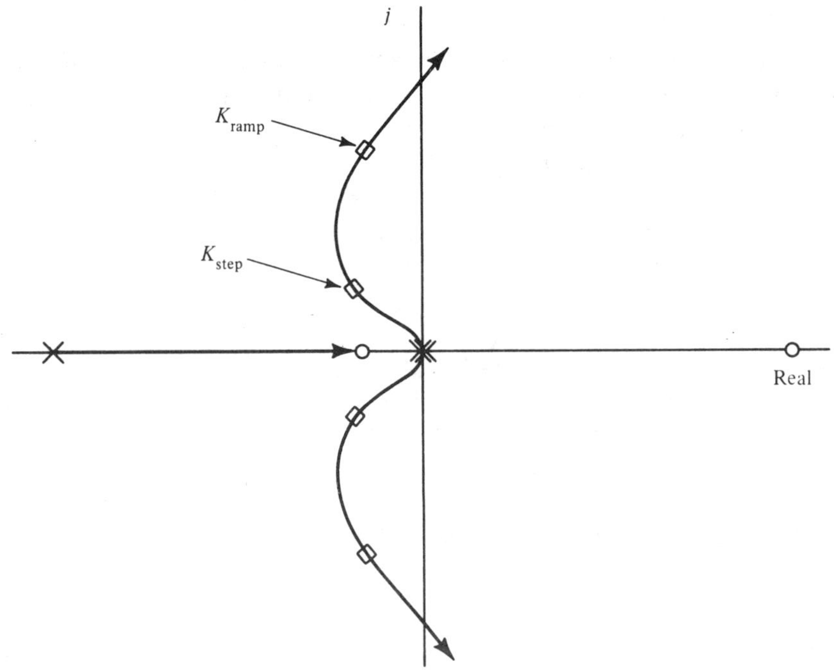

Figure 7.3-4   Root locus for $G(s) = \dfrac{K(1+s)(9-s)}{s^2(9+s)}$

Even though we may appear to be using a performance criterion here we use it only to establish an algebraic control parameter relation which allows us to determine the gain parameter K in terms of the height of the step and the slope of the ramp. This is, therefore, a parameter adaptive control system and *not* a performance adaptive control system. Obviously it is a very simple system and could be expected to perhaps not work well unless the input were truly a sequence of inputs of the form $a + bt$.

EXERCISE 7.3-1.   What are the optimum gain parameters for EXAMPLE 7.3-1 if the ITSE criterion is used?

EXERCISE 7.3-2. What are the optimum gain parameters for EXAM-PLE 7.3-1 if the error criterion is the sum of the integral square error for a step plus the sum of the integral square error for a ramp input?

A number of adaptive methods are based on various gradient or stepset descent computational algorithms. These include both parameter adaptive and performance adaptive systems. We will develop a simple gradient parameter based adaptive system first and will then conclude our development in this section with a discussion of a particular gradient performance adaptive system.

Figure (7.3-5) illustrates a simple form of gradient parameter adaptive algorithm previously investigated by a number of investigators but apparently originally by M. Margolis and C. T. Leondes. We have a physical control system with fixed plant

$$G_f(s) = \frac{\sum_{i=0}^{n-1} a_i s^i}{\sum_{i=0}^{n} b_i s^i} = \frac{Z(s)}{U_m(s)} \tag{7.3-1}$$

and we wish to determine or "learn" or identify the values of the various a and b which are unknown. We have constructed a model of the fixed plant given by

$$G_m(s) = \sum_{i=0}^{n-1} \frac{c_i s^i}{d_i s^i} = \frac{Z_m(s)}{U_f(s)} \tag{7.3-2}$$

The same signal is input to the model $U_f(s)$ that is input to the fixed plant. The difference in the output of the model and the fixed plant is

$$e(t) = z_m(t) - z(t) \tag{7.3-3}$$

$$E(s) = Z_m(s) - Z(s) = [G_m(s) - G_f(s)] U_f(s) \tag{7.3-4}$$

is obtained. We form a convex function of $e(t)$ such as the square of $e(t)$ or perhaps the time integral of square of $e(t)$. The values of the c_i and d_i which minimize this convex function are obtained by a gradient search procedure. The convex function will normally be minimized when $a_i = c_i$ and $b_i = d_i$. But if the input signal is zero $u_f(t) = 0$ then there will be no error $e(t)$ and the gradient search method may learn incorrect values of the a_i and b_i. Often this is remedied by always inserting a low level signal $u_f(t)$ and use of a smoothing operator, such as an integrator, to constitute the convex error function.

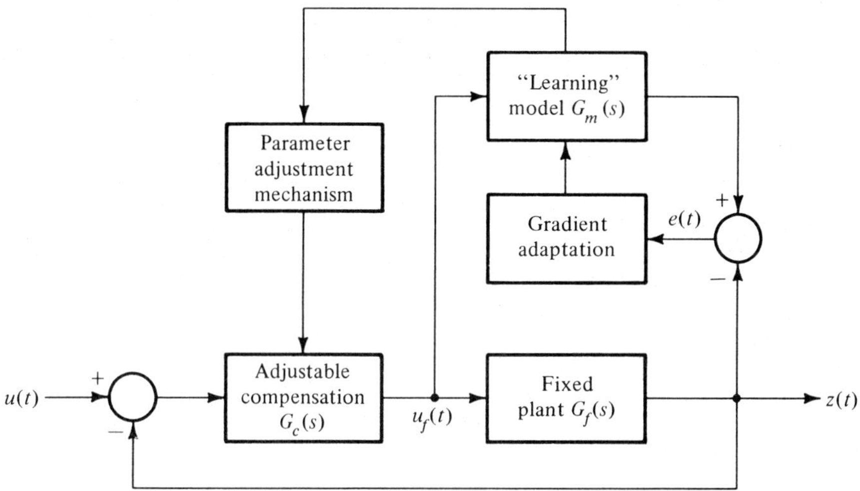

Figure 7.3-5 Gradient learning model parameter adaptive control system

For the case where we form a scalar convex function of the error $\theta[e(t)]$ we may adjust parameter d_i according to the first order relationship

$$\frac{dd_i(t)}{dt} = -K\frac{\partial\theta[e(t)]}{\partial d_i} \qquad (7.3\text{-}5)$$

or

$$d_i(t+T) = d_i(t) - KT\frac{\partial\theta[\dot{e}(t)]}{\partial d_i} \qquad (7.3\text{-}6)$$

where K is a number greater than zero. A simple example will illustrate further some of the features of this gradient parameter adaptive control system.

EXAMPLE 7.3-2. We consider the first order fixed plant described by

$$\frac{Z(s)}{U_f(s)} = G_f(s) = \frac{1}{s + 6} \tag{1}$$

We use a similar transfer function for the learning model

$$\frac{Z_m(s)}{U_f(s)} = G_m(s) = \frac{1}{s + d} \tag{2}$$

Suppose that we use the simplest possible convex error function, the error squared $e^2(t)$, as the parameter identification cost. The error equation is obtained from

$$\frac{dz}{dt} = -bz + u_f(t) \tag{3}$$

and

$$\frac{dz_m}{dt} = -dz_m + u_f(t) \tag{4}$$

by simple subtraction as

$$\frac{de(t)}{dt} = \frac{dz_m}{dt} - \frac{dz}{dt} = -dz_m + bz \tag{5}$$

It will happen, in general, for systems with no finite zeros that the formal expression for the error will not explicitly contain the input $u_f(t)$. Of course both $z_m(t)$ and $z(t)$ are functions of $u_f(t)$.

In order to implement the gradient method of Eq. (7.3-5) we need an expression for $\partial\theta[e(t)]/\partial d$. But since we have $\theta[e(t)] = e^2(t) = [z_m(t) - z(t)]^2$ and since $z(t)$ is not a function of d we have

$$\frac{\partial\theta[e(t)]}{\partial d} = 2[z_m(t) - z(t)]\,\frac{\partial z_m(t)}{\partial d}$$

$$= 2[z_m(t) - z(t)]\,\frac{\partial e(t)}{\partial d} \tag{7}$$

If we have an instrumentable expression for $\partial z_m(t)/\partial d$ or $\partial e(t)/\partial d$ we are then able to implement the gradient algorithm of Eq. (7.3-5). If we take the partial derivative of Eq. (4) we obtain

$$\frac{\partial}{\partial d}\left(\frac{dz_m}{dt}\right) = -z_m - d\,\frac{\partial z_m}{\partial d} \tag{8}$$

It is convenient to define a parameter

$$D = \frac{\partial z_m}{\partial d} \tag{9}$$

such that Eq. (8) becomes

$$\frac{dD}{dt} = -dD - z_m \tag{10}$$

Alternately we could use Eq. (5) to obtain a differential equation for $\partial e(t)/\partial d$. Solution of Eq. (10) for D allows us to implement the gradient algorithm

$$\frac{dd(t)}{dt} = -K\,\frac{\partial\theta\,[e(t)]}{\partial d} = -2K[z_m(t) - z(t)]\,D(t)$$

where we obtain $D(t)$ from solution of Eq. (10). We generally must guess at the initial value both of $d(t_0)$ and $D(t_0)$ although we will often have a reasonably good idea concerning these values. Also there is an approximation used in obtaining Eq. (8) in that the time derivative of the control signal input to the fixed plant $\partial u_f(t)/\partial d$ will not really be zero. It would be zero if no adaptive loop were present but the presence of the adaptive loop causes changes in d as d adjusts itself to identify the parameter b. It is usually true that the adaptive loop is much slower in its response characteristics than the primary control loop. Thus it is not at all unreasonable to neglect the $\partial u_f(t)/\partial d$ term.

We use the learned or identified value of b, which is the d term in our learning model, as input to a parameter adjustment computer which adjusts parameters in a compensator according to some preset algorithm. There exists the need to design the adaptive loop to insure stability and reasonable performance. Although the adaptive loop is nonlinear it can be linearized and methods presented in this text used for design purposes.

EXERCISE 7.3-3. Discuss the design of a lead network compensator and its use for the fixed plant $G_f(s) = K/s(T_s + 1)$. The time constant T is variable and a gradient parameter adaptive control system is used.

As our final type of adaptive control system we consider the special type of gradient performance adaptive system known as the model reference adaptive system and developed by a number of researchers at MIT. A block diagram of a model reference adaptive system is shown in Fig. (7.3-6). As suggested by its name, model reference adaptive control systems utilize a model, the model representing the ideal behavior and design specifications sought from the system which is to be controlled. The command input $u(t)$ is applied both to the basic control system as well as to the reference model. The control system output is subtracted from the reference model output and the error signal is obtained. Functions of the error are generated and used for parameter adjustment. Command signals for parameter adjustment are input to parameter adjustment devices in the basic control system such that the system output matches, in an optimum fashion, the output of the model. No attempt is made to identify or learn system parameters as in the previous parameter adaptive system; instead the control system is tuned such that it looks best like the model reference in the sense of minimizing an appropriately chosen error criterion.

Model reference performance adaptive systems may use either parameter adjustment (and perturbation) or control signal synthesis or a combination of these. We will discuss the parameter adjustment technique only in our efforts here. Figure (7.3-6) which illustrates the basic model reference technique shows the provision for parameter adjustment only. The adaptive loop functions quasi-independently of the basic control system and the basic control system should still function in a nonadaptive way if the adaptive loop fails. The error signal generated is input to a performance computation subsystem to cause optimal compensating network parameter changes to cause the system output to be closely matched to the output of the model reference system which system, as mentioned previously, does not change.

All adaptive systems are inherently nonlinear and the model reference system is no exception. An adaptive system is generally much too complex for completely analytical treatment. However with some simplifying assumptions basic analytical design features may be developed and the resulting design subjected to simulation and test to ensure that it performs satisfactorily.

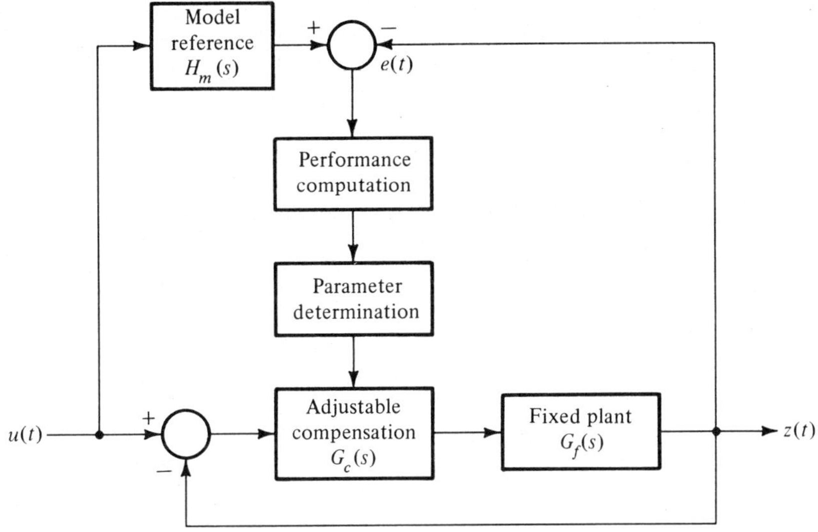

Figure 7.3-6 Gradient performance adaptive model reference parameter adjustment system

One of the principal tasks in any adaptive design philosophy is to define an appropriate measure of optimum performance and provision of an acceptable means to automatically seek this performance. We will use an integral performance measure of the form

$$J = \int_{t_1}^{t_2} \phi[e(t)] \, dt \tag{7.3-7}$$

where $\phi[e(t)]$ is an even function of $e(t)$ the error signal. Our definition of optimum performance for a system with adjustable parameter $\mathbf{p}$ requires that the gradient of the cost function of Eq. (7.3-7) with respect to the adjustable parameter be zero. Until the model reference adaptive system is in full adaptation to the model, this gradient will not be zero. This gradient is called, in model reference terminology, the error quantity

$$\mathrm{EQ} = \frac{\partial J}{\partial \mathbf{p}} = \frac{\partial}{\partial \mathbf{p}} \int_{t_1}^{t_2} \phi[e(t)] \, dt \tag{7.3-8}$$

To simplify the presentation, we shall assume that $\phi[e(t)] = e^2(t) = [z(t) - z_m(t)]^2$. We shall assume that we may exchange the operation of integration and differentiation such that we obtain

$$\text{EQ} = \int_{t_1}^{t_2} \frac{\phi[e(t)]}{\partial \mathbf{p}} \, dt = 2\int_{t_1}^{t_2} e(t) \frac{\partial e(t)}{\partial \mathbf{p}} \, dt \qquad (7.3\text{-}9)$$

A term such as $\partial e(t)/\partial p_n$ is called the error-weighting function, $W_n(t)$. Since the model output is not a function of p_n, the error-weighting function becomes $\partial z/\partial p_n$ which cannot be evaluated without identification of the dynamic characteristics of the system. Figure (7.3-7) illustrates the MIT method of explicit generation of the error quantity for a particular case in which p_n is a forward loop gain for a linearized system. In terms of Laplace transforms (for a slowly varying p_n), we have

$$W_n(s) = \frac{\partial E(s)}{\partial p_n} = \frac{\partial Z(s)}{\partial p_n} = \left[\frac{G_1(s)G_2(s)}{1 + G_1(s)p_n G_2(s)} \right]$$

$$\times \left[\frac{U(s)}{1 + G_1(s)p_n G_2(s)} \right] \qquad (7.3\text{-}10)$$

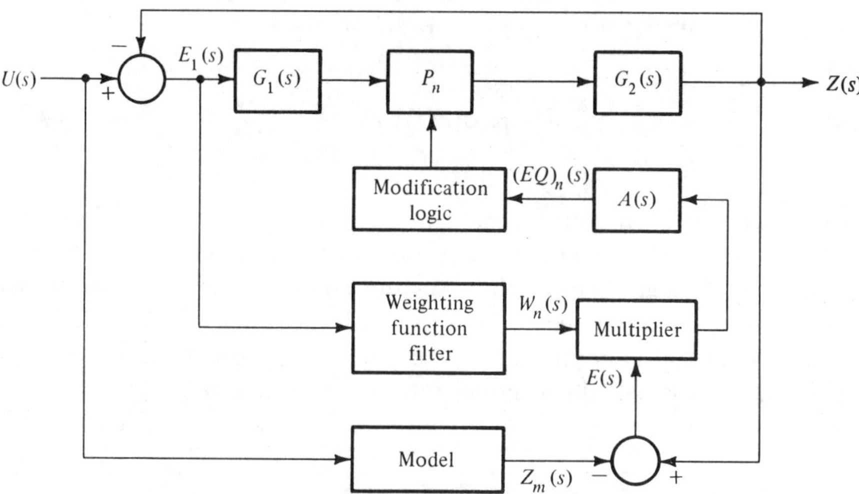

Figure 7.3-7 Signal generation in model reference performance adaptive system

The first term in this expression represents the closed loop poles and zeros of the system. The last term is the Laplace transform of the signal $e_1(t)$ which is the basic system error. It is important that we note here that $e(t)$ is not the basic system error, $e_1(t) = u(t) - z(t)$ but the performance error $z_m(t) - z(t)$. If we assume that the model is an accurate representation of the system, then the model transfer function may be used instead of the system function as a weighting function filter to yield the block diagram of Fig. (7.3-7). It can be demonstrated by appropriate simulations that this approximation works very well when an accurate model is chosen and when the unknown system parameters are near those of the model. The adjustable parameter p_n is then adjusted in accordance with an appropriate linear differential equation or possibly even a nonlinear differential equation. The basic MIT system uses a rate of change of the parameter directly proportional to the slope of the error quantity

$$\frac{\partial p_n}{\partial t} = -ke(t)W_n(t) \tag{7.3-11}$$

The response characteristics of the adaptive loop are, of course, dependent upon the adaptive loop transfer characteristics as well as the model used in the adjusting system and our input signal $u(t)$. Since the adaptive loop is nonlinear, a simple stability analysis or compensation procedure is not available. We may linearize the adaptive loop by letting $p_n(t) = p_{\text{nominal}} + \Delta p_n(t)$ in relevant systems equations and by dropping the products of dependent variables in writing the equations of the adaptive loop. This yields the linear time varying system of Fig. (7.3-8) where we assume that the model is of the same order as the system with $p_{n\,\text{model}} = p_{n\,\text{nominal}} = p_{nN}$. If the coefficient matrix or transfer function of the model is considerably different from that of the system, the linearized model of the adaptive loop is much more complicated. In this linearized model, it is apparent that adaptive loop stability is greatly dependent on the input signal. If a "fast" adaptive loop is to be designed, it seems reasonable to expect that we could use a standard test input to advantage in the adaptive loop.

EXAMPLE 7.3-3. We will use a simple example to demonstrate the design technique and also to point up the availability requirements of signals for parameter adjustment. Figure (7.3-9) is the block diagram of

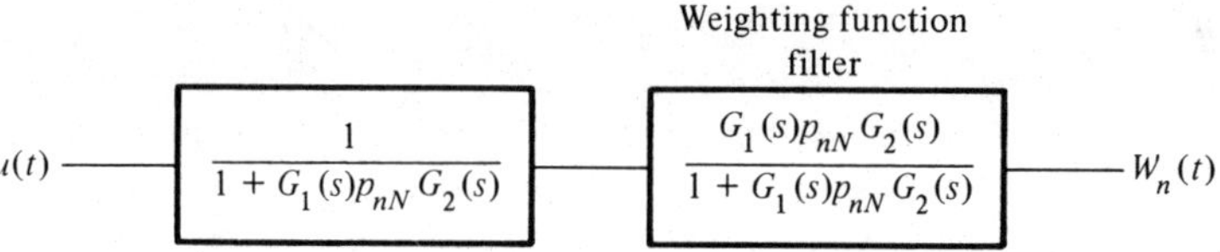

(a) Weighting function generation model

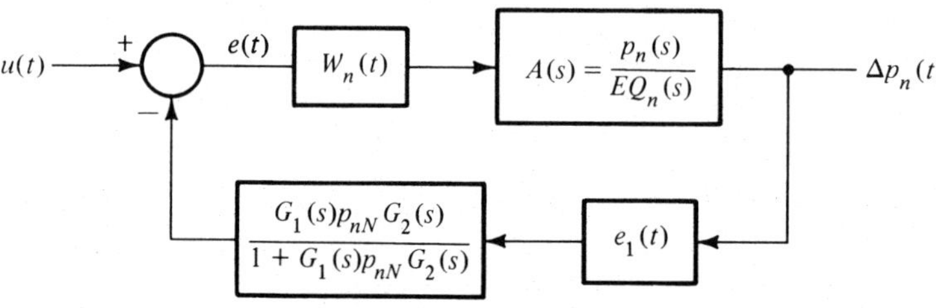

(b) Dynamic parameter change model

Figure 7.3-8   Linearized adaptive loop

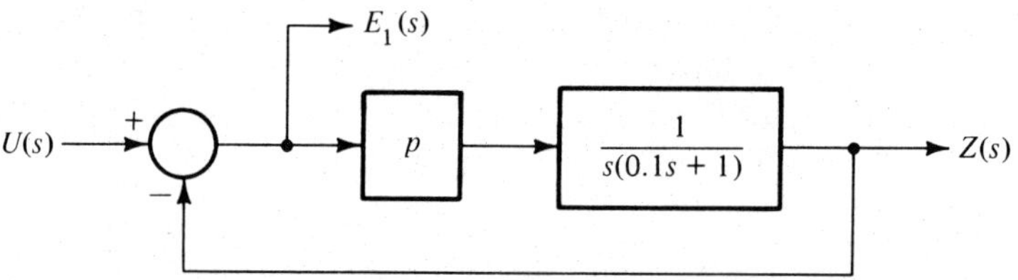

Figure 7.3-9   Second order system with adjustable gain

a second order system with an adjustable parameter in the feedforward path. We desire to adjust p by using a reference model with the open loop transfer function

$$G(s) = \frac{10}{s(0.1s + 1)} \tag{1}$$

The design procedure calls for the generation of $\partial Z(s)/\partial p$. For this system we have

$$Z(s) = \frac{p}{s(0.1s + 1) + p} U(s) \tag{2}$$

The required partial derivative is found to be

$$\frac{\partial Z(s)}{\partial p} = \frac{s(0.1s + 1)}{[s(0.1s + 1) + p]^2} U(s)$$

$$= (\text{TF}_{\text{sys}}) \frac{1}{p} \frac{s(0.1s + 1)}{s(0.1s + 1) + p} U(s) \tag{3}$$

where TF_{sys} is the closed loop transfer function of the controlled system. In terms of the input signal to p, we have

$$\frac{\partial Z(s)}{\partial p} = (\text{TF}_{\text{sys}}) \frac{1}{p} E_1(s) \tag{4}$$

At this point we replace the typically not-so-well-known transfer function of the controlled system by the transfer function of the reference model. The error-weighting function is now written, using this approximation, as

$$W(s) = (\text{TF}_{\text{mod}}) \frac{1}{p} E_1(s) \tag{5}$$

We find it advantageous to replace the $1/p$ term in the equation by unity. This does not significantly affect the adaptive response, and division by a varying quantity is not necessary. This modification is employed here and we obtain

$$W(s) = (\text{TF})_{\text{mod}} E_1(s) \tag{6}$$

Parameter adjustment calls for integration of the product of the response error and the error-weighting function. Our design procedure was derived with the assumption that parameter variation during adaptive response by the system is zero. The nonlinear nature of the system requires this assumption, which is obviously not true in the actual control system. It turns out however that the parameter variation seen by the adaptive system is an apparent parameter variation which is very small compared to the actual parameter variation, thus establishing the validity of the assumption. Figure (7.3-10) is a block diagram depicting the elements of the example system with its adaptive control loop. Figure (7.3-11) illustrates the adaptive response of the system for two offset values of the gain parameter p. Despite the approximation for the purposes of analysis, the model reference adaptive concept does work well and has been extensively investigated and proposed for a variety of applications. Among these are the control of large flexible boosters and the startup of a nuclear rocket engine.*

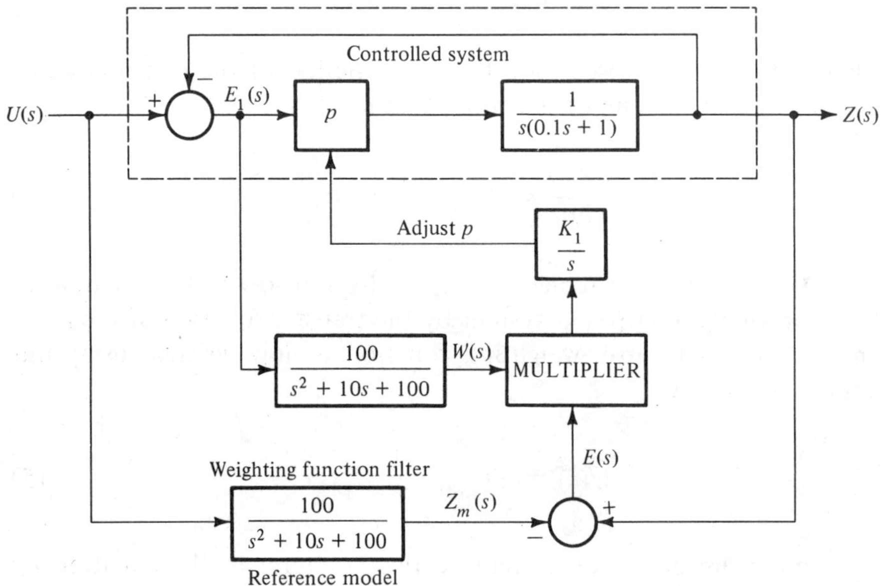

Figure 7.3-10 Model reference performance adaptive block diagram

In this section we have indicated some simple approaches to the adaptive control of systems. Three simple examples of the basic performance adaptive and parameter adaptive systems have been discussed. As we have seen in our examples there will often be considerable complexity inherent in an adaptive system and one should not normally attempt control by adaptive means unless it has been shown that con-

---

*See for instance: Kezer, A., Hofman, L. G., and Engle, A. G., "Application of Model Reference Adaptive Control Techniques to Provide Improved Bending Response of Large Flexible Missiles," *Proceedings Sixth Symposium on Ballistic Missile and Space Technology*, Academic Press, Inc., New York, 1961, pp. 113–151, and Humphries, J. T., Uhrig, R. E., and Sage, A. P., "A Model Reference Adaptive Control System for a Nuclear Rocket Engineer," *Proc. Joint Automatic Control Conf.*, Aug. 1966, pp. 492–499.

ventional means fail. There have been some past successful practical applications of adaptive control, notably to autopilot control systems, and the topic is currently an active research area as pattern recognition, speech processing and recognition, self organizing control, and other areas all use systems based on the adaptive concept, and of course the human body and human machine systems are all inherently adaptive.

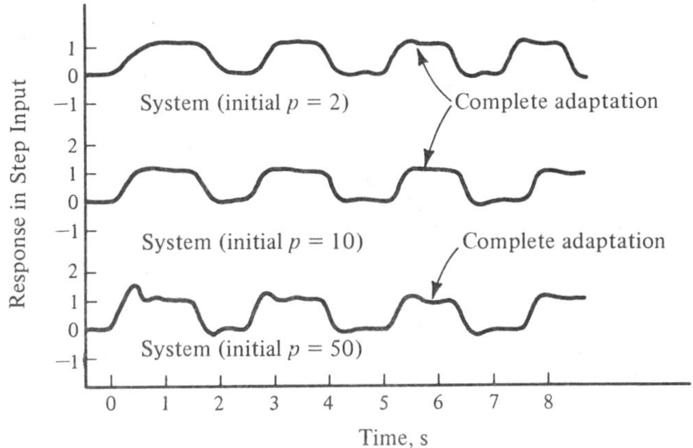

Figure 7.3-11　Response of adaptive system to a sequence of step inputs [Note: $p = 10$ in reference model.]

7.4 SUMMARY

Three topics have been examined in this, our concluding chapter of Linear Systems Control. We have decided to aggregate together the topical areas of digital, optimal, and adaptive control primarily because they are areas which represent extensions of the basic theory presented in earlier chapters. Also these three areas represent the beginning of a transition from the so-called classical control to the so-called modern control. The names classical and modern are inappropriate names but they are the names often used.

They each compliment the other as we have attempted to indicate at several points in this chapter. Knowledge of both is essential to enable most appropriate selection of the most beneficial tools with which to approach a particular systems control problem.

7.5 REFERENCES

There are a number of references which we might cite as the topics mentioned in this chapter have been the subject of intensive research over the past 20 years and at the present time. As appropriate to an introductory text we will only cite a selection of textbook references which often expand upon our coverage here. These textbooks contain full citations to the original literature and authors responsible for the original developments.

Some of the earlier textbooks on sampled data and discrete time control remain the very best. The following texts present much discussion of sampled data, discrete time, and digital control concepts:

Ragazzini, J. R. and Franklin, G. F., *Sampled Data Control Systems*, McGraw-Hill Book Co., New York, 1958.

Tou, J. T., *Digital and Sampled Data Control Systems*, McGraw-Hill Book Co., New York, 1959.

Jury, E. I., *Sampled Data Control Systems*, John Wiley & Sons, Inc., New York, 1958.

Also there are chapters on sampled data and digital control in a number of the basic textbooks referenced in earlier chapters. The following texts present a discussion both of discrete time systems and optimum systems; and present discussions of adaptive systems as well.

Tou, J. T., *Modern Control Theory*, McGraw-Hill Book Co., New York, 1964.

Kuo, B. C., *Discrete Data Control Systems*, Prentice Hall Book Co., Englewood Cliffs, N. J., 1970.

Gibson, J. E., *Nonlinear Automatic Control*, McGraw-Hill Book Co., New York, 1963.

Many recent texts are devoted more or less exclusively to the subject of optimum control and related topics of estimation, system identification and stochastic adaptive control. A selection among these which the reader may find helpful includes:

Newton, G. C., Jr., Gould, L. A., and Kaiser, J. F., *Analytical Design of Linear Feedback Controls*, John Wiley & Sons, New York, 1957.

Eveleigh, V. W., *Adaptive Control and Optimization Techniques*, McGraw-Hill Book Co., New York, 1967.

both of which contain detailed tables of integrals useful in evaluating integral square error as well as extensive discussions of the frequency domain approach to linear system optimization. Also of considerable interest are:

Astrom, K. J., *Introduction to Stochastic Control Theory*, Academic Press, New York, 1970.

Athans, M. and Falb, P. L., *Optimal Control*, McGraw-Hill Book Co., New York, 1966.

Bryson, A. E. and Ho, Y. C., *Applied Optimal Control*, Blaisdell, Boston, 1969.

Meditch, J. S., *Stochastic Optimal Linear Estimation and Control*, McGraw-Hill Book Co., New York, 1969.

Sage, A. P. and Melsa, J. L., *Estimation Theory with Applications to Communications and Control*, McGraw-Hill Book Co., New York, 1971.

Sage, A. P. and Melsa, J. L., *System Identification*, Academic Press, New York, 1971.

Sage, A. P. and White, C. C., *Optimum Systems Control*, Second Edition, Prentice Hall, Inc., Englewood Cliffs, N. J., 1977.

A very complete summary of progress in adaptive, learning and self organizing control is available in:

Saridis, G. N., *Self Organizing Control of Stochastic Systems*, Marcell Dekker Inc., New York, 1977.

7.6 PROBLEMS

1. For the block diagram of Fig. (7.1-1) suppose that

$$G(s) = \frac{1 - e^{-sT}}{s} \frac{K}{s(s + 1)}$$

where the sample period is 1 second. Determine:

   a. the closed loop transfer functions

   b. a Nyquist plot of the system

   c. a root locus diagram of the system

   d.   the phase margin of the system as a function of K

   e.   the step response of the system for several representative values of K.

2. A method sometimes used to analyze sampled data systems is that of using the first two terms of the starred transform approximation of Eq. (7.1-13)

$$G^*(s) = \frac{1}{T}G(s) + \frac{1}{T}G(s - j\omega_s) + \frac{1}{T}G(s + j\omega_s)$$

Use this approximation to plot a Nyquist diagram for a system with the block diagram of Fig. (7.1-1) with $T = 1$ and

$$G(s) = \frac{K}{s(1 + s)}$$

Compare with the true Nyquist diagram for the sampled system.

3. We may develop a general method for data reconstruction as follows. The general purpose of a data reconstructor is to predict an input τ seconds into the future from the last sample point. That is to say accomplish the operation $e^{s\tau}$. Now $e^{s\tau}$ is an unrealizable transfer function and must be approximated in order to make it realizable. We can rewrite this expression as $[e^{-sT}]^{-\tau/T}$ where e^{-sT} is realizable. We expand this further as

$$e^{s\tau} = [e^{-sT}]^{-\tau/T} = [1 - (1 - e^{-sT})]^{-\tau/T}$$

and expand this in a Taylor series about $1 - e^{-sT} = \eta = 0$. We obtain

$$e^{s\tau} \cong 1 + (1 - e^{-sT})\frac{\tau}{T} + \frac{1}{2}(1 - e^{-sT})^2 \frac{\tau + T}{T}\frac{\tau}{T} + \cdots$$

   a.   Using this relationship show that the input output relationship for the hold circuit is

$$z(nT + \tau) = u(nT) + \frac{\tau}{T}[u(nT) - u(\overline{n - 1}\,T)]$$

$$+ \frac{(\tau + T)}{2T^2}[u(nT) - 2u(\overline{n - 1}\,T) + u(\overline{n - 2}\,T)] + \cdots$$

    b. What is the next term in the approximation?

    c. Derive the zero, first and second order hold circuits transfer functions from this relationship.

    d. Find the response of the three hold circuits to a step input, a ramp input, and a t^2 input.

4. Show that the static and dynamic error coefficient approach discussed in Chap. 4 is directly applicable to sampled data systems. Derive these error coefficients for sampled data systems of the form of Fig. (7.1-6c) where $G_2(s) = 1$.

5. A suggestion has been made that a sampled data system in the form of Fig. (7.1-1) with $G(s) = G_{Ho}(s)G_c(s)G_f(s)$ where $G_{Ho}(s)$ represents a zero order hold, $G_c(s)$ represents a compensating network transfer function, and $G_f(s)$ the fixed plant, behaves essentially as a continuous time system containing $G_c(s)G_f(s)$ but where the sample hold device is replaced by a time delay of half a sampling period or a finite pole zero approximation to this. Investigate this assertion using the system of Problem 7.1.

6. A "deadbeat" second order system is one that has the response for a sample sequence ramp input, $u(nT) = nT$, $n \geqslant 0$, the output $z(nT) = 0$, $n < 2$, and $z(nT) = nT$, $n \geqslant 2$. Find the z transfer function of a closed loop system which realizes this and the open loop transfer function $G(z)$ of a unity ratio feedback system. What is the phase margin of this second order deadbeat system? What is the step response?

7. Determine root loci for the system of Fig. (7.1-1) where the open loop system $G(s)$ consists of a transfer function

$$G_f(s) = \frac{K}{s(1 + s)}$$

The sample period is 5 seconds and the fixed plant is preceded by:

    a. no hold circuit,

    b. a zero order hold, and

    c. a first order hold

What value of gain K gives the most reasonable phase margin and closed loop poles?

8. For the system of Fig. (7.1-9) with

$$G_{\text{Ho}}(s) = \frac{1 - e^{-sT}}{s}$$

$$G_f'(s) = \frac{10}{s(1 + s/10)^2}$$

Find suitable $G_c(z)$ to yield a 45° phase margin and 0.707 damping. Use EXAMPLE 7.1-11 as a guide to your analysis and design efforts.

9. Standard integration formulae, routinely obtainable from Newton's formula, are:

$$\frac{Tz^{-1}}{1 - z^{-1}} \qquad\qquad \text{Rectangular Rule}$$

$$\frac{T(1 + z^{-1})}{2(1 - z^{-1})} \qquad\qquad \text{Trapezoidal Rule}$$

$$\frac{T}{2}\frac{(z^{-1})(4 - 3z^{-1} + z^2)}{(1 - z^{-1})} \qquad\qquad \text{Three Point Extrapolation}$$

$$\frac{T}{3}\frac{1 + 4z^{-1} + z^{-2}}{1 - z^2} \qquad\qquad \text{Simpson's Rule}$$

$$\frac{T}{3}\frac{(z^{-1})(8 - 5z^{-1} + 4z^{-2} - z^{-3})}{(1 - z^{-2})} \qquad\qquad \text{Four Point Rule}$$

Rather than develop expressions for s^{-n} we can write $s^{-n} = (s^{-1})^n$ and then cascade n integrators from the above table. Discuss the use of these and, for a numerical example, consider simulation of $\dfrac{\omega_0^2}{s^2 + \omega_0^2}$, a sinusoidal oscillator.

10. If a system is to be designed to have zero steady state error for an input $U(s) = 1/s^{n+1}$ then the closed loop normalized system transfer function must be of the form

$$H_{n,k}(s) = \frac{1 + a_1 s + a_2 s^2 + \ldots + a_n s^n + b_{n+1} s^{n+1} + \ldots}{1 + a_1 s + a_2 s^2 + \ldots + a_n s^n + a_{n+1} s^{n+1} + \ldots + s^{n+k}}$$

The particular case where $k = 1$ will be of interest in this example. Find the coefficients for $n = 1, 2, 3$ to yield minimum integral square error. What is the step response and phase margin of the resulting systems?

11. Find the Tustin z form representation of the system of Problem 7.10. Please investigate the possibility of digital system design by finding a continuous time optimum system and discretizing it using a z form representation. In particular does the presence of a fixed continuous time element pose a problem?

12. Find the response of the following system using the suggested z form approximations

$$G(s) = \frac{2^{1/2}(1 + s)}{s^2}$$

$$\frac{Z(s)}{U(s)} = H(s) = \frac{G(s)}{1 + G(s)}$$

Evaluate the sum of the error squared and compare it with the integral square error.

13. Determine the structure of the linear feedback controller to minimize

$$J = \frac{1}{2} x^T(t_f) S x(t_f) + \frac{1}{2} \int_{t_0}^{t_f} (x^T Q x + x^T M^T u + u^T M x + u^T R u) \, dt$$

for the linear system

$$\dot{x} = Ax + Bu \qquad x(t_0) = x_0$$

14. A unity feedback ratio system has the open loop transfer function

$$G(s) = \frac{K(s+1)}{s(s-1)(s+25)}$$

Find the best K for minimum

a. ISE

b. ITSE

c. ISTSE

d. IEMSE

Determine the phase margin and root locus as a function of K as well as the dominant pole damping ratio. Compare your analytical design results with the classical results.

15. For the fixed plant described by

$$G_f(s) = \frac{\sqrt{2}\,(s+1)}{s^2} = \frac{Z(s)}{U_f(s)}$$

Find the best control $u_f(t)$ to minimize

$$J = \frac{1}{2}\int_0^\infty [z^2(t) + Ru_f^2(t)]\ dt$$

where $z(0) = 1$ and $\dot{z}(0) = 0$. Do this by considering the final time as some fixed value and then let t_f approach infinity in the optimal control solution. Also obtain the closed loop control $u_f(t) = \mathbf{K}\mathbf{x}(t)$ for the infinite time interval case and then obtain the open loop control from this regulator solution. Contrast and compare the effort involved in the two procedures. Compare your infinite time results with those obtained from the integral square error frequency domain approach. Show appropriate block diagram representations for all infinite time results.

16. Discuss the design of a parameter adaptive system to control a unity feedback ratio system with

$$G_f(s) = \frac{K}{s^2}$$

The nominal gain K is 100 but this varies and cannot be controlled. An adjustable lead network

$$G_c(s) = \frac{(1 + s/\omega_1)}{(1 + s/9\omega_1)}$$

is used where the adaptive adjustment device seeks to adjust ω_1 for maximum phase margin. If a convenient simulation facility is available simulate your results and observe the step response under a variety of conditions.

17. Repeat Problem P.16 using a performance adaptive model reference system where the reference model is

$$H_m(s) = \frac{100 + 14.14s}{100 + 14.14s + s^2}$$

If a simulation facility is available simulate your results and observe the step response under a variety of conditions.

APPENDIX A

It will be convenient throughout this text to use a DELTA Chart method to graphically portray various linear control system design methods. DELTA is an anacronym devised by Warfield and Hill from the first letter of five graphic elements used in constructing a DELTA Chart. These elements are the decision box, event box, logic box, time arrow, and activity box. Two other graphic elements are often used, a connection matrix to depict a large fanout of activities and a ground or end termination to denote the termination of a sequence of preceding elements.

The seven symbols are shown in Fig. (A.1). These can be arranged in a flow chart configuration to present an overall program or design concepts.

REFERENCE

Warfield, J. N. and Hill, J. D., "The DELTA Chart—A Method for R&D Project Portrayal," *IEEE Trans. on Engr. Mgt.*, Vol. EM-18, No. 4, November 1971.

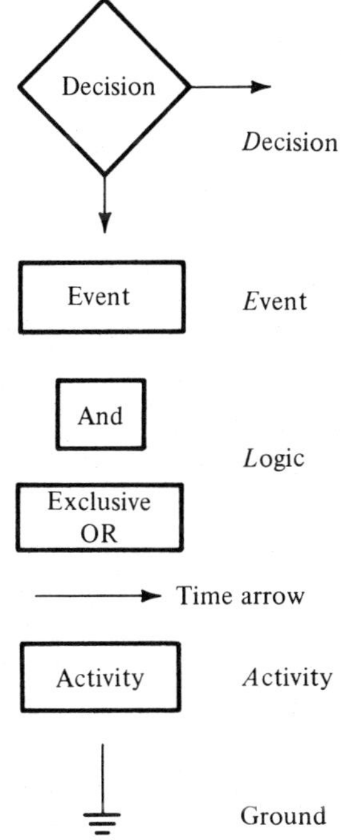

Figure A.1 Symbol for DELTA Charts

INDEX